STREPTOCOCCAL GENETICS

STREPTOCOCCAL GENETICS

Editors

JOSEPH J. FERRETTI

University of Oklahoma Health Sciences Center
Oklahoma City, Oklahoma

ROY CURTISS III

Washington University
St. Louis, Missouri

AMERICAN SOCIETY FOR MICROBIOLOGY
Washington, D.C.
1987

Library of Congress Cataloging-in-Publication Data

ASM Conference on Streptococcal Genetics (2nd : 1986 :
  Miami Beach, Fla.)

  Streptococcal genetics.

  Includes indexes.
    1. Streptococcus—Congresses. 2. Bacterial genetics—
Congresses. I. Ferretti, Joseph J. II. Curtiss, Roy.
III. American Society for Microbiology. IV. Title.
[DNLM: 1. Streptococcus—genetics—congresses.
W3 AS443 2nd 1986s / QW 142.5.C6 A836 1986s]
QR82.S78A86   1986   589.9'5   87-1468

ISBN 0-914826-93-X

Contents

C. Pneumococci

IV. Oral Streptococci

V. Lactic Acid Streptococci

Introduction

The Second ASM Conference on Streptococcal Genetics was held in Miami, Florida, 21–24 May 1986. Approximately 200 researchers from 16 different countries were in attendance for the eight oral and poster sessions. The proceedings of the first conference, held 5 years previously, were published in the ASM volume *Microbiology—1982*. Since then, there has been a dramatic increase in interest and activity as investigators from all disciplines have embraced the new approaches and tools that genetic studies afford. Initially, streptococcal genetics research centered on the study of gene transfer, antibiotic resistance, and plasmid biology. However, in recent years there has been an emphasis on genetic aspects of streptococcal virulence, pathogenicity, and metabolism. These studies are directed towards the major health problems associated with streptococcal diseases, namely, rheumatic heart disease, glomerulonephritis, dental caries, neonatal meningitis and septicemia, pneumonia, and skin and throat infections. Additionally, basic studies aimed at the elucidation of streptococcal fermentation pathways are of prime importance for food-processing and dairy industries.

This volume is divided into five main sections, consisting of papers of interest from both the oral and poster sessions of the conference. Attempts have been made to provide the reader with adequate introductions to present a historical perspective for each of the topics. We hope that this volume will be of interest to individuals unfamiliar with streptococcal genetics as well as those actively involved in this area of research. Appendixes of useful streptococcal cloning vectors and recently sequenced genes have been included as a reference resource.

We thank the National Institute of Allergy and Infectious Diseases and the National Institute of Dental Research, National Institutes of Health, for providing funds for young investigators to participate in the conference. We also thank the ASM Conference Committee, Meetings Board, and Publications Department for their contributions in making the conference a success and this volume a reality.

We also thank the following companies for their generous contributions in support of the Second ASM Conference on Streptococcal Genetics: Bristol-Myers Co.; General Mills; Mallinckrodt, Inc.; Meiji Institute of Health Science; Miles Laboratories; Monsanto Co.; Neogen Corp.; Pioneer Hybrid International, Inc.; Shionogi Research Laboratories; and Upjohn Co.

JOSEPH J. FERRETTI
University of Oklahoma Health Sciences,
Oklahoma City, Okla.

ROY CURTISS III
Washington University,
St. Louis, Mo.

I. Gene Transfer

Sex Pheromones and Plasmid-Related Conjugation Phenomena in *Streptococcus faecalis*

DON B. CLEWELL,[1] ELIZABETH E. EHRENFELD,[1] FLORENCE AN,[1] ROBERT E. KESSLER,[1†]
REINHARD WIRTH,[1‡] MASAAKI MORI,[2] CHIEKO KITADA,[3] MASAHIKO FUJINO,[3]
YASUYOSHI IKE,[4] AND AKINORI SUZUKI[2]

*Departments of Oral Biology and Microbiology/Immunology, Schools of Dentistry and Medicine, and The
Dental Research Institute, The University of Michigan, Ann Arbor, Michigan 48109[1]; Department of
Agricultural Chemistry, The University of Tokyo, Bunkyo-ku, Tokyo 113, Japan[2]; Central Research Division,
Takeda Chemical Industries, Ltd., Yodogawa-ku, Osaka 532, Japan[3]; and Department of Microbiology,
Gunma University School of Medicine, Maebashi, Japan[4]*

Certain conjugative plasmids in *Streptococcus faecalis* encode mating responses to specific peptide sex pheromones excreted by potential recipient cells (1, 2). The response requires about 30 to 40 min and is characterized by the synthesis of an adherent, proteinaceous substance that uniformly covers the cell surface and facilitates the formation of mating aggregates upon random collision of donors and recipients. The newly synthesized donor surface material is referred to as "aggregation substance" (AS) and is presumed to bind to a material on the recipient designated "binding substance" (BS). BS is also present on the donor surface; thus, donors induced with culture filtrates of recipients undergo a self-aggregation and even self-mating. The donor self-aggregation or clumping response serves as the basis for quantitating a given pheromone; thus, the latter is also referred to as clumping-inducing agent.

Cell aggregates can be dissociated by exposure to EDTA, and when such cells are pelleted and resuspended in buffer (Tris, pH 7.0), reaggregation does not occur unless phosphate and divalent cations (e.g., Mg^{2+}) are added back (21). Exposure of induced, EDTA-dissociated cells to proteases, sodium dodecyl sulfate (0.05%), or heat (98°C, 5 min) destroyed the ability to reaggregate when phosphate and $MgCl_2$ were subsequently added (21). Immunoelectron microscopy studies of *S. faecalis* 39-5 (contains pPD1) revealed the appearance of a "fuzzy" material on the surface of induced

donor cells (21). Induction could also be correlated with the appearance of an immunoreactive 78,000-dalton protein (11).

Plasmid-free *S. faecalis* strains (e.g., FA2-2; OG1X) excrete multiple pheromones, specific for different conjugative plasmids (2). When a specific plasmid (e.g., pAD1) is acquired, the transconjugants shut down production of the related sex pheromone (e.g., cAD1); however, they continue to excrete unrelated pheromones. Interestingly, plasmid-containing cells excrete a peptide which behaves as a competitive inhibitor (e.g., iAD1) of the related pheromone. The inhibitor is believed to prevent induction by levels of pheromone too low to result in the generation of mating aggregates (recipients too far away to encounter by random collision) or induction by low levels of endogenous pheromone. It may also prevent self-induction by other pheromones with low levels of crossreacting activity for the plasmid-specific receptor site.

Table 1 lists some of the properties of three pheromone-responding plasmids we have investigated: pPD1, pAD1, and pAM373. The related pheromones have been isolated and characterized (13, 14; M. Mori, this volume); their amino acid sequences are shown in Fig. 1. All three are hydrophobic peptides with a single serine residue; all are generally produced by a single strain of *S. faecalis*. Whereas production of cPD1 and cAD1 by any of 11 other bacterial species tested could not be detected, a cAM373-like activity was found to be excreted by essentially all strains of *Staphylococcus aureus*, some strains of *Streptococcus sanguis*, and *Streptococcus faecium* 9790 (3). There is no evidence that the

†Present address: Bristol-Myers Co., Wallingford, CT 06492.
‡Present address: University of Munich, Munich, Federal Republic of Germany.

TABLE 1. Properties of some conjugative plasmids encoding pheromone responses

Plasmid	Size (kilobases)	Encodes mating response to:	Determines inhibitor designated:	Plasmid determinants[a]	Strain of origin	References
pAD1	56.7	cAD1	iAD1	Hly-Bac, UV^r	DS16	5, 15
pPD1	54.6	cPD1	iPD1	Bac	39–5	7, 21
pAM373	36	cAM373	iAM373		RC73	3

[a] Hly-Bac, Hemolysin-bacteriocin; UV^r, resistance to UV light.

cAM373 activity serves as a pheromone in any species other than *S. faecalis*. In connection with the production by *S. aureus*, it is interesting that no related clumping-inducing agent could be detected in filtrates of coagulase-negative staphylococci (i.e., a relatively nonpathogenic species of staphylococci).

Identification of Pheromone-Induced Surface Proteins

Antiserum was raised in rabbits against *S. faecalis* FA2-2(pAM714) cells that had been induced and fixed with glutaraldehyde by a method described elsewhere (21). (pAM714 is a pAD1::Tn917 derivative with normal conjugative behavior [9].) The antiserum was used in a Western-blot analysis (17) of surface proteins extracted from OG1X(pAM714) after exposure to pheromone for increasing lengths of time. Extraction was as previously described (11, 21) using Zwittergent 3-12 (Calbiochem-Behring, La Jolla, Calif.). Electrophoresis was performed on 10% polyacrylamide slab gels. After transfer to nitrocellulose filters, enzyme-linked immunosorbent assay of the filters was done using the directions and reagents from an Immuno-blot (GAR-HRP) assay kit (Bio-Rad Laboratories, Richmond, Calif.).

Figure 2 shows the results for extracts prepared at different times after the addition of cAD1 (a recipient filtrate). Induced and uninduced samples were paired at 15-min intervals. After 45 min of induction, four new bands were readily observed (compare lanes g and h). The

most prominent is a 74-kilodalton (kDa) band, which is accompanied by bands of lower intensity at 130, 153, and 157 kDa. These bands are referred to as AD74, AD130, AD153, and AD157. AD74 was easily resolved on Coomassie blue staining of sodium dodecyl sulfate-polyacrylamide gels (not shown); the other bands, however, could be detected only immunologically. When the same antiserum was used to similarly analyze a different pheromone-responding plasmid system pAM351 (a pPD1::Tn916 derivative; responds to cPD1) in an OG1-10 host, the same three larger bands (130, 153, and 157 kDa; designated PD130, PD153, and PD157) were detected upon induction (not shown). The previously reported 78-kDa band (PD78) of the pPD1 system (11) did

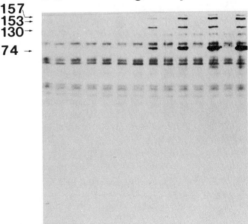

FIG. 2. Western blot analysis of *S. faecalis* OG1X(pAM714) proteins after exposure to cAD1 (culture filtrate of recipient cells). Protein samples were taken at the following time points: lanes a (uninduced) and b (induced) at 0 min; lanes c (uninduced) and d (induced) at 15 min; lanes e (uninduced) and f (induced) at 30 min; lanes g (uninduced) and h (induced) at 45 min; lanes i (uninduced) and j (induced) at 60 min; lanes k (uninduced) and l (induced) at 75 min; and lanes m (uninduced) and n (induced) at 90 min. The arrows with size estimates (in kilodaltons) point to protein bands which appear during induction.

cPD1 (912 DALTONS)

H-PHE-LEU-VAL-MET-PHE-LEU-SER-GLY-OH

cAD1 (818 DALTONS)

H-LEU-PHE-SER-LEU-VAL-LEU-ALA-GLY-OH

cAM373 (733 DALTONS)

H-ALA-ILE-PHE-ILE-LEU-ALA-SER-OH

FIG. 1. Amino acid sequences and molecular weights of the sex pheromones cAD1, cPD1, and cAM373.

not react similarly with the antiserum, although it was readily detectable by Coomassie blue staining. The reciprocal experiment generated similar results in that antiserum raised against FA2-2(pAM351) cells reacted with AD130, AD153, and AD157 of cells containing pAM714 but only slightly with AD74. Thus the two systems appear to differ with respect to a 74- and 78-kDa substance but seem to have in common the other three proteins. It seems likely that AD74 and PD78 are plasmid encoded, and their prominence in each case is suggestive that they represent AS in the two systems. If so, their difference in size and immunological specificity may relate to visible differences between these systems in the nature of the induced aggregates; the pPD1 system gives rise to more dramatic clumping (large "flakes") than the pAD1 system (a more granular appearance).

In the pCF10 plasmid system, Tortorello and Dunny (16) recently identified several new antigens appearing in the culture medium after exposure of cells to sex pheromone. It was speculated that one of the antigens, designated SA73 (a 73-kDa protein), may be similar to PD78. The other antigens had molecular weights above 100,000. The relationship between these proteins and the proteins that relate to pAD1 and pPD1 is not yet known. However, a 130-kDa protein designated C130 and believed to relate to surface exclusion was reported not to cross-react with any induced pAD1-related proteins (8).

Specificity of Plasmid Transfer from Cells Harboring Two Different Plasmids

The following experiment was conducted to determine the effect exposure to a single phero-mone (cAD1 or cPD1) has on conjugal donation from a strain simultaneously harboring pAD1 and pPD1 derivatives. OG1-10(pAM714, pAM-351) cells were induced for 1 h with pheromone preparations containing cAD1, cPD1, cAD1 +

cPD1, or no activity, prior to a short (15 min) mating with the plasmid-free recipient strain FA2-2. Table 2 shows the results of two independent experiments (trials 1 and 2). Increased levels of transfer of pAM714 occurred only after pretreatment with cAD1, and only transfer of pAM351 was enhanced after induction of the donor strain with cPD1. pAM351 transfer was somewhat increased after exposure to cAD1; the frequency, however, was at least two orders of magnitude below that which occurred upon cPD1 exposure. Cotransfer of the plasmids was seen at elevated frequencies only when the pheromone preparation containing both cAD1 and cPD1 was used. Thus the two different sex pheromones specifically induce their target plasmid systems, and there is little if any *trans* complementation with a different coresident plasmid.

Evidence that LTA May Correspond to Binding Substance

Lipoteichoic acid (LTA) is present in significant amounts in the cell wall of *S. faecalis* (19, 20); indeed, it corresponds to the Lancefield group D antigen. It was therefore of interest whether this material might be a component of BS. To explore this possibility, we extracted LTA (18, 19) from *S. faecalis* JH2-2 and tested its ability to inhibit the aggregation of induced plasmid-containing cells. *S. faecalis* OG1X (pAM714) and OG1-10(pAM351) were each induced for 90 min with an appropriate recipient culture filtrate; aggregates were dissociated by addition of EDTA (to 50 mM), washed, and resuspended in half the original volume in 50 mM Tris, pH 7.5. (The absence of phosphate ions and divalent cations prevents reaggregation at this point.) Using a microtiter plate, LTA was serially diluted (twofold) through wells containing the induced-dissociated cells. Next, a solution containing divalent cations (0.25 mM $CaCl_2$ or $MgCl_2$) and phosphate (10 mM) was added,

TABLE 2. Transfer frequencies of pAM714 and pAM351 from strain OG1X to FA2-2[a]

Trial no.	Pheromone activity	pAM714/donor	pAM351/donor	pAM714 + pAM351/donor
1	None	2.6×10^{-5}	5.1×10^{-8}	$<5.1 \times 10^{-9}$
	cAD1 + cPD1	1.8×10^{-1}	8.8×10^{-3}	9.3×10^{-5}
	cPD1	2.7×10^{-5}	1.8×10^{-2}	1.5×10^{-7}
	cAD1	2.2×10^{-1}	1.3×10^{-5}	$<1.8 \times 10^{-8}$
2	None	2.5×10^{-3}	9.8×10^{-7}	$<4.3 \times 10^{-9}$
	cAD1 + cPD1	5.1×10^{-1}	1.8×10^{-1}	1.4×10^{-2}
	cPD1	4.8×10^{-4}	1.3×10^{-1}	2.0×10^{-5}
	cAD1	1.1×10^{-1}	7.0×10^{-5}	1.7×10^{-6}

[a] Frequencies are expressed as the number of transconjugants containing the indicated plasmid per total number of donor cells. Induction was for 60 min and was followed by a 15-min mating. cPD1 was prepared from JH2-2(pAM714) and was devoid of cAD1. Conversely, cAD1 was prepared from JH2-2(pAM351) and was devoid of cPD1. cAD1 + cPD1 was prepared from the plasmid-free strain JH2-2.

and the microtiter plate was swirled at 200 rpm for 3 h. In the presence of Mg^{2+}, aggregation occurred only at LTA concentrations below 1.0 μg/ml for pAM714-containing cells and below 0.12 μg/ml for pAM351-containing cells. When Ca^{2+} ions were present, reaggregation occurred at 8 (but not 16) μg/ml and 2 (but not 4) μg/ml, respectively. (When a similar experiment was performed comparing $MgCl_2$ concentrations of 0.25 and 25.0 mM, the minimum inhibitory LTA concentrations were the same; thus, LTA was not preventing aggregation via a chelation effect on divalent cations.) This inhibition of reaggregation by relatively low concentrations of LTA supports the notion that LTA represents the receptor for aggregation substance.

It is interesting that, when pAM714 was introduced into S. faecium 9790RF (see next section), transconjugants did not exhibit a clumping-inducing-agent response. However, if these cells were exposed to cAD1 for 45 min and mated with S. faecalis JH2SS (plasmid free; resistant to streptomycin and spectinomycin [21]) for 20 min in broth, transfer occurred at a relatively high frequency (10^{-4} per donor). Transfer of pAM714 from similarly induced JH2SS(pAM714) to 9790RF did not occur in broth ($<10^{-7}$ per donor). (Filter mating is required to observe transfer in this direction; see below.) One interpretation of these data is that, while AS may be produced in S. faecium, BS is not present on the surface of this strain. It would therefore be interesting to compare the LTA components in the two strains and determine whether LTA from 9790RF can inhibit clumping of induced S. faecalis (pAM714) cells.

Plasmid-Related Inhibitor Production

The plasmid-related inhibitor peptides iAD1 and iPD1, produced by cells containing pAD1 or pPD1, respectively, have recently been isolated and characterized (12; M. Mori et al., manuscript in preparation); both were found to be hydrophobic octapeptides. In Fig. 3 their structures are compared with the related pheromones cAD1 and cPD1. The iAD1 peptide is seen to be 50% homologous with cAD1; four of the eight amino acid residues correspond. In the case of iPD1 only two of the eight residues correspond. Interestingly, the sequence -Thr-Leu-Val- is present in the C-terminal half of both iAD1 and iPD1. It was found that a threefold excess of iAD1 was necessary to inhibit cAD1 (12).

The unique structure of the inhibitor has ruled out an earlier-proposed (10) interpretation that it represented a modified form of the host-encoded pheromone, and it raises the possibility that it is plasmid encoded. In addressing this point we reasoned that, if iAD1 were encoded by pAD1, it might be possible to detect the inhibitor after

cAD1 H-Leu-Phe-Ser-Leu-Val-Leu-Ala-Gly-OH
iAD1 H-Leu-Phe-Val-Val-Thr-Leu-Val-Gly-OH

cPD1 H-Phe-Leu-Val-Met-Phe-Leu-Ser-Gly-OH
iPD1 H-Ala-Leu-Ile-Leu-Thr-Leu-Val-Ser-OH

FIG. 3. Amino acid sequences of the inhibitor peptides iAD1 and iPD1 and comparisons with their corresponding sex pheromones cAD1 and cPD1, respectively. The boxes point out similarities between inhibitor and pheromone, whereas the circled residues represent similar sequences in iAD1 and iPD1. The molecular weights of iAD1 and iPD1 are 846 and 828, respectively.

introduction of the plasmid into a bacterium that does not normally excrete cAD1 (or iAD1). S. faecium 9790 has been found to be devoid of a number of sex pheromones (including cAD1) normally excreted by S. faecalis (3). (One exception is cAM373; however, the role of this peptide as a true pheromone in this species has not been demonstrated.)

The plasmid pAM714 was introduced into S. faecium 9790RF (has mutational resistances to rifampin and fusidic acid [3]) from S. faecalis OG1X(pAM714) (9). When a filter mating procedure was used (3), transfer to S. faecium occurred at approximately 10^{-6} per donor. The production of iAD1 by S. faecium 9790RF (pAM714) is shown in Fig. 4. The activity increased with cell growth and leveled off as the culture entered stationary phase. The appearance of iAD1 is related to the presence of the plasmid, as no iAD1 activity could be detected in filtrates of a 9790RF strain devoid of pAM714. That the structure of the iAD1 activity observed here was identical to that produced by S. faecalis carrying this plasmid was strongly suggested from its characteristic retention time during fractionation by high-performance liquid chromatography (not shown). The data are consistent with the view that pAD1 determines the production of iAD1. Although the alternative possibility, that pAD1 might activate the expression of a structural determinant located on the chromosome, cannot be totally ruled out, such an occurrence seems unlikely. The absence of a cAD1 pheromone system in S. faecium 9790RF implies that there would be no apparent functional basis for a host-encoded iAD1 in this strain. It is now clear that in S. faecalis pAD1 affects the activity of cAD1 by at least two means: the inactivation or repression of endogenous cAD1 and the production of an inhibitor of this peptide.

Upon determination of the structure of iAD1 (12), we considered the interesting possibility that, in addition to its inhibitory effect on cAD1, the peptide may have a unique pheromone ac-

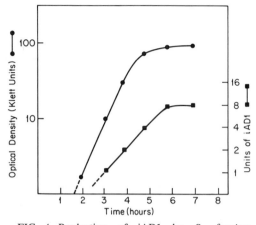

FIG. 4. Production of iAD1 by *S. faecium* 9790RF(pAM714). The medium used was Todd-Hewitt broth (Difco Laboratories, Detroit, Mich.). Cell density was measured using a Klett-Summerson colorimeter (no. 54 filter). The culture volume was 25 ml, and the activities were measured in filtrates of the sampled culture. Inhibitor activity was based on the ability to reduce the pheromone titer in a microtiter dilution assay (10). A pheromone-containing culture filtrate (from strain FA2-2) was serially diluted through 50-μl amounts of culture filtrate being tested for inhibitor activity prior to the addition of 50 μl of responder cells (DS16). The number of units of inhibitor was taken as the pheromone titer in the absence of inhibitor (broth substituted for inhibitor) divided by the titer in the presence of inhibitor.

tivity relating to a different plasmid system. However, upon screening 200 clinical isolates of *S. faecalis* (from Gunma University Hospital and Isesaki Hospital in Japan), none exhibited a clumping-inducing-agent response when exposed to synthetic iAD1 (2 ng/ml).

On the Genetic Basis of Pheromone Synthesis

Since production of sex pheromone parallels cell growth and is blocked by exposure to rifampin (2), synthesis of the peptide is believed to occur by ribosomal assembly rather than by enzymatic condensation. On the basis of the cPD1 amino acid sequence, an oligonucleotide probe corresponding to the first 14 nucleotides starting from the N terminus and representing 64 degeneracies was constructed (Genex Corp.) and used as a probe (5' labeled with ^{32}P using polynucleotide kinase) in filter-blot hybridizations to the *S. faecalis* JH2-2 chromosome. The probe mixture was found to hybridize to four *Eco*RI fragments with sizes of approximately 1.8, 2.4, 6, and 12 kilobases (not shown). Hybridization to multiple *Bcl*I (13, 6, and 4 kilobases) and *Bam*HI (11, 8, 6.5, and 1.8 kilobases) fragments was also observed. Efforts to clone and sequence the pheromone gene(s), using the

synthetic probe for screening, are currently under way.

Concluding Remarks

The total number of sex pheromones actually excreted by a single strain of *S. faecalis* is not known; conceivably it could be many. It is also likely that many strains carry more than one pheromone-responding plasmid. Indeed, *S. faecalis* DS5 has been shown to carry three such plasmids (pAMγ1, pAMγ2, and pAMγ3), encoding responses to three distinct pheromones (6). It is presumed that each plasmid encodes a pheromone-specific receptor site on the cell surface, and it is evident that the resulting signal is directed to the related plasmid and not to co-resident plasmids. The conjugation data noted above relating to a strain simultaneously harboring derivatives of both pAD1 and pPD1 are supportive of this view. The two plasmids appear to operate independently and determine in each case at least one unique protein (AD74 and PD78). Conceivably these proteins correspond to AS, and whereas AS may be unique in each case, BS (represented at least in part by LTA) may be common for all the systems.

The synthesis of AS is believed to be under negative control; in the pAD1 system, it has been possible to generate Tn917 insertion mutants in which constitutive synthesis of AS occurs (9). Such mutants map in two clusters which have been designated *traA* and *traB* (9). (AS74 is constitutively synthesized in both cases [E. Ehrenfeld, unpublished data].) The products of *traA* and *traB* appear to work together in preventing expression of AS. It is possible that a number of transfer-related functions are similarly controlled by *traA* and *traB* since plasmid transfer occurs at high frequency in the mutant derivatives. Support for this view comes with an earlier observation involving transfer between two donor strains with distinguishable derivatives of pAD1 (4). When only one of the strains was induced with cAD1 for 45 min prior to a short (20 min) mating, transfer occurred at a high frequency, but only from the induced cells to the uninduced cells. If the only function of the pheromone were to generate mating aggregates, then we would have expected transfer to occur equally well in both directions—regardless of which donor was induced prior to mating. Thus induction of donors by cAD1 also involves a "preparation for mating." Included in the response would appear to be the induction of surface exclusion functions; when both donor strains were induced prior to mating, transfer occurred in both directions but at frequencies one to two orders of magnitude lower than was the case when the observed recipient was not induced (4). Induction of a surface exclusion function has recently been reported in the pCF10 system (8).

In the bacterial world conjugative plasmids that encode mating responses to recipient-produced sex pheromones have been reported only in *S. faecalis*. However, it would seem only a matter of time before they are discovered in other species. In *S. faecalis* such plasmids are widespread and can significantly influence the transfer of various coresident plasmids (2). In addition, they can mobilize chromosomal genes and significantly enhance the transfer of chromosome-borne conjugative transposons (2). Although progress is being made in understanding the nature of these systems, much is yet to be learned about the molecular mechanisms of their behavior.

This work was supported by Public Health Service grants GM33956, AI10318, DE05916, and DE02731 from the National Institutes of Health and by a Grant-in-Aid for Scientific Research (no. 60790127) from the Ministry of Education, Science, and Culture of Japan.

We thank our laboratory colleagues for helpful discussions and advice.

LITERATURE CITED

1. **Clewell, D. B.** 1981. Plasmids, drug resistance, and gene transfer in the genus *Streptococcus*. Microbiol. Rev. **45**:409–436.
2. **Clewell, D. B.** 1985. Sex pheromones, plasmids, and conjugation in *Streptococcus faecalis*, p. 13–28. *In* H. O. Halvorson and A. Monroy (ed.), The origin and evolution of sex. Alan R. Liss, New York.
3. **Clewell, D. B., F. Y. An, B. A. White, and C. Gawron-Burke.** 1985. *Streptococcus faecalis* sex pheromone (cAM373) also produced by *Staphylococcus aureus* and identification of a conjugative transposon (Tn*918*). J. Bacteriol. **162**:1212–1220.
4. **Clewell, D. B., and B. Brown.** 1980. Sex pheromone cAD1 in *Streptococcus faecalis*: induction of a function related to plasmid transfer. J. Bacteriol. **143**:1063–1065.
5. **Clewell, D. B., P. K. Tomich, M. C. Gawron-Burke, A. E. Franke, Y. Yagi, and F. An.** 1982. Mapping of *Streptococcus faecalis* plasmids pAD1 and pAD2 and studies relating to transposition of Tn*917*. J. Bacteriol. **152**:1220–1230.
6. **Clewell, D., Y. Yagi, Y. Ike, R. Craig, B. Brown, and F. An.** 1982. Sex pheromones in *Streptococcus faecalis*: multiple pheromone systems in strain DS5, similarities of pAD1 and pAMγ1, and mutants of pAD1 altered in conjugative properties, p. 97–100. *In* D. Schlessinger (ed.), Microbiology—1982. American Society for Microbiology, Washington, D.C.
7. **Dunny, G., B. Brown, and D. B. Clewell.** 1978. Induced cell aggregation and mating in *Streptococcus faecalis*. Evidence for a bacterial sex pheromone. Proc. Natl. Acad. Sci. USA **75**:3479–3483.
8. **Dunny, G. M., D. L. Zimmerman, and M. L. Tortorello.** 1985. Induction of surface exclusion (entry exclusion) by *Streptococcus faecalis* sex pheromones: use of monoclonal antibodies to identify an inducible surface antigen involved in the exclusion process. Proc. Natl. Acad. Sci. USA **82**:8582–8586.
9. **Ike, Y., and D. B. Clewell.** 1984. Genetic analysis of the pAD1 pheromone response in *Streptococcus faecalis*, using transposon Tn*917* as an insertional mutagen. J. Bacteriol. **158**:777–783.
10. **Ike, Y., R. Craig, B. White, Y. Yagi, and D. Clewell.** 1983. Modification of *Streptococcus faecalis* after acquisition of plasmid DNA. Proc. Natl. Acad. Sci. USA **80**:5369–5373.
11. **Kessler, R. E., and Y. Yagi.** 1983. Identification and partial characterization of a pheromone-induced adhesive surface antigen of *Streptococcus faecalis*. J. Bacteriol. **155**:714–721.
12. **Mori, M., A. Isogai, Y. Sakagami, M. Fujino, C. Kitada, D. Clewell, and A. Suzuki.** 1986. Isolation and structure of *Streptococcus faecalis* sex pheromone inhibitor, iAD1, that is excreted by the donor strain harboring plasmid pAD1. Agric. Biol. Chem. **50**:539–541.
13. **Mori, M., Y. Sakagami, M. Narita, A. Isogai, M. Fujino, C. Kitada, R. Craig, D. Clewell, and A. Suzuki.** 1984. Isolation and structure of the bacterial sex pheromone, cAD1, that induces plasmid transfer in *Streptococcus faecalis*. FEBS Lett. **178**:97–100.
14. **Suzuki, A., M. Mori, Y. Sakagami, A. Isogai, M. Fujino, C. Kitada, R. Craig, and D. Clewell.** 1984. Isolation and structure of bacterial sex pheromone, cPD1. Science **226**:849–850.
15. **Tomich, P. K., F. An, S. Damle, and D. B. Clewell.** 1979. Plasmid-related transmissibility and multiple-drug resistance in *Streptococcus faecalis* subsp. *zymogenes* strain DS16. Antimicrob. Agents Chemother. **15**:828–830.
16. **Tortorello, M. L., and G. M. Dunny.** 1985. Identification of multiple cell surface antigens associated with the sex pheromone response of *Streptococcus faecalis*. J. Bacteriol. **162**:131–137.
17. **Towbin, H., T. Staehelin, and J. Gordon.** 1979. Electrophoretic transfer of proteins from polyacrylamide gels to nitrocellulose sheets: procedure and some applications. Proc. Natl. Acad. Sci. USA **76**:4350–4354.
18. **Westphal, O., O. Luderitz, and F. Bister.** 1952. Uber die Extraktion von Bakterien mit Phenol/Wasser. Naturforscher **7B**:148–155.
19. **Wicken, A. J., S. D. Elliott, and J. Baddiley.** 1963. The identity of streptococcal group D antigen with teichoic acid. J. Gen. Microbiol. **31**:231–239.
20. **Wicken, A. J., and K. W. Knox.** 1974. Lipoteichoic acids: a new class of bacterial antigens. Science **187**:1161–1167.
21. **Yagi, Y., R. Kessler, J. Shaw, D. Lopatin, F. An, and D. Clewell.** 1983. Plasmid content of *Streptococcus faecalis* strain 39-5 and identification of a pheromone (cPD1)-induced surface antigen. J. Gen. Microbiol. **129**:1207–1215.

Isolation and Structure of *Streptococcus faecalis* Sex Pheromone cAM373, Also Produced by *Staphylococcus aureus*

MASAAKI MORI

Department of Agricultural Chemistry, The University of Tokyo, Bunkyo-ku, Tokyo 113, Japan

Certain conjugative plasmids in *Streptococcus faecalis* transfer very frequently in broth culture (10^{-3} to 10^{-1} per donor). In these plasmid-transfer systems, recipient strains excrete multiple peptidal sex pheromones specific for strains harboring conjugative plasmids. Induction of donor cells by the pheromone results in the synthesis of a proteinaceous adhesin on the cell surface, which facilitates the formation of mating aggregates. Since exposure of donor cells to a cell-free culture filtrate of recipients leads to self-clumping, the sex pheromone is also called clumping-inducing agent (2–4). Recently, two sex pheromones, cPD1 and cAD1, which induce specific responses in strains harboring bacteriocin plasmid pPD1 (56.7 kilobases) and hemolysin plasmid pAD1 (54.6 kilobases), respectively, have been isolated and characterized. Their sequences were determined (9, 10) and are shown in Fig. 1. The molecular weight of cPD1 is 912, and that of cAD1 is 818.

Here the isolation, structure elucidation, and total synthesis of another sex pheromone, cAM373, are described. The sex pheromone cAM373, also excreted by recipient strains of *S. faecalis* and involved in the conjugative transfer of the plasmid pAM373, is characterized by the fact that a substance showing its activity is also excreted by certain strains of *S. sanguis* and *S. faecium* and by all *Staphylococcus aureus* strains examined (1). However, since pAM373 derivatives failed to be established in these organisms, it is conceivable that the cAM373 activity does not relate with mating in these species. It is noteworthy that almost all coagulase-negative staphylococci (e.g., *Staphylococcus epidermidis*) fail to excrete cAM373 activity (1). This raises the possibility that in the genus *Staphylococcus* the cAM373 activity might play a role in virulence.

The pheromone-producing *S. faecalis* strain JH2-2(pAM351) (7), which harbors a derivative of the conjugative plasmid pPD1, pAM351 (=pPD1::Tn*916*), but can be a recipient for plasmid pAM373, was cultured (0.25% inoculum) in THG medium (18.2 g of Oxoid Todd-Hewitt broth and 20 g of glucose in 1 liter of distilled water, 4 liters per batch) with gentle stirring at 37°C for 20 h (stationary phase). The cells were removed by centrifugation, and the supernatant was used for isolation of cAM373 (Table 1). Biological activity was assayed by means of the microtiter dilution assay (4) using pAM373-containing strain FA373 (1) as responder cells. One unit of activity was defined as the lowest amount that could induce clumping of responder cells in 100 μl of assay medium in a microtiter dilution well. During the purification, any drying treatments were avoided to protect the pheromone from insolubilization. In the final step of purification, the fractions containing active materials corresponding to four batches (16 liters of culture broth) were combined and applied to reverse-phase high-performance liquid chromatography on an SSC-ODS-262 column (Senshukagaku), and a gradient of 10 to 22% (10 min) and 22 to 28% (30 min) acetonitrile in 0.1% trifluoroacetic acid afforded 4.4 μg of pure cAM373.

Through a purification procedure consisting of eight steps, a 35×10^6-fold purification from culture broth was generated. The total weight, total activity, and specific activity of the active material in each step are summarized in Table 1. In the course of the purification, the seventh step, employing high-performance liquid chromatography on a Senshupak CN-4251-N column (Senshukagaku), was most effective and enabled a 150-fold purification. The purified cAM373 induced the self-clumping of FA373 cells at a concentration as low as 3.4 pg/100 μl (about 5×10^{-11} M) in a microtiter dilution well, a value comparable to those found for cPD1 and cAD1 (9, 10).

Since experiments on the inactivation of cAM373 with proteolytic enzymes showed the active substance to be a peptide (1), the isolated

cAM373:	H-Ala-Ile-Phe-Ile-Leu-Ala-Ser-OH
cPD1:	H-Phe-Leu-Val-Met-Phe-Leu-Ser-Gly-OH
cAD1:	H-Leu-Phe-Ser-Leu-Val-Leu-Ala-Gly-OH

FIG. 1. Amino acid sequences of *S. faecalis* sex pheromones. Coincident residues are circled (common to two of three peptides) or boxed (consensus in all three peptides).

TABLE 1. Isolation of cAM373

Purification step	Total wt (mg)	Total activity ($U \times 10^6$)	Sp act (ng/U)
Culture broth (16 liters)	610,000	5.12	119,000
Charcoal, activated	8,300	2.56	3,200
DEAE-Sephadex	1,100	2.56	430
LRP-2 (i)	180	2.56	70
LRP-2 (ii)	85	1.80	47
SSC-ODS-742 (i)	22	1.28	17
SSC-ODS-742 (ii)	1.9	1.28	1.5
CN-4251-N	0.013	1.28	0.01
SSC-ODS-262	0.0044	1.28	0.0034

sex pheromone cAM373 was subjected to sequence analysis with a gas-phase protein sequencer (model 470A, Applied Biosystems) (6) equipped with a PTH analyzer (model 120A, Applied Biosystems). As a result of analysis, the amino acid sequence H-Ala-Ile-Phe-Ile-Leu-Ala-Ser- was identified, and no amino acids were detected after the eighth step. In the fast atom bombardment mass spectrum of cAM373, measured in a matrix of glycerol containing hydrochloric acid with a JMS DX-303 mass spectrometer (JEOL), using xenon as the fast atom, the spectrum was terminated by a peak at mass-to-charge ratio 734, which was assignable to a quasi-molecular ion peak $(M + H)^+$. Thus the molecular weight of cAM373 was deduced to be 733, which explains the above-mentioned sequence with a free carboxy terminus. Accordingly, the structure of cAM373 was determined as H-Ala-Ile-Phe-Ile-Leu-Ala-Ser-OH.

The protected cAM373 was synthesized in solution by the fragment condensation between amino-terminal tripeptide and carboxy-terminal tetrapeptide. In each coupling reaction, condensation was carried out using HONB (N-hydroxy-5-norbornene-2,3-dicarboximide)-DCC (N,N'-dicyclohexylcarbodiimide) or the HONB activated ester method (5). After deblocking of all the protecting groups from the protected heptapeptide by trifluoroacetic acid treatment and hydrolysis, the desired material was collected from water. The purity of the product was checked by thin-layer chromatography with silica gel and amino acid analysis. The chromatographic behavior and clumping-inducing activity of the synthetic heptapeptide were identical with those of naturally obtained cAM373. Therefore, the chemical structure of cAM373 was unambiguously established.

Although the pheromone cAM373 is a heptapeptide and differs from the other two pheromones, both of which are octapeptides, it shares some structural characteristics with the pheromones cPD1 and cAD1: leucine as the third residue from the carboxy terminus, absence of acidic and basic residues, unusual hydrophobicity of the molecule, and Ser as the only hydrophilic residue (Fig. 1). Regardless of these similarities, each pheromone failed to induce the mating response of donors harboring plasmids related to the other pheromones at a concentration as high as 10^{-6} M. Kitada et al. (8) indicated, through synthetic studies on the structure-activity relationships of cPD1 and cAD1, that the specificity of pheromones to plasmids was determined by the amino-terminal part of the peptides. This may hold true in the case of cAM373, since the three peptides appear to differ in the amino-terminal region whereas they conserve relatively high homology in the carboxy-terminal part (Fig. 1).

The chemical structure and physiological role of the substance possessing the cAM373 activity in cultures of bacterial species other than S. faecalis remains to be determined. In this connection, the purification and characterization of the cAM373 activity present in cultures of an S. aureus strain is now in progress.

This work was supported in part by a Grant-in-Aid for Scientific Research (no. 60790127) from the Ministry of Education, Science and Culture of Japan.

LITERATURE CITED

1. **Clewell, D. B., F. Y. An, B. A. White, and C. Gawron-Burke.** 1985. *Streptococcus faecalis* sex pheromone (cAM373) also produced by *Staphylococcus aureus* and identification of a conjugative transposon (Tn918). J. Bacteriol. **162:**1212–1220.
2. **Clewell, D. B., B. A. White, Y. Ike, and F. An.** 1984. Sex pheromone and plasmid transfer in *Streptococcus faecalis*, p. 133–149. *In* R. Losick and L. Shapiro (ed.), Microbial development. Cold Spring Harbor Press, Cold Spring Harbor, N.Y.
3. **Dunny, G. M., B. L. Brown, and D. B. Clewell.** 1978. Induced cell aggregation and mating in *Streptococcus faecalis*; evidence for a bacterial sex pheromone. Proc. Natl. Acad. Sci. USA **75:**3479–3483.
4. **Dunny, G. M., R. A. Craig, R. L. Carron, and D. B. Clewell.** 1979. Plasmid transfer in *Streptococcus faecalis*: production of multiple sex pheromones by recipients. Plasmid **2:**454–465.
5. **Fujino, M., S. Kobayashi, M. Obayashi, T. Fukuda, S. Shinagawa, and O. Nishimura.** 1974. The use of N-hydroxy-5-norbornene-2,3-dicarboximide active esters in peptide synthesis. Chem. Pharm. Bull. (Tokyo) **22:**1857–1863.

6. **Hewick, R. M., M. W. Hunkapiller, L. E. Hood, and W. J. Dreyer.** 1981. A gas-liquid solid phase peptide and protein sequenator. J. Biol. Chem. **256:**7990–7997.

7. **Ike, Y., R. A. Craig, B. A. White, Y. Yagi, and D. B. Clewell.** 1983. Modification of *Streptococcus faecalis* sex pheromones after acquisition of plasmid DNA. Proc. Natl. Acad. Sci. USA **80:**5369–5373.

8. **Kitada, C., M. Fujino, M. Mori, Y. Sakagami, A. Isogai, A. Suzuki, D. Clewell, and R. Craig.** 1985. Synthesis and structure-activity relationships of *Streptococcus faecalis* sex pheromones, cPD1 and cAD1, p. 43–48. *In* N. Izumiya (ed.), Peptide chemistry 1984. Protein Research Foundation, Osaka, Japan.

9. **Mori, M., Y. Sakagami, M. Narita, A. Isogai, M. Fujino, C. Kitada, R. A. Craig, D. B. Clewell, and A. Suzuki.** 1984. Isolation and structure of the bacterial sex pheromone, cAD1, that induces plasmid transfer in *Streptococcus faecalis*. FEBS Lett. **178:**97–100.

10. **Suzuki, A., M. Mori, Y. Sakagami, A. Isogai, M. Fujino, C. Kitada, R. A. Craig, and D. B. Clewell.** 1984. Isolation and structure of bacterial sex pheromone, cPD1. Science **226:**849–850.

Streptococcus faecalis Plasmid pCF10 as a Model for Analysis of Conjugation in Streptococci and Other Gram-Positive Bacteria

G. M. DUNNY,[1] P. J. CHRISTIE,[1] M. L. TORTORELLO,[1] J. C. ADSIT,[1] R. Z. KORMAN,[2] AND S. A. ZAHLER[2]

Department of Veterinary Microbiology, New York State College of Veterinary Medicine,[1] and Section of Genetics and Development, Division of Biological Sciences,[2] Cornell University, Ithaca, New York 14853

Two types of conjugative gene transfer (where conjugation is defined as a transfer event that requires direct contact between donor and recipient cells, is insensitive to DNases, and does not involve a detectable transducing bacteriophage) have been demonstrated in streptococci and several related gram-positive genera. One type of transfer is exhibited by a number of broad-host-range plasmids and conjugative transposons (several of these are described in other papers in this volume). It involves a mechanism that generally functions at a fairly low frequency (10^{-4} to 10^{-8}) and often requires prolonged contact on a filter or agar surface. This type of transfer can occur across genus and species barriers. The second type of conjugation has, to date, been described only in *Streptococcus faecalis*. It involves a highly efficient transfer (10^{-4} to 10^{-1}) in liquid matings. The transfer functions of the plasmids that encode this second type of conjugation are inducible by peptide sex pheromones produced by recipient cells in mating mixtures. For extensive reviews of these conjugation systems and the associated plasmids and transposons, see references 3, 4a, and 14.

Several years ago we began an extensive analysis of pheromone-inducible conjugation in *S. faecalis*, employing the 58-kilobase (kb) tetracycline resistance plasmid, pCF10 (6). Several studies of conjugative hemolysin-bacteriocin plasmids by Clewell and co-workers (see their paper in this volume for a recent update) and of pCF10 by our group (7, 18) have shown that a number of biological functions and cell surface changes are induced as a result of the response of donor cells to pheromones. These include cell clumping (hence the designation of clumping-inducing agents for the peptide pheromones) (5, 15), DNA transfer functions distinct from clumping (4), surface- or entry-exclusion functions that prevent donor-donor matings (8), and synthesis of proteinaceous antigens on the cell surface (13, 18, 20). We are particularly interested in identifying, mapping, and studying the regulation of genes carried by pCF10 that are involved in these inducible processes, and also

in analyzing the cell surface changes associated with the pheromone response. In this paper we summarize the current status of this work and discuss some recent results indicating that pCF10 actually encodes two independent conjugation systems representing both of the two forms of transfer described above.

Molecular Organization of pCF10 Genes Involved in Antibiotic Resistance and Pheromone-Inducible Conjugation

Using a combination of restriction enzyme and hybridization analysis, and insertional mutagenesis with the transposon Tn917 (12, 16), we have begun to construct a physical and genetic map of pCF10 (2). Some of the important features of this map are noted in Fig. 1 and in the discussion below.

Tn917 insertions affecting pheromone response and transfer of pCF10 between *S. faecalis* strains mapped in a 20- to 30-kb region of the plasmid. Nine portions of this region, designated Tra 1 through 9, were defined by the phenotype and position of specific insertions. The portion of the plasmid carrying the Tra 4–6 regions spanned less than 2 kilobases, and insertions of Tn917 into these regions reduced or eliminated pheromone-inducible conjugation ability. Insertions of Tn917 into Tra 1–3, or Tra 7–9, resulted in higher donor potential and alterations such as constitutive clumping or differences in colony morphology. In addition, the Tra region exhibited an interesting symmetry, with a unique central Tra 5 region surrounded by pairs of regions with similar phenotypes, e.g., Tra 1 and 9, Tra 2 and 8, Tra 3 and 7, and Tra 4 and 6. It is tempting to speculate that the Tra 1–9 region may have originated from a DNA duplication event. Since the only insertions that reduced transfer occurred in the central Tra 4–6 regions, it is possible that some Tra functions of pCF10 have remained duplicated. The organization of this region also suggests that Tra 4–6 could represent an important regulatory region, such as a site of pheromone-inducible transcription initiation, with transcription proceeding

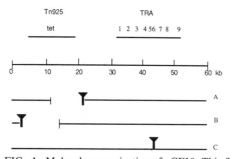

FIG. 1. Molecular organization of pCF10. This figure depicts the arrangement of the Tra and Tn925 portions of pCF10, with the plasmid linearized at an EcoRI site between the EcoRI a and d fragments, as determined by Christie and Dunny (2). A and B represent mutant derivatives of the plasmid where insertion of Tn917 (indicated by ┳) and deletion of adjacent pCF10 DNA have removed large segments of Tn925. S. faecalis strains carrying these plasmids show normal pheromone-inducible plasmid transfer to S. faecalis recipients, but fail to transfer to B. subtilis. C represents a Tn917 insertion into the Tra 5 region of the plasmid that abolishes pheromone response and plasmid transfer to S. faecalis recipients in broth matings. Strains carrying this plasmid transfer Tn925 to B. subtilis at wild-type frequencies.

outward in one or both directions. The pheromone-inducible plasmid pAD1 was subjected to a similar analysis by Ike and Clewell (12). They identified two regions of pAD1, Tra A and Tra B, that appeared to encode pheromone-inducible transfer functions. As indicated in Fig. 2, the organization of Tra A and Tra B shows some similarity to that of Tra 4–8 of pCF10. This relatedness will need to be confirmed by molecular analysis. Further molecular and genetic analyses currently under way in our laboratories include determination of the minimum amount of pCF10 DNA required for a wild-type pheromone response and transfer phenotype, determination of the extent (if any) of duplication in the Tra region, and analysis of transcription of the Tra genes in the presence and absence of pheromones.

The position of the tetracycline resistance determinant (Tet^r) of pCF10 was determined by mapping the locations of Tn917 insertions and insertions that also generated deletions of adjacent pCF10 DNA (2). The region containing this determinant showed a high degree of sequence homology to the conjugative tetracycline resistance transposon, Tn916 (9, 10). However, unlike Tn916, the Tet^r element of pCF10 showed a very stable association with the plasmid, through multiple rounds of transfer. We were never able to demonstrate transposition of this element from pCF10 to any other replicon in S. faecalis. However, two lines of evidence obtained recently have led to the conclusion that the Tet^r element of pCF10 is actually part of a transposable element that encodes a conjugal transfer system distinct from the pheromone-inducible system encoded by the Tra region.

When S. faecalis strains carrying pCF10 are mixed in broth with Bacillus subtilis strains derived from strain 168 (1), transfer of tetracycline resistance from the Streptococcus donors to the Bacillus recipients occurs at frequencies of 10^{-7} to 10^{-8} per donor. The mechanism of transfer involves direct cell-to-cell contact, and transfer is not inhibited by DNases. The tetracycline-resistant transconjugants carry only the portion of pCF10 homologous to Tn916 integrated into the chromosome at various locations. These transconjugants can serve as donors of Tet^r to other B. subtilis recipients in broth matings and to S. faecalis recipients in filter matings. S. faecalis strains that have acquired the Tet^r element can serve as donors to S. faecalis, B. subtilis, and Streptococcus agalactiae recipients. All of the donors and transconjugants in these experiments lacked plasmid DNA and carried the Tet^r element in their chromosomes. We also demonstrated that the transfer occurred normally in recombination-deficient bacterial strains. Thus, the Tet^r determinant of pCF10 seems to be a part of a conjugative transposon, now termed Tn925, that is very similar to Tn916 (10). Like Tn916, Tn925 can also insert within or near the hemolysin gene of the plasmid pAD1 (9, 10) and alter the level of hemolysin produced by the host bacterium. In a paper recently submitted for publication, we describe the transfer of Tn925 from S. faecalis to B. subtilis and present a detailed molecular and genetic characterization of Tn925 (P. Christie et al., submitted for publication). Although Tn925

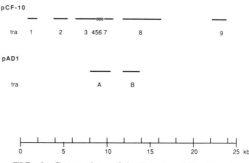

FIG. 2. Comparison of the Tra A and Tra B regions of pAD1 and the Tra 1 through 9 regions of pCF10. These maps were aligned by correlating phenotypes and map positions of Tn917 insertions into the two plasmids as reported for pAD1 by Ike and Clewell (12) and for pCF10 by Christie and Dunny (2). The actual extent of nucleotide sequence homology of these regions has not been determined.

can function separately from the rest of pCF10, we wished to learn whether this element contributes to transfer functions previously associated with the Tra region of the plasmid. In Fig. 1, the positions of several insertion and deletion derivatives of pCF10 are shown. As indicated in this figure, deletions of large portions of Tn925 that completely abolish transfer between *S. faecalis* and *B. subtilis* have no effect on pheromone-inducible transfer. Similarly, Tn917 insertions in Tra 5, which inactivate pheromone-inducible transfer of pCF10 in *S. faecalis*, have no effect on transfer of Tn925 from *S. faecalis* and *B. subtilis*.

In the course of our Tn917 mutagenesis studies of pCF10, we isolated a strain in which one or more Tn917 insertions into the plasmid resulted in loss of the plasmid and insertion into the *S. faecalis* chromosome of a derivative of Tn925 carrying Tn917 inserted about 3 kb from one end (at a position corresponding to coordinate 16 kb on the map shown in Fig. 1). Both resistance and transposition functions of this composite transposon are apparently intact. In fact, transfer of the composite Tn925::Tn917 element can be up to 100 times more efficient than that of the wild-type Tn925 (Christie et al., submitted). When Tn925::Tn917 is transferred from *S. faecalis* to *B. subtilis*, and selection is made for Tet^r, there is 100% cotransfer of the MLS^r marker of Tn917. However, when the selection is for MLS^r, about 10 to 20% of the transconjugants are tetracycline sensitive, indicating the likely transposition of Tn917 from Tn925 to some other location in the recipient genome. Thus, the composite transposon could be useful as a delivery vehicle for Tn917, as well as in its own right.

Analysis of Cell Surface Antigenic Changes Associated with the Pheromone Response

The pCF10 system represents a good model for the development of combinations of genetic, biochemical, and immunological techniques for the analysis of bacterial cell surfaces. We have used polyclonal and monoclonal antibodies to analyze the antigenic changes that occur as a result of the pheromone response encoded by pCF10. All bacterial strains used in these studies were isogenic derivatives of strain OG1 (11) that differed only in plasmid content or in the extent of pheromone induction. Rabbits, mice, or rats were immunized with whole cells, and various polyclonal and monoclonal antibody preparations obtained were tested for reactivity with surface antigens with whole-cell enzyme-linked immunoassays, or by immunoprecipitation and immunoblotting analysis of antigenic extracts. Cell surface antigens were obtained from culture supernatants, from cells extracted with Zwit-

tergent detergent (13), or from cells subjected to partial cell envelope digestion by lysozyme and mutanolysin. As described below, the use of several extraction and analytical procedures was necessary to detect the variety of antigenic changes that occur as a result of pheromone induction of *S. faecalis* cells carrying pCF10. Many of our results have been described in several publications (8, 17–19). These data are summarized below, along with some more recent experimental findings.

We have detected three major antigens that appear to be synthesized specifically during pheromone stimulation. The appearance of these antigens, as well as several chromosomally encoded antigens, is depicted in Fig. 3. Recently, we adopted the convention of referring to inducible antigenic bands on immunoblots with the designation "Tra" followed by a number indicating the apparent molecular mass in kilodaltons. As noted below, it is likely that several of these antigens appear as groups of bands that represent different molecular-weight forms of the same gene product. Therefore, we will use the numerical system to designate the bands until the relationship of the various bands to specific genes is better defined. Antigens that are chromosomally encoded are designated as "Sfa" antigens.

The antigenic bands depicted in Fig. 3 include Sfa80 (formerly SA80), the predominant chromosomal antigen detected, along with several other less intense bands that are expressed constitutively. The pheromone-inducible antigens consist of Tra73 (formerly Sa73), a proteinaceous antigen that is best extracted from induced cells with Zwittergent and that migrates as a single band, a group of bands in the 125- to 135-kilodalton range referred to as the Tra130 (formerly C130) group, and a pair of bands migrating in the 150-kilodalton range, called the Tra150 antigen. Tra150 and Tra130 can both be extracted by gentle lysozyme-mutanolysin treatment of pheromone-induced donor cells, and Tra130 is also shed into the medium in large quantities. Reactivity of the various bands with monoclonal antibodies, as well as their biochemical properties, leads us to believe that the two bands in Tra150 represent different molecular-weight forms of one protein, while several if not all of the bands in the Tra130 group represent a second protein. Tra73 appears to be a third distinct protein.

Thus far, most of our analysis of these antigens has focused on Tra130 (8). We have shown that monoclonal antibodies against this antigen develop bands in the 130-kilodalton range in extracts from pheromone-induced cells carrying pCF10. Interestingly, extracts from cells carrying a mutant derivative of pCF10 (pCF11) that

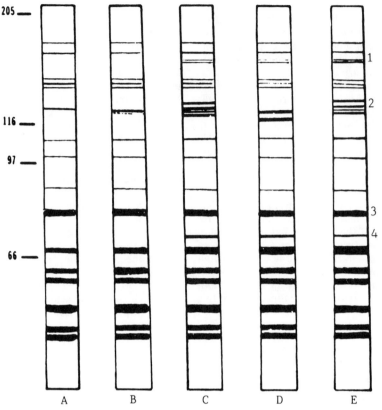

FIG. 3. Antigenic profiles of isogenic derivatives of *S. faecalis* OG1 carrying various plasmids. This drawing represents the appearance of the major antigens detected by immunoblotting analysis. The antiserum used to develop the blots was raised against pheromone-induced donor cells. The drawing represents a composite of all antigens detected in various types of extracts. As discussed in the text, not all antigens are readily detected in every blot, since they show differences in extractability with various procedures. Lane A, Antigen profile of plasmid-free cells; lane B, antigen profile of cells carrying pCF10 grown in the absence of pheromones; lane C, antigen profile of pheromone-induced cells carrying pCF10; lane D, antigen profile of cells carrying pCF11, a mutant derivative of pCF10 that confers a constitutively clumpy and elevated transfer phenotype on its host; lane E, antigen profile of cells carrying pCF11 and grown in the presence of exogenous pheromones. The numbers 1, 2, 3, and 4 indicate the position of the Tra150, Tra130, Sfa80, and Tra73 antigens, respectively.

confers a constitutive clumping and elevated transfer phenotype contain a similar group of antigenic bands, but these bands (Tra125) migrate somewhat faster in sodium dodecyl sulfate gels. Treatment of cells carrying pCF11 with exogenous pheromone preparations results in a "shift" in the banding pattern to the Tra130 pattern observed in extracts of wild-type induced donors. Addition of monoclonal antibodies against Tra130 to conventional donor-recipient mating mixtures has no effect on transfer frequencies. However, we have demonstrated that pheromones induce a surface exclusion function that inhibits mating between two cells carrying the same plasmid. Interestingly, the monoclonal antibodies against Tra130 inhibit surface exclusion. Furthermore, the strains that carry pCF11 and produce Tra125 show inducible surface exclusion that is correlated with the shift to the Tra130 banding pattern. We feel that Tra130 is involved in surface exclusion and that synthesis of an active protein may involve a posttranslational modification from a precursor protein to a higher-molecular-weight form. Apparently cells carrying pCF11 synthesize the precursor, but remain inducible for the modification activity necessary for synthesis of a functional protein. Further genetic and molecular analyses of Tra130 are currently under way to verify this model.

Monoclonal antibodies have also been useful in analysis of the Tra150 antigen. This antigen can be immunoprecipitated from lysozyme-mutanolysin extracts by polyclonal antiserum raised against induced donor cells. We derived monoclonal antibodies that show specific reactivity to pheromone-induced donor cells. One of these antibodies inhibits conjugation. The biological activity of the monoclonal antibody suggests that it is directed toward an epitope that is part of the pheromone-induced aggregation substance (13, 18, 20) that mediates contact between donor and recipient cells. This antibody has been shown recently to immunoprecipitate both of the bands of the Tra150 group. However, the precipitated antigens also include some Tra130. We believe that Tra130 and Tra150 may be a part of a large macromolecular complex on the cell surface and are coprecipitated by monoclonal antibody directed against Tra150. Previous results (18) also suggested that Tra130 was associated with a large macromolecular complex in the native state. To date, we have not been able to obtain a monoclonal antibody reactive with the Tra73 antigen, nor is any genetic evidence available to suggest a possible function for Tra73.

To better determine the role of these antigens in conjugation, we are pursuing both biochemical and genetic analyses. The availability of monoclonal antibody reagents should facilitate purification and more extensive biochemical and functional analysis. We have had some recent success in purifying Tra130 by immunoaffinity chromatography, as illustrated in Fig. 4. The purified antigens can be used for biochemical analysis, in biological assays (inhibition of mating or surface exclusion), and as immunizing reagents for production of additional monoclonal antibodies. This type of approach eventually should enable us to define specific regions of the proteins responsible for activity. In addition, the monoclonal antibodies should be useful as cytological staining reagents that may enable visualization and analysis of interactions between individual cells in mating pairs by electron microscopy. The various antibody reagents described above are also being employed in analysis of cloned fragments of pCF10 and in mutations of the plasmid obtained by transposon insertion. Thus we expect to be able to correlate specific antigens with plasmid-encoded genes and biological functions. Finally, it should be noted that transposon mutagenesis and immunochemical analysis are also being employed to analyze chromosomal genes involved in conjugation, as described by K. E. Carberry-Goh et al. (this volume).

Conclusions

The *S. faecalis* plasmid pCF10 actually encodes two different conjugation systems, a situation that has not been observed previously in a plasmid isolated from nature. Broad-host-range conjugal transfer ability is encoded by a segment of the plasmid Tn925, which is a conjugative

FIG. 4. Immunoaffinity purification of the Tra130 antigen. A monoclonal antibody against Tra130 was bound to a protein A-agarose column (Pierce Chemical), and culture supernatants from cells carrying pCF11 were applied to the column. Bound material was eluted with 50 mM Tris hydrochloride, pH 8.4. Fractions were lyophilized, desalted, and subjected to immunoblotting analysis. Lane 1, Precolumn antigen; lane 2, unbound antigen; lanes 3–5, material eluted from column.

tetracycline resistance transposon similar to Tn916. Pheromone-inducible plasmid transfer is encoded by a separate portion of the plasmid called the Tra region. This region shows a very interesting molecular organization. Pheromone induction of cells carrying pCF10 results in synthesis of at least three proteinaceous antigens that may be involved in cell aggregation, DNA transfer, and surface exclusion functions associated with the pheromone response. Genetic, molecular, and immunologic techniques have been developed that should permit a more thorough investigation of the conjugation functions encoded by pCF10.

This research was supported by Public Health Service grant AI19310 from the National Institute of Allergy and Infectious Diseases and by a grant from the Cornell Biotechnology Program.

We thank Laura Jean vanPutte for excellent assistance with the immunoblotting analysis.

LITERATURE CITED

1. **Anagnostopoulos, C., and J. Spizizen.** 1961. Requirements for transformation in *Bacillus subtilis.* J. Bacteriol. **81:**741–746.

2. **Christie, P. J., and G. M. Dunny.** 1986. Identification of regions of the *Streptococcus faecalis* plasmid pCF10 that encode antibiotic resistance and pheromone response functions. Plasmid **15:**230–241.

3. **Clewell, D. B.** 1985. Sex pheromones, plasmids, and conjugation in *Streptococcus faecalis,* p. 13–28. *In* H. O. Halvorson and A. Monroy (ed.), The origin and evolution of sex. Alan R. Liss, New York.

4. **Clewell, D. B., and B. L. Brown.** 1980. Sex pheromone cAD1 in *Streptococcus faecalis*: induction of a function related to plasmid transfer. J. Bacteriol. **143:**1063–1065.

4a. **Clewell, D. B., and C. Gawron-Burke.** 1986. Conjugative transposons and the dissemination of antibiotic resistance in streptococci. Annu. Rev. Microbiol. **40:**635–659.

5. **Dunny, G. M., B. L. Brown, and D. B. Clewell.** 1978. Induced cell aggregation and mating in *Streptococcus faecalis*: evidence for a bacterial sex pheromone. Proc. Natl. Acad. Sci. USA **75:**3475–3483.

6. **Dunny, G., C. Funk, and J. Adsit.** 1981. Direct stimulation of the transfer of antibiotic resistance by sex pheromones in *Streptococcus faecalis.* Plasmid **6:**270–278.

7. **Dunny, G., M. Yuhasz, and E. Ehrenfeld.** 1982. Genetic and physiological analysis of conjugation in *Streptococcus faecalis.* J. Bacteriol. **151:**855–859.

8. **Dunny, G. M., D. L. Zimmerman, and M. L. Tortorello.** 1985. Induction of surface exclusion by *Streptococcus faecalis* sex pheromones: use of monoclonal antibodies to identify an inducible surface antigen involved in the exclusion process. Proc. Natl. Acad. Sci. USA **82:**8582–8586.

9. **Franke, A. E., and D. B. Clewell.** 1981. Evidence for a chromosome-borne resistance transposon (Tn*916*) in *Streptococcus faecalis* that is capable of "conjugal" transfer in the absence of a conjugative plasmid. J. Bacteriol. **145:**494–502.

10. **Gawron-Burke, C., and D. B. Clewell.** 1982. A transposon in *Streptococcus faecalis* with fertility properties. Nature (London) **300:**281–284.

11. **Gold, O., H. V. Jordan, and J. van Houte.** 1975. The presence of enterococci in the human mouth and their pathogenicity in animal models. Arch. Oral Biol. **20:**473–477.

12. **Ike, Y., and D. B. Clewell.** 1984. Genetic analysis of the pAD1 pheromone response in *Streptococcus faecalis,* using transposon Tn*917* as an insertional mutagen. J. Bacteriol. **158:**777–783.

13. **Kessler, R. E., and Y. Yagi.** 1983. Identification and partial characterization of a pheromone-induced adhesive surface antigen of *Streptococcus faecalis.* J. Bacteriol. **155:**714–721.

14. **LeBlanc, D. J.** 1981. Plasmids in streptococci: a review, p. 81–90. *In* S. B. Levy, R. C. Clowes, and E. L. Koenig (ed.), Molecular biology, pathogenicity and ecology of bacterial plasmids. Plenum Publishing Corp., New York.

15. **Suzuki, A., M. Mori, Y. Sakagami, A. Isogai, M. Fugino, C. Kitada, R. A. Craig, and D. B. Clewell.** 1984. Isolation and structure of bacterial sex pheromone cPD1. Science **226:**849–850.

16. **Tomich, P. K., F. Y. An, and D. B. Clewell.** 1980. Properties of erythromycin-inducible transposon Tn*917* in *Streptococcus faecalis.* J. Bacteriol. **141:**1366–1374.

17. **Tortorello, M. L., J. C. Adsit, D. A. Krug, D. F. Antczak, and G. M. Dunny.** 1986. Monoclonal antibodies to cell surface antigens involved in sex pheromone induced matings in *Streptococcus faecalis.* J. Gen. Microbiol. **132:** 857–864.

18. **Tortorello, M. L., and G. M. Dunny.** 1985. Identification of multiple cell surface antigens associated with the sex pheromone response of *Streptococcus faecalis.* J. Bacteriol. **162:**131–137.

19. **Tortorello, M. L., D. A. Krug, and G. M. Dunny.** 1985. Identification of surface antigens associated with pheromone response in *Streptococcus faecalis,* p. 54–55. *In* Y. Kimura, S. Kotami, and Y. Shiokawa (ed.), Recent advances in streptococci and streptococcal diseases. Reedbooks, Berkshire, United Kingdom.

20. **Yagi, Y., R. E. Kessler, J. H. Shaw, D. E. Lopatin, F. An, and D. B. Clewell.** 1983. Plasmid content of *S. faecalis* strain 39-5 and identification of a pheromone (cPD1)-induced surface antigen. J. Gen. Microbiol. **129:**1207–1215.

Transposon Tn*917* Delivery Vectors for Mutagenesis in *Streptococcus faecalis*

KEITH E. WEAVER AND DON B. CLEWELL

Departments of Oral Biology and Microbiology/Immunology, Schools of Dentistry and Medicine and the Dental Research Institute, The University of Michigan, Ann Arbor, Michigan 48109-2007

Transposon-mediated insertional mutagenesis has found wide applicability for the isolation of random mutations in a variety of gram-negative procaryotes. With the discovery of transposons in gram-positive bacteria these techniques can now be extended to these organisms. Transposon Tn*551* has been used in *Staphylococcus aureus* for obtaining auxotrophic markers and for chromosome mapping (8, 9, 14). In *Streptococcus faecalis*, Tn*917* (15) has been used for genetic studies on plasmid pAD1 (7) and its hemolysin gene (2). For more efficient delivery of Tn*917*, P. Youngman (*in* K. Hardy, ed., *Plasmids: A Practical Approach*, in press) constructed a number of temperature-sensitive vectors and used them to deliver the transposon to the *Bacillus subtilis* chromosome. These vectors were derived from the *S. aureus* plasmid pE194, which is unable to replicate in *B. subtilis* at temperatures above 45°C (6). To broaden the applicability of this system to species unable to grow at these temperatures, a similar set of vectors was constructed using a mutant derivative of pE194, designated pE194(Ts) (6), which displays an even more extreme temperature-sensitive replication deficiency (Youngman, in press). This deficiency is due to a 2-base-pair change in the *rep(incA)* region of the plasmid. Some of these plasmids show promise for use in *S. faecalis*. Plasmid pTV1 is a 12.4-kilobase (kb) plasmid carrying Tn*917* and the pE194 replicon. It contains a plasmid-encoded chloramphenicol resistance and the transposon-encoded erythromycin resistance. Plasmids pTV1(Ts) and pTV-32(Ts) are identical to plasmid pTV1 except that they contain the pE194(Ts) replicon. In addition, pTV32(Ts) carries a derivative of Tn*917* containing a promoterless *lacZ* gene near one end, making it useful for constructing transcriptional gene fusions. The *lacZ* gene does not interfere with transposition or with expression of the erythromycin resistance gene (10). We have assessed the utility of these three plasmids for genetic analysis in *S. faecalis*.

Plasmid Instability in *S. faecalis*

Plasmids pTV1(Ts) and pTV32(Ts) were introduced into *S. faecalis* OG1X by protoplast transformation (17). The transformation efficiency was very low [approximately one transformant per µg of pTV1(Ts) DNA used] and required 5 to 12 days of incubation for the appearance of transformants on regeneration plates containing 5 µg of erythromycin per ml. No pTV1(Ts) or pTV32(Ts) transformants were isolated on plates containing 5 µg of chloramphenicol per ml, but chloramphenicol resistance (Cm^r) was expressed by transformants isolated on erythromycin-containing plates. The isolation of OG1X(pTV1) was previously described (12).

Youngman (in press) noted that in *B. subtilis* the double mutation present in the *rep* region of pTV1(Ts) rendered it much less stable at elevated temperature than the parental plasmid, pTV1. While pTV1 displayed a gradual decrease in copy number at increasing incubation temperatures between 35 and 45°C, pTV1(Ts) showed an abrupt replication block at 38°C and above. To determine whether these observations could be extended to *S. faecalis* we compared the rate of segregation of the two plasmids at 32 and 42°C (Fig. 1). The average half-life ($t_{1/2}$) in generations at 42°C over the extent of the experiment for pTV1 and pTV1(Ts) is 2.1 and 1.2, respectively. However, during the first four generations pTV1(Ts) has a $t_{1/2}$ of 2.5 while pTV1 shows little segregation. After the first four generations, pTV1 has a $t_{1/2}$ of 1.6 generations, while pTV1(Ts) has a $t_{1/2}$ of 1.0 (Table 1). The time at which the proportion of plasmid-containing strains reaches the 0.01% intercept (as extrapolated from the graph in Fig. 1), which would be expected to be the threshold of Tn*917* transposition frequency (18), differs by about five generations or 3 h. Plasmid pTV1(Ts) was less stable than pTV1 at 32°C (see Table 1), which may explain the difficulties experienced in isolating transformants.

The substantial difference in the behavior of the two plasmids during the first four generations of growth at the nonpermissive temperature suggested that a difference in copy number of the two plasmids might exist in this organism. The minimal inhibitory concentration of chloramphenicol for plasmid-containing strains was determined in broth inoculated with a 1:500 dilution of late-log-phase cultures grown on 25 µg of chloramphenicol per ml to maintain the

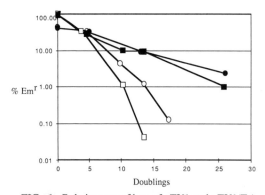

FIG. 1. Relative rate of loss of pTV1 and pTV1(Ts) at 32 and 42°C. Strains to be tested were grown overnight at the permissive temperature (32°C) in the presence of 25 µg of chloramphenicol per ml. Cells were harvested, washed, and resuspended in antibiotic-free medium. Time zero samples were taken at this time, diluted, and plated to antibiotic-free medium and to erythromycin (10 µg/ml)-containing plates for determination of the percentage Em^r. Cells were diluted 1:50 in antibiotic-free medium and incubated at the temperature indicated. Cultures were maintained in log phase by successive dilutions approximately every four generations. Samples for determination of the percentage Em^r were taken periodically and treated as at time zero. Symbols: ■, pTV1(Ts), 32°C; □, pTV1(Ts), 42°C; ●, pTV1, 32°C; ○, pTV1, 42°C.

plasmid. Test cultures were incubated for 20 h at 32°C. The minimal inhibitory concentration of chloramphenicol for OG1X[pTV1(Ts)] and OG1X (pTV1) revealed a twofold difference (62.5 and 125 µg/ml, respectively), a result consistent with a difference in copy number.

The copy number of pTV1 and pTV1(Ts) in [³H]thymidine-labeled cultures of OG1X was determined by the method of Clewell et al. (3). Assuming an *S. faecalis* genome size of 1,500 megadaltons, the estimated copy numbers of pTV1 and pTV1(Ts) were 20 and 10 copies per chromosome, respectively. This difference may account for the lag observed in the initiation of loss of pTV1 following transfer to the nonpermissive temperature. It should be noted that, while copy number was determined on log-phase cultures, stationary-phase cultures were used as inocula for determination of plasmid instability. Because of the observed instability at 32°C, copy numbers of the two plasmids may be below maximum at time zero of instability experiments as ribosomes become maximally methylated and selective pressure for maintenance of the plasmid is removed. This may explain why the plasmids begin to be lost more rapidly than would be expected from their copy number.

Table 1 is a summary of the observed rates of loss of pTV1, pTV1(Ts), and pTV32(Ts) in strain OG1X. Plasmids pTV1(Ts) and pTV32(Ts) are nearly identical in their instability characteristics in this strain. We were interested by the observation that all three plasmids displayed significant instability at temperatures below 32°C. To our knowledge no such instability has been observed in *B. subtilis*. While the basis of this instability is unknown, its practical implications are obvious. Care should be taken at all times in the storage of these strains at low temperatures.

Transposon Insertion

On the basis of the results presented above, a protocol for mutagenesis was devised. Cultures to be mutagenized were grown overnight at 32°C in Todd-Hewitt broth containing 25 µg of chloramphenicol and 0.05 µg of erythromycin per ml (the latter to induce transposition [2]). Cells were harvested, washed once in antibiotic-free medium, and resuspended in an equal volume of Todd-Hewitt broth. At time zero, the percentage of Em^r cells was determined. A 1:1,000 dilution was made in Todd-Hewitt broth, the culture was grown to stationary phase at 42°C, and a second cycle of dilution and growth at the nonpermissive temperature was performed. This growth period correlates to approximately 20 generations, after which the proportion of plasmid-bearing cells should be below the expected transposition frequency (see above and Fig. 1). Samples were diluted and plated to Todd-Hewitt broth and Todd-Hewitt broth supplemented with 10 µg of erythromycin per ml to determine the transposition frequency.

Using this procedure, we tested the ability of pTV1(Ts) to donate Tn917 to the *S. faecalis* chromosome and to two plasmids, pAD1 and pPD1. In the absence of plasmid DNA, transpo-

TABLE 1. Stability of pTV1, pTV1(Ts), and pTV32(Ts) at various temperatures

Strain[a]	Plasmid $t_{1/2}$[b]		
	22°C	32°C[c]	42°C[c]
OG1X(pTV1)	1.7	5.6	1.6
OG1X[pTV1(Ts)]	1.8	4.5	1.0
OG1X[pTV32(Ts)]	1.7	4.1	1.1

[a] Strains were treated as described in the legend of Fig. 1. Doubling times were approximately 1 h at 32°C, between 0.6 and 0.7 h at 42°C, and 1.7 h at 22°C.

[b] Plasmid stability is represented as $t_{1/2}$ at the temperature indicated; $t_{1/2}$ is defined as the number of generations required for the proportion of plasmid-containing cells to decrease by a factor of two. A plasmid unable to replicate would have a $t_{1/2}$ of 1. The $t_{1/2}$s were calculated as an average over >20 generations at 32°C and ≈10 generations at 22 and 42°C.

[c] The $t_{1/2}$s calculated at 32 and 42°C omitted the first four generations because of the difference in copy number of the two plasmids.

TABLE 2. Tn917 hybridizable fragments in chromosomal insertions[a]

Strain	Fragments[b]		Total	Less Tn917[c]
	1	2		
A16	9.4	4.2	13.6	8.4
A26	10.0	5.0	15.0	9.8
B26	8.0	3.6	11.6	6.4
B37[d]	3.9[e]	3.5[e]	7.4	2.2
	4.0	3.9[e]	7.9	2.7
C35	18.0	13.2	31.2	26.0
C48	9.3	8.2	17.5	12.3[f]
	16.0	9.3	24.3	19.1
D18[d]	13.8	3.3	17.1	11.9[f]
	3.9[e]	3.3	7.2	2.0
D48	8.0	2.8	10.8	5.6

[a] Each strain is an independent isolate of a single mutagenesis. Chromosomal DNA was purified by CsCl-ethidium bromide equilibrium gradient. Samples were cut with NcoI and fractionated on a 0.8% agarose gel. DNA was transferred to nitrocellulose by the method of Southern (13) as modified by Wahl et al. (16) and hybridized to nick-translated pAM225. Preparation of the probe and autoradiography procedures were as previously described (5).

[b] Fragment sizes are expressed as kilobases.

[c] The size of Tn917 is ≈5.2 kb.

[d] Cm[r] isolates.

[e] These bands are approximately the same size as Tn917-containing pTV1(Ts) NcoI fragments and may represent contaminating autonomous or integrated plasmid.

[f] Probably within the range of error of size determination.

sition frequencies were consistently on the order of 10^{-4} (see Table 3). This is well within the range observed by Youngman (in press) in *B. subtilis*. Approximately 1.5% of Em[r] colonies were still Cm[r], indicating that they retained a copy of the plasmid either autonomously or integrated into the chromosome.

To assess the degree of specificity of transposition, chromosomal DNA from six chloramphenicol-sensitive (Cm[s]) and two Cm[r] strains isolated after mutagenesis was purified, digested with NcoI, fractionated, transferred to nitrocellulose, and probed with radiolabeled pAM225. Plasmid pAM225 is a pBR325 derivative containing Tn917 (11). The sizes of the hybridizing fragments are shown in Table 2. Restriction enzyme NcoI was chosen because it cuts once near the center of Tn917 and twice within the pE194 section of pTV1(Ts), resulting in the formation of two fragments of approximately 4 and 3.3 kb which are recognized by a Tn917 probe. Hybridization with these fragments would indicate the presence of the plasmid.

All insertions were clearly in different sites, and all but one pair of insertions were in different size fragments. At least one Cm[s] isolate, C48, contained four hybridizing bands. Since

NcoI cuts once within Tn917, this most likely represents a double insertion event. At least one of the pTV1(Ts) specific fragments is present in each of the two Cm[r] isolates. In D18, both fragments are present but at a significantly lower intensity than the other insert in that strain. This may indicate that these bands are due to plasmid contamination in the chromosome preparation. In B37 only the larger of the two bands is present. If this is indeed a pTV1(Ts)-related band, it may indicate that plasmid integration has occurred. Interestingly, both Cm[r] strains contain inserts independent of the pTV1(Ts)-related bands.

Insertion frequencies in the presence of various plasmids are shown in Table 3. In the presence of pAD1 (56.7 kb; determines hemolysin-bacteriocin; conjugative), transposition of Tn917 shows a clear preference for plasmid DNA, as confirmed by determining the ability of mutagenized strains to donate transposon-encoded Em[r] in 2-h broth matings (1). In 20 such matings with independently isolated insertions, only one failed to transfer Em[r].

In the presence of pAM351, a derivative of pPD1 (55.7 kb; determines bacteriocin; conjugative) carrying Tn916 (4), transposition of Tn917 was severely inhibited. To determine whether Tn916 was responsible for the observed inhibition, two Tn916-bearing pAD1 derivatives (pAM210 and pAM250) (2) were tested for their ability to inhibit transposition. While transposition frequencies in the presence of these two plasmids were slightly lower than in the presence of pAD1, they were still above that observed in the absence of plasmid. As in the case of pAD1, most Tn917 insertions were into pAM210 and pAM250. Therefore, it appears that transposition inhibition is due to pPD1 rather than Tn916.

Isolation of *lacZ* Fusions with pTV32(Ts)

Youngman (in press) has described pTV-32(Ts), which contains a modified Tn917 carrying a promoterless *lacZ* gene near one end. This gene is carried along with Tn917 when it transposes, and if it is inserted downstream from an active promoter in the correct orientation, the *lacZ* gene is expressed. Perkins and Youngman (10) have observed that in *B. subtilis* approximately 15% of chromosomal inserts of this altered transposon result in production of blue pigment on plates containing 40 µg of 5-bromo-4-chloro-3-indolyl-β-D-galactoside (X-Gal) per ml, suggesting that transcriptional fusions were isolated. To determine whether this technique was applicable to *S. faecalis*, we screened several chromosomal insertions isolated after mutagenesis with pTV32(Ts) for their ability to produce pigment on X-Gal-containing plates. Ap-

TABLE 3. Plasmid preference of Tn917 transposition[a]

Strain	% Em[r] at time zero	Transposition frequency[b]	Transposition relative to OG1X[pTV1(Ts)]
OG1X[pTV1(Ts)]	43	1.32×10^{-4c}	1.0
OG1X[pTV32(Ts)]	34	2.68×10^{-5d}	0.2
OG1X[pTV1(Ts), pAD1]	≈100	$3.49 \times 10^{-3e,f}$	26.4
OG1X[pTV1(Ts), pAM351][g]	≈100	$2.0 \times 10^{-7b,c}$	0.0015
OG1X[pTV1(Ts), pAM210][h]	≈100	1.40×10^{-3f}	10.6
OG1X[pTV1(Ts), pAM250][h]	80	5.00×10^{-4f}	3.8

[a] Strains were subjected to mutagenesis as described in the text. Samples were plated at time zero to ensure that no unusual amount of plasmid loss had occurred during the initial growth.

[b] Transposition frequency is defined as the proportion of Em[r] colonies remaining after mutagenesis.

[c] Average of four experiments. Range: 3.4×10^{-5} to 2.79×10^{-4} for OG1X[pTV1(Ts)] and 1.2×10^{-7} to 4×10^{-7} for OG1X[pTV1(Ts), pAM351].

[d] Average of two experiments.

[e] Average of three experiments.

[f] As explained in the text, most transpositions occurring in these strains resulted in insertions into plasmid DNA.

[g] Plasmid pAM351 is a pPD1 derivative containing a Tn916 insert.

[h] Plasmids pAM210 and pAM250 are Tn916-containing pAD1 derivatives.

proximately 2 to 10% of Em[r] isolates showed clear expression of the *lacZ* gene above background on X-Gal (30 μg/ml)-containing plates. In addition, variability in the intensity of blue pigment produced by different clones indicates that insertions are occurring in different sites with varying levels of expression, a phenomenon also observed in *B. subtilis* (10). Detectable levels of LacZ required incubation of mutagenized colonies for 5 days on X-Gal-containing plates. In contrast, blue *B. subtilis* colonies appear overnight in most cases (10).

Discussion

We have tested the utility of three plasmids, pTV1, pTV1(Ts), and pTV32(Ts), for transpositional mutagenesis and transcriptional gene fusion in *S. faecalis*. All three plasmids are significantly temperature sensitive in plasmid replication functions in this organism. The optimum temperature for maintenance is ≈32°C, and stability decreases at incubation temperatures above and below this optimum. At 42°C a difference in stability is observed between pTV1 and pTV1(Ts). As expected from observations in *B. subtilis*, where replication of pTV1(Ts) is completely blocked above 38°C (Youngman, in press), the $t_{1/2}$ of pTV1(Ts) approaches 1 at 42°C. Since replication of pTV1 shows a gradual decrease in copy number up to 45°C, at which point replication is completely inhibited in *B. subtilis* (Youngman, in press), the difference in $t_{1/2}$ observed after the first four generations may be due to some residual replication that persists at 42°C. The difference observed between loss of pTV1 and pTV1(Ts) during the first four generations is more than likely due to a higher copy number of pTV1 at the beginning of the experiment. Thus, it takes several more generations

for segregants lacking plasmid to arise from pTV1-containing cells than from pTV1(Ts)-containing cells. It is interesting that the copy number of pTV1 in *B. subtilis* has been estimated at approximately six copies per chromosome (Youngman, in press), whereas we estimate a copy number of 20 copies per chromosome in *S. faecalis*.

Despite the observed differences in the rate of loss of pTV1 and pTV1(Ts) in *S. faecalis*, the practical significance of these differences is questionable. Both plasmids segregate rapidly at 42°C, and the time at which their frequency would be expected to fall below that of transposition differs by little more than five generations. This is probably because the incubation temperature of 42°C is so close to that at which replication of pTV1 ceases. Indeed, pTV1 has the advantage of being more stable at permissive temperature and, therefore, should be more easily transformed and stored. Plasmid pTV1(Ts) may be more useful in organisms incapable of growth above 40°C.

Transposition to the chromosome from both pTV1(Ts) and pTV32(Ts) was observed. Examination of insertions from pTV1(Ts) revealed that Tn917 has little if any site specificity. However, Youngman (in press) reported that, rather than showing preference for a specific site on the *B. subtilis* chromosome, Tn917 inserts preferentially in "hot spot" regions that may span several tens of kilobases. While the wide variety of inserts obtained using this vector system suggests that this may not be the case in *S. faecalis*, the possibility of this kind of regional specificity cannot be completely ruled out without further study.

Transposition to plasmid pAD1 from pTV1 (Ts) showed a marked preference over transposition to the chromosome. On the other hand, transposition was severely inhibited in the presence of pPD1. The nature of this inhibition is unknown at present.

LITERATURE CITED

1. **Clewell, D. B., F. Y. An, B. A. White, and C. Gawron-Burke.** 1985. *Streptococcus faecalis* sex pheromone (cAM373) also produced by *Staphylococcus aureus* and identification of a conjugative transposon (Tn*918*). J. Bacteriol. **162**:1212–1220.
2. **Clewell, D. B., P. K. Tomich, M. C. Gawron-Burke, A. E. Franke, Y. Yagi, and F. Y. An.** 1982. Mapping of the *Streptococcus faecalis* plasmids pAD1 and pAD2 and studies relating to transposition of Tn*917*. J. Bacteriol. **152**:1220–1230.
3. **Clewell, D. B., Y. Yagi, G. M. Dunny, and S. K. Schultz.** 1974. Characterization of three plasmid deoxyribonucleic acid molecules in a strain of *Streptococcus faecalis*: identification of a plasmid determining erythromycin resistance. J. Bacteriol. **117**:283–289.
4. **Gawron-Burke, C., and D. B. Clewell.** 1982. A transposon in *Streptococcus faecalis* with fertility properties. Nature (London) **300**:281–284.
5. **Gawron-Burke, C., and D. B. Clewell.** 1984. Regeneration of insertionally inactivated streptococcal DNA fragments after excision of Tn*916* in *Escherichia coli*: a strategy for targeting and cloning of genes from gram-positive bacteria. J. Bacteriol. **159**:214–221.
6. **Gryczan, T. J., J. Hahn, S. Contente, and D. Dubnau.** 1982. Replication and incompatibility properties of plasmid pE194 in *Bacillus subtilis*. J. Bacteriol. **152**:722–735.
7. **Ike, Y., and D. B. Clewell.** 1984. Genetic analysis of the pAD1 pheromone response in *Streptococcus faecalis*, using transposon Tn*917* as an insertional mutagen. J. Bacteriol. **158**:777–783.
8. **Luchanski, J. B., and P. A. Pattee.** 1984. Isolation of Tn*551* insertions near chromosomal markers of interest in *Staphylococcus aureus*. J. Bacteriol. **159**:894–899.
9. **Pattee, P. A.** 1981. Distribution of Tn*551* insertion sites responsible for auxotrophy on the *Staphylococcus aureus* chromosome. J. Bacteriol. **145**:479–488.
10. **Perkins, J. B., and P. J. Youngman.** 1986. Construction and properties of Tn*917-lac*, a transposon derivative that mediates transcriptional gene fusions in *Bacillus subtilis*. Proc. Natl. Acad. Sci. USA **83**:140–144.
11. **Shaw, J. H., and D. B. Clewell.** 1985. Complete nucleotide sequence of macrolide-lincosamide-streptogramin B resistance transposon Tn*917* in *Streptococcus faecalis*. J. Bacteriol. **164**:782–796.
12. **Smith, M. D.** 1985. Transformation and fusion of *Streptococcus faecalis* protoplasts. J. Bacteriol. **162**:92–97.
13. **Southern, E.** 1975. Detection of specific sequences among DNA fragments separated by gel electrophoresis. J. Mol. Biol. **98**:503–517.
14. **Stahl, M. L., and P. A. Pattee.** 1983. Computer-assisted chromosome mapping by protoplast fusion in *Staphylococcus aureus*. J. Bacteriol. **154**:395–405.
15. **Tomich, P. K., F. Y. An, and D. B. Clewell.** 1980. Properties of erythromycin-inducible transposon Tn*917* in *Streptococcus faecalis*. J. Bacteriol. **141**:1366–1374.
16. **Wahl, G., M. Stern, and G. Stark.** 1979. Efficient transfer of large DNA fragments from agarose gels to diazobenzyloxymethyl paper and rapid hybridization using dextran sulfate. Proc. Natl. Acad. Sci. USA **76**:3683–3687.
17. **Wirth, R., F. Y. An, and D. B. Clewell.** 1986. Highly efficient protoplast transformation system for *Streptococcus faecalis* and a new *Escherichia coli-S. faecalis* shuttle vector. J. Bacteriol. **165**:831–836.

Transposon Tn*916* Mutagenesis of Chromosomal Genes Encoding Streptococcal Cell Surface Antigens

K. E. CARBERRY-GOH, K. E. MUNKENBECK-TROTTER, A. WANGER, AND G. M. DUNNY

Department of Veterinary Microbiology, New York State College of Veterinary Medicine, Cornell University, Ithaca, New York 14853

Definitive identification of the biological role of specific bacterial components can be achieved by construction of mutant strains containing well-defined genetic lesions which block expression of a single molecular component. Transposon insertion has proved a powerful tool in the creation of such mutants (6, 7).

We have been interested in cell surface components of streptococci that may be involved in pathogenicity or in genetic transfer (2). The lack of a natural transformation or transduction system in the group D and B streptococci of interest led to the selection of Tn*916*, a tetracycline resistance, conjugative 15-kilobase transposable element (3) for insertional inactivation of chromosomal genes encoding streptococcal surface antigens.

Tn*916* was introduced by conjugation into *Streptococcus faecalis* and *Streptococcus agalactiae*. The resultant banks of strains carrying random insertions were used to study surface antigens involved in pheromone-inducible conjugation (group D) and the immune response in bovine mastitis (group B).

Tn*916* Insertions Affecting Conjugation in *S. faecalis*

Cell clumping and conjugal transfer of plasmid DNA are enhanced as a result of the response of certain *S. faecalis* donor cells to peptide sex pheromones. Our laboratory has used the conjugative plasmid pCF10 to study this phenomenon (G. M. Dunny et al., this volume). To identify chromosomal determinants involved in conjugative ability in *S. faecalis,* two banks of chromosomal Tn*916* insertional mutants were created by introducing Tn*916* by conjugation from *S. faecalis* CG180 (carrying Tn*916* on an erythromycin resistance plasmid, pAM180) (4) into plasmid-free *S. faecalis* OG1-SSp. The resulting banks were screened for resistance to the lytic phage NPV-1, a phage isolated in our laboratory from primary sewage and capable of lysing wild-type cells. It was hoped that mutants with resistance to phage would have surface alterations which would also affect conjugation ability.

Five phage-resistant colonies were isolated and purified. No evidence of lysogeny was found in any of the phage-resistant mutants. Identical phenotypes were exhibited by all phage-resistant mutants. Southern blot analysis with ^{32}P-labeled pAM120 (an *Escherichia coli* plasmid into which Tn*916* had been cloned) (5) followed by autoradiography was performed on *Eco*RI- and *Hind*III-digested chromosomal DNA of the mutant strains. This analysis showed that the phage-resistant mutant strains demonstrated multiple Tn*916* insertions into the chromosome. Four isolates from the first bank showed identical junction fragments, suggesting that they may be siblings. Data from one of these strains are used to represent all four. The isolate from the second bank showed a different junction fragment pattern.

The phage-resistant mutants were tested in a pheromone-induced conjugation system to investigate the effect of phage resistance on recipient and donor ability. Mating experiments were carried out with plasmid-free mutant strains as recipients. The mutant strains showed a 10- to 100-fold decrease in efficiency of acquiring conjugative plasmids (Table 1) in broth matings with pheromone-induced donors. Although these strains were deficient in recipient ability, it was possible to introduce (at reduced efficiency) derivatives of pCF10 into these strains. When plasmid-harboring mutant strains were tested as donors in pheromone-induced matings with normal recipients, they showed donor frequencies similar to those of the wild type (Table 1). When the same strains were tested in pure culture for clumping response to pheromone induction, they showed no clumping, whereas wild-type strains did clump.

Several monoclonal antibodies to cell surface antigens were tested for binding to mutant strains in whole-cell enzyme-linked immunosorbent assays (ELISAs). The two monoclonal antibodies, 4.5E7 and 4.7F6, have been shown to bind more strongly to plasmid-free cells than to induced plasmid-containing cells in whole-cell ELISAs. Also, addition of these antibodies to mating mixes resulted in decreased mating efficiency. In whole-cell ELISAs, both 4.5E7 and 4.7F6 bound more strongly to the mutant strains than to wild-type strains (Fig. 1).

Analysis of *S. agalactiae* Surface Antigens

A. R. Wanger and G. M. Dunny (submitted for publication) demonstrated that *S. agalactiae*

TABLE 1. Ability of chromosomal Tn916 insertion mutants in pheromone-induced matings

Strain	Ability to acquire pCF500[a] (frequency, T/R[b])	Ability to donate pC500 (frequency, T/D[c])
OG1-SSp	3.4×10^{-3}	3.1×10^{-3}
KMT-2	2.7×10^{-6}	5.5×10^{-3}
KMT-6	5.0×10^{-5}	2.4×10^{-3}

[a] A derivative of pCF10 which confers erythromycin resistance rather than tetracycline resistance but encodes a normal pheromone-inducible conjugation ability (1).

[b] T/R, Transconjugants per recipient.

[c] T/D, Transconjugants per donor.

surface proteins in the molecular-weight range of 97,000 to 104,000 elicited an antibody response when whole formalinized *S. agalactiae* strain 224 was injected into the supramammary lymph node of a dairy cow. The proteins were termed *S. agalactiae* specific (Sas97/104). Antibodies reacting with these antigens have been found in mastitic milk and are absent in normal milk. To facilitate investigation of the potential of Sas97/104 as virulence determinants in bovine mastitis, we used Tn916 mutagenesis to develop a mutant incapable of expressing Sas97/104. We used a group B Tn916 donor constructed by Wanger and Dunny (8), *S. agalactiae* W-2S$_2$ (with a Tn916 donor frequency of 3×10^{-7}).

S. agalactiae 224RF (8), derived from a strain isolated from a clinical case of bovine mastitis by selection of spontaneous mutations for rifampin and fusidic acid resistance, was used as the Tn916 recipient. Random insertions of Tn916 into 224RF chromosomal DNA were generated by a conjugation method described by Wanger and Dunny (8). Mating mixtures were plated on Todd-Hewitt base agar selective plates containing rifampin (50 μg/ml), fusidic acid (20 μg/ml), and tetracycline (10 μg/ml) (RFT plates) and incubated for 48 to 72 h. Transconjugants were harvested by flooding the plates with Todd-Hewitt broth, scraping the colonies into suspension, and incubating the harvest suspension in sterile tubes for 18 h at 37°C. The cells in this culture were immunoprecipitated with rabbit antiserum directed specifically against Sas97/104 to enrich for Sas97/104-negative mutants. The antiserum suspension was incubated with shaking for 30 min at 37°C and then centrifuged at 2,000 rpm (ca. 500 × g) for 8 min. Half of the supernatant was transferred to a sterile tube, an equal volume of Todd-Hewitt broth was added, and the suspension was incubated for 18 h at 37°C. The immunoprecipitation procedure was repeated four times. After the fourth cycle, the supernatant was incubated for 18 h and then plated on RFT selective plates and incubated for 24 to 48 h at 37°C.

Individual transconjugant colonies were isolated, grown in Todd-Hewitt broth, washed, and used in agglutination assays using rabbit antisera directed specifically against Sas97/104. One transconjugant, with an agglutination titer less than that of the parent strain 224RF, was found to lack Sas97/104 and related bands on immunoblots (Fig. 2). This Sas97/104-negative mutant was called *S. agalactiae* INY2001. ELISAs were used to further explore surface antigen differences between strain 224RF and mutant

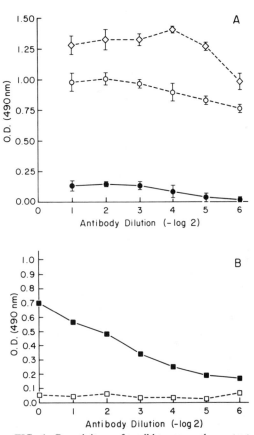

FIG. 1. Reactivity of wild-type and mutant streptococci with antibodies against specific cell surface antigens in ELISAs. Antigen was prepared by baking whole washed cells onto ELISA titer plates. Antigen was detected by adding primary antibody to plates, incubating, washing several times, applying secondary antibody (conjugated goat anti-mouse antibody), adding substrate, and reading optical densities at 490 nm (2). (A) Binding of phage-resistant mutant *S. faecalis* KMT-6 to mouse monoclonal antibody 4.5E7 (◇) and mouse monoclonal antibody 4.7F6 (○). Binding of wild-type *S. faecalis* OG1-SSp to mouse monoclonal antibody 4.7F6 (●). (Binding of monoclonal antibody 4.5E7 to OG1-SSp was identical to that of 4.7F6.) (B) Binding of mouse monoclonal antibodies directed against Sas97/104 to *S. agalactiae* 224RF (■) and to *S. agalactiae* mutant INY2001 (□).

A B

116-

97-

66-

45-

FIG. 2. Antigen profile of *S. agalactiae* 224RF and a mutant derivative (strain INY2001) containing Tn*916*. Immunoblot of culture supernatants immunoprecipitated with rabbit antiserum directed against Sas97/104, run on sodium dodecyl sulfate-polyacrylamide gel electrophoresis, transferred to nitrocellulose, and allowed to react with rabbit antiserum directed against whole *S. agalactiae* 224RF cells. Lane A, *S. agalactiae* mutant INY2001 (Sas97/104 negative); lane B, *S. agalactiae* 224RF (Sas97/104-positive parent). Numerals represent molecular masses in kilodaltons.

INY2001. Little difference in antibody binding was measured between strains 224RF and INY2001 when rabbit antiserum directed against whole-cell 224RF was used; however, lack of binding to INY2001 when compared with 224RF was demonstrated when rabbit antisera directed specifically against Sas97/104 or mouse monoclonal antibodies directed against Sas97/104 were used in whole-cell ELISAs (Fig. 1). The ELISA results suggested that mutant INY2001 did not express Sas97/104 while maintaining

many other surface antigens characteristic of the parent strain 224RF.

Autoradiography of a Southern blot (as described previously for *S. faecalis* using a Tn*916* probe) revealed a single band of the *Eco*RI-digested and two bands of the *Hin*dIII-digested chromosomal DNA of mutant INY2001. These results suggested that a single Tn916 insertion had caused the deletion of Sas97/104 expression in mutant INY2001. Cloning of the Tn916 inserted chromosomal fragments of mutant INY2001 is currently under way.

Conclusions

We have used insertional mutagenesis with Tn*916* to inactivate expression of surface antigens in group B and D streptococci. These mutants should be very useful in analysis of pheromone-inducible conjugation and of virulence determinants in bovine mastitis. The ability of Tn*916* to excise precisely from chromosomal DNA cloned into *E. coli* hosts grown in the absence of tetracycline selection (5), allowing recovery of the functional cloned genes, will also aid in further molecular and genetic analysis.

LITERATURE CITED

1. **Christie, P. J., and G. M. Dunny.** 1986. Identification of regions of the *Streptococcus faecalis* plasmid pCF-10 that encode antibiotic resistance and pheromone response functions. Plasmid **15**:230–241.
2. **Dunny, G. M., D. L. Zimmerman, and M. L. Tortorello.** 1985. Induction of surface exclusion by *Streptococcus faecalis* sex pheromones: use of monoclonal antibodies to identify an inducible surface antigen involved in the exclusion process. Proc. Natl. Acad. Sci. USA **82**:8582–8586.
3. **Franke, A. E., and D. B. Clewell.** 1981. Evidence for a chromosome-borne resistance transposon (Tn*916*) in *Streptococcus faecalis* that is capable of "conjugal" transfer in the absence of a conjugative plasmid. J. Bacteriol. **145**:494–502.
4. **Gawron-Burke, C., and D. B. Clewell.** 1982. A transposon in *Streptococcus faecalis* with fertility properties. Nature (London) **300**:281–284.
5. **Gawron-Burke, C., and D. B. Clewell.** 1984. Regeneration of insertionally inactivated streptococcal DNA fragments after excision of transposon Tn*916* in *Escherichia coli*: strategy for targeting and cloning of genes from gram-positive bacteria. J. Bacteriol. **159**:214–221.
6. **Kleckner, N., J. Roth, and D. Bolstein.** 1977. Genetic engineering in vivo using translocatable drug-resistance elements. J. Mol. Biol. **116**:125–159.
7. **Nida, K., and P. P. Cleary.** 1983. Insertional inactivation of streptolysin S expression in *Streptococcus pyogenes*. J. Bacteriol. **155**:1156–1161.
8. **Wanger, A. R., and G. M. Dunny.** 1985. Development of a system for genetic and molecular analysis of *Streptococcus agalactiae*. Res. Vet. Sci. **38**:202–208.

Highly Efficient Cloning System for *Streptococcus faecalis*: Protoplast Transformation, Shuttle Vectors, and Applications

REINHARD WIRTH,[1] FLORENCE AN,[2] AND DON B. CLEWELL[2]

Lehrstuhl für Mikrobiologie der Universität, D-8000 Munich 19, Federal Republic of Germany,[1] and Departments of Oral Biology and Microbiology/Immunology, The University of Michigan, Ann Arbor, Michigan 48109-2007[2]

Until recently, genetic analyses of *Streptococcus faecalis* were impaired by the unavailability of a transformation system in this species. In 1985, Smith (12) reported on the transformation of *S. faecalis* protoplasts; this system, however, worked only at a low frequency and in our hands was often difficult to reproduce. By systematically optimizing various parameters—growth medium, protoplasting conditions, wash buffers, transformants recovering plates, etc.—we were able to develop a system which is highly efficient (15). Here we report on the main features of this system, the construction of new *Escherichia coli-S. faecalis* shuttle vectors, and various applications of this cloning system.

Methods

A detailed description of the protoplast transformation system is given in reference 15. The main characteristics of this system are as follows.

Strains to be protoplasted are grown to early log phase in Todd-Hewitt broth containing up to 4% glycine (strain dependent) to facilitate the following protoplast formation. Todd-Hewitt broth was by far the best growth medium because of the short time needed for cell wall regeneration of transformants. This time was 2 days for Todd-Hewitt broth, 6 to 10 days for DM3 medium (3), and more than 10 days for antibiotic medium 3, brain heart infusion, and nutrient broth no. 2.

Generation of protoplasts was with the aid of lysozyme to result in at least 99.9% conversion to osmotic-labile cells; mutanolysin could be used, too. From this point on, all buffers and media were stabilized with the aid of 0.5 M sodium succinate, and 100 µg of bovine serum albumin per ml was added. Sucrose (0.5 M) could also be used as osmotic stabilizer; its use, however, resulted in a much longer time (4 to 8 days) needed for cell wall regeneration of transformants. The reason for the increased regeneration time is not totally clear; it should be noted that the use of sucrose resulted in much more acidic growth conditions for protoplasts on selective plates than the use of sodium succinate (pH 4 to 5 versus pH 7). The actual transforma-

tion was done with protoplasts washed free from lysozyme to which plasmid DNA and polyethylene glycol, to a final concentration of 40%, were added. This mixture was kept for 3 min on ice and then heat-shocked for 5 min at 37°C; the polyethylene glycol was then removed by two washing steps. For phenotypic expression of resistance, transformants were incubated for 90 min in Todd-Hewitt broth–sodium succinate–bovine serum albumin and then plated onto selective, low-percentage agar plates.

Plasmids Used for Transformation

Transformation efficiencies for some of the plasmids used in this system are shown in Table 1. We want to emphasize the following points.

Plasmid DNA prepared from *E. coli*, *S. faecalis*, or *Bacillus subtilis* transformed with equal efficiency, indicating that a significant restriction barrier does not exist between those species. Plasmid DNA linearized with an appropriate restriction enzyme transformed with at least four orders of magnitude lower efficiency than circular DNA. No size limit seems to exist for plasmids to be transformed. D. B. Clewell et al. and G. M. Dunny et al. (unpublished data) showed that pAD1 and pCF10, being about 57 and 58 kilobases (kb) in size (5, 14), could be transformed with this system; those two plasmids are the largest tested up to now for transformation. Transformation efficiencies vary from strain to strain and correlate with the ease of protoplasting these strains. The best results were obtained with strain OG1X (8) and are given here.

The usefulness of plasmids pAM910, pVA838, and pSA3 was shown in various experiments, but they might not be of general use for the following reasons: (i) pAM910 (described in reference 12) gives rise to spontaneous deletions (12); (ii) for unknown reasons pVA856, a shuttle vector very similar to pVA838 (9), could not be transformed into OG1X (unpublished data); and (iii) the transformation frequency of pSA3 (4) is very low.

In our opinion pIP501 and its derivatives should be the most useful plasmids in this system for two reasons: (i) pIP501 is a broad-host-

TABLE 1. Efficiency of transformation for various plasmids in *S. faecalis* OG1X

Plasmid	Size (kb)	Selection[a]	Transformants/μg of DNA[b]
pAM910	8.7	Tc8	6.4×10^3
pVA838	9.2	Em5	1×10^4
pSA3	10.2	Em5	2.1×10^2
pIP501	30	Cm5	6.3×10^4
pGB354	6.3	Cm5	8.9×10^5
pAM401	10.4	Cm5	1.4×10^6
pWM401	12.1	Cm5	7×10^5
pAM401	10.4	Cm5	1×10^{3c}

[a] Selection was for chloramphenicol (Cm), erythromycin (Em), or tetracycline (Tc). The numbers show the antibiotic concentration in micrograms per milliliter.

[b] Plasmid DNA isolated by isopycnic CsCl gradient centrifugation from *S. faecalis*.

[c] Plasmid DNA isolated by a modified rapid sodium dodecyl sulfate-NaOH extraction method.

range, gram-positive plasmid able to replicate, e.g., in *Staphylococcus aureus* (10), *Pediococcus* spp. (7), and *Lactobacillus casei* (6); and (ii) pIP501 and its derivatives such as pGB354 transform *S. faecalis* at efficiencies high enough to allow shotgun cloning experiments.

We therefore constructed two novel shuttle vectors, pAM401 and pWM401, to be used in the transformation system. Both vectors transform *S. faecalis* OG1X protoplasts with at least the same efficiency as CaCl₂-treated competent *E. coli* DH1. Details of the construction and maps are given in the legend of Fig. 1; in both cases the gram-negative replicon used was pACYC184

(2). The gram-positive replicon used for construction of pAM401 was pGB354 H (1), a high-copy-number derivative of pIP501; in the case of pWM401 the low-copy-number plasmid pGB351 (1), derived from pIP501, was used. Both shuttle vectors have at least nine unique cloning sites and were more stable in *E. coli* DH1 than in *S. faecalis* OG1X; therefore, they were maintained in the gram-positive host always under selective pressure.

Applications of the System

The system described here has been used by various groups with good success. Examples of experiments from our group include the following.

(i) Tn917 initially cloned into *E. coli* (11) could be introduced into *S. faecalis* OG1S protoplasts, selecting for the transposon-encoded erythromycin resistance.

(ii) In a shotgun cloning experiment we could clone the tetracycline resistance determinant of Tn919 out of *S. sanguis* FC1 chromosomal DNA (Tn919 is integrated in this strain in the chromosome) onto pAM401 (15); selection in this case was for tetracycline (Tn919-borne) plus chloramphenicol (pAM401-borne) resistance.

(iii) Various fragments of the *S. faecalis* sex pheromone plasmids pAD1, pAM373, and pPD1 were cloned into pVA838 and pAM401 (Clewell et al., unpublished data). The functions coded by these various fragments are currently being studied.

(iv) The replication region of pAM373 could be cloned into pVA891 and transformed into *S.*

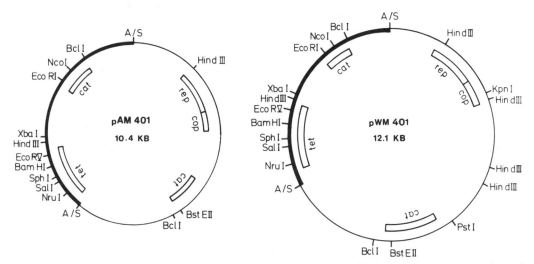

FIG. 1. Maps of the shuttle vectors pAM401 and pWM401. The heavy line represents the segment derived from pACYC184 (2); the light line represents pGB354-derived DNA (for pAM401) or pGB351-derived DNA (for pWM401). A/S indicates the location of the *Ava*I and *Sph*I sites used for joining pACYC184 with pGB354 and pGB351, respectively. A detailed description of how pAM401 was constructed is given in reference 15; the construction of pWM401 followed that protocol.

faecalis (unpublished data). pVA891 was constructed in such a way that it requires cloned replication functions for its establishment in streptococci (9).

(v) pTV1 and related plasmids could be transformed into *S. faecalis* OG1X, and the Tn*917* delivery system developed by P. Youngman for *B. subtilis* could be established in *S. faecalis* (K. E. Weaver and D. B. Clewell, this volume).

(vi) We were able to complement the gelatinase-negative phenotype of *S. faecalis* OG1X (8) by transformation with a plasmid which came from a gene bank of *S. faecalis* OG1S (a gelatinase-positive strain) chromosomal DNA in pAM401 (unpublished data). A further characterization of this plasmid is under way.

Experiments to Clone Structural Gene(s) for *S. faecalis* Sex Pheromones

Our attempts to clone a structural gene(s) for *S. faecalis* sex pheromones in high-copy-number vectors in *S. faecalis* or *E. coli* failed. Screening for such clones was by in vivo assays for pheromone production and with the aid of a cPD1-specific oligonucleotide probe (a mixture of 64 different 14mers) which could be deduced from the amino acid sequence of cPD1 (13). The use of the oligonucleotide probe enabled us to identify a 6-kb *Eco*RI fragment from chromosomal DNA of *S. faecalis* JH2-2 which should contain a cPD1-specific DNA sequence(s). Even this fragment could not be cloned in pAM401. Using *Eco*RI-cut chromosomal DNA of JH2-2 enriched for the 6-kb region (gel elution), we identified two clones hybridizing to the cPD1-specific oligonucleotide probe. These two clones were constructed in the low-copy-number shuttle vector pSA3 and transformed *E. coli* HB101 with normal frequency. Transformation into *S. faecalis* OG1X was at least three orders of magnitude less efficient than for pSA3. Surprisingly, the inserts in these two clones were not 6 kb, but 3.6 and 13.2 kb, in size. Hybridization of the cPD1-specific oligonucleotide probe to a certain region of the inserts could be demonstrated; DNA sequencing studies of these fragments are under way.

This work was supported by Public Health Service grants DE 02731, GM 33956, and AI 10318 from the National Institutes of Health to D.B.C. and by Deutsche Forschungsgemeinschaft grants Wi 731/1–1 and Wi 731/2–1 to R.W.

We thank all colleagues who contributed to this work by providing strains and plasmids. Special thanks are due to G. Wirth for typing the manuscript.

LITERATURE CITED

1. **Behnke, D., and M. S. Gilmore.** 1981. Location of antibiotic resistance determinants, copy control and replication functions on the double-selective streptococcal cloning vector pGB301. Mol. Gen. Genet. **184:**115–120.
2. **Chang, A. C. Y., and S. N. Cohen.** 1978. Construction and characterization of amplifiable multicopy deoxyribonucleic acid cloning vehicles derived from the p154 cryptic miniplasmid. J. Bacteriol. **134:**1141–1156.
3. **Chang, S., and S. N. Cohen.** 1979. High frequency transformation of *Bacillus subtilis* protoplasts by plasmid DNA. Mol. Gen. Genet. **168:**111–115.
4. **Dao, M. L., and J. J. Ferretti.** 1985. *Streptococcus-Escherichia coli* shuttle vector pSA3 and its use in the cloning of streptococcal genes. Appl. Environ. Microbiol. **49:**115–119.
5. **Dunny, G., C. Funk, and J. Adsit.** 1981. Direct stimulation of antibiotic resistance by sex pheromones in *Streptococcus faecalis* subsp. *zymogenes* strain DS16. Antimicrob. Agents Chemother. **15:**828–830.
6. **Gibson, E. M., N. M. Chace, S. B. London, and J. London.** 1979. Transfer of plasmid-mediated antibiotic resistance from streptococci to lactobacilli. J. Bacteriol. **137:**614–619.
7. **Gonzalez, C. F., and B. S. Kunka.** 1983. Plasmid transfer in *Pediococcus* spp.: intergeneric and intrageneric transfer of pIP501. Appl. Environ. Microbiol. **46:**81–89.
8. **Ike, Y., R. A. Craig, B. A. White, Y. Yagi, and D. B. Clewell.** 1983. Modification of *Streptococcus faecalis* sex pheromones after aquisition of plasmid DNA. Proc. Natl. Acad. Sci. USA **80:**5369–5373.
9. **Macrina, F. L., R. P. Evans, J. A. Tobian, D. A. Hartley, D. B. Clewell, and K. R. Jones.** 1983. Novel shuttle plasmid vehicles for *Escherichia-Streptococcus* transgeneric cloning. Gene **25:**145–150.
10. **Schaberg, D. R., D. B. Clewell, and L. Glatzer.** 1982. Conjugative transfer of R-plasmids from *Streptococcus faecalis* to *Staphylococcus aureus*. Antimicrob. Agents Chemother. **22:**204–207.
11. **Shaw, J. H., and D. B. Clewell.** 1985. Complete nucleotide sequence of macrolide-lincosamine-streptogramin B resistance transposon Tn*917* in *Streptococcus faecalis*. J. Bacteriol. **164:**782–796.
12. **Smith, M. D.** 1985. Transformation and fusion of *Streptococcus faecalis* protoplasts. J. Bacteriol. **162:**92–97.
13. **Suzuki, A., M. Mori, Y. Sakagami, A. Isogai, M. Fujino, C. Kitada, R. A. Craig, and D. B. Clewell.** 1984. Isolation and structure of bacterial sex pheromone, cPD1. Science **226:**849–850.
14. **Tomich, P. K., F. Y. An, D. P. Damle, and D. B. Clewell.** 1979. Plasmid-related transmissibility and multiple drug resistance in *Streptococcus faecalis* subsp. *zymogenes* strain DS16. Antimicrob. Agents Chemother. **15:**828–830.
15. **Wirth, R., F. Y. An, and D. B. Clewell.** 1986. Highly efficient protoplast transformation system for *Streptococcus faecalis* and a new *Escherichia coli-Streptococcus faecalis* shuttle vector. J. Bacteriol. **165:**831–836.

Aspects of Genetic Transformation in *Streptococcus mutans*

LUTHER E. LINDLER and FRANCIS L. MACRINA

Department of Microbiology and Immunology, Virginia Commonwealth University, Richmond, Virginia 23298

Streptococcus mutans is recognized as the primary cause of dental caries. The mechanism by which *S. mutans* undergoes genetic transformation has not been well characterized. The lack of knowledge about events which accompany genetic transformation of this species is partly due to the low frequency of marker recovery (3, 4). Even under the best culture conditions, the transformation frequency of *S. mutans* is much lower than that obtained with *Streptococcus sanguis*, regardless of the type of donor DNA used, i.e., chromosome or plasmid. *S. sanguis* is known to be a member of a small group of gram-positive bacteria that spontaneously develop genetic competence (6). Transformation of these organisms occurs by a mechanism that is not sequence specific and involves conversion of donor DNA to a single-stranded molecule during uptake. The development of genetic competence in *S. sanguis* is inducible. We set out to determine whether transformation in *S. mutans* was operationally similar to that in other gram-positive organisms. Because much interest has been focused on the homologous cloning and characterization of *S. mutans* virulence determinants, we considered it important to investigate the nature of the transformation process in this organism.

To artificially raise the transformation frequency usually obtained with *S. mutans* recipients, we developed a marker rescue system (1; Fig. 1). The system involved the recombinational rescue of an *Escherichia coli* plasmid (pVA981) by a resident streptococcal plasmid (pVA1208). Plasmid pVA1208 conferred erythromycin resistance (Em^r) but not tetracycline resistance (Tc^s) and was derived from streptococcal plasmid pVA982 (1, 7). During transformational marker rescue, tetracycline resistance (Tc^r) is restored by recombination between pVA1208 and pVA981. pVA981 cannot replicate in streptococcal hosts, but contains a functional Tc^r determinant that is flanked by DNA homologous to areas present on pVA1208 (Fig. 1). The advantages associated with this system lie in the use of homogeneous donor DNA and the higher copy number of target sequences compared with a chromosomal locus within recipients. Transformation frequencies obtained using the Tc^r rescue system were found to be 8.4-and 160-fold

higher than transformation with chromosomal and plasmid markers, respectively (1, 3).

We used the marker rescue system to study competence development by *S. mutans* HS6, GS5, and MT557, which are members of Bratthall serotypes a, c, and f, respectively. Using the culture conditions defined by Perry et al. (4), we found that all the above strains demonstrated a sequential rise and fall of Tc^r recombinants similar to that shown in Fig. 2 for *S. mutans* GS5. *S. mutans* HS6 and GS5 developed maximal competence after 3 h of incubation, while the serotype f (MT557) strain required only 2 h to reach peak competence. Thus, it appeared that competence was inducible in *S. mutans*.

Since we were interested in comparing the transformation process in *S. mutans* and *S. sanguis*, we examined the frequency of recovery of Tc^r recombinants obtained when we used the marker rescue system (1). pVA981 was added to competent cells, which were then treated with DNase I to destroy exogenous DNA. Tc^r transformants were expressed at essentially the same rate and reached maximal levels within 15 min after addition of donor DNA. Marker expression was rapid, with 40% of the Tc^r transformants occurring after only 5 min of exposure to donor DNA. These results were similar to those obtained with *S. sanguis* by Raina and Ravin (5), who used differentially labeled donor and recipient DNAs followed by velocity sedimentation analysis.

None of the *S. mutans* strains examined required a specific sequence for binding of exogenous DNA (1). When various amounts of heterologous nontransforming DNA were mixed with a constant amount of pVA981, the transformation frequency to Tc^r decreased as a function of increasing competing DNA concentration. Also, we investigated the possibility that *S. sanguis* Challis competence factor might activate *S. mutans* genetic competence. Preparations of competence factor which activated noncompetent *S. sanguis* cultures had no effect on *S. mutans* transformation.

The last element of *S. mutans* transformation that we examined was the fate of bound donor DNA. We allowed radioactive (^{3}H-labeled) plasmid DNA to bind to competent *S. mutans* cells for 2 min followed by DNase I treatment for 1

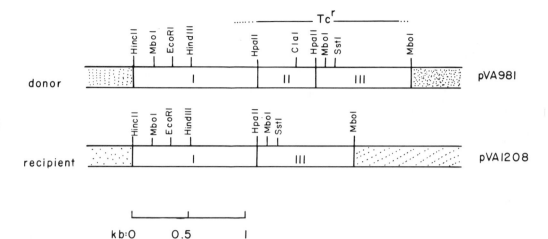

FIG. 1. Homology between pVA981 and pVA1208. Donor plasmid pVA981 restores Tc^r of recipient *Streptococcus* spp. harboring the pVA1208 construct. Recombination is initiated within areas labeled I and III and directs the insertion of area II into the streptococcal replicon pVA1208, thus restoring Tc^r. Each stippled area represents contiguous but nonhomologous regions on the two plasmids.

min to render unbound DNA acid soluble. A sample was removed immediately to determine 100% acid-insoluble radioactivity bound by competent cells, followed by removal of samples at later times to determine the percentage of remaining acid-insoluble material. Approximately 50% of the initially bound donor plasmid DNA remained in an acid-insoluble form after 30 min of incubation at 37°C (Fig. 3). From these results it appeared that *S. mutans* degraded bound donor DNA into a single-stranded form before or during uptake. In *S. sanguis* (2), this fact has a significant effect on the dose-response

kinetics expected for monomeric versus multimeric plasmid molecules. We separated monomeric and multimeric plasmid DNA by the method of Macrina et al. (2) and examined the kinetics of transformant recovery with various subsaturating concentrations of each form (1). Multimeric plasmid DNA yielded transformants with first-order kinetics; however, monomeric plasmids transformed with second-order kinetics, suggesting the requirement for entry of two

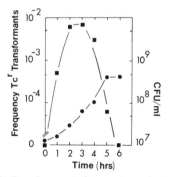

FIG. 2. Development of competence in *S. mutans* GS5 monitored by using the marker rescue system. Samples were removed at the times indicated for addition of pVA981 (1 μg/ml) and determination of CFU per milliliter. After 10 min of incubation, donor DNA uptake was terminated by addition of 10 μg of DNase I per ml. The sample was then incubated for a total of 60 min before the selection of Tc^r recombinants. CFU were determined at the time of donor DNA addition. Symbols: ■, transformation frequency; ●, CFU per milliliter.

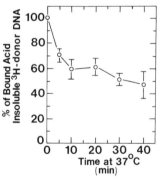

FIG. 3. Solubilization of donor plasmid DNA by competent *S. mutans* GS5. Radioactive plasmid DNA was allowed to bind to competent cells for 2 min, followed by treatment with 10 μg of DNase I per ml for 1 min. A 50-μl sample was removed and deposited on a Whatman GF/A glass-filter. The filter was immersed in 250 ml of ice-cold 10% trichloroacetic acid for 5 min. This step was repeated, followed by two washes in 250 ml of ice-cold ethanol. This initial sample was' regarded as 100% acid insoluble, and subsequent samples treated similarly were compared with it. *S. mutans* bound 50.7% of the input donor DNA counts per minute in 2 min.

molecules to yield a viable transformant. Furthermore, chromosomal markers and marker rescue (Tc^r) occurred as a first-order process. Thus, kinetics of transformant recovery were consistent with our findings from the DNA solubilization assay.

Our work has expanded our understanding of the mechanism by which *S. mutans* undergoes genetic transformation. From an applied point of view, our results indicate that researchers should employ shuttle vectors or marker rescue when conducting experiments involving cloning of *S. mutans* genes. From the basic science aspect, our results demonstrate that *S. mutans* should be included with other gram-positive, naturally transformable bacteria (6). The common niche occupied by *S. sanguis* and *S. mutans* in the oral cavity may indicate a common evolution of genes necessary for transformation of these organisms. Studies are presently under way in our laboratory to investigate this possibility.

This work was supported by Public Health Service grant DE04224 from the National Institute of Dental Research.

LITERATURE CITED

1. **Lindler, L. E., and F. L. Macrina.** 1986. Characterization of genetic transformation in *Streptococcus mutans* by using a novel high-efficiency plasmid marker rescue system. J. Bacteriol. **166:**658–665.
2. **Macrina, F. L., K. R. Jones, and R. A. Welch.** 1981. Transformation of *Streptococcus sanguis* with monomeric pVA736 plasmid deoxyribonucleic acid. J. Bacteriol. **146:**826–830.
3. **Perry, D., and H. K. Kuramitsu.** 1981. Genetic transformation of *Streptococcus mutans.* Infect. Immun. **32:**1295–1297.
4. **Perry, D., L. M. Wondruck, and H. K. Kuramitsu.** 1983. Genetic transformation of putative cariogenic properties in *Streptococcus mutans.* Infect. Immun. **41:**722–727.
5. **Raina, J. L., and A. W. Ravin.** 1978. Fate of homospecific transforming deoxyribonucleic acid bound to *Streptococcus sanguis.* J. Bacteriol. **133:**1212–1223.
6. **Smith, H. O., and D. B. Danner.** 1981. Genetic transformation. Annu. Rev. Biochem. **50:**41–68.
7. **Tobian, J. A., M. L. Cline, and F. L. Macrina.** 1984. Characterization and expression of a cloned tetracycline resistance determinant from the chromosome of *Streptococcus mutans.* J. Bacteriol. **160:**556–563.

Genetics of the Complementary Restriction Systems *Dpn*I and *Dpn*II Revealed by Cloning and Recombination in *Streptococcus pneumoniae*

SANFORD A. LACKS, BRUNO M. MANNARELLI,† SYLVIA S. SPRINGHORN, BILL GREENBERG, AND ADELA G. DE LA CAMPA

Biology Department, Brookhaven National Laboratory, Upton, New York 11973

Restriction enzymes, because they are able to recognize and cleave specific sequences in DNA, have had an enormous impact on genetic analysis and engineering. This review is concerned with some unusual restriction enzyme systems found in *Streptococcus pneumoniae*. The typical restriction endonuclease cleaves DNA at or near a particular nucleotide sequence; cells that make the enzyme protect their own DNA from cleavage by methylating a base in that sequence. Not long after the discovery of sequence-specific restriction enzyme cleavage, however, a different kind of restriction enzyme was found in *S. pneumoniae* (21). This endonuclease, *Dpn*I, cleaves only methylated DNA at the sequence 5'-GmeATC-3' (22). In contrast to all other restriction enzymes, the methylated sequence is susceptible to cleavage and the unmethylated sequence is resistant. Cells that produce *Dpn*I, therefore, require no modification of their DNA.

Only certain strains of *S. pneumoniae* produce *Dpn*I. Other strains, which were also isolated from patients with pneumonia, were found to contain, instead, a different restriction system, *Dpn*II (31). The latter is a typical restriction/modification system with an endonuclease that cleaves an unmethylated site and a DNA methylase that modifies that site to protect it from cleavage. The curious aspect of the *Dpn*II endonuclease is that it is complementary to *Dpn*I. It cleaves the sequence 5'-GATC-3' only when it is not methylated (22). The systems are complementary in that *Dpn*I cleaves DNA from a *Dpn*II strain and vice versa.

Transformation and cloning of the *Dpn*I and *Dpn*II endonuclease genes have clarified the genetic basis of the two restriction systems. Molecular cloning was carried out in the gram-positive *S. pneumoniae* host/vector system (37). Cloned chromosomal fragments from both *Dpn*I- and *Dpn*II-producing strains were subjected to nucleotide sequence determination and were used as probes for DNA hybridization

† Present address: U.S. Department of Agriculture, Agricultural Research Service, Peoria, IL 61604.

analysis (23a). It was shown that the restriction enzyme phenotype of *S. pneumoniae* depended on an intercellular genetic cassette mechanism (23a). In this review some aspects of the evolution of restriction systems in *S. pneumoniae* and other bacteria are discussed.

Aside from revealing the novel genetic basis of the restriction systems and pointing out some interesting problems in the regulation of gene expression, the work under review has made available cloned DNA containing the restriction enzyme genes. These clones and derivatives of them should prove useful for producing large amounts of the enzymes for the benefit of both laboratory and commerce. The availability of cloned streptococcal restriction genes, with the prospect of introducing them into bacterial cultures of commercial importance, may prove useful, especially to the dairy industry.

Properties of the Restriction Enzymes

The *Dpn*I endonuclease will cleave double-stranded DNA only at sites in which the 5'-GATC-3' palindrome is methylated on both strands. It will not act on methylated single-stranded DNA, nor will it cleave hemimethylated sites even in the methylated strand (39). The enzyme cuts between meA and T in each strand of the duplex target to give blunt-ended fragments (13). *Dpn*I endonuclease has been partially purified by gel filtration (19) and ion-exchange chromatography (13). From its behavior in gel filtration the native enzyme protein appears to have a molecular weight of 20,000.

Although it acts only on unmethylated 5'-GATC-3' sites, the *Dpn*II endonuclease behaves similarly to *Dpn*I in that it cleaves double-stranded DNA sites only when neither strand is methylated. Also, it will not cleave single-stranded DNA. *Dpn*II cuts the sequence on each strand at its 5' end to produce overhanging 5'-GATC tails (20). This enzyme has also been partially purified by gel filtration and ion-exchange chromatography (19), and the native enzyme exhibited a molecular weight of 70,000 in gel filtration.

The *Dpn*II methylase transfers methyl groups from *S*-adenosylmethionine to the N_6 position of adenine in 5'-GATC-3' sequences of DNA. The native enzyme was partially purified by gel filtration in which it exhibited a molecular weight of approximately 60,000 (22). Inasmuch as the cloned gene encoding the methylase can specify a polypeptide no greater than 33,000 in molecular weight, it appears that the methylase is a dimer composed of identical subunits.

Restriction Phenotypes of S. pneumoniae

Wild strains of *S. pneumoniae* appear to contain either the *Dpn*I endonuclease (phenotype I) or the *Dpn*II endonuclease and methylase (phenotype II). Their DNA is, therefore, respectively unmethylated and methylated at 5'-GATC-3' sites. Also, as a result of the restriction enzymes that they contain, cells of phenotype I restrict bacteriophage grown on phenotype II and vice versa.

A strain of *S. pneumoniae* called Rx, which was grown for some years in the laboratory, exhibited a null phenotype in that it made neither the *Dpn*I nor the *Dpn*II endonuclease, nor did it methylate its DNA (31). This strain is susceptible to phage grown on either phenotype I or II. Genetically, it is very similar to a *Dpn*I strain, but it carries a single base deletion in the gene encoding the endonuclease (23a; unpublished data).

The restriction phenotype of a strain can be determined by (i) analysis of its DNA methylation pattern using purified *Dpn*I and *Dpn*II endonucleases, (ii) testing its susceptibility to phage grown on phenotypes I and II, or (iii) examining extracts directly for *Dpn*I and *Dpn*II endonuclease activity (31). Direct examination of crude extracts can be accomplished, however, only in strains containing an *end* mutation, which reduces the major cellular DNase activity (23) so that it does not interfere with the restriction endonuclease measurement.

Restriction Effects on DNA Introduced into Cells

Bernheimer found that a bacteriophage, HB-3, which she isolated from a lysogenic strain of *S. pneumoniae*, varied, depending on the strain in which it was grown, in its infectivity toward indicator strains (5). This variation was shown to result from restriction and modification by the *Dpn*I and *Dpn*II systems (31). Plaque formation by HB-3 grown on a *Dpn*I strain was reduced by a factor of 5×10^{-6} when plated on a *Dpn*II strain compared with a *Dpn*I strain. Restriction of phage grown on a phenotype II strain by *Dpn*I cells was similarly found to give a reduction to 10^{-6}. In phage infection double-stranded DNA is presumably injected into the cell. Such DNA is a suitable substrate for the restriction endonucleases. It would appear,

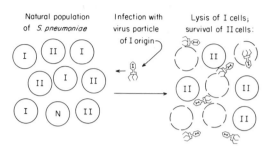

FIG. 1. Survival value of complementary restriction enzyme systems. I and II, Cells of phenotypes making *Dpn*I and *Dpn*II, respectively; N, cell of null phenotype.

then, that the main purpose of the *Dpn*I and *Dpn*II restriction systems is protection of the cell against viral infection. Supporting this notion is the presence of abnormal DNA bases in some pneumococcal phages (32). These abnormal bases render the phage DNA resistant to both *Dpn*I and *Dpn*II (22); they presumably evolved to protect the virus against such restriction systems.

Why do strains of *S. pneumoniae* contain complementary restriction systems? The presence of complementary systems in different cells among natural populations could allow the survival of a remnant of a population after infection initiated in a cell of one particular restriction phenotype (22, 23a). This concept is illustrated in Fig. 1. The complementary restriction systems would thus protect bacterial populations from viral epidemics and thereby enhance survival of the species.

Although the restriction systems of *S. pneumoniae* act very effectively on infecting phage DNA, they have no effect on chromosomal transformation. DNA methylated at 5'-GATC-3' sites transforms *Dpn*I strains as well as unmethylated DNA (21), and the same is true for *Dpn*II-containing recipients (39). The reason that chromosomal transformation is unaffected by restriction is related to the molecular fate of transforming DNA during uptake by the cell (reviewed in reference 18). Transforming DNA is taken up by the cell as single strands (16) and is integrated into the chromosome to give a donor/recipient heteroduplex (12). Neither *Dpn*I nor *Dpn*II will attack either the single-stranded DNA or the hemimethylated heteroduplex DNA (39). These enzymes appear to be cleverly designed to protect the *S. pneumoniae* cell against injected viral DNA without interfering with transformation, which is the natural process of genetic exchange in this species.

Plasmid transfer is somewhat affected by restriction. Although the mechanism of DNA entry in this case is the same as that for chromosomal transformation, two independently entering complementary strand segments of the donor plasmid must interact in the recipient cell to establish a circular replicon. Parts of this initial plasmid are newly synthesized, but enough of the original donor strands would be present to provide the double-stranded substrate susceptible to the restriction enzymes. Thus it was found that plasmids containing 7 or 10 potential restriction sites were reduced in the cross-transformation (i.e., plasmid from *Dpn*I strain introduced into *Dpn*II strain, or vice versa) to 40% of the level observed without restriction effects (25).

Transformation of Restriction Phenotype

What could be the genetic basis of the different restriction phenotypes of *S. pneumoniae*? Because strains of different phenotype were very similar in other characters, and they could exchange genetic information freely, they did not appear to constitute subspecies. One possibility was that genes for both systems were present in all strains, but since they were mutually exclusive, they were regulated so that one system was turned off when the other was turned on. It was conceivable even that DNA methylation in *Dpn*II strains played a role in repressing *Dpn*I gene function (22). An alternative possibility was that genes of the complementary restriction system were not normally present but that one set of genes could substitute for the other in any given cell. Analysis of the transformation of restriction phenotypes by chromosomal DNA suggested that the latter possibility was correct; cloning of the genes and DNA hybridization analysis showed this definitely to be the case.

Transformation of the null strain with chromosomal DNA from a *Dpn*I strain yielded *Dpn*I-containing transformants, which were selected by their resistance to phage grown on a *Dpn*II strain as opposed to the null recipient cells which were killed by this phage (31). The frequency of this transformation was as high as that of a single-site drug-resistance marker used for reference, which indicated that the restriction phenotype transformation involved a relatively small genetic change. This was later corroborated by a comparison of the DNA sequence of the null and *Dpn*I strains, which showed them to differ by a single nucleotide in the gene encoding the *Dpn*I endonuclease (unpublished data).

Transformation of the null strain with chromosomal DNA from a *Dpn*II strain gave transformants to the *Dpn*II phenotype, but the frequency of such transformants was only 20% of that shown by the single-site reference marker (31). In this transformation the selection was for the endonuclease, but the methylase gene would also have to be transferred to protect the cellular DNA of the transformant. The significance of the observed transformation frequency, as determined from previous work on genetic analysis by transformation (17), is that the endonuclease and methylase genes must be linked on the chromosome. If they had been present on separate pieces of donor DNA, the transformation requiring in this case entry of two separate pieces of DNA into the cell would have occurred at much lower frequency. That the observed frequency was considerably less than that given by a single-site marker indicates that the two genes were transferred as a linked cluster on a piece of DNA several kilobases (kb) in length. This supposition was also corroborated by the cloning work, which showed that a nonhomologous segment of DNA 2.9 kb in length was introduced into the null strain by this transformation (23a).

The restriction phenotype of the transformants always was that of the donor DNA. Attempts to select *Dpn*I transformants, for example, when the donor DNA was from a *Dpn*II strain all met with failure (31). This result suggested that unexpressed genes from the complementary system were not present in cells expressing a particular restriction system.

Cloning and Characterization of the Restriction Genes

Cloning of the methylase gene. The first gene cloned was that encoding a DNA methylase of the *Dpn*II system (24). It was cloned on a 3.7-kb *Bam*HI chromosomal fragment inserted into the *Bgl*II site of the vector plasmid pMP5 (11). All cloning was carried out with streptococcal plasmid vectors in the *S. pneumoniae* cloning system as described (37). To select for methylase clones, we used the strategy proposed by Mann et al. (28) and realized by Szomolanyi et al. (38), as well as by us. To wit, after transformation with a shotgun ligation mixture of chromosomal and plasmid DNA, the pooled plasmid population was treated with *Dpn*II endonuclease to destroy all plasmids not methylated at 5'-GATC-3' sites, so that only the remaining methylase-encoding plasmids were obtained in a subsequent transformation.

Characterization of the methylase gene. The *Bam*HI fragment was cloned in both orientations to give plasmids pMP8 (Fig. 2) and pMP10, both of which conferred the ability to make the *Dpn*II DNA methylase in a null strain at a level five times greater than that conferred by the chromosomal gene in a *Dpn*II strain (24). The DNA sequence of the methylase gene was determined with pMP7, a shortened version of pMP8 (29). Similar levels of methylase activity with the insert in opposite orientations suggested that the promoter for the methylase gene accompanied it in the insert. This was borne out by the DNA sequence in that a near-consensus promoter sequence (34), TTGAtA..18 nucleotides..TAa-AAT, was found just upstream of the methylase structural gene. The structural gene itself was localized to an open reading frame contained completely within a 1.5-kb *Hae*III restriction fragment (Fig. 3). Proof that it was the structural gene was obtained by transfer of a plasmid containing the gene to a foreign host, *Bacillus subtilis*, which then expressed the methylase activity (29).

Recombinant plasmids containing the methylase gene could be transferred to *Dpn*II and null strains of *S. pneumoniae* without any problem. However, attempting to transfer a plasmid such as pMP8 to a *Dpn*I strain runs into the problem that establishment of the plasmid would set up a suicidal situation: methylation of the cellular DNA would allow it to be cleaved by the *Dpn*I endonuclease made by the cell. When such a transfer was attempted, very few transformants containing plasmids were obtained (25). Nevertheless, a frequency 0.2% of that obtained in a null strain was observed. Analysis of the plasmids in these transformants showed that half of them contained a deleted donor plasmid that no longer encoded a methylase. The other half, however, contained intact pMP8. Analysis of cured cells indicated that these isolates had lost *Dpn*I activity. The alteration causing this loss of *Dpn*I remains to be determined. Its frequency, approximately 0.1%, is higher than would be expected for a spontaneous mutation in the gene encoding *Dpn*I. Perhaps it corresponds to a regulatory event that, by turning *Dpn*I off, would allow a transition from a *Dpn*I to a *Dpn*II system.

Cloning of the *Dpn*II endonuclease. The 3.7-kb *Bam*HI chromosomal fragment cloned in pMP8 that contained the methylase gene did not confer the ability to make the *Dpn*II endonuclease. Inasmuch as the genes for these enzymes appeared to be linked, a larger chromosomal piece, cloned similarly by selection for the methylase, might be expected to yield both products. A 17.4-kb *Bgl*II fragment inserted into the *Bcl*I site of the vector pLS101 (2) also failed to give *Dpn*II production. Although this plasmid, pLS130 (Fig. 2), extended the cloned region 13 kb in one direction, it extended it only 0.4 kb in the other. Restriction mapping of pLS130 and DNA hybridization analysis assisted in the choice of chromosomal fragments for further cloning in the latter direction. Attempts to clone an 8.6-kb *Hin*dIII fragment were unsuccessful, presumably because the fragment contained a gene deleterious in a multicopy plasmid. However, a 4.6-kb *Bst*EII fragment was cloned in pLS101 to give pLS201 (Fig. 2), which conferred the ability to make

*Dpn*II, as indicated by phage restriction patterns (Table 1) and enzyme activity in extracts (data not shown). A 2.3-kb *Bst*EII fragment from elsewhere in the chromosome was adventitiously cloned together with the 4.6-kb fragment in pLS201. That the fragments were not adjacent in the chromosome was shown by DNA hybridization analysis (unpublished data). That the expression of *Dpn*II did not require the smaller fragment was shown by its removal to give pLS202, which retained the ability to make *Dpn*II.

Characterization of the *Dpn*II genes. The nucleotide sequence of a 3.5-kb segment of pLS201 containing the *Dpn*II genes was determined (23a). A detailed restriction map and some of the sequencing strategy are indicated in the lower portion of Fig. 3. The sequence of the entire stretch has now been determined on both strands. In the upper portion of Fig. 3 are indicated open reading frames deduced from the DNA sequence along with potential ATG start codons. An open reading frame leading into the left end of the segment is observed in phase 1, followed by an extensive open frame in the same phase. Continuing from left to right, an open reading frame is observed in phase 3, and then another one in phase 2. The only extensive open reading frame from right to left is found leading into the right end of the segment in phase 3. The first complete gene, *dpnM*, was identified as the structural gene for the *Dpn*II DNA methylase (29). If translation begins at the first ATG start codon, the molecular weight of the methylase polypeptide would be 33,000. The gene associated with the second full open reading frame is designated *dpnA*. No function is known for this gene. Depending on whether its translation begins at the first or second start codon (see below), its product would be a polypeptide of molecular weight 31,000 or 30,000. The next open reading frame could encode a polypeptide of molecular weight 34,000. Preliminary results of site-directed insertion mutagenesis indicate that the product of this third gene, *dpnB*, corresponds to the *Dpn*II endonuclease.

In addition to the aforementioned promoter, several features that could potentially correspond to control signals were observed in the DNA sequence of the *Dpn*II genes (23a, 29). No typical ribosome binding site (36) precedes the methylase gene. However, the sequence 5'-AATTCT..4 or 5 nucleotides..TATA..9 or 10 nucleotides.. precedes the first ATG start codon in both the *dpnM* and *dpnA* open reading frames (29). This sequence, perhaps in conjunction with a cellular protein, may conceivably act as an alternative ribosome binding site. A typical Shine-Dalgarno sequence, 5'-GGAGGTG-3', is associated with the second ATG codon in the *dpnA* open reading frame. Two different pro-

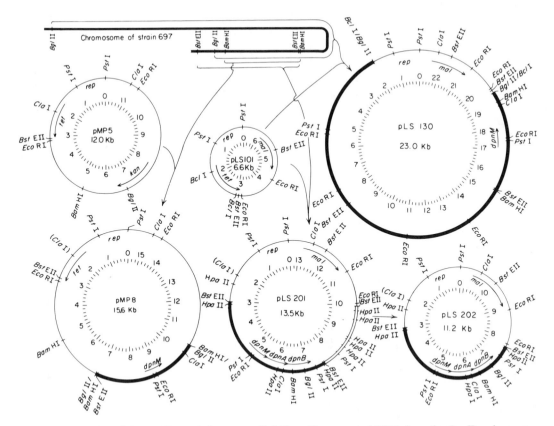

FIG. 2. Cloning of the *Dpn*II restriction genes. Solid bar, Chromosomal DNA from the *Dpn*II region; open bar, adventitious chromosomal DNA; thin line, vector. Arrows between plasmids indicate derivation. The region essential for replication of the plasmids, *rep*, and the extent and direction of transcription of intact plasmid-borne genes are indicated.

TABLE 1. Restriction endonuclease phenotype in strains of *S. pneumoniae* carrying plasmids with chromosomal insertions

Strain	Derivation and features	Resistance[a] to infection by phage HB-3 grown in		Restriction endonuclease phenotype
		*Dpn*I strain	*Dpn*II strain	
193	R6Δ*mal*	−	+	*Dpn*I
533	R6 *str*	−	+	*Dpn*I
697	HB264	+	−	*Dpn*II
707	R36NC	−	+	*Dpn*I
762	Rx	−	−	Null
763	Rx trI	−	+	*Dpn*I
764	Rx trII	+	−	*Dpn*II
777	RxΔ*mal*	−	−	Null
762(pMP8)		−	−	Null
777(pLS130)		−	−	Null
777(pLS201)		+	−	*Dpn*II
777(pLS202)		+	−	*Dpn*II
777(pLS203)		−	−	Null
777(pLS207)		−	+	*Dpn*I

[a] Plus and minus signs indicate resistance and lack of resistance, respectively.

teins could possibly be made by this gene. A strong ribosome binding site, 5'-GGAGG..7 nucleotides..ATG, is found at the start of the *dpn*B gene encoding the endonuclease (23a). The coding region for this polypeptide, however, overlaps that of the prior one by 11 bases. Finally, on the 3' side of the last gene in the *Dpn*II segment, a sequence corresponding to a transcription terminator is found (23a). It appears that the three *Dpn*II restriction genes are expressed from a single mRNA transcript.

We have recently transferred the *Dpn*II restriction gene segment to a hyperexpression vector, which has allowed the isolation of protein products of the *dpn*M, *dpn*A, and *dpn*B genes (unpublished data). Analysis of these proteins by gel filtration and sodium dodecyl sulfate-polyacrylamide gel electrophoresis showed that all three of them are dimers composed of identical subunits and that these subunits correspond approximately in size to the polypeptides predicted from the DNA sequence (unpublished data).

Cloning of the *Dpn*I genes by chromosomal facilitation of plasmid establishment. When plas-

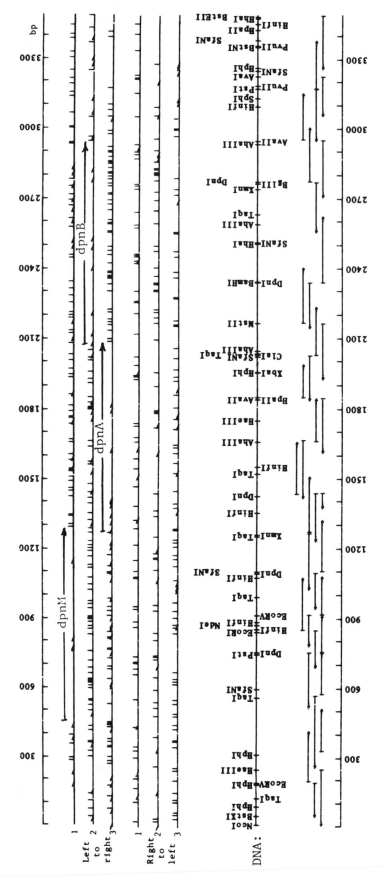

FIG. 3. Restriction map of *Dpn*II gene region showing DNA sequencing strategy and open reading frames. Segments for which nucleotide sequence was determined are shown below the restriction site map by arrows indicating the strand sequenced: vertical marks, 5'-labeled terminus; arrowhead, end of reading. (Further sequencing has now entirely covered both strands.) Open reading frames are indicated above the map for all three phases in both directions: vertical marks, termination codons; oblique marks, potential ATG start sites. *Dpn*II-specific genes are indicated by arrows.

36

mid DNA is introduced into a pneumococcal cell, if it contains DNA homologous to the recipient chromosome, interaction with the chromosome can facilitate plasmid establishment by an entering single-strand fragment (27). During such facilitation genetic information from the chromosome may be transferred to the newly established plasmid at a frequency ranging from 10 to 90% (3, 26, 27). Such a chromosomal facilitation event, in which a pLS202 strand fragment lacking the central region of its chromosomal insert interacted with the recipient chromosome at homologies bracketing that region to pick up a new, shorter central region to give a 9.9-kb plasmid, apparently occurred when pLS202 was transferred to the null strain 777 (23a). Although three of eight plasmids examined were established as the 11.2-kb pLS202, the other five were only 9.9 kb in length. In particular, the 4.6-kb *Bst*EII fragment containing the *Dpn*II genes was replaced by a 3.3-kb fragment. The new plasmid, pLS203, conferred neither *Dpn*I- nor *Dpn*II-producing ability, in keeping with its origin from the null strain chromosome.

When pLS203 was transferred to the *Dpn*I-producing strain 193, all Mal$^+$ transformants contained 9.9-kb plasmids. Plasmid preparations from four such isolates were used to transform strain 777. A number of transformants, which all contained 9.9-kb plasmids, from each of the four crosses were tested for *Dpn*I or *Dpn*II production by the phage restriction assay. In one case, only null transformants were observed, presumably because pLS203 was originally established in strain 193. In the other three cases approximately half of the transformants produced *Dpn*I, presumably because in these cases the plasmid established in 193 had benefited from chromo-

somal facilitation during which an active *Dpn*I gene was transferred from the chromosome to the plasmid (23a). The active plasmid, called pLS207, again on account of chromosomal facilitation, established itself intact in only half of the transformants of strain 777. In the other half, the null chromosomal configuration was substituted to convert pLS207 back to pLS203.

When pLS202 is introduced directly into the *Dpn*I-producing strain, chromosomal facilitation of its establishment gives rise to pLS207 in a single step (Fig. 4). It therefore appears that the *Dpn*I and *Dpn*II gene segments are located at the same position in the chromosome. Table 1 indicates the restriction patterns observed for strains bearing the various plasmids discussed.

Characterization of *Dpn*I genes. A 1.7-kb segment of pLS207 containing *Dpn*I-specific genes was subjected to sequence determination (23a). Its detailed restriction map, some of the sequencing strategy employed, and open reading frames deduced from the DNA sequence are indicated in Fig. 5. The sequence of the entire stretch has now been determined on both strands. Only two extensive open frames are apparent. Both read from left to right and are complete within the segment. The first is found in phase 3, and the second, shorter one is in phase 1. The first open reading frame could encode a polypeptide of 30,000 molecular weight. However, the next potential ATG start site, which is associated with a reasonable Shine-Dalgarno sequence, AAtGatGGTG (upper case indicates complementarity to the terminus of 16S rRNA from *B. subtilis*), would give rise to a polypeptide of 20,000 molecular weight (23a). Preliminary results of site-directed insertion mutagenesis indicate that this open reading frame encodes the *Dpn*I endonuclease. The molecular weight of the

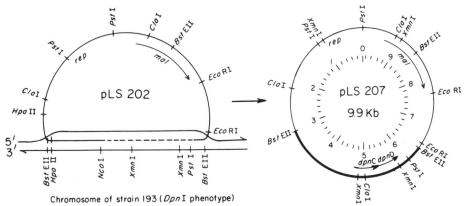

Chromosome of strain 193 (*Dpn* I phenotype)

FIG. 4. Cloning of *Dpn*I genes by chromosomal facilitation of plasmid establishment: replacement during facilitated transfer of pLS202 of its *Dpn*II gene segment by the *Dpn*I gene segment from the chromosome of strain 193 to give pLS207. A single-strand fragment of the donor plasmid interacts by circular synapsis with homologous chromosomal DNA to give a D-loop structure. Dashed line, Portion of plasmid strand lost during entry that is resynthesized from the chromosomal template.

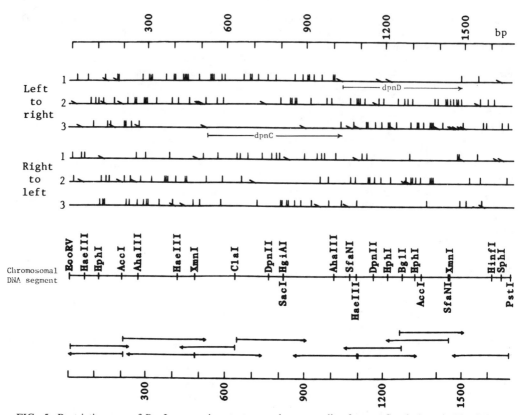

FIG. 5. Restriction map of *Dpn*I sequencing strategy and open reading frames. Symbols as in Fig. 3 except arrows indicate *Dpn*I-specific genes.

native enzyme was previously estimated to be 20,000 (19), so the corresponding *dpnC* gene was tentatively ascribed to the coding region beginning at the second ATG start site (23a).

Between the termination of the open reading frame that enters the segment and the start of the *Dpn*I endonuclease gene, several potential transcription promoters occur. The one most similar to *Escherichia coli* RNA polymerase binding sites (34), TTGcCc..19 nucleotides..TtaAAT (upper case corresponds to the *E. coli* consensus sequence), falls between the two translational start sites mentioned above. Transcription from this promoter could account for production of the shorter polypeptide from the *dpnC* open reading frame. The coding region for the next open reading frame overlaps the prior endonuclease gene by a single base. It, too, has a reasonable Shine-Dalgarno site, AgAGGAaagtATC. No function has been ascribed to the polypeptide of molecular weight 18,000 that this *dpnD* gene could encode.

Cassette Mechanism of Restriction Gene Transfer

Although the sequenced regions of the *Dpn*I and *Dpn*II segments differ markedly in their central portions, which include the structural genes described above, the ends of the segments are highly homologous, with >90% correspondence of the nucleotide sequence in the parts examined (23a). On each side of the restriction gene segments homologous proteins are encoded in opposite directions, as indicated by the fragmented box arrows in Fig. 6. Figure 6, which compares the chromosomal structure surrounding the restriction genes in *Dpn*I and *Dpn*II strains, summarizes data from restriction site mapping of cloned fragments and from DNA hybridization analysis of chromosomal DNA fragments with various probes (23a). These data show that the restriction enzyme genes are found in cassettes located in the same position of the chromosome of *S. pneumoniae* (23a).

The DNA hybridization analysis clearly showed that unexpressed *Dpn*I and *Dpn*II genes were not present in strains of the complementary phenotype (23a). This negates the original hypothesis that both sets of genes are present in all strains but are regulated in their phenotypic expression (22). The determination of restriction phenotype, rather than being due to a regulatory mechanism, appears to be based on a mecha-

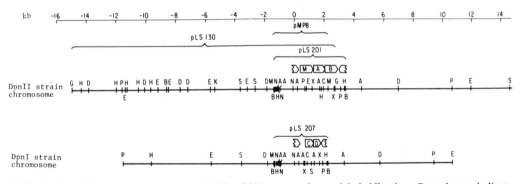

FIG. 6. Restriction gene cassettes revealed by DNA sequencing and hybridization. Open boxes indicate polypeptides encoded by open reading frames; pointed ends show direction of transcription and letters indicate *dpn* gene products. Restriction sites were determined by cleavage of plasmid inserts or hybridization with chromosomal fragments. Key: A, *Hae*III; B, *Bst*EII; C, *Cla*I; D, *Hind*III; E, *Eco*RI; G, *Bgl*II; H, *Hpa*II; K, *Kpn*I; M, *Bam*HI; N, *Nco*I; P, *Pst*I; S, *Sac*I; X, *Xmn*I.

nism of intercellular exchange of genetic cassettes (23a). The presence of homologous DNA on both sides of the restriction gene cassettes would allow transfer of restriction cassettes from one strain to another by transformation-mediated recombination. Replacement of a nonfunctional *Dpn*I cassette by a *Dpn*II cassette has already been demonstrated (31).

The regulation of endonuclease expression relative to the methylase is a general problem both for restriction systems carried on chromosomal cassettes and for those borne by plasmids. On introduction into a new host, expression of the endonuclease prior to appropriate modification of the cellular DNA would be lethal. Several bacterial type II restriction systems have now been cloned and analyzed. The study of these clones has not yet revealed how the enzyme expression is controlled, but several different patterns of restriction gene organization have been found. With *Pae*R7 (14) and *Hha*II (35) the methylase gene is transcribed before the endonuclease gene, as it is with *Dpn*II; with *Eco*RI this order is reversed (6); and with *Eco*RV (8), *Pst*I (35, 41), and *Pvu*II (7) the genes are divergently transcribed. Further exploration and comparison of these cloned systems should reveal whether differences in their regulatory mechanisms can be ascribed to a chromosomal or plasmid origin and whether members of a complementary system, such as *Dpn*I and *Dpn*II, exhibit unique regulatory features.

Are other genetic systems based on intercellularly transferred cassettes? A possible candidate is capsular determination in *S. pneumoniae*. Over 80 different polysaccharide capsule types occur in this species, and in several cases DNA-mediated chromosomal transformation has substituted genes determining one capsule type for genes determining another type (1).

Evolution of Restriction Systems in *E. coli*

In *E. coli*, strand targeting during DNA mismatch repair by the Mut system can be directed by hypomethylation at 5'-GATC-3' sites (33, 40; reviewed in reference 10). To accommodate this feature to mismatch repair in other bacteria, such as *S. pneumoniae*, in which strand breaks appear to be responsible for targeting in mismatch repair, it was suggested that hemimethylation gives rise to single-strand breaks (20). Inasmuch as the *dam* gene (30) product, which is responsible for 5'-GATC-3' methylation in *E. coli* (22), shares considerable homology with the *Dpn*II methylase, it was further suggested that a *Dpn*II-like restriction system in an ancestor of *E. coli* evolved to function in mismatch repair, with the DNA methylase retaining its original activity and the endonuclease converted to an enzyme that nicks DNA at unmethylated sites (29). It will therefore be interesting to see whether any of the Mut system components (reviewed in reference 10) are homologous to the *Dpn*II endonuclease.

Considerations of the possible evolution of restriction enzymes in *S. pneumoniae* and *E. coli* are summarized in Fig. 7. In *S. pneumoniae*, the *Dpn*I system presumably evolved in an ancestral species that already contained the more common *Dpn*II type of restriction system. Evolution in *E. coli* of its ancestral type II restriction system to a new function may have prompted this species to evolve its current chromosomally determined *Eco*K and *Eco*B restriction systems, which are of type I. Such type I systems have been found so far only in gram-negative bacteria. Even though they are not complementary, but because they are allelic (9), the *Eco*K and *Eco*B restriction systems could enhance viral resistance of *E. coli* in a manner similar to the cassette systems of *S. pneumoniae*.

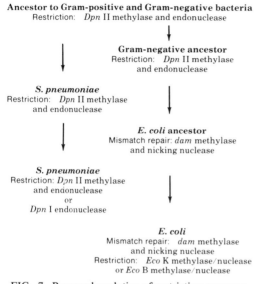

Ancestor to Gram-positive and Gram-negative bacteria
Restriction: *Dpn* II methylase and endonuclease

Gram-negative ancestor
Restriction: *Dpn* II methylase
and endonuclease

S. pneumoniae
Restriction: *Dpn* II methylase
and endonuclease

E. coli ancestor
Mismatch repair: *dam* methylase
and nicking nuclease

S. pneumoniae
Restriction: *Dpn* II methylase
and endonuclease
or
Dpn I endonuclease

E. coli
Mismatch repair: *dam* methylase
and nicking nuclease
Restriction: *Eco* K methylase/nuclease
or *Eco* B methylase/nuclease

FIG. 7. Proposed evolution of restriction enzymes.

Evolution of Methylation at 5'-GATC-3' Sites in DNA

The pneumococcal *Dpn*II DNA methylase and the *E. coli dam* methylase are identical in 30% of their amino acid residues (29). This correspondence, which occurs in several tracts distributed throughout the protein, attests to their common ancestry. Since one species is gram positive and the other is gram negative, they must have diverged from their common ancestor over a billion years ago. The suggestion that 5'-GATC-3' methylation arose relatively recently in gram-negative bacteria (4) appears to be erroneous. Several other DNA methylases found in *E. coli* and its viruses share homology with the *Dpn*II methylase. These include T-even phage-encoded enzymes (15) and, curiously, *Eco*RV methylase (8; R. Roberts, personal communication), which methylates 5'-GATATC-3' rather than 5'-GATC-3'.

So far the ability of a restriction enzyme to recognize a methylated DNA sequence is confined to 5'-GmeATC-3'. However, in addition to *Dpn*I such enzymes have been found in *Neisseria* and *Caulobacter* species (I. Schildkraut, personal communication). Will restriction endonucleases specific for other methylated sites be found? We venture to predict that they will, provided that they are sought in the right place and in the right manner. Other bacterial species should also be able to profit from the evolutionary advantage conferred by complementary restriction systems similar to those in *S. pneumoniae*. Therefore, the place to seek the methyl-specific endonuclease is in different strains of a species in which at least one strain methylates its DNA to protect it from the ordinary restriction endonuclease that it makes. The appropriate substrate with which to seek the methyl-specific enzyme, then, is the methylated DNA from the latter strain.

We thank T. S. Balganesh and P. Kale for their experimental contributions to the project.

This research was conducted at Brookhaven National Laboratory under the auspices of the U.S. Department of Energy Office of Health and Environmental Research. It was supported by U.S. Public Health Service grants AI14885 and GMAI29721 from the National Institutes of Health.

LITERATURE CITED

1. **Austrian, R., H. P. Bernheimer, E. E. B. Smith, and G. T. Mills.** 1959. Simultaneous production of two capsular polysaccharides by pneumococcus. II. The genetic and biochemical bases of binary capsulation. J. Exp. Med. **110:**585–602.
2. **Balganesh, T. S., and S. A. Lacks.** 1984. Plasmid vector for cloning in *Streptococcus pneumoniae* and strategies for enrichment for recombinant plasmids. Gene **29:**221–230.
3. **Balganesh, T. S., and S. A. Lacks.** 1985. Heteroduplex DNA mismatch repair system of *Streptococcus pneumoniae*: cloning and expression of the *hexA* gene. J. Bacteriol. **162:**979–984.
4. **Barbeyron, T., K. Kean, and P. Forterre.** 1984. DNA adenine methylation of GATC sequences appeared recently in the *Escherichia coli* lineage. J. Bacteriol. **160:**586–590.
5. **Bernheimer, H. P.** 1979. Lysogenic pneumococci and their bacteriophages. J. Bacteriol. **138:**618–624.
6. **Betlach, M., V. Hershfield, L. Chow, W. Brown, H. M. Goodman, and H. W. Boyer.** 1976. A restriction endonuclease analysis of the bacterial plasmid controlling the *Eco*RI restriction and modification of DNA. Fed. Proc. **35:**2037–2043.
7. **Blumenthal, R. M., S. A. Gregory, and J. S. Cooperider.** 1985. Cloning of a restriction-modification system from *Proteus vulgaris* and its use in analyzing a methylase-sensitive phenotype in *Escherichia coli*. J. Bacteriol. **164:**501–509.
8. **Bougueleret, M., M. Schwarzstein, A. Tsugita, and M. Zabeau.** 1984. Characterization of the genes coding for the *Eco*RV restriction and modification system of *Escherichia coli*. Nucleic Acids Res. **12:**3659–3677.
9. **Boyer, H.** 1964. Genetic control of restriction and modification in *Escherichia coli*. J. Bacteriol. **88:**1652–1660.
10. **Claverys, J. P., and S. A. Lacks.** 1986. Heteroduplex DNA base mismatch repair in bacteria. Microbiol. Rev. **50:**133–165.
11. **Espinosa, M., P. Lopez, M. T. Perez-Urena, and S. A. Lacks.** 1982. Interspecific plasmid transfer between *Streptococcus pneumoniae* and *Bacillus subtilis*. Mol. Gen. Genet. **188:**195–201.
12. **Fox, M. S., and M. K. Allen.** 1964. On the mechanism of deoxyribonucleate integration in pneumococcal transformation. Proc. Natl. Acad. Sci. USA **52:**412–419.
13. **Geier, G. E., and P. Modrich.** 1979. Recognition sequence of the *dam* methylase of *Escherichia coli* K12 and mode of cleavage of *Dpn*I endonuclease. J. Biol. Chem. **254:**1408–1413.
14. **Gingeras, T. R., and J. E. Brooks.** 1983. Cloned restriction modification system from *Pseudomonas aeruginosa*. Proc. Natl. Acad. Sci. USA **80:**402–406.
15. **Hattman, S., J. Wilkinson, D. Swinton, S. Schlagman, P. M. MacDonald, and G. Mosig.** 1985. Common evolutionary origin of the phage T4 *dam* and host *Escherichia coli dam* DNA-adenine methyltransferase genes. J. Bacteriol. **164:**932–937.
16. **Lacks, S.** 1962. Molecular fate of DNA in genetic transformation of pneumococcus. J. Mol. Biol. **5:**119–131.

17. **Lacks, S.** 1966. Integration efficiency and genetic recombination in pneumococcal transformation. Genetics **53:**207–235.

18. **Lacks, S. A.** 1977. Binding and entry of DNA in pneumococcal transformation, p. 179–232. *In* J. Reissig (ed.), Microbial interactions. Chapman and Hall, London.

19. **Lacks, S. A.** 1980. Purification and properties of the complementary endonucleases *Dpn*I and *Dpn*II. Methods Enzymol. **65:**138–146.

20. **Lacks, S. A., J. J. Dunn, and B. Greenberg.** 1982. Identification of base mismatches recognized by the heteroduplex-DNA-repair system of *Streptococcus pneumoniae*. Cell **31:**327–336.

21. **Lacks, S., and B. Greenberg.** 1975. A deoxyribonuclease of *Diplococcus pneumoniae* specific for methylated DNA. J. Biol. Chem. **250:**4060–4066.

22. **Lacks, S., and B. Greenberg.** 1977. Complementary specificity of restriction endonucleases of *Diplococcus pneumoniae* with respect to DNA methylation. J. Mol. Biol. **114:**153–168.

23. **Lacks, S., B. Greenberg, and M. Neuberger.** 1975. Identification of a deoxyribonuclease implicated in genetic transformation of *Diplococcus pneumoniae*. J. Bacteriol. **123:**222–232.

23a.**Lacks, S. A., B. M. Mannarelli, S. S. Springhorn, and B. Greenberg.** 1986. Genetic basis of the complementary *Dpn*I and *Dpn*II restriction systems of *Streptococcus pneumoniae*: an intercellular cassette mechanism. Cell **46:**993–1000.

24. **Lacks, S. A., and S. S. Springhorn.** 1984. Cloning in *Streptococcus pneumoniae* of the gene for *Dpn*II DNA methylase. J. Bacteriol. **157:**934–936.

25. **Lacks, S. A., and S. S. Springhorn.** 1984. Transfer of recombinant plasmids containing the gene for *Dpn*II DNA methylase into strains of *Streptococcus pneumoniae* that produce *Dpn*I and *Dpn*II restriction nucleases. J. Bacteriol. **158:**905–909.

26. **Lopez, P., M. Espinosa, and S. A. Lacks.** 1984. Physical structure and genetic expression of the sulfonamide-resistance plasmid pLS80 and its derivatives in *Streptococcus pneumoniae*. Mol. Gen. Genet. **195:**402–410.

27. **Lopez, P., M. Espinosa, D. L. Stassi, and S. A. Lacks.** 1982. Facilitation of plasmid transfer in *Streptococcus pneumoniae* by chromosomal homology. J. Bacteriol. **150:**692–701.

28. **Mann, M. B., R. N. Rao, and H. O. Smith.** 1978. Cloning of restriction and modification genes in *E. coli*: the *Hha*II system from *Haemophilus haemolyticus*. Gene **3:**97–112.

29. **Mannarelli, B. M., T. S. Balganesh, B. Greenberg, S. S. Springhorn, and S. A. Lacks.** 1985. Nucleotide sequence of the *Dpn*II DNA methylase gene of *Streptococcus pneumoniae* and its relationship to the *dam* gene of *Escherichia coli*. Proc. Natl. Acad. Sci. USA **82:**4468–4472.

30. **Marinus, M. G., and N. R. Morris.** 1973. Isolation of deoxyribonucleic acid methylase mutants of *Escherichia coli* K-12. J. Bacteriol. **114:**1143–1150.

31. **Muckerman, C. C., S. S. Springhorn, B. Greenberg, and S. A. Lacks.** 1982. Transformation of restriction endonuclease phenotype in *Streptococcus pneumoniae*. J. Bacteriol. **152:**183–190.

32. **Porter, R. D., and W. R. Guild.** 1976. Characterization of some pneumococcal bacteriophages. J. Virol. **19:**659–667.

33. **Pukkila, P. J., J. Peterson, G. Herman, P. Modrich, and M. Meselson.** 1983. Effects of high levels of DNA adenine methylation on methyl-directed mismatch repair in *Escherichia coli*. Genetics **104:**571–582.

34. **Rosenberg, M., and D. Court.** 1979. Regulatory sequences involved in the promotion and termination of RNA transcription. Annu. Rev. Genet. **13:**319–353.

35. **Schoner, B., S. Kelly, and H. O. Smith.** 1983. The nucleotide sequence of the *Hha*II restriction and modification genes from *Haemophilus haemolyticus*. Gene **24:**227–236.

36. **Shine, J., and L. Dalgarno.** 1975. Determinant of cistron specificity in bacterial ribosomes. Nature (London) **254:**34–38.

37. **Stassi, D. L., P. Lopez, M. Espinosa, and S. A. Lacks.** 1981. Cloning of chromosomal genes in *Streptococcus pneumoniae*. Proc. Natl. Acad. Sci. USA **78:**7028–7032.

38. **Szomolanyi, E., A. Kiss, and P. Venetianer.** 1980. Cloning the modification methylase gene of *Bacillus sphaericus* R in *Escherichia coli*. Gene **10:**219–225.

39. **Vovis, G. F., and S. Lacks.** 1977. Complementary action of restriction enzymes Endo R.*Dpn*I and Endo R.*Dpn*II on bacteriophage f1 DNA. J. Mol. Biol. **115:**525–538.

40. **Wagner, R., C. Dohet, M. Jones, M.-P. Doutriaux, F. Hutchinson, and M. Radman.** 1984. Involvement of *Escherichia coli* mismatch repair in DNA replication and recombination. Cold Spring Harbor Symp. Quant. Biol. **49:**611–615.

41. **Walder, R. Y., J. L. Hartley, J. E. Donelson, and J. A. Walder.** 1981. Cloning and expression of the *Pst*I restriction-modification system in *Escherichia coli*. Proc. Natl. Acad. Sci. USA **78:**1503–1507.

II. Antibiotic Resistance

Dissemination of Streptococcal Antibiotic Resistance Determinants in the Natural Environment

DONALD J. LeBLANC,[1] LINDA N. LEE,[1] MENACHEM BANAI,[2] LARRY D. ROLLINS,[3] AND JULIA M. INAMINE[4]

Bacterial Virulence Section, Laboratory of Molecular Microbiology, National Institute of Allergy and Infectious Diseases, Fort Detrick, Frederick, Maryland 21701-1013[1]; International Genetic Sciences Partnership, Jerusalem 91042, Israel[2]; Division of Therapeutic Drugs for Food Animals, Center for Veterinary Medicine, Food and Drug Administration, Rockville, Maryland 20857[3]; and Department of Microbiology and Immunology, University of North Carolina at Chapel Hill, Chapel Hill, North Carolina 27514[4]

Virtually all groups of bacteria capable of causing human and animal diseases have become increasingly resistant to antibiotics over the past 40 years. The streptococci are no exception. Reports of tetracycline (Tc)-resistant (33, 42) and erythromycin (Em)-resistant (34) isolates of *Streptococcus pyogenes* from human clinical sources first appeared in the 1950s. By the late 1960s and early 1970s, group A streptococci resistant to lincomycin (17, 38, 53) and chloramphenicol (Cm; 38) were also emerging. From 1969 to 1979, increases in antibiotic resistance among several groups of streptococci were being reported (18, 36, 40, 59). Among the group D streptococci, resistance to high levels ($>2,000$ µg/ml) of the aminoglycoside antibiotics, such as neomycin, kanamycin (Km), and streptomycin (Sm), has become quite common (4, 9, 39, 41). More recently, group D strains resistant to high levels of gentamicin (Gm; 26, 37) and spectinomycin (Sp; this report) have been isolated. The first report of a streptococcal isolate producing a transferable β-lactamase activity is of a strain of *Streptococcus faecalis* obtained from a human patient (44). Thus, streptococci resistant to Tc, Cm, the MLS group of antibiotics (macrolides such as Em, lincosamides such as lincomycin, and streptogramin B-type antibiotics), virtually all clinically relevant aminoglycosides, and penicillin (as a result of β-lactamase activity) have now been isolated. Many of these strains have been shown to be resistant to more than one antibiotic.

The antibiotic resistance traits of streptococci are usually encoded by plasmids. The first report of plasmid-mediated antibiotic resistance in *Streptococcus* species appeared in 1972 (14).

Just 2 years later, Jacob and Hobbs (30) described a multiple antibiotic resistance plasmid from a strain of *S. faecalis* that was transferable to a plasmid-free strain of *S. faecalis* by a mechanism resembling conjugation. Since then, a large number of transferable plasmids, many encoding several antibiotic resistance traits, have been described in natural isolates of streptococci (see reference 12 for review). The most promiscuous of the streptococcal conjugative R plasmids, relative to interspecies and intergeneric transfer, are the so-called MLS resistance plasmids. These represent a group of closely related replicons which may encode resistance to other antibiotics as well as the MLS group of antibiotics. The MLS resistance plasmids have been transferred, in nature or in the laboratory, among virtually all species of streptococci and to other genera of gram-positive bacteria including *Lactobacillus, Bacillus*, and *Staphylococcus* (see reference 24 for review). In addition, Weisblum and co-workers (64) demonstrated DNA sequence homology among MLS resistance plasmids isolated from streptococci and staphylococci found in nature. In retrospect, the broad host range of the MLS resistance plasmids predicted that other streptococcal antibiotic resistance determinants would also be found outside this genus.

The transfer of antibiotic resistance determinants in streptococci is not always associated with plasmids. There have been several reports on the conjugative transfer of antibiotic resistance among strains of *Streptococcus pneumoniae* (8, 56, 57) for which no plasmid association could be demonstrated. In 1981, Horodniceanu et al. (25) described the en bloc transfer of Tc,

Cm, and MLS resistance in the absence of detectable plasmid DNA from clinical isolates of group A, B, F, and G streptococci. These same investigators obtained similar results regarding the transfer of high-level aminoglycoside resistance from clinical isolates of group A, B, G, D (*Streptococcus bovis*), and viridans streptococci (27). Inamine and Burdett (28) characterized a large (67-kilobase-pair [kbp]) conjugative element mediating resistance to Cm, Em, and Tc from the chromosome of a group B streptococcal isolate. Clewell and associates (20–22; J. M. Jones et al., this volume) have provided a molecular basis for plasmid-free resistance transfer with the characterization of a Tc resistance transposon from *S. faecalis*, Tn916. Tn916 encodes its own conjugative functions and thus can mediate its own transfer from one bacterial strain to another in the absence of plasmid DNA. Although the existence of this type of transposon may not explain all of the presumed non-plasmid-mediated antibiotic resistance transfers referred to above, other conjugative transposons have since been described (19, 23).

By 1982 it was obvious that the streptococci were carrying several antibiotic resistance determinants and that many of these traits were transferable, singly or in groups, by a number of different mechanisms. Although most of the studies dealt with resistance in human clinical isolates, we and others (11) had observed many of the same phenotypic traits in streptococci obtained from animal sources. These observations, coupled with the reports of MLS resistance plasmid transfer to several genera of gram-positive bacteria (reviewed in reference 24), the shared DNA sequence homology between such plasmids from streptococci and staphylococci (64), and reports of similar mechanisms for aminoglycoside resistance in these two genera (10, 15), prompted us to initiate a comprehensive study of antibiotic resistance determinants in streptococci. Our major objectives were to assess the magnitude of the antibiotic resistance gene pool shared by human and animal streptococcal strains, and to determine the extent and mechanisms of its dissemination in the natural environment. In this report we summarize our results to date. We also include data from our laboratory, as well as others, relative to the dissemination of the streptococcal antibiotic resistance gene pool among diverse bacterial genera.

Molecular and Genetic Analysis of Streptococcal R Plasmid pJH1

The first streptococcal conjugative R plasmid described in the literature, pJH1 (30), encodes resistance to Tc, Km, Sm, and Em, the last being representative of the MLS resistance phenotype. Since these phenotypes represent the most common antibiotic resistance traits observed among streptococci of human and animal origin (4, 9, 12, 18, 37, 52, 61), we chose pJH1 as a prototype for further studies. The original clinical isolate of *S. faecalis* that was shown to harbor this plasmid, strain JH1, also contained a second conjugative plasmid, pJH2, which encoded a hemolysin-bacteriocin (Hly-Bcn) determinant (29). Therefore, it was necessary to separate these two plasmids by conjugation, using the plasmid-free strain of *S. faecalis*, JH2-2 (30), as a recipient. A transconjugant that exhibited the pJH1-associated resistance phenotype, strain DL77, was the source of pure pJH1 DNA used to construct a restriction endonuclease map (2; Fig. 1). Also included in Fig. 1 is a summary of results, to be described below and obtained from a series of Southern blot hybridizations, heteroduplex studies, deletion analyses, transformations, and cloning experiments, that provided evidence for the presence in pJH1 of portions of two different R plasmids originally isolated from other strains of *S. faecalis*.

The conjugation experiments used to derive strain DL77 also produced transconjugants exhibiting a variety of resistance phenotypes (2). These included the expression of all four of the

FIG. 1. Molecular, genetic, and functional properties of pJH1. The largest circular restriction endonuclease map is that of pJH1. Smaller circular maps illustrate the region(s) of homology between pJH1 and pAD2 (13) and their locations on the map of pJH1 (solid bars), or between pJH1 and pAMα1δ1 (47) and its location on the map of pJH1 (cross-hatched bars). The Tc resistance plasmid derived from pJH1, pDL316 (1), is represented by the cross-hatched bar plus the flanking open bar regions on the map of pJH1 and by the cross-hatched plus open bar circle superimposed on the map of pAMα1δ1. The open bar represents a *Taq*I fragment of pDL316 that did not exhibit any homology to pAMα1δ1. The linear restriction endonuclease map at the bottom of the figure represents a virtually identical 11-kbp region of pJH1 and pAD2 (31). The locations of the structural genes for resistance to Sm, Km, Em, and Tc are indicated by *str, kan, erm*, and *tet*, respectively. The origin of replication on the map of pAMα1δ1 is indicated by *ori*. Capital letters on the maps designate sites for cleavage by restriction endonucleases: A, *Ava*I; B, *Bam*HI; C, *Cla*I; E, *Eco*RI; H, *Hin*dIII; T, *Taq*I; X, *Xba*I. F412, F413, and F414 designate the cloned fragments from pJH1 encoding resistance to Sm, Km, and Em, respectively, that were used as hybridization probes to detect homologous sequences in DNA from multiply resistant streptococcal isolates (31). Tn917 (13) and Tn3871 (3) are similar, if not identical, Em resistance transposons located on pAD2 and pJH1, respectively.

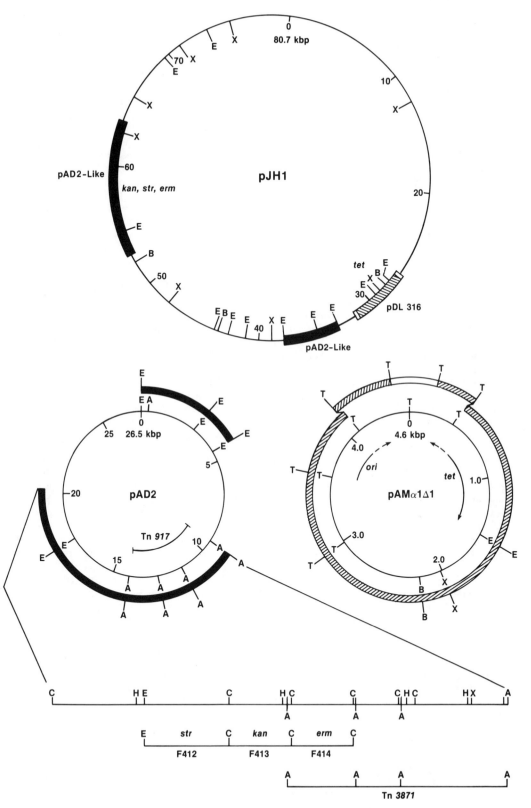

pJH1-associated resistance traits: resistance to Km, Sm, and Em only; Em and Tc only; Em alone; and Tc alone. The vast majority of the transconjugants also expressed the pJH2-associated Hly-Bcn trait. These transconjugants could be divided into six categories with respect to plasmid content: just pJH1, intact pJH1 and pJH2, intact pJH2 and a deleted pJH1 molecule, a cointegrate plasmid consisting of pJH1 and pJH2, pJH2 with an insert composed of a segment of pJH1 DNA, or a recombinant plasmid composed of portions of pJH1 and pJH2. Plasmid DNA from several of these transconjugants was used in the genetic, molecular, and functional analysis of pJH1.

Analysis of the Tc resistance region (*tet*) of pJH1 was initiated with a study of one of the transconjugants described above, strain DL310, which was resistant to Tc and expressed the Hly-Bcn phenotype of pJH2. Strain DL310 contained an intact pJH2 plasmid and a second replicon with sequences from pJH1 and pJH2. Plasmid DNA from this strain was used to transform a competent culture of *Streptococcus sanguis* strain Wicky (46), and a Tc-resistant transformant, strain DL316, was chosen for further study (1). This transformant contained a 4.7-kbp plasmid, pDL316, that was composed entirely of pJH1-derived sequences. Restriction endonuclease digests, Southern blot hybridizations, and heteroduplex analyses were used to determine the origin of pDL316 on pJH1 and to show that it was nearly identical to another Tc resistance plasmid, pAMα1δ1, a 4.6-kbp replicon also of streptococcal origin (47). The location of pDL316 on pJH1 is represented by the cross-hatched bar in Fig. 1, as is its relationship to pAMα1δ1. A restriction endonuclease map of pDL316 could almost be superimposed on a previously published map of pAMα1δ1 (47). Only a small region of pDL316, represented by the open bar at the top of the smaller circular map, and the open bars flanking the cross-hatched bar on the map of pJH1, showed no homology to pAMα1δ1. The homology between pDL316 and pAMα1δ1 included the Tc resistance region and origin of replication of the latter plasmid.

The Km, Sm, and Em resistance region (*kan str erm*) of pJH1 was located by using several of the transconjugants obtained from the matings between strain JH1 and strain JH2-2 which exhibited Hly-Bcn activity and were resistant to Km, Sm, and Em, but not Tc. Most of these transconjugants contained a single plasmid composed of intact pJH2 with a segment of pJH1 DNA inserted into it. In addition, a spontaneous deletion derivative of pJH1 was isolated which mediated resistance to Tc, but no longer expressed resistance to Km, Sm, or Em; this derivative was smaller than pJH1 by approximately 16 kbp. The pJH2-pJH1 hybrid

plasmids and the pJH1 deletion derivative were used in restriction endonuclease analyses and Southern blot hybridizations to provide approximate locations for these determinants on the map of pJH1 (2), as indicated in Fig. 1.

Further study of the *erm* region was prompted by the observation of a high frequency (~20%) of transconjugants from the JH1 × JH2-2 matings which were positive for the Hly-Bcn trait and resistant only to Em. The single plasmid species harbored by these transconjugants were pJH2 molecules with inserts from pJH1 of 5 to 6 kbp (2). Additional studies provided evidence that the Em resistance determinant of pJH1 resided on a transposon, Tn*3871* (3), that was virtually identical to Tn*917*, a 5.1-kbp transposon originally shown to be present in pAD2, which is a 26.5-kbp R plasmid from *S. faecalis* DS16 (13, 61). Plasmid pAD2 also encodes resistance to Km and Sm. Further analysis showed that the relationship between pJH1 and pAD2 extended beyond the respective Em resistance transposons (31).

We compared the resistance regions of pJH1 and pAD2 by cloning the resistance determinants of both plasmids in the Challis strain of *S. sanguis*, using the cryptic plasmid pVA380-1 (35) as a vector and the restriction endonucleases *Eco*RI, *Hin*dIII, and *Cla*I, singly and in combination (31). These cloned fragments were used to construct a map of an apparently identical 11-kbp segment of DNA from pJH1 and pAD2 which includes the structural genes for Km and Sm resistance, as well as the Em resistance transposon. The location of this DNA segment on the map of pJH1 is indicated by the larger of the two solid bars in Fig. 1. A detailed linear map of this segment, as well as its location on a circular map of pAD2, is also illustrated in Fig. 1. Southern blot hybridizations provided evidence for the presence of a second segment of pAD2 DNA in pJH1. This pAD2 DNA region includes one of two long inverted repeat sequences (13), and its location in pJH1 and pAD2 is indicated by the smaller of the two solid bars drawn on the respective maps in Fig. 1. Whether either of the two pAD2-associated DNA sequences present in pJH1 includes an origin of replication from pAD2 has not been determined. However, it is clear from all of the results presented in this section that pJH1 may very likely represent a composite plasmid containing DNA from at least two other streptococcal replicons, plasmids pAD2 and pAMα1δ1. It is also possible that pAD2 and pAMα1δ1 may have originated, at least in part, from pJH1.

Dissemination of Streptococcal Animoglycoside Resistance Determinants among Gram-Positive Cocci of Human and Animal Origin

The extent to which homologous resistance genes have been disseminated among group D streptococci collected from human and animal sources was evaluated by a series of blot hybridizations (31). The smallest cloned pJH1 DNA fragments expressing resistance to Sm (fragment F412) and Km (fragment F413) which were used as molecular probes are shown on the linear map at the bottom of Fig. 1. We examined DNA samples from a total of 91 multiply resistant isolates obtained from healthy pigs and chickens from U.S. farms (52), and from human clinical sources in four different countries (31, 52). Among the pig isolates obtained from seven states, 76 and 52% of the DNA samples exhibited homology to fragments F412 and F413, respectively. With samples from the chicken isolates from six states, the percentages were 72 and 70, respectively. The human clinical isolates were obtained from Houston, Texas, Washington, D.C., and Ann Arbor, Michigan, in the United States, and from England, Chile, and Thailand. Among the DNA samples from these strains, 83% hybridized to the Sm resistance determinant probe, and 79% hybridized to the Km resistance probe.

H. Ounissi and P. Courvalin (this volume) used intragenic fragments from the pJH1-derived Sm and Km resistance determinants as probes to examine DNA from 500 multiply resistant human clinical streptococcal isolates obtained from 15 hospitals in France. They reported that nearly 78 and 88% of the DNA samples examined exhibited homology to the Sm and Km resistance probes, respectively. In addition, they showed that these same Sm and Km resistance probes hybridized to nearly 79 and 89% of the DNA samples, respectively, from 500 multiply resistant staphylococcal isolates. These investigators also used an intragenic fragment from a streptococcal Km-Gm resistance determinant as a probe in hybridizations with the DNA samples from the same group of streptococcal and staphylococcal isolates. This probe hybridized to 98 and 99% of the DNA samples from the streptococcal and staphylococcal strains, respectively.

There is clearly a common evolutionary origin for the antibiotic resistance genes found in streptococci isolated from humans and those isolated from animals. Similar genes for the high-level Sm, Km, and Km-Gm resistance traits originally cloned from streptococcal isolates have also been disseminated among strains of *Staphylococcus* species obtained from human sources. Further studies on the evolutionary relationships among these genes from different sources will be aided by the availability of the nucleotide sequence of the pJH1-encoded Km resistance structural gene (62) and more recently of the pJH1-encoded Sm resistance gene (Ounissi and Courvalin, this volume). In this regard, P. Courvalin, C. Carlier, and F. Caillaud (this volume) have shown that the nucleotide sequences of the structural genes for Km resistance from pJH1, a transposon from a strain of *S. pneumoniae*, and a plasmid from a gram-negative intestinal pathogen, *Campylobacter coli*, are identical. The dissemination of closely related antibiotic resistance determinants among gram-positive, gram-negative, and cell wall-less bacterial genera is discussed in a later section of this report.

Dissemination of the MLS Resistance Determinant of pJH1 among Group D Streptococci of Human and Animal Origin

Since previous studies had demonstrated the presence of an MLS resistance determinant on transposon Tn*3871* in pJH1 (3) and on Tn*917* in pAD2 (13, 61), we decided to search for similar elements in the collection of human and animal isolates described in the previous section. Initially, we used the smallest cloned fragment from pJH1 that was still able to express resistance to the MLS group of antibiotics, fragment F414 on the linear map at the bottom of Fig. 1, as a probe to identify homologous sequences in the DNA samples from the collection of isolates. This probe hybridized to 81, 74, and 83% of the samples from pigs, chickens, and humans, respectively. Several of these strains were able to serve as antibiotic resistance donors in conjugation experiments. Among the transconjugants obtained from these matings, eight contained a single plasmid encoding the same resistance phenotype as pJH1, i.e., resistance to Tc, Km, Sm, and Em, while a single plasmid from two others exhibited the same phenotype as pAD2, i.e., resistance to Km, Sm, and Em, but not Tc. These plasmids were purified from the respective transconjugants, and their *Ava*I fragment patterns were examined by agarose gel electrophoresis. The rationale for this experiment can be explained from the linear map shown at the bottom of Fig. 1. There are four cleavage sites for *Ava*I within transposon Tn*3871*. This is also true for Tn*917*, as indicated on the circular map of pAD2. Thus, any plasmid containing a transposon that is closely related to Tn*3871* or Tn*917* will yield three distinct fragments when digested with *Ava*I. The recently published sequence of Tn*917* (48, 54) indicated that these three fragments account for greater than 99% of this transposon. The *Ava*I digestion patterns of the plasmids examined in this experiment indicated that, in addition to pJH1 and pAD2, plasmids from two additional human isolates, one pig isolate, and two chicken isolates contained the three distinct Tn*917*-associated *Ava*I fragments. Two of the Tn*917*-like *Ava*I fragments were also present on a plasmid from a second pig isolate.

TABLE 1. Relationship of pJH1 to selected group D streptococcal plasmids

Plasmid	Size (kbp)	Source of original strain	Geographical origin of strain	Extent of homology to pJH1 (kbp)	Tn917-related AvaI fragments	Transposition of Em resistance demonstrated
pJH1	81	Human	London	81	3	+
pAD2	26	Human	Ann Arbor	11	3	+
pLDR505	105	Human	Washington, D.C.	11	3	NT[a]
pLDR506	105	Human	Washington, D.C.	11	3	+
pLDR512	53	Pig	Illinois	11	3	+
pLDR504	63	Pig	Nebraska	9	2	−
pLDR167	?	Chicken	North Carolina	+[b]	3	NT
pLDR191	?	Chicken	Arkansas	+	3	NT

[a] NT, Not tested.

[b] +, Exact amount of homology to pJH1 is not known beyond the Km and Sm resistance determinants and the three Tn3871-like AvaI fragments.

Homology between these fragments and authentic Tn917-specific DNA was confirmed by Southern blot hybridization. These results, as well as the results of additional experiments performed with these plasmids, are summarized in Table 1. The strains from which the plasmids containing Tn917-like sequences had been transferred represented a broad geographical distribution, and the plasmids themselves exhibited a wide range of sizes. Thus far, transposition of the MLS resistance phenotype and the presence of the three distinct AvaI fragments has been demonstrated with four of the plasmids. An Em resistance transposon, Tn551, very similar to Tn917, has also been described in a strain of Staphylococcus aureus (45).

Dissemination of Plasmid DNA among Gram-Positive Bacteria

The presence of similar or identical antibiotic resistance determinants in streptococcal isolates obtained from numerous geographical areas throughout the world, in strains of human or animal origin, and in different genera of gram-positive bacteria can be explained by several mechanisms. When such determinants are plasmid borne, it seems likely that the genes are at least initially transferred from organism to organism with their associated plasmids. When the plasmid is able to replicate in the new host to which it has been transferred, the resistance determinants may be maintained on that plasmid. If one or more of the plasmid-mediated resistance genes are encoded on a transposon, then these genes may relocate to a different plasmid present in the new host, or to that host's chromosome, whether or not the incoming plasmid can replicate in the new host. If the transferred plasmid cannot replicate in the new host, then the expression of any of its resistance determinants will depend on the integration of the genes into a resident plasmid or the host's chromosome, either by transposition or by recombination. Evidence has been accumulating that indicates the involvement of all of these mechanisms in the distribution of antibiotic resistance genes in the natural environment.

With the exception of pLDR505 and pLDR506, which appear to be identical, all of the plasmids listed in Table 1 provided very different EcoRI restriction patterns in agarose gels. Despite this apparent heterogeneity, the first six plasmids in Table 1 all contained the three pJH1-related resistance determinants on a single EcoRI fragment ranging in size between 11 and 20 kbp (31). From a series of Southern blot hybridizations and subcloning experiments, we have shown that five of the plasmids, including pJH1, were indistinguishable over an 11-kbp region that included the Tn917-associated AvaI fragments. The sixth plasmid, pLDR504, was identical to the other five over a 9-kbp region but lacked a portion of the largest of the three Tn917-like AvaI fragments illustrated at the right-hand side of the linear map at the bottom of Fig. 1. Whether the presence of a common segment of plasmid DNA on at least five otherwise different plasmids represents the result of recombinational events between a resident plasmid and a newly acquired plasmid, or a transpositional event, has not been determined.

The antibiotic resistance phenotype most widely disseminated among streptococci, and bacteria in general, is resistance to Tc. Burdett and co-workers (6) have described three different Tc resistance determinants associated with streptococci: tetL, which has been found exclusively on plasmids (6); tetM, which has been located on the chromosome of numerous streptococcal species (6, 7, 23, 60); and tetN, for which there is currently only one known example, and this is on a plasmid (6). Burdett et al. (6, 7) demonstrated the presence of tetL on pJH1 and on the small S. faecalis Tc resistance plasmid, pAMα1. As described in an earlier section, smaller Tc resistance plasmid derivatives of these plasmids appear to be almost identical; pDL316 from pJH1 (1) and pAMα1δ1 from pAMα1 (47) share the same replication origin and Tc resistance determinant. On the basis of data from several laboratories, it now appears that pAMα1δ1-like replicons have not only been disseminated among streptococci, but among *Bacillus* species as well (Table 2). Both pBC16, from a strain of *B. cereus* (5), and pNS1981, from a strain of *B. subtilis* (55), are virtually identical to pAMα1δ1 (47). Thus, in these instances, the *tet* determinant has been widely distributed among gram-positive bacteria via a common plasmid. Furthermore, this replicon has also been responsible for the dissemination of at least one other antibiotic resistance determinant to a gram-positive organism. Plasmid pUB110, isolated from a strain of *S. aureus*, is virtually identical to pBC16 except that the Tc resistance gene has been replaced by a Km resistance determinant (49).

Dissemination of Antibiotic Resistance Determinants among Gram-Positive, Gram-Negative, and Cell Wall-Less Bacteria

It now appears that *tetM* may be the most widely distributed of all the streptococcal antibiotic resistance genes, having been found in gram-positive, gram-negative, and cell wall-less bacterial genera. The *tetM* determinant was first cloned in *Escherichia coli* from a Tc-resistant strain of *Streptococcus agalactiae* and was shown by hybridization analysis to be present in strains of *S. faecalis* and *S. pneumoniae* by Burdett and associates (6). In 1982 we reported the presence of a second Tc resistance determinant in *S. faecalis* JH1, the strain originally shown to harbor pJH1 (32). This determinant was in the chromosome of JH1, but was capable of transposing onto the Hly-Bcn plasmid, pJH2. The transposon carrying this Tc resistance gene is very similar to Tn*916*, the first conjugative transposon to be identified (20, 21). The Tc resistance encoded by both of these streptococcal transposons is due to the presence of *tetM* (6, 7; unpublished data). Tobian and Macrina (60) then showed that *tetM* was present in a Tc-resistant clinical isolate of *S. mutans*. In 1984, in

collaboration with Hartley et al. (23), we showed that this determinant was responsible for the expression of Tc resistance by several species of oral streptococci (*S. sanguis* I and II, *S. salivarius*, and *S. mitis*) and that in some cases the gene appeared to be located on a conjugative transposon. The following year, in collaboration with Roberts and co-workers, we demonstrated the presence of *tetM* among clinical isolates of the cell wall-less pathogen *Mycoplasma hominis* (51), and earlier this year Roberts and Kenny (50) extended these observations to include a second cell wall-less species, *Ureaplasma urealyticum*. In the latter study, the presence of *tetM* in Tc-resistant strains of *Gardnerella vaginalis*, a urogenital organism of uncertain taxonomic status, was also reported. Taylor (58) presented evidence for the presence of *tetM* in *Campylobacter jejuni*; it appears, however, that the homology detected between cloned *tetM*-containing DNA from *S. agalactiae* and cloned DNA from a Tc resistance plasmid from *C. jejuni* may have been due to DNA sequences flanking the *tetM* gene (P. Courvalin, 2nd ASM Conference on Streptococcal Genetics). M. Roberts (personal communication) has recently detected *tetM* sequences in DNA from Tc-resistant isolates of the gram-negative pathogen *Neisseria gonorrhoeae*. All of the accumulated data on the distribution of *tetM* in the natural environment indicate that it can be ranked among the most promiscuous of bacterial genetic determinants and that it fails to recognize any procaryotic boundaries. The transfer of antibiotic resistance genes between gram-positive and gram-negative bacteria is not confined to *tetM*. Trieu-Cuot et al. (63) have shown that the nucleotide sequences of the structural genes for Km resistance from pJH1 and from the *C. coli* plasmid, pIP1433, are identical. More recently, these investigators (P. Trieu-Cuot, M. Arthur, and P. Courvalin, this volume) provided evidence for an identical 220-base-pair region in the structural gene for MLS resistance in *C. coli* and in the streptococcal transposon Tn*917*.

Molecular Analysis of a Novel Spectinomycin Resistance Determinant from *S. faecalis*

Bacterial adenylylating enzymes that modify Sp are usually active on Sm (16). During a recent

TABLE 2. Dissemination of pAMα1δ1-like replicons among gram-positive bacteria

Replicon	Original host	Original replicon	Discovered in	Reference
pAMα1δ1	*S. faecalis*	pAMα1	*B. subtilis*	47
pDL316	*S. faecalis*	pJH1	*S. sanguis*	1
pBC16	*B. cereus*	pBC16	*B. cereus*	5
pNS1981	*B. subtilis*	Chromosome	*B. subtilis*	55
pUB110	*S. aureus*	pUB110	*S. aureus*	49

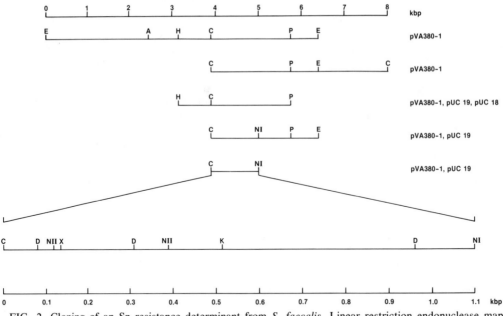

FIG. 2. Cloning of an Sp resistance determinant from *S. faecalis*. Linear restriction endonuclease maps represent fragments cloned from pDL55, a 26-kbp plasmid encoding resistance to Sp that originated in a human clinical isolate of *S. faecalis*. All fragments represented enabled recombinant clones of *S. sanguis* or *E. coli* to grow in the presence of high levels of Sp when ligated to vector plasmid pVA380-1 or pUC19 (or pUC18), respectively, as indicated at the right of each fragment. Capital letters designate sites for cleavage by restriction endonucleases: A, *Ava*I; C, *Cla*I; D, *Dra*I; E, *Eco*RI; H, *Hin*dIII; K, *Kpn*I; NI, *Nde*I; NII, *Nde*II; P, *Pst*I; X, *Xba*I.

survey of group D streptococci, we observed that 1 of 26 human isolates and 59 of 225 animal isolates were resistant to high levels of Sp, and that of these, the human strain and eight of the animal isolates did not express resistance to Sm (unpublished data). Since this aminoglycoside resistance phenotype had been reported for only one other bacterial isolate, a strain of *S. aureus* (16), we decided to examine the human *S. faecalis* strain further (a detailed description of this study will be reported elsewhere). This isolate, strain LDR55, transferred the Sp resistance to a plasmid-free *S. faecalis* recipient, but the transconjugants were unable to serve as donors in subsequent mating experiments. A 26-kbp plasmid, pDL55, was isolated from one of the transconjugants and used in cloning experiments (Fig. 2). Initially, *Eco*RI- or *Cla*I-digested pDL55 DNA was ligated to appropriately digested pVA380-1 vector DNA and used to transform the Challis strain of *S. sanguis*. Sp-resistant transformants containing hybrid plasmids with a 6.4-kbp *Eco*RI or a 4.2-kbp *Cla*I insert from pDL55 were obtained. After constructing restriction endonuclease maps of these two cloned fragments, we subcloned a 2.5-kbp *Hin*dIII-*Pst*I fragment and a 2.5-kbp *Cla*I-*Eco*RI fragment into pVA380-1. *S. sanguis* recombinant clones containing any of the four pDL55-

derived fragments were resistant to greater than 10,000 µg of Sp per ml. The two smaller fragments were subcloned onto pUC plasmids and used to transform *E. coli*. These recombinant clones were resistant to greater than 50,000 µg of Sp per ml. Finally, a 1.1-kbp *Cla*I-*Nde*I fragment was subcloned into *S. sanguis* and *E. coli* with appropriate vectors, and this fragment also conferred resistance to Sp in both strains. The nucleotide base sequence of the 1.1-kbp fragment was determined and compared with the published sequence of a similar Sp resistance determinant from *S. aureus* (43). The Sp resistance gene from *S. aureus* was cloned from Tn554, which also encodes an Em resistance determinant (43), while the human *S. faecalis* isolate carried the Sp resistance gene on plasmid pDL55. Strain LDR55 is also resistant to Em and Tc, but the location of these resistance traits has not been established. Minimal inhibitory concentration of Sp for the *S. aureus* strain from which the Sp resistance determinant was cloned was 4,000 µg/ml, while the transconjugant from which pDL55 was isolated was resistant to greater than 25,000 µg of Sp per ml. Strains carrying either determinant possessed an aminoglycoside adenylylating enzyme active on Sp but not on Sm, presumably an AAD(9). The staphylococcal sequence contained a 783-base open reading frame corresponding to a protein of nearly 29,000 daltons, while the streptococcal sequence contained a 753-base open reading frame corresponding to a protein slightly larger than 29,000 daltons in size. The two open reading frames shared nearly 50% base sequence homology, and the predicted proteins shared approximately 40% amino acid homology, which was expanded to 50% when conserved amino acid changes were included.

Southern blots were obtained from agarose gels containing restriction endonuclease-digested DNA from Sp-resistant *S. faecalis* animal isolates and hybridized to cloned fragments from pDL55. The results indicated that the eight Sp-resistant, Sm-sensitive animal isolates contained the same Sp resistance determinant as the human strain, LDR55. We have also obtained preliminary evidence for homology between the cloned Sp resistance genes from both *S. aureus* and *S. faecalis* and chromosomal DNA from *N. gonorrhoeae* isolates exhibiting resistance to greater than 5,000 μg of Sp per ml.

Conclusions

The results presented in this report clearly indicate that closely related antibiotic resistance genes have been widely disseminated over a broad geographical range among diverse groups of bacteria. The genetic determinants for four of the resistance traits examined, i.e., resistance to Sm, Em, Km, and Tc, were first shown to be carried on a conjugative plasmid, pJH1, from a strain of *S. faecalis* (30). DNAs from the majority of multiply resistant streptococci of human (this report; Ounissi et al., this volume) and animal (34; this report) origin, as well as human staphylococcal isolates (Ounissi et al., this volume), exhibit homology to the pJH1-associated Sm, Em, and Km resistance determinants. The last determinant was also implicated in the recent emergence of Km resistance in *C. coli* (63), a gram-negative enteric pathogen. The Tc resistance determinant of pJH1, *tetL* (6), has been distributed among streptococci (1, 47) and *Bacillus* species (5, 55) via a similar, and in some cases identical, replicon. This same replicon, carrying a Km resistance determinant in place of the *tetL* gene, was also isolated from a strain of *S. aureus* (49). The most widely disseminated of the "streptococcal" antibiotic resistance determinants appears to be *tetM* (6), which has been identified in DNA from Tc-resistant strains of nearly all *Streptococcus* species (6, 23, 60), cell wall-less *M. hominis* (51) and *U. urealyticum* (50) isolates, the gram-negative organism *N. gonorrhoeae* (Roberts, personal communication), and the urogenital bacterium *G. vaginalis* (50). With the exception of the streptococci, the acquisition of Tc resistance by these microorga-

nisms appears to be relatively recent. Sp-specific adenylylating enzyme activity was reported in 1978 in a single strain of *S. aureus* (16). All other known bacterial aminoglycoside adenylylating enzymes that modified Sp were also active on Sm. We have recently obtained nine Sp-resistant, Sm-susceptible strains of *S. faecalis* which were isolated between 1979 and 1982. Eight of these isolates were from animal sources and one was from a human patient. All of the streptococcal strains appear to contain the same Sp resistance gene, which shares at least 50% DNA sequence homology with the Sp resistance gene from *S. aureus* (43). Recent isolates of *N. gonorrhoeae* exhibiting high-level resistance to Sp appear to carry a related genetic determinant. High-level resistance to Gm is also a recently acquired trait in streptococci (26, 37), and a cloned Km-Gm resistance determinant from one such strain was shown to hybridize to nearly all Gm-resistant streptococcal and staphylococcal strains examined (Ounissi and Courvalin, this volume). All of these results suggest that there is constant exchange of genetic information among *Streptococcus* and *Staphylococcus* species, and perhaps also *Bacillus* species, in the natural environment. Furthermore, these gram-positive bacteria may be the sources of newly emerging antibiotic resistance traits among several gram-negative and cell wall-less bacterial species.

J.M.I. was the recipient of National Research Service Award DE05346 from the National Institute of Dental Research.

LITERATURE CITED

1. **Banai, M., M. A. Gonda, J. M. Ranhand, and D. J. LeBlanc.** 1985. *Streptococcus faecalis* R plasmid pJH1 contains a pAMα1δ1-like replicon. J. Bacteriol. **164**:626–632.
2. **Banai, M., and D. J. LeBlanc.** 1983. Genetic, molecular, and functional analysis of *Streptococcus faecalis* R plasmid pJH1. J. Bacteriol. **155**:1094–1104.
3. **Banai, M., and D. J. LeBlanc.** 1984. *Streptococcus faecalis* R plasmid pJH1 contains an erythromycin resistance transposon (Tn*3871*) similar to transposon Tn*917*. J. Bacteriol. **158**:1172–1174.
4. **Basker, M. J., B. Slocomber, and R. Sutherland.** 1977. Aminoglycoside-resistant enterococci. J. Clin. Pathol. **30**:375–380.
5. **Bernhard, K., H. Schrempf, and W. Goebel.** 1978. Bacteriocin and antibiotic resistance plasmids in *Bacillus cereus* and *Bacillus subtilis*. J. Bacteriol. **138**:897–903.
6. **Burdett, V., J. Inamine, and S. Rajagopalan.** 1982. Heterogenicity of tetracycline resistance determinants in *Streptococcus*. J. Bacteriol. **149**:995–1004.
7. **Burdett, V., J. Inamine, and S. Rajagopalan.** 1982. Multiple tetracycline resistance determinants in *Streptococcus*, p. 155–158. *In* D. Schessinger (ed.), Microbiology—1982. American Society for Microbiology, Washington, D.C.
8. **Buu-Hoi, A., and T. Horodniceanu.** 1980. Conjugative transfer of multiple antibiotic-resistance markers in *Streptococcus pneumoniae*. J. Bacteriol. **143**:313–320.
9. **Calderwood, S. A., C. Wennersten, R. C. Moellering, Jr., L. J. Kunz, and D. J. Krogstad.** 1977. Resistance to six aminoglycosidic aminocyclitol antibiotics among entero-

cocci: prevalence, evolution, and relationship to synergism with penicillin. Antimicrob. Agents Chemother. **12:**401–405.

10. **Carlier, C., and P. Courvalin.** 1982. Resistance of streptococci to aminoglycoside-aminocyclitol antibiotics, p. 162–166. *In* D. Schlessinger (ed.), Microbiology—1982. American Society for Microbiology, Washington, D.C.

11. **Christie, P. J., and G. M. Dunny.** 1984. Antibiotic selection pressure resulting in multiple antibiotic resistance and localization of resistance determinants to conjugative plasmids in streptococci. J. Infect. Dis. **149:**74–82.

12. **Clewell, D. B.** 1981. Plasmids, drug resistance, and gene transfer in the genus *Streptococcus.* Microbiol. Rev. **45:**409–436.

13. **Clewell, D. B., P. K. Tomich, M. C. Gawron-Burke, A. Franke, Y. Yagi, and F. Y. An.** 1982. Mapping of *Streptococcus faecalis* plasmids pAD1 and pAD2 and studies relating to transposition of Tn*917.* J. Bacteriol. **152:**1220–1230.

14. **Courvalin, P. M., C. Carlier, and Y. A. Chabbert.** 1972. Plasmid inherited resistance in group D "*Streptococcus*". Ann. Inst. Pasteur (Paris) **123:**755–759.

15. **Courvalin, P., C. Carlier, and E. Collatz.** 1980. Plasmid-mediated resistance to aminocyclitol antibiotics in group D streptococci. J. Bacteriol. **143:**541–555.

16. **Davies, J., and D. I. Smith.** 1978. Plasmid-determined resistance to antimicrobial agents. Annu. Rev. Microbiol. **32:**469–518.

17. **Dixon, J. M. S.** 1968. Group A streptococcus resistant to erythromycin and lincomycin. Can. Med. Assoc. J. **99:**1093–1094.

18. **Finland, M.** 1979. Emergence of antibiotic resistance in hospitals, 1935–1975. Rev. Infect. Dis. **1:**4–21.

19. **Fitzgerald, G. F., and D. B. Clewell.** 1985. A conjugative transposon (Tn*919*) in *Streptococcus sanquis*. Infect. Immun. **47:**415–420.

20. **Franke, A. E., and D. B. Clewell.** 1980. Evidence for conjugal transfer of a *Streptococcus faecalis* transposon (Tn*916*) from a chromosomal site in the absence of plasmid DNA. Cold Spring Harbor Symp. Quant. Biol. **45:**77–80.

21. **Franke, A. E., and D. B. Clewell.** 1981. Evidence for a chromosome-borne resistance transposon (Tn*916*) in *Streptococcus faecalis* that is capable of "conjugal" transfer in the absence of a conjugative plasmid. J. Bacteriol. **145:**494–502.

22. **Gawron-Burke, C., and D. B. Clewell.** 1982. A transposon in *Streptococcus faecalis* with fertility properties. Nature (London) **300:**281–284.

23. **Hartley, D. L., K. R. Jones, J. A. Tobian, D. J. LeBlanc, and F. L. Macrina.** 1984. Disseminated tetracycline resistance in oral streptococci: implication of a conjugative transposon. Infect. Immun. **45:**13–17.

24. **Horaud, T., C. Le Bouguenec, and K. Pepper.** 1985. Molecular genetics of resistance to macrolides, lincosamides and streptogramin B (MLS) in streptococci. J. Antimicrob. Chemother. **16:**111–135.

25. **Horodniceanu, T., L. Bougueleret, and G. Bieth.** 1981. Conjugative transfer of multiple-antibiotic resistance markers in beta-hemolytic group A, B, F, and G streptococci in the absence of extrachromosomal deoxyribonucleic acid. Plasmid **5:**127–137.

26. **Horodniceanu, T., L. Bougueleret, N. El-Solh, G. Bieth, and F. Delbos.** 1979. High-level, plasmid-borne resistance to gentamicin in *Streptococcus faecalis* subsp. *zymogenes*. Antimicrob. Agents Chemother. **16:**686–689.

27. **Horodniceanu, T., A. Buu-Hoi, F. Delbos, and G. Bieth.** 1982. High-level aminoglycoside resistance in group A, B, G, D (*Streptococcus bovis*), and viridans streptococci. Antimicrob. Agents Chemother. **21:**176–179.

28. **Inamine, J. M., and V. Burdett.** 1985. Structural organization of a 67-kilobase streptococcal conjugative element mediating multiple antibiotic resistance. J. Bacteriol. **161:**620–626.

29. **Jacob, A. E., G. J. Douglas, and S. J. Hobbs.** 1975. Self-transferable plasmids determining the hemolysin and bacteriocin of *Streptococcus faecalis* var. *zymogenes*. J. Bacteriol. **121:**863–872.

30. **Jacob, A. E., and S. J. Hobbs.** 1974. Conjugal transfer of plasmid-borne multiple antibiotic resistance in *Streptococcus faecalis* subsp. *zymogenes*. J. Bacteriol. **117:**360–372.

31. **LeBlanc, D. J., J. M. Inamine, and L. N. Lee.** 1986. Broad geographical distribution of homologous erythromycin, kanamycin, and streptomycin resistance determinants among group D streptococci of human and animal origin. Antimicrob. Agents Chemother. **29:**549–555.

32. **LeBlanc, D. J., and L. N. Lee.** 1982. Characterization of two tetracycline resistance determinants in *Streptococcus faecalis* JH1. J. Bacteriol. **150:**835–843.

33. **Lowbury, E. J. L., and J. S. Cason.** 1954. Aureomycin and erythromycin therapy for *Str. pyogenes* in burns. Br. Med. J. **2:**914–915.

34. **Lowbury, E. J. L., and L. Hurst.** 1959. The sensitivity of staphylococci and other wound bacteria to erythromycin, oleandomycin, and spiramycin. J. Clin. Pathol. **12:**163–169.

35. **Macrina, F. L., S. S. Virgili, and C. L. Scott.** 1978. Extrachromosomal gene systems in *Streptococcus mutans*. Adv. Exp. Med. Biol. **107:**859–868.

36. **Matsen, J. M., and C. B. Coghlan.** 1972. Antibiotic testing and susceptibility patterns of streptococci, p. 189–204. *In* L. W. Wannamaker and J. M. Matsen (ed.), Streptococci and streptococcal diseases. Academic Press, Inc., New York.

37. **Mederski-Samoraj, B. D., and B. E. Murray.** 1983. High-level resistance to gentamicin in clinical isolates of enterococci. J. Infect. Dis. **147:**751–757.

38. **Miyamoto, Y., K. Takizawa, A. Matsushima, Y. Asai, and S. Nakatsuka.** 1978. Stepwise acquisition of multiple drug resistance by beta-hemolytic streptococci and difference in resistance pattern by type. Antimicrob. Agents Chemother. **13:**399–404.

39. **Moellering, R. C., Jr., O. M. Korzeniowski, M. A. Sande, and C. B. Wennersten.** 1979. Species-specific resistance to antimicrobial synergism in *Streptococcus faecium* and *Streptococcus faecalis*. J. Infect. Dis. **140:**203–208.

40. **Moellering, R. C., Jr., and D. J. Krogstad.** 1979. Antibiotic resistance in enterococci, p. 293–298. *In* D. Schlessinger (ed.), Microbiology—1979. American Society for Microbiology, Washington, D.C.

41. **Moellering, R. C., Jr., C. Wennersten, T. Medrek, and A. N. Weinberg.** 1971. Prevalence of high-level resistance to aminoglycosides in clinical isolates of enterococci, p. 335–340. Antimicrob. Agents Chemother. 1970.

42. **Mogabgab, W. J., and W. Pelon.** 1958. An outbreak of pharyngitis due to tetracycline-resistant group A, type 12 streptococci. Am. J. Dis. Child. **96:**696–698.

43. **Murphy, E.** 1985. Nucleotide sequence of a spectinomycin andenyltransferase AAD(9) determinant from *Staphylococcus aureus* and its relationship to AAD(3″) (9). Mol. Gen. Genet. **200:**33–39.

44. **Murray, B. E., and B. Mederski-Samoraj.** 1983. Transferable beta-lactamase: a new mechanism for *in vitro* resistance in *Streptococcus faecalis*. J. Clin. Invest. **72:**1168–1171.

45. **Novick, R. P., I. Edelman, M. Schwesinger, A. Gruss, E. Swanson, and P. A. Pattee.** 1979. Genetic translocation in *Staphylococcus aureus*. Proc. Natl. Acad. Sci. USA **76:**400–404.

46. **Pakula, R., and W. Walczak.** 1963. On the nature of competence of transformable streptococci. J. Gen. Microbiol. **31:**125–133.

47. **Perkins, J. B., and P. Youngman.** 1983. *Streptococcus* plasmid pAMα1 is a composite of two separable replicons, one of which is closely related to *Bacillus* plasmid pBC16. J. Bacteriol. **155:**607–615.

48. **Perkins, J. B., and P. J. Youngman.** 1984. A physical and functional analysis of Tn*917*, a *Streptococcus* transposon in the Tn*3* family that functions in *Bacillus*. Plasmid **12:**119–138.

49. **Polack, J., and R. P. Novick.** 1982. Closely related plasmids from *Staphylococcus aureus* and soil bacteria. Plasmid **7**:152–162.

50. **Roberts, M. C., and G. E. Kenny.** 1986. Dissemination of the *tetM* tetracycline resistance determinant to *Ureaplasma urealyticum.* Antimicrob. Agents Chemother. **29**:350–352.

51. **Roberts, M. C., L. A. Koutsky, K. K. Holmes, D. J. LeBlanc, and G. E. Kenny.** 1985. Tetracycline-resistant *Mycoplasma hominis* strains contain streptococcal *tetM* sequences. Antimicrob. Agents Chemother. **28**:141–143.

52. **Rollins, L. D., L. N. Lee, and D. J. LeBlanc.** 1985. Evidence for a disseminated erythromycin resistance determinant mediated by Tn*917*-like sequences among group D streptococci isolated from pigs, chickens, and humans. Antimicrob. Agents Chemother. **27**:439–444.

53. **Sanders, E., M. T. Foster, and D. Scott.** 1968. Group A beta-hemolytic streptococci resistant to erythromycin and lincomycin. N. Engl. J. Med. **278**:538–540.

54. **Shaw, J. H., and D. B. Clewell.** 1985. Complete nucleotide sequence of macrolide-lincosamide-streptogramin B-resistance transposon Tn*917* in *Streptococcus faecalis.* J. Bacteriol. **164**:782–796.

55. **Shishoda, K., and Y. Tanaka.** 1984. A restriction map of *Bacillus subtilis* tetracycline-resistance plasmid pNS1981. Plasmid **12**:65–66.

56. **Shoemaker, N. B., M. D. Smith, and W. R. Guild.** 1979. Organization and transfer of heterologous chloramphenicol and tetracycline resistance genes in *Pneumococcus.* J. Bacteriol. **139**:432–441.

57. **Shoemaker, N. B., M. D. Smith, and W. R. Guild.** 1980. DNase-resistant transfer of chromosomal *cat* and *tet* insertions by filter mating in *Pneumococcus.* Plasmid **3**:80–87.

58. **Taylor, D. E.** 1986. Plasmid-mediated tetracycline resistance in *Campylobacter jejuni*: expression in *Escherichia coli* and identification of homology with streptococcal class M determinant. J. Bacteriol. **165**:1037–1039.

59. **Toala, P., A. McDonald, C. Wilcox, and M. Finland.** 1969. Susceptibility of group D *Streptococcus* (enterococcus) to 21 antibiotics *in vitro*, with special reference to species differences. Am. J. Med. Sci. **258**:416–430.

60. **Tobian, J. A., and F. L. Macrina.** 1982. Helper plasmid cloning in *Streptococcus sanguis*: cloning of a tetracycline resistance determinant from the *Streptococcus mutans* chromosome. J. Bacteriol. **152**:215–222.

61. **Tomich, P. K., F. Y. An, S. P. Damle, and D. B. Clewell.** 1979. Plasmid-mediated transmissibility of multiple drug resistance in *Streptococcus faecalis* subsp. *zymogenes* strain DS16. Antimicrob. Agents Chemother. **15**:828–830.

62. **Trieu-Cuot, P., and P. Courvalin.** 1983. Nucleotide sequence of the *Streptococcus faecalis* plasmid gene coding the 3′,5″-aminoglycoside phosphotransferase type III. Gene **23**:331–341.

63. **Trieu-Cuot, P., G. Gerbaud, T. Lambert, and P. Courvalin.** 1985. *In vivo* transfer of genetic information between Gram-positive and Gram-negative bacteria. EMBO J. **4**:3583–3587.

64. **Weisblum, B., S. B. Holder, and S. M. Halling.** 1979. Deoxyribonucleic acid sequence common to staphylococcal and streptococcal plasmids which specify erythromycin resistance. J. Bacteriol. **138**:990–998.

Structural and Genetic Studies of the Conjugative Transposon Tn916

JOANNE M. JONES, CYNTHIA GAWRON-BURKE,† SUSAN E. FLANNAGAN,
MITSUYO YAMAMOTO,‡ ELISABETH SENGHAS, and DON B. CLEWELL

Departments of Oral Biology and Microbiology/Immunology, Schools of Dentistry and Medicine, and The Dental Research Institute, The University of Michigan, Ann Arbor, Michigan 48109

Tn916 is a 16.4-kilobase (kb) transposon originally identified on the chromosome of the multiply resistant *Streptococcus faecalis* strain DS16 (9–11). The transposon encodes tetracycline resistance, as well as functions which facilitate its conjugal transfer to recipient cells in the absence of mobilizing plasmids. Tn916 has been called a "conjugative transposon" and typifies a number of similarly behaving elements (some encoding multiple resistance traits) that have now been observed in several different species of streptococci (4). Tn918, from *S. faecalis* RC73 (3), and Tn919, from *S. sanguis* FC1 (8), are closely related to Tn916.

Matings between plasmid-free strains are generally conducted on filter membranes, and transfer gives rise to insertions at different sites in the recipient chromosome (10). The frequency of transfer ranges from about 10^{-8} to 10^{-5} per donor and is characteristic for the particular donor strain. The location of Tn916 in the chromosome and possible influence from adjacent sequences may affect the frequency of transfer. There is a quantitative correlation between the conjugative donor potential and the frequency of transposition to a subsequently introduced plasmid such as pAD1 (56.7 kb; encodes hemolysin; conjugative [6]); the two phenomena therefore appear to share a common step. In addition, transposition to a resident plasmid and conjugative transfer in the absence of plasmid DNA are both Rec-independent events (9). It was previously proposed that movement of Tn916 occurs by an excision/insertion mechanism (Fig. 1) and that conjugative transfer simply represents a transposition event where the donor and recipient replicons are located in different cells (9, 10). (At a size of 16.4 kb, there would seem to be ample room for fertility genes necessary for intercellular transfer.) Excision is viewed as the rate-limiting step which, in turn, triggers an efficient expression of functions necessary for insertion and conjugation.

Support for an excision/insertion mechanism is based on the behavior of Tn916 when residing

on a conjugative plasmid such as pAD1 or the broad-host-range erythromycin resistance plasmid pAM81 (26 kb) (10). Transfer of the plasmid results in a "zygotic induction" in the recipient, which leads to an excision of the element from the plasmid and its subsequent loss (segregation) or insertion into the chromosome. The excision appears precise, since in the case of a pAD1::Tn916 derivative it restored expression of an insertionally inactivated hemolysin gene. Zygotically induced transposition was also observed when a plasmid bearing Tn916 was introduced into *S. sanguis* (Challis) by transformation (11). The majority of transformants from such experiments contain Tn916 inserts at different sites in the chromosome.

Tn916 has been cloned into a plasmid vector, pGL101 (a derivative of pBR322; encodes resistance to ampicillin), in *Escherichia coli* DH1 by selecting for expression of tetracycline resistance (11). In the absence of selective pressure, however, the element excises (RecA independent) from the plasmid at high frequency and segregates. (Overnight growth in the absence of tetracycline results in greater than 90% of the cells becoming sensitive to tetracycline.) Insertions into the *E. coli* chromosome are rare (10^{-7}). The excision gives rise to plasmid DNA in which the sequences that flanked the transposon are spliced together. It is conceivable that the propensity for Tn916 to excise in *E. coli* represents an aberrant effort by the element to transpose; in the *E. coli* host, a fully balanced expression of all the required genes for transposition and its control may not occur. The ability to clone Tn916 in *E. coli*, however, has greatly facilitated its characterization, and recent progress in this regard is described below.

Structure of Tn916

Figure 2 shows a restriction map of Tn916 based on analyses of a chimeric molecule, pAM120, representing an *Eco*RI restriction fragment of a pAD1::Tn916 derivative ligated into pGL101. The insert was in *Eco*RI fragment F, a segment located within the hemolysin determinant of pAD1 (6). A comparison of restriction patterns derived from pAM120 with those ob-

† Present address: Ecogen, Inc., Langhorne, PA 19047.

‡ Present address: Institute of Medical Science Advance, Sagamihara Kanagawa 229, Japan.

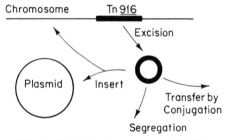

FIG. 1. Model for the behavior of Tn916.

tained from a derivative of pAM120 from which Tn916 excised (pAM120LT) allowed the identification of Tn916-specific fragments. The tetracycline resistance determinant (tet) is believed to contain the single HindIII site, since insertion of a fragment into this site results in loss of resistance (11). Subcloning experiments localizing tetracy-

cline resistance in the 4.8-kb HincII fragment of pAM120 (see Fig. 2) are consistent with this view (C. Gawron-Burke and J. Jones, unpublished data). In other experiments in which the two HindIII transposon-junction fragments (HindIII A and HindIII B) of the streptococcal plasmid pAM180 (pAM81::Tn916) were each cloned into the E. coli vector pACYC177, neither of the two clones expressed tetracycline resistance (11).

The termini of Tn916 have been sequenced using the following strategy. A chimera designated pAM160 (pGL101 with the pAD1 EcoRI H fragment containing a Tn916 insert) was used as a source of DNA for subcloning into M13 vectors (M13mp18 and M13mp19). Plaques containing inserts were screened for the presence of Tn916 sequences (using an appropriate Tn916 probe; ^{32}P-labeled by nick translation) as well as pAD1 sequences (probing with pAM717 [12], a pAD1 derivative containing a large deletion not

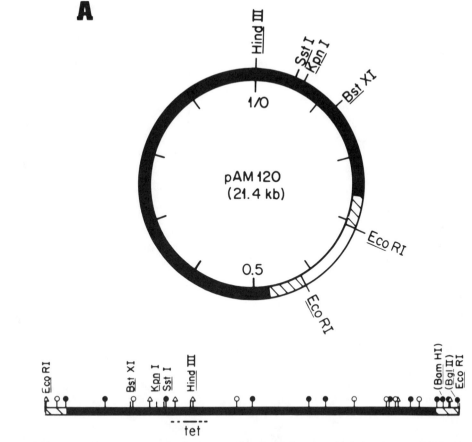

FIG. 2. Restriction map of Tn916. (A) pAM120 representing EcoRI fragment F::Tn916 from pAD1::Tn916 (pAM211) cloned into the E. coli vector pGL101. The dark segment represents Tn916. The hatched segment represents the EcoRI F fragment sequences of pAD1. The remainder represents pGL101. (B) More detailed map of Tn916. Restriction enzymes having single sites are named on the diagram. Multiple-site enzymes are shown as follows: ●, Sau3A; ○, HincII; △, HpaII.

A

```
            10        20        30        40        50        60        70
5'...GTGAAACAAATGATAATATTATT[AAACTAAA]CAAAGTATAAATTTCTAATTATCTTTTTATATTTTCTTAAATGCTCG


        80        90       100       110       120       130       140       150
TAAAGCCTTATTCTATGTGCTTTCGAGTATTTTTACTGTAGGAAGATACTTCACGTTTCTTTGCATATTTCCTCATGTCT


             170       180       190       200       210       220       230
TAGCTGTCAGAAGTGGTAAATAAGTAGTAAATTCATTTGTACTACTAAGCAACAAGACGCTCCTGTTGCTTCTCTTTATT
               "DR-2"      DR-2


            250       260
CAAGCGTTTCATTTCTGCCATT... 3'
```

B

```
            10        20        30        40        50        60        70        80
5'...TGATTCTTGATTTTTTGTTTTCATAAGTTCACTTCCTTTCAAAATCGGGTAAAAAAAATAGACACCTCATTTTTTGAAGTG


        90       100       110       120       130       140       150       160
TCATCCTATTAAATATTCAAATTTTATTGGAAGTATCTTTATATCTTCACTTTTCAAGGATAAATCGTCGTATCAAAGCT


            170       180       190       200       210       220       230       240
CATTCATAAGTAGTAAATTAGTAGTAAATTGAGTGGTTTTGACCTTGATAAAGTGTGATAAGTCCAGTTTTTATGCGGAT
        DR-2        DR-2      ******                          ******  DR-1
                                -35                              -10


            250       260       270       280       290       300
AACTAGATTTTTATGCTATTTTT[AACTAAAA]GAAATATCTTTTGAATTTTGTAAAAATAA... 3'
DR-1
```

FIG. 3. Nucleotide sequences of left (A) and right (B) ends of Tn916. The junction sequences are boxed. The direct repeat sequences DR-1, DR-2, and "DR-2" are underlined, and potential promoter hexamers (−10 and −35) within the right end are indicated by asterisks.

affecting the *Eco*RI H fragment). A chimera of pACYC184 containing a pAD1 *Bam*HI-*Sal*I fragment which included the *Eco*RI H fragment was used as a source of DNA containing the original target for insertion. The pAD1 *Eco*RI H fragment was cloned from this source into M13, and sequence comparisons with the clones harboring fragments of Tn916 facilitated identification of the transposon termini. The DNA sequencing protocol was as described elsewhere (16), using M13 universal primers (New England BioLabs, Inc., Beverly, Mass.) and synthetic primers (Systec, Inc., Minneapolis, Minn.).

The sequences of the Tn916 termini are shown in Fig. 3. In the left end, the 229 nucleotides had an overall guanine-plus-cytosine content of 32%

(18% in the first 50 base pairs). At the right end, 263 bases were sequenced and had an overall guanine-plus-cytosine content of 28%. Within the right terminus are two sets of short direct repeats designated DR-1 and DR-2. The two DR-1 sequences are 9 base pairs long and are separated by 11 base pairs; these repeats are close to the end of the transposon. The DR-2 sequences are 11 nucleotides long and are contiguous. Another set of contiguous DR-2 repeats appear in the left end. However, the first segment (indicated as "DR-2") differs by two base pairs. Several potential outwardly reading promoter sites in the right end were also revealed, one of which is shown in Fig. 3 with a −35 hexamer sequence overlapping DR-2.

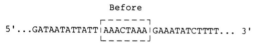

```
                   Before
                 ┌ ─ ─ ─ ─ ─ ┐
5'...GATAATATTATT│AAACTAAA│GAAATATCTTTT... 3'
                 └ ─ ─ ─ ─ ─ ┘


                   After
                 ┌ ─ ─ ─ ─ ─ ┐
5'...GATAATATTATT│AACTAAAA│GAAATATCTTTT... 3'
                 └ ─ ─ ─ ─ ─ ┘
                 ┌ ─ ─ ─ ─ ─ ┐
5'...GATAATATTATT│AAACTAAA│GAAATATCTTTT... 3'
                 └ ─ ─ ─ ─ ─ ┘
```

FIG. 4. Sequence of target DNA prior to insertion
(Before) and after excision of Tn916 in E. coli (After).
Two different sequences were observed for two inde-
pendently obtained excision products.

At the left end of Tn916 the sequence
AAACTAAA (boxed in Fig. 3) was found to
correspond to a sequence in the target DNA
(Fig. 4, Before). This sequence is not duplicated
at the right end of the insertion; however, a
similar sequence AACTAAAA (boxed in Fig. 3)
is present there. The only difference is an absence of a single A at one end of the octamer and
the presence of an additional A at the other end.
This difference is particularly interesting in the
light of additional sequencing data on DNA
corresponding to the target region after spontaneous excision of the transposon in E. coli. The
latter DNA was subcloned in M13 from
pAM160LT, a plasmid from a tetracycline-sensitive segregant of pAM160 containing the regenerated EcoRI H fragment of pAD1. Two
independently derived M13 clones yielded two
different sequences (Fig. 4, After). Interestingly,
one of them contains AAACTAAA, and the
other contains AACTAAAA. The data are suggestive of a nonreplicative insertion mechanism
where Tn916 contains a sequence at least partially
homologous with the target sequence which facilitates insertion of a circular intermediate of the
transposon via a reciprocal recombination. To
generate the configuration shown in Fig. 3, however, the recombination event must occur outside
the octamer sequences. During excision in E. coli,
which may resemble the first step of a transposition event, either of the two octamers is lost with
equal efficiency. In this regard, when a sequencing
analysis was performed directly on pAM160LT,
using a method reported by Chen and Seeburg (2)
and a synthetic primer corresponding to a sequence adjacent to the target site, we observed an
approximately equal mixture of the two octamer
sequences (not shown).

Genetic Analysis of Tn916 Using Tn5 as an Insertional Mutagen

A genetic analysis of Tn916 behavior was
performed using the following approach. First,
the EcoRI fragment containing Tn916 in

pAM120 (Fig. 2) was subcloned to the single
EcoRI site within the chloramphenicol resistance determinant of the vector pVA891 (13).
The latter plasmid is a deletion derivative of the
E. coli-streptococcus shuttle vector pVA838 (14)
which is no longer able to replicate in streptococci. It contains a streptococcal erythromycin
resistance determinant. For the Tn916-contain-
ing chimera, designated pAM620, it was found
that spontaneous excision of the transposon in
the absence of selective pressure was significantly lower than was the case for pAM120.
(Overnight growth in the absence of tetracycline
resulted in only 5%, or less, of the cells segre-
gating Tn916.) When pAM620 was used to transform S. faecalis OG1X (plasmid-free) using a
recently developed protoplast transformation
system (18), tetracycline-resistant transformants
arose at a frequency of about 30 per μg of DNA.
The transformants were erythromycin sensitive,
indicating that zygotically induced insertions
had occurred. Transformants were able to donate Tn916 in filter matings, and when pAD1
was introduced it could be observed to acquire
transpositions from the chromosome. The
pAM620 derivative serves as a useful system to
investigate the genetics of transposition; mutants of Tn916 could be examined for their
ability to (i) excise in E. coli, (ii) transform S.
faecalis and exhibit zygotically induced transposition, (iii) transfer by conjugation from transformants, and (iv) transpose to subsequently
introduced pAD1.

The transposon Tn5 (encodes resistance to
kanamycin) was used to generate insertions in
pAM620 by a modification of the procedure of
de Bruijn and Lupski (7). For this purpose, a
lambda lysogen of the DH1 strain harboring
pAM620 was constructed. (The DH1 genotype is
F⁻ recA1 endA1 gyrA96 thi-1 hsdR17 supE44.)
The prophage served to block phage replication
of a Tn5 delivery vehicle, lambda 467 (lambda
b221 rex::Tn5 cI857 Oam29 Pam80). Lambda::Tn5 was used at a multiplicity of infection of
1, and after the cells were shaken for 2 h at 30°C
they were plated on LB agar (11) containing
kanamycin (20 μg/ml) and tetracycline (4
μg/ml). The plates were incubated at 30°C overnight, and the resistant transductants were
washed off the plates by adding 5 ml of 25%
sucrose in TES (30 mM Tris hydrochloride [pH
8.0], 5 mM EDTA, and 50 mM NaCl) to each
plate and using a sterile glass spreading rod. For
each plate the cells were then pelleted and
resuspended in 1.5 ml of 25% sucrose in TES.
Plasmid DNA was isolated by a cleared-lysate/
ethidium bromide-CsCl method (5) and used to
transform competent E. coli DH1 cells selecting
for resistance to kanamycin and tetracycline. All
transformants examined had Tn5 inserted into

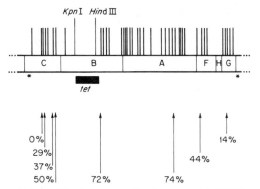

FIG. 5. Map locations of Tn*5* inserts in Tn*916*. The insertions are indicated by the vertical lines. The letters identify *Hinc*II fragments corresponding to those in pAM120. The sizes, in kilobases, of the latter fragments are: A, 5.5; B, 4.8; C, 3.6; F, 1.6; G, 1.1; and H, 0.4. Mapping was by restriction analyses. The arrows point out the position of inserts in derivatives that exhibited various degrees of segregation (indicated as the percentage of cells that became tetracycline sensitive) during growth overnight (about 10 generations) in the absence of tetracycline.

plasmid DNA, with 58% of the Tn*5* insertions in Tn*916*.

Restriction enzyme analyses facilitated the mapping of 56 Tn*5* inserts in Tn*916*; 40 of these insertions are shown in Fig. 5. Insertions were observed over most of Tn*916* except in the region of *tet*; this was expected since the original transductants were selected on media containing tetracycline. In addition, no Tn*5* insertions have yet been found in the Tn*916* 0.4-kb *Hinc*II H fragment. The *Eco*RI fragments containing the Tn*916*::Tn*5* inserts from 31 derivatives were subcloned to pGL101 to examine the relative rates of excision. (As noted above, excision from the pAM620 plasmid generally occurred at much lower frequencies, making measurements less convenient.) All but two exhibited significant excision as judged by the frequency of appearance of tetracycline-sensitive (ampicillin-resistant) derivatives after growth overnight in the absence of tetracycline. No excision was detected in the case of two derivatives, both of which mapped at the left end. Three others, also mapping near the left end, excised at 7% or less. Representative excision frequencies are indicated in Fig. 5 for insertions at various positions in Tn*916*.

The same 31 pAM620::Tn*5* plasmid derivatives which were used for subcloning into pGL101 were also used to transform protoplasts of *S. faecalis* OG1X. All but five derivatives gave rise to tetracycline-resistant transformants (the frequency ranged from 2 to 37 transformants per μg of plasmid DNA), and an absence of erythromycin resistance was consistent with an

establishment of Tn*916* via zygotic induction. Southern blot hybridization (17) using pVA891 as a probe showed that vector sequences were absent. The five Tn*5* insertions that did not transform mapped at the left end of Tn*916* and corresponded to those derivatives that excised in *E. coli* at 0 to 5% (i.e., when subcloned to the pGL101 vector). The region of these Tn*5* inserts is indicated in Fig. 6; the phenotype is designated PT⁻ (for protoplast transformation negative).

Protoplast transformants representing the other 26 derivatives were purified and tested for their ability to donate Tn*916* by conjugation. Since donor potential can vary greatly depending on the location of the insertion in the *S. faecalis* chromosome (see the introduction), three different transformants representing each introduced transposon derivative were tested for the ability to transfer tetracycline resistance from the OG1X (chromosomal mutation to streptomycin resistance) host to *S. faecalis* FA2-2 (plasmid-free; chromosomal mutations for fusidic acid and rifampin resistances) in filter matings. Twenty-four derivatives failed to yield tetracycline-resistant transconjugants (frequency $<10^{-9}$ per donor); they are located in the region designated Tra⁻ in Fig. 6. Most of these, however, were able to transpose to a subsequently introduced pAD1. Transpositions from the chromosome of OG1X to the highly conjugative pAD1 were detected on the basis of matings with FA2-2 whereby recipients acquired pAD1::Tn*916*::Tn*5* at frequencies comparable to wild-type Tn*916* (up to 10^{-6} per cell per generation under these conditions). Six of the Tra⁻ derivatives were defective in transposition and transposed to pAD1 at much lower or undetectable frequencies; these are indicated as Tn⁻ (Fig. 6). (Three transposed at no greater than 10^{-8} per cell per generation; the other three were at less than 10^{-9}.) Overall, only two Tn*5* insertions in

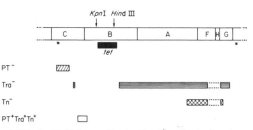

FIG. 6. Tn*916* map illustrating the effect of various Tn*5* inserts. Insertion derivatives that affect different behaviors are indicated as: PT⁻, unable to transform *S. faecalis* OG1X protoplasts; Tra⁻, unable to transfer by conjugation; Tn⁻, defective in transposition to pAD1; PT⁺ Tra⁺ Tn⁺, behavior similar to wild-type Tn*916*. The asterisks indicate the ends of the transposon.

Tn916 appeared completely normal (PT⁺ Tra⁺ Tn⁺ in Fig. 6) and resembled the wild-type situation corresponding to transformants arising from pAM620. It is apparent that a region to the left of *tet* may be nonessential for normal behavior of Tn916.

It is interesting that some derivatives are impaired in their ability to transpose to pAD1 from an established site on the *S. faecalis* chromosome despite their ability to generate transformants (of protoplasts) by a zygotically induced transposition event. Conceivably, the control mechanisms that initiate zygotically induced transposition bypass certain steps that are necessary in the case of the already established element.

Concluding Remarks

The information presented above, relating to the structure and behavior of Tn916, begins to provide some insight into the nature of this interesting transposon. Unlike many other transposons, Tn916 does not have homologous sequences on its two ends, nor does it appear to generate a duplication of the target site. (In this regard, however, there is a resemblance to the staphylococcal transposon Tn554 [15].) Transposition would appear to more closely resemble a lambdalike mechanism in which a circular intermediate bears a sequence (an octamer?) at least partially homologous with a sequence on the target DNA. Insertion would generate an appearance of these two sequences at opposite ends of the element, and a subsequent transposition would begin with an excision leading to an intermediate that could contain either of the two sequences. The specific flanking sequence to accompany the intermediate would depend on the side of the "homologous pairing" on which the reciprocal recombination event occurs. The sequencing data reported above are supportive of such a behavior.

The genetic data clearly show that there are regions on Tn916 encoding products which are necessary for conjugation but which are not necessary for intracellular transposition. Other regions are necessary for transposition, and conjugative transfer is, as would be expected, dependent on such regions. A region near the left end of Tn916 encodes a product(s) necessary for excision in *E. coli*; the fact that mutants in this region also do not insert during transformation of *S. faecalis* protoplasts is consistent with the view that transposition is dependent on the ability to excise. A segment within the Tra region (Fig. 6) of Tn916 appears to encode products which facilitate transposition. It is interesting that such products are not required during a transposition which occurs as a result of zygotic induction.

Conjugative transposons are ubiquitous in *Streptococcus* species and appear to have played a major role in the spread of antibiotic resistance in this genus (4). In some species of streptococci (e.g., *S. pneumoniae, S. pyogenes*, and the viridans streptococci), they may be more prevalent than conjugative R plasmids. Recent studies of Tn1545, a multiple resistance element from *S. pneumoniae*, show remarkable similarities to Tn916 (P. Courvalin et al., this volume). It is likely that many of the conjugative nonplasmid elements of streptococci have a common origin. Consistent with this view is the observation that all such elements reported thus far contain a tetracycline resistance determinant of the *tetM* class (1).

This work was supported by Public Health Service grant AI10318 from the National Institute of Allergy and Infectious Diseases. C.G.-B. was the recipient of a Junior Faculty Research Award from the American Cancer Society.

LITERATURE CITED

1. Burdett, V., J. Inamine, and S. Rajagopalan. 1982. Heterogeneity of tetracycline resistance determinants in *Streptococcus*. J. Bacteriol. 149:995–1004.
2. Chen, E. Y., and P. H. Seeburg. 1985. Supercoil sequencing: a fast and simple method for sequencing plasmid DNA. DNA 4:165–170.
3. Clewell, D. B., F. Y. An, B. A. White, and C. Gawron-Burke. 1985. *Streptococcus faecalis* sex pheromone (cAM373) also produced by *Staphylococcus aureus* and identification of a conjugative transposon (Tn918). J. Bacteriol. 162:1212–1220.
4. Clewell, D. B., and C. Gawron-Burke. 1986. Conjugative transposons and the dissemination of antibiotic resistance in streptococci. Annu. Rev. Microbiol. 40:635–659.
5. Clewell, D. B., and D. R. Helinski. 1970. Properties of deoxyribonucleic acid-protein relaxation complex and strand specificity of the relaxation event. Biochemistry 9:4428–4440.
6. Clewell, D. B., P. K. Tomich, M. C. Gawron-Burke, A. E. Franke, Y. Yagi, and F. Y. An. 1982. Mapping of *Streptococcus faecalis* plasmids pAD1 and pAD2 and studies relating to transposition of Tn917. J. Bacteriol. 152:1220–1230.
7. de Bruijn, F. J., and J. R. Lupski. 1984. The use of transposon Tn5 mutagenesis in the rapid generation of correlated physical and genetic maps of DNA segments cloned into multicopy plasmids—a review. Gene 27:131–149.
8. Fitzgerald, G. F., and D. B. Clewell. 1985. A conjugative transposon (Tn919) in *Streptococcus sanguis*. Infect. Immun. 47:415–420.
9. Franke, A E., and D. B. Clewell. 1981. Evidence for a chromosome-borne resistance transposon (Tn916) in *Streptococcus faecalis* that is capable of "conjugal" transfer in the absence of a conjugative plasmid. J. Bacteriol. 145:494–502.
10. Gawron-Burke, C., and D. B. Clewell. 1982. A transposon in *Streptococcus faecalis* with fertility properties. Nature (London) 300:281–284.
11. Gawron-Burke, C., and D. B. Clewell. 1984. Regeneration of insertionally inactivated streptococcal DNA fragments after excision of Tn916 in *Escherichia coli*: strategy for targeting and cloning genes from gram-positive bacteria. J. Bacteriol. 159:214–221.
12. Ike, Y., and D. B. Clewell. 1984. Genetic analysis of the pAD1 pheromone response in *Streptococcus faecalis*, using transposon Tn917 as an insertional mutagen. J. Bacteriol. 158:777–783.

13. **Macrina, F. L., R. P. Evans, J. A. Tobian, D. L. Hartley, D. B. Clewell, and K. R. Jones.** 1983. Novel shuttle plasmid vehicles for *Escherichia-Streptococcus* transgeneric cloning. Gene **25:**145–150.

14. **Macrina, F. L., J. A. Tobian, K. R. Jones, R. P. Evans, and D. B. Clewell.** 1982. A cloning vector able to replicate in *Escherichia coli* and *Streptococcus sanguis*. Gene **19:** 345–353.

15. **Murphy, E., and S. Lofdahl.** 1984. Transposition of Tn*554* does not generate a target duplication. Nature (London) **307:**292–294.

16. **Shaw, J. H., and D. B. Clewell.** 1985. Complete nucleotide sequence of macrolide-lincosamide-streptogramin B-resistance transposon Tn*917* in *Streptococcus faecalis*. J. Bacteriol. **164:**782–796.

17. **Southern, E. M.** 1975. Detection of specific sequences among DNA fragments separated by gel electrophoresis. J. Mol. Biol. **98:**503–517.

18. **Wirth, R., F. Y. An, and D. B. Clewell.** 1986. A highly efficient protoplast transformation system for *Streptococcus faecalis* and a new *Escherichia coli-S. faecalis* shuttle vector. J. Bacteriol. **165:**831–836.

Functional Anatomy of the Conjugative Shuttle Transposon Tn1545

PATRICE COURVALIN, CÉCILE CARLIER, AND FRÉDÉRIC CAILLAUD

Unité des Agents Antibactériens, Centre National de la Recherche Scientifique U.A. 271, Institut Pasteur, 75724 Paris Cedex 15, France

Since the initial detection of plasmids in streptococci (5), it has become evident that the majority of the genes mediating resistance to antibiotics in this genus are plasmid borne. This type of resistance has spread in recent years among clinical isolates of all the streptococcal species studied, with the remarkable exception of *Streptococcus pneumoniae*. Multiple antibiotic resistance has emerged in *S. pneumoniae* since 1977 (9). The acquisition of resistance in this species is intriguing because of a novel form of conjugative transfer in the apparent absence of extrachromosomal DNA involvement (1, 13). In an attempt to elucidate this sudden acquisition of resistance in the absence of plasmids, we detected Tn1545 (2) in the chromosome of the clinical isolate BM4200 (P. Courvalin, C. Carlier, and E. Collatz, Program Abstr. 11th Lunteren Lecture, abstr. no. 85, Lunteren, The Netherlands, 1979).

Transposition of Em, Km, and Tc Resistances of BM4200

S. pneumoniae BM4200 is resistant to chloramphenicol, high levels of kanamycin and structurally related aminoglycosides, macrolide-lincosamide-streptogramin B-type (MLS) antibiotics, penicillin, sulfonamide, tetracycline and its lipophilic analogs minocycline and chelocardin, and trimethoprim (Courvalin et al., 11th Lunteren Lecture). All the resistance determinants are located in the chromosome (1, 2, 4, 8). In BM4200, a chromosomal sequence containing a block of four resistances (Cm, Em, Km, and Tc) transfers as a unit during a conjugationlike process (1, 8). Resistance to kanamycin is due to the presence of an *aphA3* gene encoding a 3'-aminoglycoside phosphotransferase type III (2–4), resistance to MLS antibiotics is mediated by an *ermAM* gene (11), and resistance to tetracycline is due to a *tetM* gene (P. Martin, P. Trieu-Cuot, and P. Courvalin, this volume). The conjugative resistances of BM4200 were transferred at low frequencies from BM4200 to the chromosome of the plasmid-free *Streptococcus faecalis* strain UV202 deficient in general recombination. Plasmid pIP964 (Tra$^+$, Hly, 57.5 kilobases [kb]) was then introduced into the transcipients by conjugation with high efficiency. The strains constructed were then used as donors in conjugation experiments. The Em, Km, and Tc determinants were transferred en bloc. Three phenotypic classes of resistant transconjugants, with regard to hemolysin production, were obtained. Strains belonging to the major class were hemolytic (Hly$^+$) and harbored a plasmid with a size indistinguishable from that of pIP964. In retransfer experiments, transcipients of this class could conjugate the Em, Km, and Tc resistances en bloc to *S. faecalis* at frequencies similar to those of the primary mating. The two minor classes of transconjugants were nonhemolytic (Hly$^-$) or hyperhemolytic (Hly^{+2}), respectively. Acquisition, by clones of both classes, of the three resistance determinants resulted from the transfer of plasmids with a size increase of approximately 25 kb relative to that of the donor strains. Transconjugants of these two classes could retransfer the Em, Km, and Tc resistances to JH2-2 at high frequencies.

Restriction Endonuclease Analysis of Plasmid DNA

Plasmid DNA from the three classes of transconjugants was digested with *Eco*RI and analyzed by agarose gel electrophoresis (Fig. 1). Plasmid pIP964 had 11 *Eco*RI-generated DNA fragments which were numbered in order of decreasing size. In plasmids (pIP806–pIP808) from Hly^{+2} transconjugants, DNA fragment 5 (5.54 kb) was absent and a new fragment (31 kb) was present. In Hly$^-$ plasmids (pIP804 and pIP805) *Eco*RI fragment 9 (1.3 kb) was replaced by an extra fragment (27 kb). The latter plasmids gave rise to spontaneous Hly$^+$ revertants (pIP804-1 and -2 and pIP805-1 and -2, respectively) which had *Eco*RI-generated fragment patterns identical to that of pIP964. Plasmids from Hly$^+$ transconjugants were also indistinguishable from pIP964. The presence on pIP964 of the Em, Km, and Tc resistance genes of BM4200 resulted, therefore, from the acquisition, at various sites, of a DNA fragment of 25.3 kb. The translocatable element which can induce insertional inactivation and can be lost after apparently clean excision was designated Tn1545 (10).

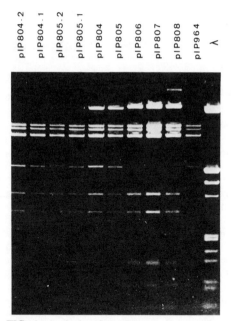

FIG. 1. Analysis of *Eco*RI-generated patterns of plasmid DNA by gel electrophoresis. Plasmid DNA was digested with *Eco*RI, and the resulting fragments were fractionated by electrophoresis in agarose gel. Fragments obtained by digestion of bacteriophage λ DNA with restriction endonucleases *Eco*RI and *Hin*dIII were used as molecular size standards.

Host Range of Tn*1545*

The resistances encoded by Tn*1545* have been shown to transfer, by a mechanism which fits the working definition of conjugation, from *S. pneumoniae* BM4200 to other, nonencapsulated, pneumococcal strains and to *S. faecalis* (1, 8). In addition to *S. faecalis*, we were able to transfer Tn*1545* by conjugation at low frequencies to *Streptococcus cremoris*, *S. diacetylactis*, *S. lactis*, *S. sanguis* Challis, *Staphylococcus aureus* (restriction deficient), and *Listeria monocytogenes*. Plasmid DNA could not be detected after analysis of transcipient lysates. Transposon Tn*1545* could be retransferred to *S. faecalis* only from the *L. monocytogenes* transconjugants. We failed to transfer the element to *Bacillus subtilis*, *B. thuringiensis*, *Corynebacterium diphtheriae*, or the gram-negative organisms *Escherichia coli* and *Campylobacter jejuni*.

Cloning of Tn*1545* in *E. coli* (Fig. 2)

Total DNA from *S. pneumoniae* BM4200 and pSF2124 DNA were mixed, digested with *Eco*RI, ligated, and introduced by transformation into *E. coli*. In the transformants, two plasmids (40.9 and 15.6 kb in size) were present in various amounts. Plasmid DNA from one transformant was purified and analyzed by

agarose gel electrophoresis after digestion with *Eco*RI. Plasmid pAT10 consisted of pSF2124 containing a 29.8-kb *Eco*RI fragment designated original insert, whereas pAT10-1 consisted of pSF2124 containing a 4.5-kb *Eco*RI fragment designated residual insert. The original insert of plasmid pAT10 was purified, cloned into pBR325, and transformed into *E. coli* HB101 (*recA*). Two plasmids, with sizes of 36.9 and 11.6 kb, were present in every transformant. Plasmid DNA from a transformant, BM2910, was purified and compared with that of pAT10 and pAT10-1. As expected, plasmid pAT11 consisted of pBR325 containing a 29.8-kb *Eco*RI fragment indistinguishable from the insert in pAT10 and also designated original insert. Plasmid pAT11-1 consisted of a 4.5-kb *Eco*RI fragment with an electrophoretic mobility identical to that of the insert of pAT11-1 and therefore also designated residual insert, cloned into pBR325. This observation suggests that the residual insert results from a deletion in the original insert.

Apparently Clean Excision of Tn*1545* in *E. coli*

E. coli BM2910 grown in the presence of ampicillin spontaneously gave rise, at very high frequencies, to derivatives which were susceptible to kanamycin and erythromycin but retained resistance to ampicillin and tetracycline. Analysis of the plasmid content of these strains

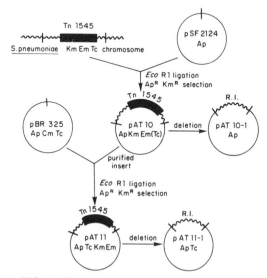

FIG. 2. Cloning of Tn*1545* in *E. coli*. Heavy bar, Tn*1545;* wavy line, BM4200 chromosomal DNA; I, *Eco*RI recognition site; R.I., residual insert; Ap, ampicillin resistance; Cm, chloramphenicol resistance; Em, erythromycin resistance; Km, kanamycin resistance; Tc, tetracycline resistance; (Tc), low-level tetracycline resistance. Not drawn to scale.

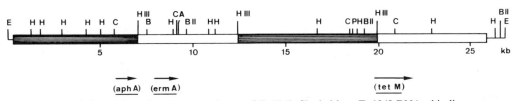

FIG. 3. Restriction endonuclease-generated map of Tn*1545*. Shaded bar, Tn*1545* DNA; thin line, pneumococcal DNA; A, *Alu*I; B, *Bam*HI; BII, *Bgl*II; C, *Cla*I; E, *Eco*RI; H, *Hinc*II; HIII, *Hind*III; P, *Pst*I; *aphA*, kanamycin resistance gene; *ermAM*, erythromycin resistance gene; *tetM*, tetracycline resistance gene. Arrows indicate direction and extent of transcription. Sizes are specified in kilobase pairs.

indicated that they harbored only pAT11-1. This observation suggested that the resistance genes of Tn*1545* were carried by the original insert present in pAT11. The relationship between pAT11 and pAT11-1 was studied by DNA-DNA hybridization, and the homology observed between the original insert and the residual insert confirmed the relationship between these two fragments. It therefore appears that pAT10-1 and pAT11-1 result from loss, after apparently clean excision at very high frequency, of Tn*1545* from pAT10 and pAT11, respectively.

Successive Transpositions of Tn*1545* in *E. coli*

Transposition from pAT11 to the chromosome. *E. coli* BM2910, when subcultured in the presence of kanamycin, spontaneously gave rise to derivatives which had lost ampicillin and tetracycline resistance. The remaining resistance determinants were tentatively assigned to the chromosome of these strains since plasmid DNA was not detected. The presence of Tn*1545* at various sites in the HB101 chromosome was confirmed by Southern hybridization using a pAT11 probe.

Transposition from the chromosome to pIP135-1. Plasmid pIP135-1 (Inc 7-M, Tc, Hg) was introduced in strain HB101::Tn*1545* and exconjugated into *E. coli* DB10. The transposition frequency of Tn*1545* was 10^{-8} to 10^{-9} and acquisition of Tn*1545* resistance determinants by the transconjugants resulted from the transfer of a plasmid with a size increase of approximately 25 kb relative to that of the donor strain. In each experiment, certain transconjugants became spontaneously susceptible to kanamycin and erythromycin after loss of Tn*1545* by apparently clean excision.

Transposition of Tn*1545* into *B. subtilis* Chromosome

Plasmid pAT11 was introduced by transformation into *B. subtilis*, and plasmid DNA was not detected after analysis of the transformants.

Physical Analysis of Tn*1545*

Analysis of the element by restriction endonucleases, molecular cloning, electron micros-

copy of heteroduplexes, DNA hybridization, and sequencing allowed us to (i) establish a physical map of Tn*1545*, (ii) localize the R genes, determine their direction of transcription (Fig. 3), and compare them with other characterized R determinants, (iii) show that Tn*1545* is not flanked by terminal repeated sequences in either direct or opposite orientation, (iv) show that Tn*1545* does not generate duplication of the target DNA upon insertion, (v) detect a target consensus sequence, and (vi) confirm that Tn*1545* is capable of clean excision.

Conclusion

S. pneumoniae BM4200, isolated from a clinical specimen, is multiply resistant to antibiotics, and all the resistance determinants are located in the chromosome (1, 2, 4, 8). The data reported here indicate that the *aphA-3*, *ermAM*, and *tetM* genes are located on a conjugative transposon designated Tn*1545*. The element is 25.3 kb in size and can transpose, in a host devoid of homologous recombination, to at least two different sites on the self-transferable hemolysin plasmid pIP964 (Fig. 1). Transposon Tn*1545* can induce mutations upon insertion and can be lost by apparently clean excision (Fig. 1). Nonhemolytic transconjugants probably result from insertion-inactivation, whereas hyperhemolytic clones could be due to insertion of the element in a regulatory region or to expression of hemolysin under the control of a Tn*1545*-associated promoter. The transposition frequency of the element onto pIP964 is 10^{-5} to 10^{-6}, a value similar to those already reported for other conjugative transposable elements (6).

Transposon Tn*1545* exhibits a broad host range. It can conjugate to and transposes in various species of streptococci, including group N lactic streptococci of industrial importance, and in the phylogenetically remote pathogenic *S. aureus* and *L. monocytogenes*. This element, which constitutes, to our knowledge, the sole genetic system so far available in the latter species, was used successfully as a mutagen to study the role of hemolysin in the virulence of *Listeria* species (7).

The element was cloned in its entirety in *E. coli* and was found to be unstable in the new host. On the basis of the size of the restriction plasmid DNA fragments, Tn*1545* was lost, as in gram-positive bacteria, by apparently clean excision. However, the instability of the element paralleled the copy number of the bearing replicon. Tn*1545* was extremely unstable when present on high-copy-number plasmids (Fig. 2). The element was relatively stable after transposition on the low-copy number plasmid pIP135-1 and very stable when integrated into the host chromosome. The reason for this plasmid-copy-number effect on the instability of Tn*1545* in *E. coli* is not known.

Transposon Tn*1545* is not self-transferable to and among gram-negative organisms but readily transposes in gram-negative bacteria. We were able to transpose the element, in a host devoid of homologous recombination, to numerous sites of the chromosome and to two different loci of the conjugative resistance plasmid pIP135-1. However, the transposition frequency (10^{-8} to 10^{-9}) of the element when present in the *E. coli* chromosome is 10^{-3} lower than that in a gram-positive background (P. Courvalin and C. Carlier, submitted for publication).

Resistance to numerous antibiotics has appeared recently in clinical isolates of *S. pneumoniae* (9). No plasmids have been detected in these strains, although pneumococci can stably replicate streptococcal plasmids introduced by transformation (12) or by conjugation (14). The existence of conjugative transposons such as Tn*1545* clearly accounts for the recent emergence, rapid dissemination, and stabilization of multiple antibiotic resistance in *S. pneumoniae* in the absence of plasmids.

Direct transfer of *aphA3* and *ermAM* genes from gram-positive to gram-negative bacteria under natural conditions has recently been reported (16). Since these microorganisms are extremely distantly related, transfer has probably occurred by transformation followed by transposition (15). Transposon Tn*1545*, or related elements, are the most likely candidates to perform the second step in this process.

We thank F. Goldstein for the gift of strain BM4200 and Y. A. Chabbert for his interest in this work and for material support.

This work was supported by grant A.T.P. 955548 from the Centre National de la Recherche Scientifique and by a grant from the Caisse Nationale Assurance Maladie Travailleurs Salariés.

LITERATURE CITED

1. **Buu-Hoï, A., and T. Horodniceanu.** 1980. Conjugative transfer of multiple antibiotic resistance markers in *Streptococcus pneumoniae*. J. Bacteriol. **143:**313–320.
2. **Carlier, C., and P. Courvalin.** 1982. Resistance of streptococci to aminoglycoside-aminocyclitol antibiotics, p. 162–166. *In* D. Schlessinger (ed.), Microbiology—1982. American Society for Microbiology, Washington, D.C.
3. **Collatz, E., C. Carlier, and P. Courvalin.** 1983. The chromosomal 3',5"-aminoglycoside phosphotransferase in *Streptococcus pneumoniae* is closely related to its plasmid-coded homologs in *Streptococcus faecalis* and *Staphylococcus aureus*. J. Bacteriol. **156:**1373–1377.
4. **Collatz, E., C. Carlier, and P. Courvalin.** 1984. Characterization of high-level aminoglycoside resistance in a strain of *Streptococcus pneumoniae*. J. Gen. Microbiol. **130:** 1665–1671.
5. **Courvalin, P. M., C. Carlier, and Y. A. Chabbert.** 1972. Plasmid-linked tetracycline and erythromycin resistance in group D "*Streptococcus.*" Ann. Inst. Pasteur. (Paris) **123:**755–759.
6. **Franke, A. E., and D. B. Clewell.** 1981. Evidence for a chromosome-borne resistance transposon (Tn*916*) in *Streptococcus faecalis* that is capable of "conjugal" transfer in the absence of a conjugative plasmid. J. Bacteriol. **145:**494–502.
7. **Gaillard, J. L., P. Berche, and P. Sansonetti.** 1986. Transposon mutagenesis as a tool to study the role of hemolysin in the virulence of *Listeria monocytogenes*. Infect. Immun. **52:**50–55.
8. **Guild, W. R., M. D. Smith, and N. B. Shoemaker.** 1982. Conjugative transfer of chromosomal R determinants in *Streptococcus pneumoniae*, p. 88–92. *In* D. Schlessinger (ed.), Microbiology—1982. American Society for Microbiology, Washington, D.C.
9. **Jacobs, M. R., H. J. Koornhof, R. M. Robins-Browne, C. M. Stevenson, I. Freiman, G. B. Miller, M. A. Witcomb, M. Isaacson, J. I. Ward, and R. Austrian.** 1978. Emergence of multiply resistant pneumonocci. N. Engl. J. Med. **299:**735–740.
10. **Lederberg, E. M.** 1981. Plasmid reference center registry of transposon (Tn) allocations through July 1981. Gene **16:** 59–61.
11. **Ounissi, H., and P. Courvalin.** 1982. Heterogeneity of macrolide-lincosamide-streptogramin B-type antibiotic resistance determinants, p. 167–169. *In* D. Schlessinger (ed.), Microbiology—1982. American Society for Microbiology, Washington, D.C.
12. **Shoemaker, N. B., M. D. Smith, and W. R. Guild.** 1979. Organization and transfer of heterologous chloramphenicol and tetracycline resistance genes in *Pneumococcus*. J. Bacteriol. **139:**432–441.
13. **Shoemaker, N. B., M. D. Smith, and W. R. Guild.** 1980. DNase-resistant transfer of chromosomal *cat* and *tet* insertions by filter mating in *Pneumococcus*. Plasmid **3:** 80–87.
14. **Smith, M. D., N. B. Shoemaker, V. Burdett, and W. R. Guild.** 1980. Transfer of plasmids by conjugation in *Streptococcus pneumoniae*. Plasmid **3:**70–79.
15. **Trieu-Cuot, P., and P. Courvalin.** 1986. Evolution and transfer of aminoglycoside resistance genes under natural conditions. J. Antimicrob. Chemother. **18**(Suppl. C):93–102.
16. **Trieu-Cuot, P., G. Gerbaud, T. Lambert, and P. Courvalin.** 1985. *In vivo* transfer of genetic information between Gram-positive and Gram-negative bacteria. EMBO J. **4:**3583–3587.

Transfer of Genetic Information between Gram-Positive and Gram-Negative Bacteria under Natural Conditions

P. TRIEU-CUOT, M. ARTHUR, AND P. COURVALIN

Unité des Agents Antibactériens, Centre National de la Recherche Scientifique U.A. 271, Institut Pasteur, 75724 Paris Cedex 15, France

Since their clinical introduction in the mid 1940s, antibiotics have been found to be therapeutically effective agents, but their use has often been limited by the emergence of antibiotic-resistant pathogens. One of the major causes of the evolution of bacterial resistance to antibiotics is the spreading of already known genes into "new" bacterial genera, i.e., bacteria which were previously uniformly susceptible. This evolution is due to the fact that most resistance genes are located on plasmids or transposons and can be readily disseminated. Changes in antibiotic susceptibility among different bacterial species have resulted in increased patient morbidity, necessitating rapid changes in antibiotic usage (10a). It is therefore important to determine the extent of gene transfer among bacteria under natural conditions to anticipate the dissemination of antibiotic resistance. Here we present recent results from our laboratory which indicate that gram-positive cocci (staphylococci and streptococci) can serve as a reservoir of resistance genes for gram-negative bacteria.

Kanamycin Resistance in *Campylobacter* Species

Aminoglycosides constitute a large and clinically important family of antibiotics useful in the treatment of severe infections due to gram-negative and gram-positive organisms. Bacterial resistance to aminoglycosides is frequently mediated by aminoglycoside-modifying enzymes, which are classified according to the reaction catalyzed (N acetylation, O nucleotidylation, and O phosphorylation) and the site on the antibiotic molecule which they modify (8). The 3'-aminoglycoside phosphotransferases, APH(3'), catalyze the phosphorylation of the hydroxyl group in position 3' of aminohexose I of kanamycin and structurally related antibiotics. In pathogenic bacteria, three types of APH(3') can be distinguished, in particular on the basis of their substrate range in vitro (Table 1). The respective genes, although thought to diverge from a common ancestor (14a), do not cross-hybridize (7). Until now, types I and II were specific for gram-negative bacteria, whereas type III, which can be distinguished in that it modifies butirosin, lividomycin, and amikacin in vitro (6), was confined to gram-positive cocci (5).

Campylobacter coli and *Campylobacter jejuni* are gram-negative bacteria frequently responsible for bacterial acute gastroenteritis including traveler's diarrhea, the so-called "tourista," in humans. *C. coli* BM2509, resistant to ampicillin, chloramphenicol, erythromycin, kanamycin, spectinomycin, streptomycin, and tetracycline, was isolated in 1983 at the Hôpital Saint-Joseph in Paris from the feces of a 78-year-old diabetic patient with hospital-acquired diarrhea (10). Kanamycin and tetracycline resistance are carried by pIP1433, a 47.2-kilobase plasmid self-transferable to other *Campylobacter* cells, but not to *Escherichia coli*. Resistance to kanamycin is novel in the genus *Campylobacter*. Phosphocellulose paper-binding assays indicated that resistance to kanamycin and structurally related antibiotics in strain BM2509 was due to the synthesis of an APH(3') of type III, an enzyme not detected previously in a gram-negative bacterium. DNA annealing studies indicated a close structural relationship between the APH(3')-III gene of *C. coli* BM2509 and that representative of this type of resistance determinant in gram-positive cocci. This similarity was confirmed by determination of the nucleotide sequence of a 1,427-base-pair pIP1433 DNA fragment conferring resistance to kanamycin that was cloned in *E. coli* (15). The resistance gene was located in an open reading frame of 792 bp and appeared to be identical to the corresponding genes of gram-positive cocci (Fig. 1). Furthermore, the DNA sequences upstream from the structural genes originating in *Campylobacter* spp., *Streptococcus* spp., and pneumococci were also closely related (Fig. 1). The finding of identical genes in evolutionarily distant bacteria can only be interpreted as the result of a horizontal transfer of genetic material, and we conclude that emergence of resistance to kanamycin in *Campylobacter* spp. is due to acquisition in vivo of a gene or a plasmid from gram-positive bacteria. Plasmid pIP1433 is extremely stable in its original host, *C. coli* BM2509, and transfers at a high frequency to other *Campylobacter* spp., where it is also extremely stable. This behavior suggests that the replication apparatus and the

TABLE 1. Substrate specificities of
APH(3′) types I to III

Antibiotic	APH(3′)		
	I	II	III
Neomycin	+[a]	+	+
Butirosin	−	+	+
Lividomycin	+	−	+
Amikacin	−	−	(+)

[a] +, Substrate; −, nonsubstrate; (+), substrate in
vitro, but cells remain susceptible in vivo.

transfer machinery of this plasmid are particularly well tuned to the genus *Campylobacter*.
Moreover, restriction endonuclease analysis and
DNA annealing studies indicated that plasmid
pIP1433 shares extensive sequence homology
with plasmid pMAK175, a representative of the
tetracycline resistance plasmids of *Campylobacter* spp. (13). Taken together, these observations
suggest that kanamycin resistance in *Campylobacter* spp. results from the acquisition of a gene
rather than that of a replicon en bloc. This
represents the first example of genetic exchange
between gram-positive and gram-negative bacteria under natural conditions.

MLS Antibiotic Resistance in *E. coli*

Among gram-positive bacteria, resistance
to macrolide-lincosamide-streptogramin B-type
(MLS) antibiotics is widespread and is generally
due to the presence of a methylase which modifies a specific adenine residue in 23S rRNA (14,
15). Gram-negative bacteria are resistant to low
levels of MLS antibiotics, probably by imperme-

ability (1, 2). This intrinsic resistance precludes
the use of these drugs in systemic infections.
However, because of high local concentrations
achieved, erythromycin has recently found therapeutic applications in the modulation of the
gram-negative flora of the intestinal tract (1, 2).
Gram-negative organisms highly resistant to
erythromycin can be isolated, usually in association with previous intake of the drug (1, 3).

We recently described *E. coli* BM2570, the
first gram-negative clinical isolate resistant to
high levels of MLS antibiotics (4a). This phenotype is due to the presence on a 150-kilobase
plasmid, pIP1527, of two genes, *ereB* and *erxA*,
which contribute cooperatively to erythromycin
resistance by two different mechanisms. The
ereB gene encodes an erythromycin esterase
type II (4), and *erxA* confers resistance to MLS
without inactivation of the drugs. The latter gene
was cloned into pUC8 on a 1.8-kilobase *Hind*III-
*Pst*I DNA fragment of pIP1527, and the resulting
plasmid was designated pAT69 (4a). The sequence of 375 nucleotides at the *Hind*III end of
the pAT69 insert was determined (Fig. 2). A
computer search indicated the existence of
nearly identical sequences in transposons Tn*917*
(12) and Tn*1545* (6; F. Caillaud and P.
Courvalin, unpublished data) and in plasmid
pAM77 (9) (Fig. 2). These three genetic structures originate in streptococci and carry
ermAM, which encodes an rRNA methylase.
The sequence of the last 216 nucleotides of *erxA*
is identical to that encoding the last 72 amino
acids of the methylase in Tn*917*, Tn*1545*, and
pAM77. The *erxA* gene therefore encodes an
rRNA methylase identical, or closely related, to

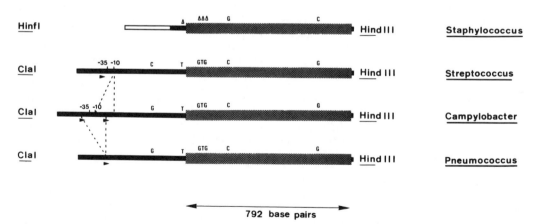

FIG. 1. Comparison of nucleotide sequences of APH(3′)-III genes. Large shaded segments correspond to the
structural gene for the APH(3′)-III. Capital letters (A, C, G, T) and Δ above the sequences specify nucleotide
substitutions or deletions, respectively. Heavy black lines depict homology between the DNA sequences
upstream from the structural genes. The −35 recognition site and −10 Pribnow boxes are indicated when known.
Deletion-insertion events leading to rearrangements in the promoter regions are represented by dashed lines.
Small arrows below the DNA fragment of *Campylobacter* spp. represent two 12-base-pair long direct repeats.

```
pIP1527  |AAACTTACCC GCCATACCAC AGATGTTCCA GATAAATATT GGAAGCTATA TACGTACTTT|  60
Tn917    |++++++++++ ++++++++++ ++++++++++ ++++++++++ ++++++++++ ++++++++++|
pAM77    |++++++++++ ++++++++++ ++++++++++ ++++++++++ ++++++++++ ++++++++++|

pIP1527  |GTTTCAAAAT GGGTCAATCG AGAATATCGT CAACTGTTTA CTAAAAATCA GTTTCATCAA| 120
Tn917    |++++++++++ ++++++++++ ++++++++++ ++++++++++ ++++++++++ ++++++++++|
pAM77    |++++++++++ ++++++++++ ++++++++++ ++++++++++ ++++++++++ ++++++++++|

pIP1527  |GCAATGAAAC ACGCCAAAGT AAACAATTTA AGTACCGTTA CTTATGAGCA AGTATTGTCT| 180
Tn917    |++++++++++ ++++++++++ ++++++++++ ++++++++++ ++++++++++ ++++++++++|
pAM77    |++++++++++ ++++++++++ ++++++++++ ++++++++++ ++++++++++ ++++++++++|

pIP1527  |ATTTTTAATA GTTATCTATT ATTTAACGGG AGGAAA|TAAT TCTATGAGTC GCTTTTGTAA| 240
Tn917    |++++++++++ ++++++++++ ++++++++++ ++++++|++++ ++++++++++ ++++++++++|
pAM77    |++++++++++ ++++++++++ ++++++++++ ++++++|++++ ++++++++++ ++++++++++|

pIP1527   ATTTGGAAAG TTACACGTTA CTAAAGGGAA TGTAGATAAA TTATTAGGTA TACTACTGAC  300
Tn917     ++++++++++ ++++++++++ ++++++++++ ++++++++++ ++++++++++ ++++++++++
pAM77     ++++++++++ ++++++++++ ++++++++++ ++++++++++ ++++++++++ ++++++++++

pIP1527   AGCTTCCAAG GAGCTAAAGA GGTCCCTAGA CTAGCAAGAA GTACACAAGA AGCCTTAAAG  360
Tn917     ++++++++++ ++++++++++ +++++++++C GCCTACGGGG AATTTGTATC GATAAGGAAT
pAM77     ++++++++++ A+++++++++ ++++++++++ ++++++++++ ++++++++++ +++++++GGT

pIP1527   ATTATAGAAA AGCTT                                                  375
Tn917     AGATTTAAAA ATTTCGCTGT
pAM77     ACAAATTCCC ACTAAGCGCT
```

FIG. 2. Nucleotide sequences of the DNA fragments containing part of the *erxA* gene from plasmid pIP1527 (*E. coli*) and the *ermAM* gene from Tn917 and pAM77 (*Streptococcus*). Homology with the DNA sequences of *E. coli* is indicated by plus signs. The structural region for the rRNA methylase is boxed by solid lines. The putative resolution site of Tn917 is underlined.

that coded for by *ermAM*. The identity of DNA sequences originating in streptococci and *E. coli* provides evidence for an in vivo horizontal transfer of genetic information between these phylogenetically remote pathogenic bacteria. The guanine-plus-cytosine content of the *ermAM* gene is 33 mol% (9). This value, similar to that of the *Streptococcus* genome (11), is significantly different from the 50 mol% content of the *E. coli* chromosome (11). Consequently, acquisition of a gene from a gram-positive bacterium by *E. coli* seems more likely than the opposite transfer. These results indicate a recent extension of the gene pool of gram-positive cocci to enterobacteria. The distribution of nucleotide sequences homologous to the streptococcal *ermAM* gene was studied by colony hybridization in 21 clinical isolates of enterobacteria highly resistant to erythromycin. A positive hybridization was detected with three strains of *E. coli* belonging to different biotypes and one strain of *Kiebsiella pneumoniae*, which indicates that the streptococcal *ermAM* gene has already disseminated into enterobacteria.

Conclusions

Antibiotic resistance determinants from gram-positive organisms are generally expressed in gram-negative bacteria, whereas the reverse is uncommon. Therefore, the only apparent barrier to the acquisition of genes from a gram-positive by a gram-negative bacterium lies in the transfer process and in the stable replication of exogenous DNA. Transfer of genetic information by conjugation or transduction between gram-positive and gram-negative organisms has never been obtained under laboratory conditions. Although this observation does not rule out the possibility that such a transfer could occur under natural conditions, which is an almost impossible claim to substantiate, exchange of genetic material between distantly related species is most probably mediated by a transformationlike process. It remains to be demonstrated that *Campylobacter* spp. and *E. coli* can acquire exogenous DNA by transformation in their natural environment. Since, in most cases, plasmids from gram-positive bacteria cannot be stably maintained in gram-negative organisms, the second important aspect is the integration, upon entry, of the foreign DNA into the genome of the new host. The absence of homology between the genomes of phylogenetically remote bacteria makes illegitimate recombination a likely step for in vivo stabilization of exogenous DNA in procaryotes. The resistance gene flux from gram-positive to gram-negative bacteria is therefore probably facilitated by the presence of the genes on transposable elements. The fact that the conjugative streptococcal shuttle transposon Tn1545 is also active in *E. coli* (5) is consistent with this notion. Thus, a valuable feature for the kanamycin and MLS resistance genes detected in *Campylobacter* spp. and in *E. coli*, respectively, would be their location on transposable elements. This is currently being tested.

We thank F. Caillaud for communicating results prior to publication.

LITERATURE CITED

1. **Andremont, A., H. Sancho-Garnier, and C. Tancrede.** 1986. Epidemiology of intestinal colonization by members of the family *Enterobacteriaceae* highly resistant to erythromycin in a hematology-oncology unit. Antimicrob. Agents Chemother. **29:**1104–1107.
2. **Andremont, A., and C. Tancrede.** 1981. Reduction of the aerobic gram-negative bacterial flora of the gastro-intestinal tract and prevention of traveler's diarrhea using oral erythromycin. Ann. Microbiol. (Paris) **132B:**419–427.
3. **Arthur, M., A. Andremont, and P. Courvalin.** 1986. Heterogeneity of genes conferring high-level resistance to erythromycin by inactivation in enterobacteria. Ann. Microbiol. (Paris) **137A:**125–134.
4. **Arthur, M., D. Autissier, and P. Courvalin.** 1986. Analysis of the nucleotide sequence of the gene *ereB* encoding the erythromycin esterase type II. Nucleic Acids Res. **14:**4987–4999.
4a.**Arthur, M., and P. Courvalin.** 1986. Contribution of two different mechanisms to erythromycin resistance in *Escherichia coli*. Antimicrob. Agents Chemother. **30:**694-700.
5. **Carlier, C., and P. Courvalin.** 1982. Resistance of streptococci to aminoglycoside-aminocyclitol antibiotics, p. 162–166. *In* D. Schlessinger (ed.), Microbiology—1982. American Society for Microbiology, Washington, D.C.
6. **Courvalin, P., and J. Davies.** 1977. Plasmid-mediated aminoglycoside phosphotransferase of broad substrate range that phosphorylates amikacin. Antimicrob. Agents Chemother. **11:**619–624.
7. **Courvalin, P., M. Fiandt, and J. Davies.** 1978. DNA relationships between genes coding for aminoglycoside-modifying enzymes from antibiotic-producing bacteria and R-plasmids, p. 262–266. *In* D. Schlessinger (ed.), Microbiology—1978. American Society for Microbiology, Washington, D.C.
8. **Davies, J., and D. Smith.** 1978. Plasmid-determined resistance to antimicrobial agents. Annu. Rev. Microbiol. **32:**469–518.
9. **Horinouchi, S., W. H. Byeon, and B. Weisblum.** 1983. A complex attenuator regulates inducible resistance to macrolides, lincosamides, and streptogramin type B antibiotics in *Streptococcus sanguis*. J. Bacteriol. **154:**1252–1262.
10. **Lambert, T., G. Gerbaud, P. Trieu-Cuot, and P. Courvalin.** 1985. Structural relationship between the genes encoding 3'-aminoglycoside phosphotransferases in *Campylobacter* and in gram-positive cocci. Ann. Microbiol. (Paris) **136B:**135–150.
10a.**Meyer, K. H.** 1986. The epidemiology of antibiotic resistance in hospitals. J. Antimicrob. Chemother. **18**(Suppl. C):223–233.
11. **Normore, W. M.** 1976. Guanine-plus-cytosine (GC) composition of the DNA of bacteria, fungi, algae and protozoa, p. 65–240. *In* G. D. Fasman (ed.), Handbook of biochemistry and molecular biology, vol. 2, Nucleic acids, 3rd ed. CRC Press, Cleveland.
12. **Shaw, J. H., and D. B. Clewell.** 1985. Complete nucleotide sequence of macrolide-lincosamide-streptogramin B-resistance transposon Tn*917* in *Streptococcus faecalis*. J. Bacteriol. **164:**782–796.
13. **Taylor, D. E., R. S. Garner, and B. J. Allan.** 1983. Characterization of tetracycline resistance plasmids from *Campylobacter jejuni* and *Campylobacter coli*. Antimicrob. Agents Chemother. **24:**930–935.
14. **Thakker-Varia, S., A. C. Ranzini, and D. T. Dubin.** 1985. Ribosomal RNA methylation in *Staphylococcus aureus* and *Escherichia coli*: effect of the "MLS" (erythromycin resistance) methylase. Plasmid **14:**152–161.
14a.**Trieu-Cuot, P., and P. Courvalin.** 1986. Evolution and transfer of aminoglycoside resistance genes under natural conditions. J. Antimicrob. Chemother. **18**(Suppl. C):93–102.
15. **Trieu-Cuot, P., G. Gerbaud, T. Lambert, and P. Courvalin.** 1985. *In vivo* transfer of genetic information between gram-positive and gram-negative bacteria. EMBO J. **4:**3583–3587.
16. **Weisblum, B.** 1985. Inducible resistance to macrolides, lincosamides, and streptogramin type B antibiotics: the resistance phenotype, its biological diversity, and structural elements that regulate expression—a review. J. Antimicrob. Chemother. **16**(Suppl. A):63–90.

Molecular Analysis of the Gene Specifying the Bifunctional 6'-Aminoglycoside Acetyltransferase–2''-Aminoglycoside Phosphotransferase Enzyme in *Streptococcus faecalis*

JOSEPH J. FERRETTI,[1] KEETA S. GILMORE,[1] AND PATRICE COURVALIN[2]

Department of Microbiology and Immunology, University of Oklahoma Health Sciences Center, Oklahoma City, Oklahoma 73190,[1] and Unité des Agents Antibactériens, Institut Pasteur, 75724 Paris, France[2]

Multiresistant strains of gram-positive organisms have been reported with increasing frequency and are primarily the result of plasmid- or transposon-encoded enzymes. In view of the facile transfer of many of these plasmids and transposons to a wide variety of hosts, there is ample opportunity for genetic rearrangements to occur. Such may have been the case with plasmids found in *Streptococcus faecalis* and *Staphylococcus aureus* strains which encode both 6'-aminoglycoside acetyltransferase [AAC(6')] and 2''-aminoglycoside phosphotransferase [APH(2'')] activities and which are thought to be specified by a bifunctional enzyme (6, 12–14). The possibility that a gene fusion event may have resulted in the AAC(6')-APH(2'') resistance determinant is supported by recent information concerning the molecular weight of a purified bifunctional AAC(6'')-APH(2'') enzyme from *S. aureus*, reported to be 56,000 (22), which is approximately twice the size of individual AACs (1, 5, 10, 11, 19) and APHs (2, 9, 17, 20, 21) that have been sequenced. We have previously cloned an AAC(6')-APH(2'') resistance determinant from *S. faecalis* plasmid pIP800 (3), and its availability made possible a further investigation into its sequence and the product it encodes, and provided insight into its possible evolutionary rearrangements.

Nucleotide Sequence of the AAC(6')-APH(2'') Gene

The complete nucleotide sequence of the AAC(6')-APH(2'') gene from *S. faecalis* was determined by the dideoxy method of Sanger et al. (18) and is presented in Appendix B of this volume and elsewhere (7). The AAC(6')-APH(2'') gene was located in a 1.5-kilobase *Alu*I fragment that was cloned into plasmid pUC8 to form the plasmid pSF815A. The bifunctional enzyme expressed both AAC and APH activities in *Escherichia coli*, and minicell experiments demonstrated the presence of a protein with an apparent molecular weight of 56,000. A single open reading frame containing 1,437 base pairs codes for the AAC(6')-APH(2'') resistance protein, and the deduced protein contains 479 amino acids with a molecular weight of 56,850.

Amino Acid Sequence and Homology

The deduced amino acid sequence of the AAC(6')-APH(2'') resistance gene product was shown to have a partial homology with the sequence of two other known resistance proteins, as shown in Fig. 1. In the N-terminal region, there are 30 amino acids identical to and 52 amino acids similar to the chloramphenicol acetyltransferase specified by the *cat-86* gene of *Bacillus pumilus* (10). In the C-terminal region, there are 28 amino acids identical to and 42 amino acids similar to the APH(3') of *Streptomyces fradiae* (20). These results tentatively identified two domains of activity and suggested that the AAC(6')-APH(2'') resistance determinant arose as a fusion between two genes specifying the individual activities.

Subcloning of the AAC(6') and APH(2'') Specifying Gene Regions

Subcloning of the regions specifying the putative AAC(6') and APH(2'') activities was made possible because of a convenient *Sca*I site in the middle of the 1.5-kilobase *Alu*I fragment. Each gene segment, as shown in Fig. 2, was subcloned into plasmid pUC8, and strains containing the recombinant plasmids were assayed for the presence of AAC(6') and APH(2'') activities. The results, also presented in Fig. 2, confirm that gene segments containing the individual activities can be obtained separately from one another.

Discussion

We have shown that the bifunctional AAC-APH protein contains two independent domains, each responsible for an individual resistance enzyme activity. The demonstration of two domains of activity within a single bifunctional enzyme also suggests that the formation of the AAC-APH resistance gene came about as a result of a gene fusion event since the individual resistance proteins of known sequence are all about one-half the size of the AAC-APH resistance protein.

```
AAC/APH    1  MNIVENEICIRTLIDDDFPLMLKWLTDERVLEFYGGRDKKYTLESLKKHYTEPWEDEVFR

cat-86    31                                        *  MDQ  *E  *YW            **
AAC/APH   61  VIIEYNNVPIGYGQIYKMYDELYTDYHYPKTDEIVYGMDQFIGEPNYWSKGIGTRYIKLI

cat-86    85  ***L*KE ** * I *P **N **   *** * **F* **L P*****E** * * L
AAC/APH  121  FEFLKKERNANAVILDPHKNN-PRAIRAY--QKSGFRIIEDL-PEHELHEGKKEDCYLME

cat-86   145    **D * NV**     YL* * ***FKV** *II *    V * V **Y
AAC/APH  177  YRYDDNATNVKA-MKYLIEHY-FDNFKVDSIEIIGSGYDSVAYLVNNEYIFKTKFSTNKK


AAC/APH  235  KGYAKEKAIYNFLNTNLETNVKIPNIEYSYISDELSILGYKEIKGTFLTPEIYSTMSEEE

APH      109                                                              * *
AAC/APH  295  QNLLKRDIASFLRQMHGLDYTDISECTIDNKQNVLEEYILLRETIYNDLTDIEKDYIESF

APH      169  ** L* T    ***  *CH*D** N**LLD G* R*TG*ID G  G* D * D*      *
AAC/APH  355  MERLNATTVFEGKKCLCHNDFSCNHLLLD-GNNRLTGIIDFGDSGIIDEYCDFIYLLED-

APH      231    *E*  *G****E *L *YG*  **K*K** ****E**
AAC/APH  413  --SEEE-IGTNFGEDILRMYGNIDIEKAKEYQDIVEEYYPIETIVYGIKNIKQEFIENGR
```

AAC/APH 470 KEIYKRTYKD

FIG. 1. Homology of the predicted amino acid sequence of the bifunctional AAC(6')-APH(2'') resistance determinant with the *cat-86* gene product from *B. pumilus* (10) and the APH gene product from *S. fradiae* (14). The PRTALN program of Wilbur and Lipman (24) was used to align the sequences. Identical amino acids are indicated by the corresponding one-letter amino acid symbol, and an asterisk indicates similar amino acids. The numbering system at the beginning of each amino acid sequence is that given by the original investigators.

The discovery of the two domains of activity in the bifunctional AAC-APH protein was made possible because of the amino acid sequence homologies found with two other sequenced resistance proteins, i.e., the chloramphenicol acetyltransferase specified by the *cat-86* gene of

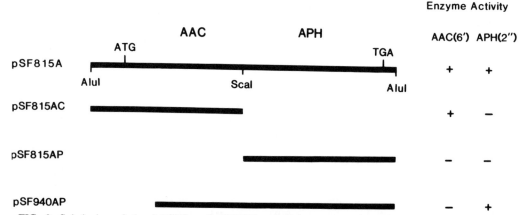

FIG. 2. Subcloning of the AAC(6') and APH(2'') specifying gene regions. The recombinant plasmids designated on the left contain gene segments (heavy line) cloned into the plasmid pUC8. Enzyme assays performed for each clone are indicated on the right.

B. pumilus (10) and the APH of *S. fradiae* (20). These homologies most likely represent similarities in important regions, such as active sites, since they are conserved in these distantly related organisms. It was of interest that there was greater homology of the APH region of the bifunctional enzyme with the APH of *S. fradiae* than with reported sequences from streptococci (21) or *E. coli* (2, 17). This finding is consistent with previous suggestions by Walker and Walker (23) and Benveniste and Davies (4) that bacterial antibiotic resistance genes may have originated in antibiotic-producing *Streptomyces* spp. Additionally, it appears that the APH gene may have been passed directly from the *Streptomyces* spp. to *S. faecalis* rather than through other organisms, but it is difficult to ascertain where in evolution the gene fusion event occurred between the AAC and APH genes. Perhaps some insight into this question will be obtained when additional sequence information is available for an AAC gene from *Streptomyces* spp.

There are numerous reports of a bifunctional AAC(6')-APH(2'') enzyme from *S. aureus* (11–16) which appears to possess properties similar to those of the bifunctional enzyme of this study. For example, Ubukata et al. (22) purified a bifunctional AAC(6')-APH(2'') enzyme that had a molecular weight of 56,000, and Martel et al. (16) performed kinetic studies to demonstrate that there was no interaction between the APH and AAC activities. More recently, Lyon et al. (15) described a 4.5-kilobase transposon, Tn4001, which mediates resistance to gentamicin, tobramycin, and kanamycin by AAC(6')-APH(2'') activity. Within the central portion of Tn4001 is a 2.5-kilobase *Hin*dIII fragment which appears to be common among *S. aureus* strains possessing gentamicin resistance plasmids (8). A similar size *Hin*dIII fragment is present in streptococcal plasmid pIP800, although it is not yet known whether this fragment contains the region specifying the AAC(6')-APH(2'') activity.

This work was supported by a National Science Foundation United States-France Cooperative Grant and a Fondation pour la Recherche Medicale Award.

We thank David R. Lorenz for technical expertise, innovative ideas, and helpful discussions.

LITERATURE CITED

1. **Allmansberger, R., B. Brau, and W. Piepersberg.** 1985. Genes for gentamicin-(3)-N-acetyl-transferases III and IV. II. Nucleotide sequences of three AAC(3)-III genes and evolutionary aspects. Mol. Gen. Genet. **198:**514–520.
2. **Beck, E., G. Ludwig, E. A. Auerswald, B. Reiss, and H. Schaller.** 1982. Nucleotide sequence and exact localization of the neomycin phosphotransferase gene from transposon Tn5. Gene **19:**327–336.
3. **Behnke, D., M. S. Gilmore, and J. J. Ferretti.** 1981. Plasmid pGB301, a new multiple resistance streptococcal cloning vehicle and its use in cloning of a gentamicin/kanamycin resistance determinant. Mol. Gen. Genet. **182:**414–421.
4. **Benveniste, R., and J. Davies.** 1973. Aminoglycoside antibiotic inactivating enzymes in Actinomyces similar to those present in clinical isolates of antibiotic resistant bacteria. Proc. Natl. Acad. Sci. USA **70:**2276–2280.
5. **Brau, B., U. Pilz, and W. Piepersberg.** 1983. Genes for gentamicin-(3)-N-acetyltransferases III and IV. I. Nucleotide sequence of the AAC(3)IV gene and possible involvement of an IS140 element in its expression. Mol. Gen. Genet. **193:**179–187.
6. **Courvalin, P., C. Carlier, and E. Collatz.** 1980. Plasmid-mediated resistance to aminocyclitol antibiotics in group D streptococci. J. Bacteriol. **143:**541–551.
7. **Ferretti, J. J., K. S. Gilmore, and P. Courvalin.** 1986. Nucleotide sequence analysis of the gene specifying the bifunctional 6'-aminoglycoside acetyltransferase-2''-aminoglycoside phosphotransferase enzyme in *Streptococcus faecalis* and identification and cloning of gene regions specifying the two activities. J. Bacteriol. **167:**631–638.
8. **Gillespie, M. L., and R. A. Skurray.** 1986. Plasmids in multiresistant *Staphylococcus aureus*. Microbiol. Sci. **3:** 53–58.
9. **Gray, S. G., and W. M. Fitch.** 1983. Evolution of antibiotic resistance genes: the DNA sequence of a kanamycin resistance gene from *Staphylococcus aureus*. Mol. Biol. Evol. **1:**57–66.
10. **Harwood, C. R., D. M. Williams, and P. S. Lovett.** 1983. Nucleotide sequence of a *Bacillus pumilus* gene specifying chloramphenicol acetyltransferase. Gene **24:**163–169.
11. **Horinouchi, S., and B. Weisblum.** 1982. Nucleotide sequence and functional map of pC194, a plasmid that specifies chloramphenicol resistance. J. Bacteriol. **150:** 812–825.
12. **Le Goffic, F.** 1977. The resistance of *Staphylococcus aureus* to aminoglycoside antibiotics and pristinamycins in France in 1976–1977. J. Antibiot. **30**(Suppl.):286–291.
13. **Le Goffic, F., A. Martel, N. Moreau, M. L. Capmau, C. J. Soussy, and J. Duval.** 1977. 2''-O-Phosphorylation of gentamicin components by a *Staphylococcus aureus* strain carrying a plasmid. Antimicrob. Agents Chemother. **12:**26–30.
14. **Le Goffic, F., N. Moreau, and M. Masson.** 1977. Ann. Microbiol. (Paris) **128B:**465–469.
15. **Lyon, B. R., J. W. May, and R. A. Skurray.** 1984. Tn4001: a gentamicin and kanamycin resistance transposon in *Staphylococcus aureus*. Mol. Gen. Genet. **193:**554–556.
16. **Martel, A., M. Masson, N. Moreau, and F. LeGoffic.** 1983. Kinetic studies of aminoglycoside acetyltransferase and phosphotransferase from *Staphylococcus aureus* RPAL. Eur. J. Biochem. **133:**515–521.
17. **Oka, A., H. Sugisaki, and M. Takanami.** 1981. Nucleotide sequence of the kanamycin resistance transposon Tn903. J. Mol. Biol. **147:**217–226.
18. **Sanger, F., S. Nicklen, and A. R. Coulson.** 1977. DNA sequencing with chain terminating inhibitors. Proc. Natl. Acad. Sci. USA **74:**5463–5467.
19. **Shaw, W. V.** 1983. Chloramphenicol acetyltransferase: enzymology and molecular biology. Crit. Rev. Biochem. **14:**1–43.
20. **Thompson, C. J., and G. S. Gray.** 1983. Nucleotide sequence of a streptomycete aminoglycoside phosphotransferase gene and its relationship to phosphotransferases encoded by resistance plasmids. Proc. Natl. Acad. Sci. USA **80:**5190–5194.
21. **Trieu-Cuot, P., and P. Courvalin.** 1983. Nucleotide sequence of the plasmid gene encoding the aminoglycoside phosphotransferase APH(3')(5'')-III in *Streptococcus faecalis*. Gene **23:**331–341.
22. **Ubukata, K., N. Yamashita, A. Gotoh, and M. Konno.** 1984. Purification and characterization of aminoglycoside-modifying enzymes from *Staphylococcus aureus* and *Staphylococcus epidermidis*. Antimicrob. Agents Chemo-

ther. **25:**754–759.

23. **Walker, M. S., and J. B. Walker.** 1970. Streptomycin biosynthesis and metabolism. J. Biol. Chem. **245:**6683–6689.

24. **Wilbur, W. J., and D. J. Lipman.** 1983. Rapid similarity searches of nucleic acid and protein data banks. Proc. Natl. Acad. Sci. USA **80:**726–730.

Nucleotide Sequence of the Gene, *tetM*, Conferring Resistance to Tetracycline-Minocycline in Gram-Positive Cocci

P. MARTIN, P. TRIEU-CUOT, AND P. COURVALIN

Unité des Agents Antibactériens, Centre National de la Recherche Scientifique U.A. 271, Institut Pasteur, 75724 Paris Cedex 15, France

The nucleotide sequence of the tetracycline resistance gene *tetM*, encoded by the streptococcal conjugative shuttle transposon Tn*1545*, has been determined. The resistance gene was located by analysis of the initiation and termination codons and by in vitro inactivation into an open reading frame of 1,917 base pairs corresponding to a protein with an M_r of 72,500. This value is in good agreement with that (68,000) estimated by sodium dodecyl sulfate-polyacrylamide gel electrophoresis of extracts of *Escherichia coli* minicells. The *tetM* gene product does not exhibit any sequence homology with either the gram-negative (*tetA, tetB,* and *tetC*) or the *Bacillus* spp. and *Staphylococcus* spp. tetracycline resistance proteins. The average hydropathy value of the *tetM* gene product (-0.21) contrasts with those calculated for the other TET proteins, which are markedly hydrophobic (0.76 to 0.93). Hybridization experiments performed with an intragenic *tetM* probe do not support the claim (1) that tetracycline resistance in *Campylobacter* spp. is due to acquisition of *tetM*.

The nucleotide sequence of the *tetM* gene is presented in Appendix B of this volume. The first base pair in the sequence, as presented, is defined as position 1. The presumed ribosome binding site, optimally spaced (8 base pairs) before an initiation codon, is boxed, and the 8 of 10 bases complementary with the 3'-OH terminus of the 16S rRNA of *B. subtilis* are indicated in small letters. Arrows indicate the 21-base-pair perfect diad symmetry regions. The sequence of pBR322 is framed. The deduced amino acid sequence of the TETM protein is indicated.

LITERATURE CITED

1. **Taylor, D. E.** 1986. Plasmid-mediated tetracycline resistance in *Campylobacter jejuni*: expression in *Escherichia coli* and identification of homology with streptococcal class M determinant. J. Bacteriol. **165**:1037–1039.

Genetic and Molecular Analysis of Streptococcal and Enterococcal Chromosome-Borne Antibiotic Resistance Markers

THEA HORAUD, CHANTAL LE BOUGUÉNEC, AND GILDA DE CÉSPÈDES

Laboratoire des Staphylocoques et des Streptocoques, Institut Pasteur, 75724 Paris Cedex 15, France

Chromosome-borne resistance to tetracycline has been described for *Enterococcus faecalis* (8), *Enterococcus faecium* (19), and *Streptococcus sanguis* (9), and chromosomal resistance to various antibiotics, including tetracycline, has been described for different streptococci (12). Over the past few years, 83 multiply resistant strains of groups A, B, C, D (*Streptococcus bovis*), F, and G, *Streptococcus pneumoniae*, and viridans streptococci have been examined in our laboratory for transfer of resistance markers and plasmid content. Of these, only nine, belonging to group A, B, C, or G, were found to carry R plasmids (2, 4, 13, 16). The markers borne by these plasmids are erythromycin-lincomycin-streptogramin B, with or without chloramphenicol. All nine strains are also resistant to tetracycline, but this resistance is neither conjugative nor associated with plasmids (16). In the remaining 74 strains (90%), no extra-chromosomal DNA could be detected. Twenty-four (32.4%) of these plasmid-free strains transfer their antibiotic resistance markers en bloc by conjugation at a low frequency (10^{-4} to 10^{-9} transconjugants per donor cell) into several streptococcal and enterococcal recipients (5, 6, 14–16). Therefore, it appears that a chromosomal location is fairly common for antibiotic resistance determinants in streptococci.

(In accordance with the proposal of Schleifer and Kilpper-Bälz [22] that *S. faecalis* and *S. faecium* be transferred to the genus *Enterococcus*, we have used the designations *Enterococcus faecalis* and *Enterococcus faecium* throughout this paper.)

Chromosomal conjugative elements coding for tetracycline in *E. faecalis* (8) and *S. sanguis* (9) and coding for multiple antibiotic resistance in *Streptococcus pyogenes* (18), *S. agalactiae* (17), and *S. pneumoniae* (7, 25) have been reported. In this report the element of *S. pyogenes* A454 is characterized in greater detail. One strain of *E. faecalis* and one of *E. faecium* harboring chromosomal conjugative elements are also described.

Analysis of the Chromosomal Conjugative Element Carried by A454

The chromosomal conjugative element (CCE) harbored by the plasmid-free *S. pyogenes* strain A454 (14) encodes resistance to erythromycin (Em^r; MLS phenotype) and tetracycline (Tc^r). This element transfers by conjugation from chromosome to chromosome, translocates onto two hemolysin-bacteriocin (Hly-Bcn) plasmids, pIP964 (1) and pAD1 (24), and spontaneously excises from the chromosome of A454 during prolonged broth culture (18).

Chromosomal transfer of the element of A454. The CCE transfers from A454 by conjugation (only by filter mating), at frequencies of 10^{-4} to 10^{-5} transconjugants per donor, into six streptococcal recipients (groups A, B, C, and G, *S. sanguis*, and *S. pneumoniae*) and at a frequency of 10^{-7} per donor into Rec$^+$ JH2-2 or Rec$^-$ UV202 *E. faecalis* recipients. All transconjugants were Em^r Tc^r. Plasmid DNA was not detected in any of the new hosts (18). No Em^r Tc^r transconjugants were detected when A454 was crossed with *S. bovis*, *E. faecium*, *E. durans*, *Staphylococcus aureus*, or *Listeria innocua* recipients (frequencies $< 10^{-8}$ per donor) (5, 16).

Translocation of the chromosomal conjugative element onto Hly-Bcn plasmids. The CCE was transferred, by filter mating, either directly from its wild-type host, A454, into *E. faecalis* Rec$^+$ JH2-2(pIP964) and Rec$^-$ UV202(pIP964) or from a JH2-2(pIP964) transconjugant into a plasmid-free *E. faecalis* recipient, BM133. The transconjugants obtained from these matings were resistant to both erythromycin and tetracycline, except for four clones which were resistant to only one of the two antibiotics. Some of the beta-hemolytic (Hly) transconjugants obtained segregated in the course of purification into Hly and nonhemolytic (NH) clones. Other transconjugants displayed a hyperhemolytic (H-Hly) phenotype; in subsequent cultures on tetracycline-containing media these clones continued to be H-Hly but they were NH on antibiotic-free or erythromycin-containing media. Similar results were obtained when the element translocated onto pAD1 (18).

Eleven plasmids examined, harbored by six NH, two H-Hly, and three Hly clones, had restriction patterns different from those of pIP964 with either *Eco*RI or *Hin*dIII. The mod-

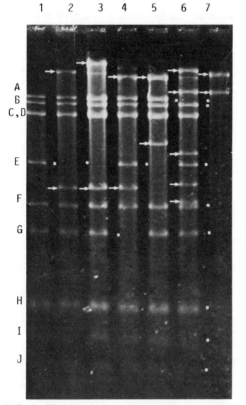

1 2 3 4 5 6 7

A
B
C,D

E

F

G

H

I

J

FIG. 1. Electrophoretic pattern of EcoRI-digested plasmids. The restriction fragments were separated on a 0.7% agarose gel. Lane 1, Cleavage of pIP964 generated 10 fragments A–J. Lanes 2–7, Em^r Tc^r plasmids derived from pIP964: 2, pIP1077; 3, pIP1115; 4, pIP1117; 5, pIP1118; 6, pIP1116; 7, pIP1114. Dots indicate positions of missing bands, and arrows indicate new bands containing the insertions.

Tc^r, pIP1037 codes for only Em^r, and pIP1038 codes for only Tc^r (18). In the largest derivative plasmid, pIP1116, the EcoRI A fragment of pIP964 was replaced by five additional fragments, A'1 to A'5, corresponding to an insertion of 45 kb. ^{32}P-labeled pIP964 was used to probe pIP1116 digested by EcoRI; hybridization was detected only with the additional fragments A'2 and A'3 (besides the pIP964 fragments), indicating that these are junction fragments.

For the localization of the Tc^r determinant the single additional EcoRI fragment of pIP1038, G' (19.5 kb), was used to probe the five plasmids. Sequence homologies were found on the additional bands E'1 (25.1 kb) of pIP1077 and A'1 (28.2 kb) of pIP1116 and with the largest fragment (25.1 kb) of pIP1114. Moreover, each of the five derived plasmids, as well as pIP964, was used to probe HincII-digested pAD1 and pAD1::Tn916 (8). Sequence homologies were detected on the first four HincII fragments of Tn916 with all plasmids except pIP1037. These results and those obtained by comparison of the HincII profiles of both the modified plasmids and Tn916 indicated that the tetM gene (3) and a structure similar to Tn916 are present in the CCE of A454 (C. Le Bouguénec, G. de Céspèdes, and T. Horaud, manuscript in preparation).

To localize the Em^r determinant, each of the five derived plasmids was used to probe pAD2, carrying Tn917 (23), digested with either AvaI or HpaI-KpnI. Sequences homologous to the erm gene of Tn917 (21) were detected with all the plasmids except pIP1038. pAD2 was then used to probe the five derived plasmids. Sequences homologous to the erm gene of Tn917 were found on the same additional bands as the Tc^r determinant for pIP1077, pIP1116, and pIP1114 and on the E'1 (8.1 kb) fragment of pIP1037. No homology was detected between pAD2 and pIP1038.

Chromosomal insertion of the element of A454. ^{32}P-labeled pIP1077 was used to probe the EcoRI-digested cellular DNAs of A454, its spontaneous antibiotic-susceptible derivative (A467), the streptococcal transconjugants, and the corresponding recipients. In each case, except for the antibiotic-susceptible strains, two hybridizing EcoRI fragments were visible, which were the same sizes in both the wild-type and the new streptococcal hosts, about 28.2 and 10.6 kb (18). To verify that the CCE has a preferential insertion site in the streptococcal chromosomes, pIP1116 was used to probe the EcoRI-digested cellular DNAs of 10 transconjugants independently obtained by conjugative transfer of CCE from A454 into a group B Streptococcus recipient. In each of the 10 chromosomes, as well as in that of A454, the same five EcoRI fragments hybridized with the probe (Fig. 2).

ified profiles consisted of insertions of 7 to 45 kilobases (kb) into the EcoRI A, E, or G fragment of pIP964, generating one, two, or more additional bands (Fig. 1). The insertions into the EcoRI E or G fragment of pIP964 modified the expression of the genes involved in beta-hemolysis, yielding H-Hly or NH phenotypes, whereas insertion into the EcoRI A fragment left the Hly phenotype unaltered.

Restoration of both the normal function of Hly genes and the restriction profile of pIP964 was observed when two of the derivative plasmids, which confer the NH phenotype, were transferred to a new recipient, indicating that the element had retranslocated onto the chromosome in the new host (18).

Location of antibiotic resistance markers on the plasmids derived by translocation of the CCE of A454 onto pIP964. Of the five derived plasmids used for further studies, pIP1077, pIP1116, and pIP1114 (Fig. 1, lanes 2, 6, and 7) code for Em^r

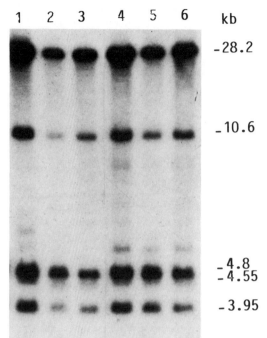

FIG. 2. Filter blot hybridization of *Eco*RI-digested cellular DNA with ³²P-labeled pIP1116. Lane 1, A454; lanes 2–6, independent group B transconjugants. The sizes of the five hybridizing fragments were the same for every strain.

Similar hybridization experiments were performed, in which the modified plasmids were used to probe the *Eco*RI-digested cellular DNAs of 10 independent transconjugants having an *E. faecalis* strain as host. The results were different from those obtained with the streptococcal chromosomes and suggested that in *E. faecalis* the integration of the CCE of A454 occurs at several sites (Le Bouguenec et al., manuscript in preparation).

Comparison of the A454 CCE with Tn*916*. The A454 CCE and Tn*916* are closely related, insofar as both transfer by filter-mating conjugation from chromosome to chromosome without the intervention of a vehicle plasmid; for both of them this transfer is Rec independent. Moreover, the two elements may also jump from the chromosome onto Hly-Bcn plasmids and vice versa, and both insert at several specific sites of the vector plasmid. *Eco*RI cleaves the CCE of A454 but not Tn*916*.

The two elements differ considerably with respect to their location on the host chromosome. In streptococcal hosts (18), the CCE of A454 inserts site specifically, whereas Tn*916* can insert in various sites and, therefore, multiple copies may be present in the same strain (8, 20).

For Tn*916*, no variation in insertion size in the course of transposition onto Hly-Bcn plasmids has been reported (10, 11), whereas with the CCE of A454 insertions of different sizes in the course of translocation were obtained (18). In fact, it is not clear whether the largest insertion obtained (45 kb) actually represents the entire element. The variety of insertion sizes in pIP964 may be the result of molecular rearrangements, which occur, after translocation, during stabilization of the plasmids. Another possibility could be that the insertions in pIP964 represent partial excisions of the element from the chromosome of A454.

That the CCE of A454 was observed to spontaneously excise from its original host as well as from *S. pyogenes* and *E. faecalis* transconjugants distinguishes it from Tn*916*. Neither Tn*916* nor the CCE of A454 could be cured.

Chromosomal Conjugative Tetracycline Resistance Elements in Multiply Antibiotic-Resistant *E. faecalis* and *E. faecium* Strains

In a recent study of the genetic basis of antibiotic resistance of 12 *E. faecalis* strains resistant to tetracycline and erythromycin (MLS phenotype) and chloramphenicol, it was found that in all cases Tc^r transferred by filter-mating conjugation from the chromosomes of wild-type donors into plasmid-free *E. faecalis* (JH2-2) and group A, B, C, and G, *S. sanguis*, and *S. pneumoniae* streptococcal recipients. In these strains, the Tc^r determinant was plasmid borne in only one of the wild-type strains, D397, and in its corresponding *E. faecalis* transconjugant (K, Pepper, T. Horaud, C. Le Bouguénec, and G. de Céspèdes, manuscript in preparation).

Mating the *E. faecalis* wild-type strain D395 with JH2-2(pIP964), with selection on tetracycline-containing media, yielded 35 Hly Tc^r transconjugants. One of these segregated during purification into clones having either the H-Hly or the normal Hly phenotype. Analysis of *Eco*RI-digested plasmids isolated from one of each type of clone revealed modification of the parental plasmid, pIP964. In each case the modification consisted of an insertion of 18 kb in *Eco*RI fragment A (Hly phenotype; pIP1132) or E (H-Hly phenotype; pIP1131), which generated two additional fragments (Fig. 3).

E. faecium D344, encoding resistance to tetracycline and erythromycin and high-level kanamycin and streptomycin, harbors several plasmids (Fig. 4, lane 1) (19). Tc^r transferred by filter-mating conjugation from D344 into JH2-2. No plasmids were detected in the new host, BM5503. When this strain was mated with JH2-2(pIP964), the results were similar to those obtained with D395. One of the H-Hly clones examined harbored a modified plasmid, the

*Eco*RI restriction profile of which was similar to that of pIP1131 (Fig. 3).

Erythromycin Resistance of *E. faecium* D344

Em[r] of D344 transferred by filter-mating conjugation into both *E. faecalis* JH2-2 and *E. faecium* BM115 recipients at frequencies of 10^{-7}

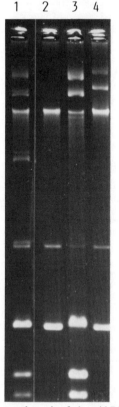

FIG. 4. Electrophoresis of plasmid DNAs. Lane 1, D344; lane 2, *E. faecium* recipient strain BM115; lane 3, BM5539; lane 4, BM5540 (harboring pIP1137).

and 10^{-4} per donor, respectively. No plasmid DNA was detected in the *E. faecalis* transconjugant, BM5502. In the *E. faecium* transconjugant BM5539, in addition to the resident cryptic plasmids (Fig. 4, lane 2), several plasmids, identical in size to plasmids of D344, were found (Fig. 1, lanes 1 and 3). Retransfer of Em[r] from the plasmid-free *E. faecalis* transconjugant BM5502 into an appropriate *E. faecium* recipient, BM110, yielded transconjugants at a frequency of 5×10^{-3} per donor. One of these, BM5540, was examined and found to harbor a plasmid, pIP1137 (Fig. 4, lane 4). Sequences homologous to the *erm* gene of Tn*917* were found in pIP1137 as well as in one *Eco*RI chromosomal fragment of BM5502. Further studies are in progress in our laboratory to elucidate this phenomenon.

We thank Karen Pepper for criticism of the manuscript and O. Rouelland for secretarial assistance. We are grateful to D. Clewell for sending us strains harboring pAD2, pAD1, pAD1::Tn*916*, and pAM170.

This work was supported by grants from the Caisse Nationale de l'Assurance Maladie des Travailleurs Salariés to T.H.

FIG. 3. Electrophoretic pattern of *Eco*RI-digested plasmids. Lane 1, pIP964; lane 2, pIP1131; lane 3, pIP1132. Circles, Positions of missing bands; arrows, new bands containing the insertions.

LITERATURE CITED

1. **Borderon, E., G. Bieth, and T. Horodniceanu.** 1982. Genetic and physical studies of *Streptococcus faecalis* hemolysin plasmids. FEMS Microbiol. Lett. **14:**51–55.

2. **Bougueleret, L., G. Bieth, and T. Horodniceanu.** 1981. Conjugative R plasmids in group C and G streptococci. J. Bacteriol. **145:**1102–1105.

3. **Burdett, V., J. Inamine, and S. Rajagopalan.** 1982. Heterogeneity of tetracycline resistance determinants in *Streptococcus.* J. Bacteriol. **149:**995–1004.

4. **Buu-Hoï, A., G. de Céspèdes, and T. Horaud.** 1985. Deoxyribonuclease-sensitive transfer of an R plasmid in *Streptococcus pyogenes* (group A). FEMS Microbiol. Lett. **30:**407–410.

5. **Buu-Hoï, A., and T. Horaud.** 1985. Genetic basis of antibiotic resistance in group A, C and G streptococci, p. 231–232. *In* Y. Kimura, S. Kotami, and Y. Shiokawa (ed.), Recent advances in streptococci and streptococcal diseases. Reedbooks, Bracknell, England.

6. **Buu-Hoï, A., and T. Horodniceanu.** 1980. Conjugative transfer of multiple antibiotic resistance markers in *Streptococcus pneumoniae.* J. Bacteriol. **142:**313–320.

7. **Carlier, C., and P. Courvalin.** 1982. Resistance of streptococci to aminoglycoside-aminocyclitol antibiotics, p. 162–166. *In* D. Schlessinger (ed.), Microbiology—1982. American Society for Microbiology, Washington, D.C.

8. **Clewell, D. B., G. F. Fitzgerald, L. Dempsey, L. E. Pearce, F. Y. An, B. A. White, Y. Yagi, and C. Gawron-Burke.** 1984. Streptococcal conjugation: plasmids, sex pheromones, and conjugative transposons, p. 194–203. *In* S. E. Mergenhagen and B. Rosan (ed.), Molecular basis of oral microbial adhesion. American Society for Microbiology, Washington, D.C.

9. **Fitzgerald, G. F., and D. B. Clewell.** 1985. A conjugative transposon (Tn*919*) in *Streptococcus sanguis.* Infect. Immun. **47:**415–420.

10. **Franke, A. E., and D. B. Clewell.** 1981. Evidence for a chromosome-borne resistance transposon (Tn*916*) in *Streptococcus faecalis* that is capable of "conjugal" transfer in the absence of a conjugative plasmid. J. Bacteriol. **145:**494–502.

11. **Gawron-Burke, C., and D. B. Clewell.** 1982. A transposon in *Streptococcus faecalis* with fertility properties. Nature (London) **300:**281–284.

12. **Horaud, T., C. Le Bouguenec, and K. Pepper.** 1985. Molecular genetics of resistance to macrolides, lincosamides and streptogramin B (MLS) in streptococci (a review). J. Antimicrob. Chemother. **16**(Suppl A):111–135.

13. **Horodniceanu, T., D. Bouanchaud, G. Bieth, and Y. A. Chabbert.** 1976. R plasmids in *Streptococcus agalactiae* (group B). Antimicrob. Agents Chemother. **10:**795–801.

14. **Horodniceanu, T., L. Bougueleret, and G. Bieth.** 1981. Conjugative transfer of multiple-antibiotic resistance markers in beta-hemolytic group A, B, F, and G streptococci in the absence of extrachromosomal deoxyribonucleic acid. Plasmid **5:**127–137.

15. **Horodniceanu, T., A. Buu-Hoï, F. Delbos, and G. Bieth.** 1982. High-level aminoglycoside resistance in group A, B, G, D (*Streptococcus bovis*), and viridans streptococci. Antimicrob. Agents Chemother. **21:**176–179.

16. **Horodniceanu, T., C. Le Bouguenec, A., Buu-Hoï, and G. Bieth.** 1982. Conjugative transfer of antibiotic resistance markers in beta-hemolytic streptococci in the presence and absence of plasmid DNA, p. 105–108. *In* D. Schlessinger (ed.), Microbiology—1982. American Society for Microbiology, Washington, D.C.

17. **Inamine, J. M., and V. Burdett.** 1985. Structural organization of a 67-kilobase streptococcal conjugative element mediating multiple antibiotic resistance. J. Bacteriol. **161:** 620–626.

18. **Le Bouguenec, C., T. Horaud, G. Bieth, R. Colimon, and C. Dauguet.** 1984. Translocation of antibiotic resistance markers of a plasmid-free *Streptococcus pyogenes* (group A) strain into different streptococcal hemolysin plasmids. Mol. Gen. Genet. **194:**377–387.

19. **Le Bouguenec, C., and T. Horodniceanu.** 1982. Conjugative R plasmids in *Streptococcus faecium* (group D). Antimicrob. Agents Chemother. **21:**698–705.

20. **Nida, K., and P. Cleary.** 1983. Insertional inactivation of streptolysin S expression in *Streptococcus pyogenes.* J. Bacteriol. **155,**1156–1161.

21. **Perkins, J. B., and J. Youngman.** 1984. A physical and functional analysis of Tn*917*, a *Streptococcus* transposon in the Tn*3* family that functions in *Bacillus.* Plasmid **12:** 119–138.

22. **Schleifer, K. H., and R. Kilpper-Bälz.** 1984. Transfer of *Streptococcus faecalis* and *Streptococcus faecium* to the genus *Enterococcus* nom. rev. as *Enterococcus faecalis* comb. nov. and *Enterococcus faecium* comb. nov. Int. J. Syst. Bacteriol. **34:**31–34.

23. **Tomich, P. K., F. Y. An, and D. B. Clewell.** 1978. A transposon (Tn*917*) in *Streptococcus faecalis* that exhibits enhanced transposition during induction of drug resistance. Cold Spring Harbor Symp. Quant. Biol. **43:**1217–1221.

24. **Tomich, P. K., F. Y. An, S. P. Damle, and D. B. Clewell.** 1979. Plasmid-related transmissibility and multiple drug resistance in *Streptococcus faecalis* subsp. *zymogenes* strain DS16. Antimicrob. Agents Chemother. **15:**828–830.

25. **Vijayakumar, M. N., S. D. Priebe, and W. R. Guild.** 1986. Structure of a conjugative element in *Streptococcus pneumoniae.* J. Bacteriol. **166:**978–984.

Dissemination of a Plasmid-Borne Chloramphenicol Resistance Gene in Streptococcal and Enterococcal Clinical Isolates

KAREN PEPPER, CHANTAL LE BOUGUÉNEC, GILDA DE CÉSPÈDES, ISABELLE COLMAR, AND THEA HORAUD

Laboratoire des Staphylocoques et des Streptocoques, Institut Pasteur, 75724 Paris Cedex 15, France

In clinical isolates of various streptococcal and enterococcal species isolated over a 7-year period, resistance to chloramphenicol (Cm^r) was observed at the following frequencies: 0.5% of 486 group A, B, and G streptococci, 4% of 197 *Streptococcus pneumoniae* strains, 4% of 775 viridans streptococci, 40% of 230 *Enterococcus faecalis* strains, and 50% of 70 *Enterococcus faecium* strains. None of the group C strains nor any of the *Streptococcus bovis, Enterococcus durans*, or *Enterococcus avium* strains examined was resistant to chloramphenicol (statistics of the French National Reference Center for Streptococci, Institut Pasteur, Paris).

In the streptococci and the enterococci, Cm^r determinants may be either chromosomal or plasmid borne. We were interested in learning whether the dispersal of a single gene might account for the appearance of this marker on plasmids and chromosomes of diverse origin. Thirty-four strains resistant to various antibiotics, including chloramphenicol, were tested by DNA/DNA hybridization with the Cm^r gene of pIP501, a 30-kilobase (kb) plasmid originally isolated from *Streptococcus agalactiae* B96 (10). In half of these strains, the Cm^r gene is on the chromosome and in half it is plasmid borne.

The Cm^r gene of pIP501 is situated on a 6.3-kb *Hin*dIII fragment, which contains a site for *Bst*EII just within the gene and a site for *Hpa*II about 200 base pairs beyond the other end of the gene (1, 8). This 6.3-kb *Hin*dIII fragment was cloned into pBR322, and the hybrid plasmid thus obtained was designated pIP1322. The 6.3-kb *Hin*dIII fragment was recovered in large quantity and digested with *Bst*EII and *Hpa*II. The 1.6-kb fragment bounded by these sites was then recovered from low-melting-point agarose gels and used as a probe (17a). Expression of the Cm^r determinant of pIP501 was observed in the *Escherichia coli* host (strain HVC45) harboring pIP1322, in accordance with results reported by Evans et al. (9). The minimal inhibitory concentration of chloramphenicol for this strain was 64 μg ml^{-1}, the same as that of wild-type strain B96.

Two series of hybridization experiments were carried out. In the first series, 17 *Hin*dIII-di-gested plasmids encoding Cm^r were probed with the 1.6-kb *Bst*EII-*Hpa*II fragment derived from pIP501 (Table 1). In the second series of experiments, the cellular DNA of 17 plasmid-free strains encoding Cm^r (5, 6, 11, 13a) was tested with the same probe by dot-blot hybridization. These strains are listed in Table 2.

The results of the first series of experiments are presented in Table 1 and Fig. 1. Two plasmids of group B origin and one of group G origin, with *Hin*dIII profiles similar to that of pIP501, had fragments of 6.3 kb that hybridized to the probe. Hybridizing bands of the same size were detected as well with two plasmids of *E. faecalis* and two of *E. faecium*, though for one of the *E. faecium* plasmids another fragment (9.0 kb) also hybridized. In two other *E. faecalis* plasmids, it was not clear whether hybridization occurred on a fragment of 6 or 6.3 kb, as these fragments were poorly resolved by electrophoresis. One *E. faecalis* plasmid had two hybridizing fragments (8.2 and 3.1 kb). Hybridization was also observed with the 7.4-kb *Hin*dIII fragment of the *Streptococcus pyogenes* plasmid pIP955. One plasmid isolated from *Staphylococcus aureus*, pC221, also hybridized with the probe. Since this plasmid has only a single site for *Hin*dIII, it was digested with *Bst*EII and *Hpa*II (20). A 1.6-kb fragment generated by this digestion hybridized with the pIP501-derived probe; according to the published sequence of pC221 (19), this fragment spans most of the Cm^r gene, indicating a structural similarity with the gene of pIP501 and possibly a common origin.

Five of the plasmids tested did not hybridize with the probe: pC02, of group B origin; pIP655, pIP683, and pIP1326 originating from *E. faecalis*; and the *S. aureus* plasmid pC194 (i.e., the 2.7-kb *Hin*dIII fragment of pHV33; see Table 1).

pHV33 was subsequently used to probe the following plasmids: pIP955, pIP612, pIP635, pIP920, pIP1321, pIP1436, pIP716, and pIP717 (Table 1). pIP955 was the only one of these plasmids in which sequences homologous to pC194 were detected. No hybridization was detected with a control probe consisting of pBR322.

TABLE 1. Plasmids and hybridization results

Plasmid designation (reference)	Phenotype[a]	Wild-type host or construction	Size (kb) of HindIII fragment(s) hybridizing to 1.6-kb pIP501-derived probe
pIP955 (4)	Cm MLS	*S. pyogenes*	7.4
pIP501 (10)	Cm MLS	*S. agalactiae*	6.3
pIP612 (7)	Cm MLS	*S. agalactiae*	6.3
pIP635 (13)	Cm MLS	*S. agalactiae*	6.3
pC02 (16)	Cm MLS	*S. agalactiae*	None
pIP920 (3)	Cm MLS	*Streptococcus* sp. (group G)	6.3
pIP655 (12)	Cm Gm Km	*E. faecalis*	None
pIP683 (12)	Cm Gm Km	*E. faecalis*	None
pIP1321 (2)	Cm MLS Tc	*E. faecalis*	6.0 or 6.3
pIP1326 (17a)	Cm	*E. faecalis*	None
pIP1435 (2)	Cm MLS	*E. faecalis*	6.3
pIP1436 (17a)	Cm MLS	*E. faecalis*	6.0 or 6.3
pIP1438 (2)	Cm MLS	*E. faecalis*	3.1, 8.2
pIP1503 (17a)	Cm MLS	*E. faecalis*	6.3
pIP716 (15)	Cm MLS Km Sm Tc	*E. faecium*	6.3, 9.0
pIP717 (15)	Cm MLS Km Sm Tc	*E. faecium*	6.3
pC221 (17)	Cm	*Staphylococcus aureus*	1.6
pHV33 (18)	Cm Ap	pC194 (14)-pBR322 in *E. coli* HVC45	None
pIP1322 (17a)	Cm Ap	6.3-kb HindIII pIP501-pBR322 in *E. coli* HVC45	6.3

[a] Abbreviations: Resistance to Ap, ampicillin; Cm, chloramphenicol; Gm, gentamicin; Km, kanamycin; MLS, macrolides-lincosamides-streptogramin B; Sm, streptomycin; Tc, tetracycline.

The cellular DNA of one of the three group A strains, three of the eight group B strains, and one of the three group G strains in which Cm[r] is chromosomally located hybridized with the probe in dot-blot experiments (Fig. 2). We are presently attempting to localize the Cm[r] genes of these strains by probing their HindIII-digested chromosomes with the pIP501-derived 1.6-kb fragment. None of the three *S. pneumoniae* strains tested carried DNA sequences homologous to the probe. There was no apparent relationship between hybridization with the pIP501-derived probe and the conjugative transfer of antibiotic resistance markers in these strains

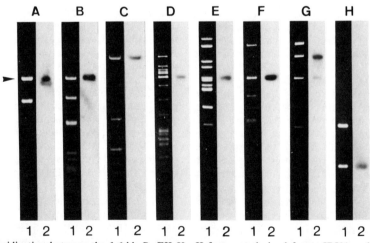

FIG. 1. Hybridization between the 1.6-kb BstEII-HpaII fragment derived from pIP501 and representative streptococcal, enterococcal, and staphylococcal plasmids. Lane 1: Plasmids were digested with HindIII (except for pC221, which was digested with BstEII and HpaII) and run on 0.8% agarose gels. Lane 2: Autoradiograms of the same DNA after transfer to nitrocellulose filters and hybridization. (A) pIP1322, (B) pIP501 (the same HindIII restriction pattern and hybridization result were observed with pIP612, pIP635, and pI920), (C) pIP955, (D) pIP1503, (E) pIP1321, (F) pIP717, (G) pIP716, (H) pC221. Arrow: 6.3 kb.

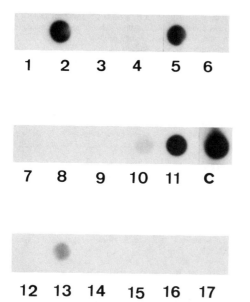

FIG. 2. Dot-blot hybridization between the 1.6-kb fragment derived from pIP501 and the cellular DNA of plasmid-free, chloramphenicol-resistant streptococci. Numbers correspond to strains presented in Table 2. C: pIP501 (control).

(Table 2). Whether any of the chromosomal Cm^r genes that failed to hybridize with the pIP501-derived probe bears homology to pC194 is presently being investigated (see Addendum).

We conclude from this study that the Cm^r gene of the *S. agalactiae* plasmid pIP501, or a closely related gene, is widely distributed among streptococcal and enterococcal plasmids and is also present on the chromosomes in several streptococcal strains of groups A, B, and G. The absence of homology between several plasmids and the probe indicates that there is at least one other plasmid-borne gene coding for Cm^r in the streptococci and in the enterococci. There is also at least one other chromosomal Cm^r determinant in the streptococci which, if the sample studied here is representative of chloramphenicol-resistant, plasmid-free strains, appears to be predominant.

This work was supported by grants from the Caisse Nationale de l'Assurance Maladie des Travailleurs Salariés to T.H. and by a French government student grant to K.P.

ADDENDUM

Since this work was submitted, the chromosomal DNA of the strains listed in Table 2 was probed with the

TABLE 2. Dot-blot hybridization with cellular DNA of plasmid-free streptococci

Strain designation[a] (reference)	Markers[b]	Conjugative transfer[c]	Hybridization[d]	
			pIP501-derived probe	pC194[e]
Group A				
1. A451 (5)	Cm Em Tc	−	−	+
2. A453 (13a)	Cm Em Km Sm	−	+	−
3. A456 (5)	Cm Em	−	−	−
Group B				
4. B109 (11,13)	Cm Em Tc	+	−	+
5. B116 (13)	Cm Em	−	+	−
6. B117 (11,13)	Cm Tc	+	−	−
7. B118 (13)	Cm	−	−	−
8. B120 (13)	Cm	−	−	−
9. B125 (this study)	Cm Tc	NT	−	+
10. B126 (this study)	Cm Em Tc	NT	−	−
11. B127 (this study)	Cm Em Tc	NT	+	−
Group G				
12. G44 (11)	Cm Em Tc	+	−	−
13. G52 (13a)	Cm Em Tc Km Sm	−	+	−
14. G54 (this study)	Cm	NT	−	−
S. pneumoniae				
15. BM4200 (6)	Cm Em Tc Km Pc Su Tp	+	−	+
16. BM6012 (6)	Cm Tc	+	−	+
17. BM6025 (this study)	Cm Tc	−	−	+

[a] Numbers preceding strain designations correspond to those used on Fig. 2.

[b] Pc, Resistance to penicillin; Su, resistance to sulfonamides; Tp, resistance to trimethoprim. For other abbreviations, see Table 1, footnote *a*.

[c] +, Present; −, absent; NT, not tested.

[d] No hybridization was detected with either probe or chloramphenicol-susceptible control strains of groups A and B and *S. pneumoniae*.

[e] See Addendum.

staphylococcal Cm^r plasmid pC194. Sequence homology was detected between this plasmid and the DNA of streptococcal strains A451 (group A), B109 and B125 (group B), and BM4200, BM6012, and BM6025 (*S. pneumoniae*). In light of these results, we would suggest that there are at least three chromosomal genes encoding Cm^r in the streptococci.

LITERATURE CITED

1. **Behnke, D., and M. S. Gilmore.** 1985. Location of antibiotic resistance determinants, copy control, and replication functions on the double-selective streptococcal cloning vector pGB301. Mol. Gen. Genet. **184**:115–120.
2. **Bieth, G., K. Pepper, C. Le Bouguénec, and T. Horaud.** 1985. Genetic and molecular characterization of group B and D streptococcal plasmids, p. 224–226. *In* Y. Kimura, S. Kotami, and Y. Shiokawa (ed.), Recent advances in streptococci and streptococcal disease. Reedbooks, Bracknell, England.
3. **Bougueleret, L., G. Bieth, and T. Horodniceanu.** 1981. Conjugative R plasmids in group C and G streptococci. J. Bacteriol. **145**:1102–1105.
4. **Buu-Hoï, A., G. de Céspèdes, and T. Horaud.** 1985. Deoxyribonuclease sensitive transfer of an R plasmid in *Streptococcus pyogenes* (group A). FEMS Microbiol. Lett. **30**:407–410.
5. **Buu-Hoï, A., and T. Horaud.** 1985. Genetic basis of antibiotic resistance in group A, C and G streptococci, p. 231–232. *In* Y. Kimura, K. Kotami, and Y. Shiokawa (ed.), Recent advances in streptococci and streptococcal diseases. Reedbooks, Bracknell, England.
6. **Buu-Hoï, A., and T. Horodniceanu.** 1980. Conjugative transfer of multiple antibiotic resistance markers in *Streptococcus pneumoniae*. J. Bacteriol. **143**:313–320.
7. **El Solh, N., D. H. Bouanchaud, T. Horodniceanu, A. Roussel, and Y. A. Chabbert.** 1978. Molecular studies and possible relatedness between R plasmids from groups B and D streptococci. Antimicrob. Agents Chemother. **14**:19–23.
8. **Evans, R. P., and F. L. Macrina.** 1983. Streptococcal R plasmid pIP501: endonuclease site map, resistance determinants location, and construction of novel derivatives. J. Bacteriol. **154**:1347–1355.
9. **Evans, R. P., R. B. Winter, and F. L. Macrina.** 1985. Molecular cloning of a pIP501 derivative yields a model replicon for the study of streptococcal conjugation. J. Gen. Microbiol. **131**:145–153.
10. **Horodniceanu, T., D. H. Bouanchaud, G. Bieth, and Y. A. Chabbert.** 1976. R plasmids in *Streptococcus agalactiae*

(group B). Antimicrob. Agents Chemother. **10**:795–801.
11. **Horodniceanu, T., L. Bougueleret, and G. Bieth.** 1981. Conjugative transfer of multiple-antibiotic resistance markers in beta hemolytic group A, B, F, and G streptococci in the absence of extrachromosomal deoxyribonucleic acid. Plasmid **5**:127–137.
12. **Horodniceanu, T., L. Bougueleret, N. El Solh, G. Bieth, and F. Delbos.** 1979. High-level, plasmid-borne resistance to gentamicin in *Streptococcus faecalis* subsp. *zymogenes*. Antimicrob. Agents Chemother. **16**:686–689.
13. **Horodniceanu, T., L. Bougueleret, N. El Solh, D. H. Bouanchaud, and Y. A. Chabbert.** 1979. Conjugative R plasmids in *Streptococcus agalactiae* (group B). Plasmid **2**:197–206.
13a. **Horodniceanu, T., A. Buu-Hoï, F. Delbos, and G. Bieth.** 1982. High-level aminoglycoside resistance in groups A, B, G, D (*Streptococcus bovis*), and viridans streptococci. Antimicrob. Agents Chemother. **21**:176–179.
14. **Iordanescu, S.** 1976. Three distinct plasmids originating in the same *Staphylococcus aureus* strain. Arch. Roum. Pathol. Exp. Microbiol. **35**:111–118.
15. **Le Bouguenec, C., and T. Horodniceanu.** 1982. Conjugative R plasmids in *Streptococcus faecium* (group D). Antimicrob. Agents Chemother. **21**:698–705.
16. **Lütticken, R., and R. Laufs.** 1979. Characterization and use for transformation of plasmid DNA from two Group B streptococcal strains, p. 279–281. *In* M. T. Parker (ed.), Pathogenic streptococci. Reedbooks, Chertsey, England.
17. **Novick, R. P., and D. H. Bouanchaud.** 1971. Extrachromosomal nature of drug resistance in *Staphylococcus aureus*. Ann. N.Y. Acad. Sci. **182**:279–294.
17a. **Pepper, K., C. Le Bouguénec, G. de Céspèdes, and T. Horaud.** 1986. Dispersal of a plasmid-borne chloramphenicol resistance gene in streptococcal and enterococcal plasmids. Plasmid **16**:195–203.
18. **Primrose, S. B., and S. D. Ehrlich.** 1981. Isolation of plasmid deletion mutants and study of their instability. Plasmid **6**:193–201.
19. **Shaw, W. V., D. G. Brenner, S. F. J. Le Grice, S. E. Skinner, and A. R. Hawkins.** 1985. Chloramphenicol acetyltransferase gene of staphylococcal plasmid pC221. Nucleotide sequence analysis and expression studies. FEBS Lett. **179**:101–106.
20. **Wilson, C. R., S. E. Skinner, and W. V. Shaw.** 1981. Analysis of two chloramphenicol resistance plasmids from *Staphylococcus aureus*: insertional inactivation of Cm resistance, mapping of restriction sites, and construction of cloning vehicles. Plasmid **5**:245–258.

Plasmid-Mediated Beta-Lactamase in *Enterococcus faecalis*

BARBARA E. MURRAY

*Department of Medicine and Program in Infectious Diseases and Clinical Microbiology,
University of Texas Medical School, Houston, Texas 77030*

Enterococci are among the most resistant members of the family *Streptococcaceae*. They are intrinsically resistant to a number of commonly used antimicrobial agents such as clindamycin, aminoglycosides, the penicillinase-resistant semisynthetic penicillins, and cephalosporins (1, 14, 15). Although ampicillin and penicillin are adequate for soft tissue infections, minimal inhibitory concentrations average 1 to 4 µg/ml, which is 100 to 1,000 times that for other streptococci. Acquired resistance to erythromycin, tetracycline, and chloramphenicol and to high levels (>2,000 µg/ml) of aminoglycosides is also common among enterococci. The latter resistance eliminates the synergistic bactericidal effect of adding an aminoglycoside to penicillin, an effect which is needed when treating endocarditis. As with aminoglycoside resistance among other organisms, high-level resistance to streptomycin and kanamycin appeared first (10). A report from France in 1979 documented the emergence of high-level resistance to gentamicin and tobramycin, and in the early 1980s we reported strains with high-level resistance to all known aminoglycosides (6, 9, 13). It was during one of these studies that we encountered a strain that not only had the new aminoglycoside resistance traits but also produced β-lactamase (Bla⁺), a new property in this genus (11). A second β-lactamase-producing multiresistant enterococcus has subsequently been isolated in Pennsylvania and sent to us for further analysis. This paper describes studies to date on both isolates.

Susceptibility and Enzyme Studies

The initial Bla⁺ enterococcus (HH22) was isolated from the urine of a patient in Houston, Texas, in 1980 (11); the second strain (PA) was isolated from the bloodstream of a patient in Philadelphia, Pennsylvania, in 1983 by M. Levison, M. Ingerman, and E. Abrutyn (10a). Both strains were resistant to tetracycline and erythromycin, were highly resistant to all aminoglycosides, and produced β-lactamase; strain PA was also resistant to chloramphenicol. The production of β-lactamase was initially detected by using the chromogenic cephalosporin nitrocefin; it was confirmed by demonstration of inactivation of penicillin by bioassay and by microiodometry (11). The effect of the β-lacta-

mase was studied in two ways and is shown in Table 1. The susceptibility of the Bla⁺ strains to various antibiotics was determined with low inocula (10^3 CFU/ml) and high inocula (10^7 CFU/ml). In addition, a partially purified β-lactamase preparation from HH22 was used to determine the hydrolysis of these compounds. There was no hydrolysis and little or no inoculum effect with methicillin and related penicillinase-resistant penicillins, cephalosporins, imipenem (a new β-lactamase resistant carbapenem), or vancomycin, a non-β-lactam agent (10a, 12). There was rapid hydrolysis and a 500-fold or greater difference in the minimal inhibitory concentrations with high and low inocula for penicillin G, ampicillin, and the newer penicillins such as piperacillin. There was an intermediate inoculum effect and partial hydrolysis of ticarcillin. These results are similar to those seen with penicillinase-producing staphylococci (2–5). When β-lactamase inhibitors were studied, clavulanic acid inhibited hydrolysis at 1 µg/ml; sulbactam was partially inhibitory at 4 µg/ml and was maximally inhibitory at 25 to 50 µg/ml (10a, 12).

Resistance Transfers

Strains HH22 and PA both transferred gentamicin resistance to the enterococcal recipient strain JH2-7 (10a, 11). The frequency of transfer with HH22 as the donor was 5×10^{-5} transconjugants per donor in broth and 4×10^{-3} on filters; PA transferred gentamicin resistance at approximately 100-fold lower frequency in each system. One hundred percent of gentamicin-resistant transconjugants from each donor produced β-lactamase. The cotransfer of other resistances from HH22 and PA, respectively, was as follows: erythromycin, 95% and 90%; streptomycin, 1% and 95%; tetracycline, 1% and 92%; chloramphenicol, 98% for PA (HH22 was not chloramphenicol resistant).

DNA Studies

We have previously shown that an 840-base-pair fragment known to encode the leader sequence and 80% of the staphylococcal β-lactamase structural gene (8) hybridized to a 5.1-kilobase (kb) *Eco*RI fragment contained within the 56-megadalton plasmid of HH22 (12). This

TABLE 1. Substrate profile of enterococcal β-lactamase and effect of inoculum on the susceptibility to various agents

Antibiotic	% Hydrolysis[a]	Minimal inhibitory concn (μg/ml)			
		Strain PA		Strain HH22	
		10^3 CFU/ml	10^7 CFU/ml	10^3 CFU/ml	10^7 CFU/ml
Penicillin[b]	100	2	>1,000	2	>1,000
Ampicillin	157	4	>1,000	2	>1,000
Piperacillin	175	2	>1,000	2	>1,000
Mezlocillin	200	NT[c]	NT	1	>1,000
Ticarcillin	29	64	500	64	500
Methicillin	3	32	32	32	32
Nafcillin	<1	NT	NT	8	8
Cephalothin	<1	64	64	32	64
Moxalactam	<1	500	500	500	500
Imipenem	<1	≤1	≤1	≤1	≤1
Vancomycin	<1	4	8	4	4

[a] Relative to the rate of hydrolysis of penicillin G.
[b] Data for penicillin are in units per milliliter.
[c] Not tested.

plasmid has also been shown to hybridize to a 2.3-kb HindIII fragment mediating gentamicin resistance which was cloned from the staphylococcal plasmid pSH6 (7; B. Mederski-Samoraj, personal communication). The 5.1-kb EcoRI enterococcal fragment has now been cloned into pACYC184 and shown to encode β-lactamase (10a).

EcoRI-digested plasmid DNAs from transconjugants of HH22 and PA were compared (Fig. 1). No bands were shared between transconjugants from the different strains. The 5.1-kb EcoRI fragment cloned from HH22 was then hybridized to EcoRI and EcoRI-HindIII-digested plasmid DNA from the transconjugants. This fragment hybridized to itself and to a 10.1-kb EcoRI fragment from transconjugants from strain PA. The EcoRI-HindIII digests also confirmed that different-sized fragments hybridized to this probe.

Enzyme Studies

The hybridization and substrate profile data indicate that the enterococcal β-lactamase is similar to that from staphylococci. Certain differences, however, were found. In staphylococci, the production of β-lactamase is typically inducible (4). With the enterococcal strains, repeated passage on antibiotic-free media generated organisms that produced as much β-lactamase as organisms grown in the presence of penicillin or in subinhibitory concentrations of methicillin, a typical inducer of the staphylococcal enzyme. A disk approximation test using penicillin, cefoxitin, and methicillin disks failed to show any diminution in the zones of inhibition around penicillin. Another difference between the enterococcal and staphylococcal enzymes is that neither of the enterococci released detect-

able amounts of enzyme into the extracellular medium, a typical feature of staphylococci (4). Even when supernatants of cultures were concentrated 100-fold with a Diaflo Ultra Filter PM10, no enzyme activity was detectable in the supernatants. Pellets obtained with HH22 were sonicated extensively and subjected to a series of filtrations and centrifugations. The β-lactamase activity was found to pellet with the mem-

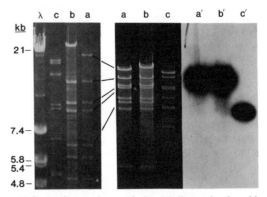

FIG. 1. Comparison of EcoRI-digested plasmid DNAs from transconjugants of strains HH22 and PA. The left and center panels show agarose gels of EcoRI-digested plasmid DNA electrophoresed for 24 and 8 h, respectively. The panel on the right shows an autoradiograph made from the DNA in the middle panel after hybridization with a 5.1-kb ^{32}P-labeled bla-specifying fragment from pBEM1. This fragment was originally cloned from XH22 (a transconjugant of JH2-7 and HH22), and the hybridization shown in lane c represents hybridization to the original 5.1-kb enterococcal fragment. Lanes a and a' show EcoRI-digested plasmid DNA from XPA-a (a transconjugant of JH2-7 and PA); lanes b and b', XPA-b (another transconjugant of JH2-7 and PA); and lanes c and c', XH22. Bacteriophage lambda has been digested with EcoRI.

brane fractions (12). The material obtained after centrifugation at 100,000 × g for 90 min was passed over a Biogel A 0.5 column, and the β-lactamase activity eluted in the void volume, indicating a molecular weight greater than 500,000; the β-lactamases from staphylococci are in the range of 30,000 molecular weight, adding further evidence that the enterococcal enzyme is bound to membrane fragments.

Discussion

Production of β-lactamase is the most common mechanism that bacteria use to overcome the effects of β-lactam antibiotics such as penicillin. Despite four decades of penicillin use, the *Streptococcaceae* have only recently been shown to produce β-lactamase. The first strain was isolated in 1981, and despite screening over 1,000 human isolates subsequent to this, no further β-lactamase producers have been found in the same institution (unpublished data). In addition, 120 isolates from animal sources were screened for β-lactamase production and none was positive (A. Wanger, personal communication). Both of the enterococci that produce β-lactamase are multiresistant and have the other new resistance marker in enterococci, namely, high-level gentamicin resistance. The production of β-lactamase obviously complicates therapy of infections caused by these organisms since the enterococcal penicillinase is most active against those penicillins (penicillin G, ampicillin, piperacillin) that are normally the most effective against these organisms. Methicillin and the cephalosporins are resistant to the penicillinase but have low intrinsic activity against enterococci and are inadequate for therapy (14, 15). Compounds active in vitro include imipenem, vancomycin, and penicillin plus clavulanic acid. Although likely to be effective for many infections, none of these compounds is bactericidal, and they are not likely to be adequate for endocarditis. Since most strains are also highly resistant to all aminoglycosides, there is no known bactericidal regimen for these strains.

The substrate profile of the β-lactamase of strain HH22 as well as the inoculum effect on susceptibility testing is like that seen with β-lactamase-producing staphylococci (2–5). The suggested relationship was confirmed by demonstrating hybridization to a staphylococcal penicillinase gene probe. Since staphylococci and streptococci can exchange certain plasmids by conjugation and are known to share other resistance genes, movement of this enzyme from staphylococci to enterococci is not particularly surprising. What is more surprising is that it has not occurred sooner. Differences between the enterococcal and staphylococcal β-lactamases include the fact that in enterococci the enzyme is

cell bound whereas it is typically extracellularly released in staphylococci. In enterococci, the production of the enzyme is constitutive whereas in staphylococci it is generally inducible. Whether these changes were necessary before the enzyme could be expressed in enterococci is unknown.

The emergence of penicillinase in enterococci is the first demonstration of this resistance mechanism in streptococci. This enzyme is encoded on a transferable or at least mobilizable plasmid in both instances; moreover, the genes for penicillinase appear to be on different-sized fragments of different or differently arranged plasmids, emphasizing the potential for rearrangement and possible spread. Whether this new resistance trait will spread to other streptococci or remain restricted to enterococci remains to be seen.

This work was supported in part by Public Health Service grant AI 19011 from the National Institute of Allergy and Infectious Diseases and in part by grants from Eli Lilly & Co. and Merck Sharp & Dohme.

LITERATURE CITED

1. **Atkinson, B. A., and V. Lorian.** 1984. Antimicrobial agent susceptibility patterns of bacteria in hospitals from 1971 to 1982. J. Clin. Microbiol. **20**:791–796.
2. **Barrett, F. F., J. I. Casey, C. Wilcox, and M. Finland.** 1970. Bacteriophage types and antibiotic susceptibility of *Staphylococcus aureus.* Arch. Intern. Med. **25**:867–873.
3. **Basker, M. J., R. A. E. Edmondson, and R. Sutherland.** 1979. Comparative antibacterial activity of azlocillin, mezlocillin, carbenicillin and ticarcillin and relative stability to beta-lactamases. Infection **7**:67–73.
4. **Citri, N., and R. Pollock.** 1966. The biochemistry and function of B-lactamase (penicillinase). Adv. Enzymol. **28**:237–320.
5. **Farrar, W. E., and P. K. Gramling.** 1976. Antistaphylococcal activity and B-lactamase resistance of newer cephalosporins. J. Infect. Dis. **133**:691–695.
6. **Horodniceanu, T., G. Bougueleret, N. El-Sohl, B. Bieth, and F. Delbos.** 1979. High-level, plasmid-borne resistance to gentamicin in *Streptococcus faecalis* subsp. *zymogenes.* Antimicrob. Agents Chemother. **16**:686–689.
7. **McDonnell, R. W., H. M. Sweeney, and S. Cohen.** 1983. Conjugational transfer of gentamicin resistance plasmids intra- and interspecifically in *Staphylococcus aureus* and *Staphylococcus epidermidis.* Antimicrob. Agents Chemother. **23**:151–160.
8. **McLaughlin, J. R., C. L. Murray, and J. C. Rabinowitz.** 1981. Unique features of the ribosomal binding site sequence of gram-positive *Staphylococcus aureus* beta-lactamase gene. J. Biol. Chem. **256**:11283–11291.
9. **Mederski-Samoraj, B. D., and B. E. Murray.** 1983. High-level resistance to gentamicin in clinical isolates of enterococci. J. Infect. Dis. **147**:751–757.
10. **Moellering, R. C., C. Wennersten, T. Medrek, and A. N. Weinberg.** 1971. Prevalence of high level resistance to aminoglycosides in clinical isolates of enterococci, p. 335–340. Antimicrob. Agents Chemother. 1970.
10a. **Murray, B. E., D. A. Church, A. Wanger, K. Zscheck, M. E. Levison, M. J. Ingerman, E. Abrutyn, and B. Mederski-Samoraj.** 1986. Comparison of two β-lactamase-producing strains of *Streptococcus faecalis.* Antimicrob. Agents Chemother. **30**:861–864.
11. **Murray, B. E., and B. Mederski-Samoraj.** 1983. Transfer-

able beta-lactamase: a new mechanism for *in vitro* penicillin resistance in *Streptococcus faecalis*. J. Clin. Invest. **72**:1168–1171.

12. **Murray, B. E., B. Mederski-Samoraj, S. J. Foster, J. L. Burton, and P. Harford.** 1986. *In vitro* studies of plasmid-mediated penicillinase from *Streptococcus faecalis* suggest a staphylococcal origin. J. Clin. Invest. **77**:289–293.

13. **Murray, B. E., J. Tsao, and J. Panida.** 1983. Enterococci from Bangkok, Thailand, with high-level resistance to currently available aminoglycosides. Antimicrob. Agents Chemother. **23**:799–802.

14. **Toala, P., A. McDonald, C. Wilcox, and M. Finland.** 1969. Susceptibility of group D streptococcus (enterococcus) to 21 antibiotics *in vitro* with special reference to species differences. Am. J. Med. Sci. **258**:416.

15. **Verbist, L., and J. Verhaegen.** 1981. In vitro activity of *N*-formimidoyl thienamycin in comparison with cefotaxime, moxalactam, and ceftazidime. Antimicrob. Agents Chemother. **19**:402–406.

Biochemistry and Genetics of Penicillin Resistance in Pneumococci

ALEXANDER TOMASZ

The Rockefeller University, New York, New York 10021

Penicillin resistance in pneumococci involves alterations in a set of membrane-bound proteins (penicillin-binding proteins, PBPs) that are presumed to be enzymes catalyzing terminal stages of cell wall assembly. While PBP-linked antibiotic resistance has already been known among laboratory mutants for some time, this type of resistance represents a relatively new mechanism by which clinical isolates of bacteria respond to the selective pressure of antibiotic use. The species among which PBP-linked resistance has been detected now include virtually all the major human invasive pathogens (for a recent review, see reference 13a).

Studies on the mechanism of penicillin resistance among clinical isolates of pneumococci promise to provide interesting clues about a number of basic questions in procaryotic cell biology. What are the factors that contribute to the emergence and spread of antibiotic resistance genes in natural populations of a major human pathogen? Do the alterations in the PBPs of resistant pneumococci affect the kinetic properties of these enzymes? How close to the active center of the PBPs do the molecular alterations related to resistance take place? And how do the resistant cells reconcile the lower reactivity of PBPs for the substrate-analog penicillin with an efficient catalytic functioning of the same proteins with their natural substrates in cell wall biosynthesis? Studies addressing these questions should provide important insights into the mechanism of the least well understood (terminal) stages of cell wall assembly and may also open up a new approach to a better understanding of the relationship between the structure of β-lactam antibiotics and their antibacterial effectiveness.

Epidemiology of Penicillin Resistance in Pneumococci

To appreciate the potential clinical impact of the emerging penicillin-resistant pneumococcal strains, it may be useful to provide a brief historical backdrop (for a brief review, see reference 18). At the time of the introduction of penicillin into chemotherapeutic practice (in the late 1940s), clinical isolates of pneumococci were uniformly and highly susceptible to this antibiotic with minimal inhibitory concentrations (MICs) in the order of 6 to 8 ng/ml. The introduction of penicillin therapy has led to a massive decline in the mortality from pneumococcal disease (for example, from 40% to 0.1% in the case of pneumococcal lung disease). Nevertheless, the incidence of pneumococcal disease has remained virtually unchanged, and these bacteria remained major etiological agents of such diseases as otitis media (estimated to affect about 50% of all children within the first 10 years of life), lung disease, and meningitis (with a morbidity rate of 1.5 to 2.5 cases per 100,000 population per year and with a mortality rate in children as high as 20 to 30%). The main potential targets of pneumococcal disease are infants and children, i.e., the very population for which the recently introduced polyvalent pneumococcal vaccine seems to be relatively ineffective. Antibiotic therapy, primarily with penicillin and other β-lactam antibiotics, remains the intervention of choice.

It seems that the first penicillin-resistant clinical isolate (with an 80- to 100-fold increase in MIC) was described in 1967 (6), i.e., about 18 to 20 years after the introduction of penicillin into therapeutic practice. The capacity of such strains to cause serious human disease, complications, and failure in chemotherapy was strikingly demonstrated by the South African epidemics in Soweto, Durban, and some other hospitals in 1977 (8). The MICs for the most highly resistant South African strains were 6 to 12 μg/ml for benzylpenicillin (to be compared with an MIC of 6 to 8 ng/ml for the more typical susceptible isolates). Several of these strains not only were resistant but were also tolerant to penicillin (10), and they carried traits of resistance to a number of other antibiotics also. Sporadic reports on the detection of moderately and highly penicillin-resistant pneumococcal isolates at a wide variety of geographical locales have continued to appear in the clinical literature, including a report on such strains in New York in 1985 (MIC of penicillin, 0.8 to 1 μg/ml) (12). An important recent survey by Klugman and his colleagues documented the ability of penicillin-resistant pneumococcal strains to become a quantitatively significant fraction of the nasopharyngeal flora of healthy children in some day-care centers in rural South Africa (9). The

TABLE 1. Susceptibility of *Streptococcus pneumoniae* isolates to penicillin[a]

Year	Location	No. of isolates	% of isolates requiring MIC (μg/ml) of:				
			>1.0	0.1–1	0.06	0.03	<0.015
1953–1955	Boston, Mass.	114	0	0	0	12	88
1961–1964	United States	50	0	0	0	30	70
1976–1977	United States	50	0	0	0	87	11
1977	Oklahoma	103	0	16	5	79	ND[b]
1977	Switzerland	100	0	3	2	5	90
1980	Federal Republic of Germany	206	0	7	25	39	29
1981	Spain	318	1	8	ND	91	ND
1981	Spain	200	2	9	5	12	73
1981	Colorado	101	0	7	1	92	ND
1981	Israel	229	7	21	72	0	0
1984–1985	Soweto, South Africa	124	0	35	10	43	12

[a] Reproduced with permission from reference 4.
[b] ND, Not done.

available evidence indicates that the penicillin resistance trait can be harbored in pneumococci of several capsule types as a stable, chromosomal trait fully compatible with normal growth rates, survival in natural environments, and virulence.

An examination of available data on the penicillin susceptibilities of pneumococcal strains that were isolated in the 1950s, 1960s, and more recently in the 1980s suggests that a significant fraction of strains in natural populations may have already acquired low-level penicillin resistance traits (4). A compilation of such data is reproduced in Table 1. An upward shift in the more recent isolates from the MIC range of <0.015 μg/ml is apparent in each of the populations examined. Strains with intermediate-level resistance (>0.1 μg/ml) are relatively rare (with the exception of the amazingly high incidence of such strains in the South African survey), and none of the resistance levels observed would require change in the currently used antibiotic regimen. Nevertheless, the isolates with the somewhat elevated MICs clearly represent "changed" bacteria: DNA isolated from such strains can transform susceptible pneumococci to the penicillin resistance level of the DNA donor strains, and the antibiotic-binding capacity of PBPs in such low-level resistant strains is clearly decreased relative to the binding capacity of sensitive strains isolated within the same geographical area (4).

A likely scenario for the emergence of penicillin resistance among natural isolates of pneumococci involves the selective pressure of clinical or prophylactic use of penicillin. Recent experiments have clearly shown that sustained exposure of penicillin-susceptible pneumococci to gradually increasing levels of penicillin in the laboratory environment results in the selection of stable resistant mutants exhibiting stepwise increases in MIC and a parallel decrease in the

penicillin-binding capacity of PBPs (4). In this respect it is probably highly significant that the highest incidence of resistant strains appears to be in areas of extensive prophylactic use of penicillin, such as Papua, New Guinea (1), some Indian reservations in the United States (13), and some areas of South Africa (8). It is not clear where the ecological source of these strains may be. The high degree of colonization of infants and young children by pneumococci (7, 11) suggests that the original resistant mutants may have emerged within such carrier floras.

Mechanism of Penicillin Resistance in Pneumococci

The currently used fluorographic technique detects five PBPs in typical penicillin-susceptible pneumococci, such as the Rockefeller University laboratory strain R6 and a number of wild-type isolates from geographically diverse locations and of different capsular types. The molecular sizes of the PBPs (in kilodaltons [kDa]) are estimated as 98 (PBP 1A), 96 (PBP 1B), 80 (PBP 2A), 78 (PBP 2B), and 52 (PBP 3). The relative "reactivity" of these proteins with benzyl penicillin follows the order of PBP 3 > 1A > 2A > 1B > 2B, and the approximate numbers of these membrane-bound proteins per cell are about 5,000 (PBP 1A), 1,000 (PBP 1B), 3,000 (PBP 2A), 3,500 (PBP 2B), and 5,000 (PBP 3) (15, 16). In contrast to the penicillin-susceptible strains, the two highly resistant South African strains (8249 and D20) contain a radically different pattern of PBPs: the resistant bacteria contain only three PBPs. Two of these, PBPs 2A and 3, appear to have the same molecular sizes as the corresponding PBPs in the susceptible strains, but the third, PBP 1C, is a novel species with a molecular size of 92 kDa and with an extremely low reactivity to penicillin (17). In addition to these differences, PBP 2A of the resistant bacteria has greatly diminished reactivity to penicillin compared with that of the same PBP from the susceptible cells. No differences between the PBPs 3 of the susceptible and resistant cells have been detected so far. Genetic crosses (using transforming DNA from strain 8249 or D20 and the susceptible R6 strain as recipient) have established that the strikingly different PBP profile of the resistant strains is indeed related to penicillin-resistant phenotype since transformants selected for high resistance also showed the PBP pattern characteristic of the DNA donor cells. Thus, the interpretation of the PBP pattern of resistant strains as altered or remodeled forms of the proteins found in the susceptible strain appears to be a valid one. Since four of the five PBPs resolvable by the fluorographic method undergo alterations in parallel with acquisition of resistance, we concluded that these four proteins perform physiologically essential functions (17). The construction of isogenic transformant strains differing only in levels of penicillin resistance allowed a detailed comparison of the antibiotic-binding capacities of PBPs of these strains. The basic finding of these studies, to be summarized briefly next, was that the penicillin resistance mutations cause a remodeling of PBPs in a way that changes their kinetic properties.

The kinetics of interaction between several penicillin-sensitive bacterial transpeptidase/carboxypeptidase enzymes and β-lactam antibiotics or model peptide substrates has been formulated by the sequence of reactions (2) shown in Fig. 1A, in which Enzyme-Penicillin represents the (reversible) inhibitor-enzyme complex which is converted to the covalent complex of the acylated (penicilloyl) enzyme. It is this latter product that can be detected by the penicillin-binding technique as a PBP. Finally, the penicilloyl-enzyme complex decomposes to degraded antibiotic, and active enzyme is regenerated. A similar sequence of reactions (Fig. 1B) can describe the interaction of the same (penicillin-sensitive) enzymes with their natural substrates in cell wall synthesis.

In the live bacterial cell encountering penicillin in the environment, reactions A and B (Fig. 1) compete with each other, and the antibacterial effects of penicillin start to manifest themselves only as the antibiotic has managed to "trap" a critical fraction of the enzyme in the inactive (PBP) form (2).

These reaction sequences offer, in principle, several alternative strategies for the bacterial cell as mechanisms for decreased penicillin susceptibility: (i) alteration (increase) may occur in the cellular concentration of natural substrate, which would then act as a more effective endogenous competitor against the antibiotic molecules; (ii) the number of enzyme molecules (rates of biosynthesis) may be increased to compensate for the fraction of the protein entrapped in the inactive antibiotic complex; (iii) the affinity of the enzyme for penicillin (k_2/k_1) may decrease; (iv) the rate of acylation (k_3) may decrease; or (v) the rate of deacylation (k_4) may increase. In addition, penicillin resistance may involve (vi) the acquisition of a β-lactamase or (vii) a barrier function (making the access of PBPs more difficult). Analyses of the penicillin-resistant South African strains have excluded mechanisms vi and vii as the basis of resistance. Also, at least in the two resistant strains (8249 and D20) and their genetic transformant derivatives examined in detail, mechanism i cannot be the major factor affecting resistance since the radically changed PBP profiles of those bacteria can also be demonstrated in vitro, in membranes prepared from the cells. More detailed kinetic studies with the PBPs of susceptible and resistant strains have also excluded mechanism v (5). It is interesting to note that mechanism v (i.e., a greatly increased deacylation rate, k_4) would not only provide a mechanism for the faster recovery of an essential enzyme but could also, in effect, achieve a net inactivation of the antibiotic molecule. Such a β-lactamase-like mechanism has apparently not been "invented" by pneumococci, while in staphylococci, enzymatic drug destruction was the first natural penicillin resist-

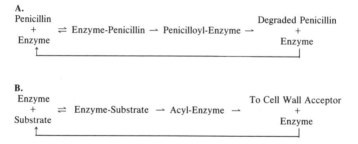

FIG. 1. Kinetics of interaction of cell wall synthetic enzymes with penicillin (A) and with their natural substrates (B).

ance mechanism to emerge, greatly limiting the chemotherapeutic usefulness of β-lactamase-sensitive penicillin by the mid-1950s. The next round of β-lactam antibiotic resistance (methicillin resistance) which also involved PBP alteration first appeared in staphylococci in the early 1960s, just a few years before the first penicillin-resistant pneumococcus was described.

A careful quantitative evaluation of the kinetics of interaction between the PBPs and radioactive penicillin has revealed that the basis of penicillin resistance in the South African pneumococcal strain 8249 involves mechanism iii or iv or both. PBPs 1A, 2A, and 2B (and probably 1B as well) were shown to undergo structural changes resulting in gradual decrease in affinity or acylation rate or both toward the antibiotic molecule (5). Figure 2 shows the results of such PBP titrations for the susceptible strain R6 and the resistant strain 8249. Recent collaborative studies with Hakenbeck's group at the Max Planck Institute for Genetics, Berlin, have shown that the "disappearance" of PBP 2B from the fluorograms of highly resistant bacteria is actually a reflection of greatly decreased reactivity with penicillin such that this protein is no longer detectable by the binding technique. Nevertheless, PBP 2B is present in the resistant bacteria, as evidenced by the specific antibody labeling technique developed by Hakenbeck's group (3). The introduction of an antibody blotting technique has also helped to clarify another puzzling aspect of PBP alterations in pneumococci: the disappearance of PBP 1A from the fluorograms of highly resistant cells and the simultaneous appearance of a new PBP (PBP 1C) of extremely low antibiotic-binding capacity that is not detectable in the sensitive bacteria

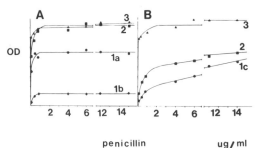

FIG. 2. Quantitative evaluation of the penicillin-binding capacities of the PBPs of penicillin-susceptible (part A) and penicillin-resistant (part B) pneumococci. Membranes prepared from susceptible (strain R6) and resistant (strain 8249) pneumococci were exposed to different concentrations (micrograms per milliliter) of radioactive penicillin for 10 min; fluorograms were prepared, and the amounts of penicillin bound to PBPs were quantitated by densitometry. OD, Optical density. (Reproduced with permission from reference 5.)

and that has a somewhat smaller molecular size than PBP 1A. Development of the sodium dodecyl sulfate-polyacrylamide gels of membrane preparations from resistant cells with antibody prepared against purified PBP 1A detected a single strongly reactive band at the position of PBP 1C. No serologically reactive material was detectable at the position expected for PBP 1A. In membrane gels from sensitive bacteria the anti-PBP 1A antiserum has detected a single reactive band at the exact position of PBP 1A. These results indicate that PBP 1A is not present in the membranes of resistant cells; instead, they contain another binding protein that is immunologically related to PBP 1A. Earlier studies with the penicillin-binding method showed that PBP 1A and 1C were never present in the same cell and that the apparent "replacement" of PBP 1A with 1C occurs at a sharply defined level of penicillin resistance (17). The immunological relatedness of PBP 1C and PBP 1A is consistent with (but does not prove) the proposal that PBP 1C may be a truncated form of PBP 1A. Hakenbeck and her colleagues have also tested with their anti-PBP 1A antiserum membranes prepared from some additional resistant pneumococcal strains, one also isolated in South Africa (A95211) and another from New Guinea (P2955). Neither of these strains contains PBP 1A (98 kDa) detectable by the penicillin-binding method. Instead, strain A95211 contains a new protein of about 96 kDa while strain P2955 contains a somewhat smaller new protein band (about 94 kDa). The immunoblotting technique has fully confirmed these observations. Just as in the case of PBP 1C of strain 8249, both the 96- and 94-kDa proteins gave strong immunological reaction with the antiserum against PBP 1A; no reactive band was detectable at the position expected for PBP 1A (3). The 96- and 94-kDa proteins may also represent truncated versions of PBP 1A. Alternatively, amino acid replacements may have resulted in a changed net charge and sodium dodecyl sulfate binding capacity of the proteins, which in turn may have caused an apparent change in molecular size.

The picture that emerges from these studies is that of extensive structural alterations in several of the cell wall synthetic enzymes such that the kinetic properties of the proteins are changed. Recent peptide mapping studies performed in collaboration with Hakenbeck's group indicate that at least some of these structural alterations (amino acid replacements) occur in the vicinity of the β-lactam-binding site (i.e., enzymatic active center) of the PBP molecules (17). These amino acid replacements are presumably the consequence of point mutations in the structural genes for PBPs. Genetic studies suggest that the genes of both PBP groups 1 and 2 contain multiple mutations, each set of which may correspond to a distinct level of penicillin resistance (i.e., MIC). In the case of the PBP 1A gene the accumulation of a certain number of the amino acid replacements may produce some sort of processing signal (proteolytic cutting site) leading to the truncation of the 98-kDa PBP 1A molecules to the 92-kDa PBP 1C.

Genetic Transformation of Penicillin Resistance

Genetic transformation of penicillin resistance has produced three types of puzzling observations (14, 17).

(i) The high degree of penicillin resistance characteristic of the South African strains 8249 or D20 could be built up in sensitive recipient cells only by multiple (six to eight) "rounds" of transformation, in which low and, subsequently, intermediate level transformants had to be used in succession as recipients. This suggests the involvement of multiple genetic determinants. The finding is understandable if one accepts the notion that pneumococci contain at least four distinct penicillin target proteins.

(ii) A second puzzling finding of the transformation studies was that the absolute frequency of transformation declined rapidly with increasing levels of resistance in spite of the fact that the recipients used in these crosses were transformants from the previous round of DNA exposure. This finding was clearly related to some feature of the penicillin resistance markers, since there was no decline in the general competence of the cells to transform other genetic markers. A plausible explanation of the observation would be as follows. Since susceptible pneumococci contain four essential PBPs with different relative affinities for the antibiotic (in the decreasing order of PBPs 1A > 2A > 1B > 2B), the acquisition of the first (low) incremental resistance level may involve "tuning down" the antibiotic affinity of a single protein (PBP 1A). The relative penicillin affinity of the altered PBP 1A may now be comparable to that of PBP 2A, i.e., the protein that is next in reactivity toward penicillin to that of the original PBP 1A. Thus, acquisition of the next level of resistance would then require change in two genetic elements (i.e., a further alteration in the PBP 1A gene plus change in the gene for PBP 2A as well). The third and higher levels of resistance may involve parallel changes in three or more determinants, thus causing a decline in transformation frequencies. This type of mechanism could also explain, at least in part, the observed directionality of PBP alterations.

(iii) A third surprising observation of genetic transformation studies was the apparent directionality or orderliness of PBP changes. Transformants to a distinct resistance level appeared always to contain the same unique set of PBPs, giving the impression of gradual, orderly change from the PBP pattern of susceptible bacteria towards the PBP pattern of the DNA donor (resistant) cells. The reason for such an orderly process is not clear. It is not clear, for instance, why transformants of low or intermediate resistance level never express the most extensively altered form of a PBP, in spite of the fact that the genetic determinants corresponding to the most extensively altered forms of PBPs are clearly present in the DNA preparation used to produce the transformants with low and intermediate resistance levels. The exclusion of most possible combinations of PBP alterations from the patterns observed in transformants of low or intermediate resistance suggests that only a limited number of PBP alterations produce viable penicillin-resistant transformants. This could be expected if PBPs were to perform their catalytic functions in a cooperative manner. If one assumes that the penicillin affinity of a PBP is also a reflection of its affinity for the natural substrate, then one may expect that PBPs in the pneumococci with various resistance levels have a corresponding spectrum of somewhat lower catalytic efficiencies in cell wall synthesis. If all PBPs function as a single cooperative unit or as a kind of "assembly line," then it is possible that the substrate affinities of individual PBPs must also be "tuned" relative to one another for normal wall synthesis to continue. For instance, a molecule of PBP 1A with the maximum degree of affinity decrease would not be able to function normally in the company of the other PBPs if the latter had been changed to only a limited degree, required for a low level of penicillin resistance. The topographic complexity of wall assembly would make the cooperative functioning of PBPs as parts of a single organelle an attractive possibility, and the results of the genetic crosses discussed above certainly support such a model.

Perhaps the most intriguing aspect of PBP-linked β-lactam antibiotic resistance is the fact that such mutants exist at all. If the β-lactam ring is indeed an active site-directed acylating agent with structural analogy to the D-alanyl-D-alanine carboxy terminus of cell wall building blocks, then decreased affinity for substrate analog (as seen in resistant mutants) should also lower catalytic efficiency in the physiologically important reactions catalyzed by these proteins. Thus, mutants of this type could be expected to have abnormalities in wall assembly and growth. Yet, this clearly need not be the case, as shown by the emergence of PBP-linked resistant mutants in several species of bacteria. It is conceivable that such mutants contain additional mutations to compensate for the reduced catalytic efficiency of PBPs. Alternatively, PBPs may be remodeled in such a way that a distinction between substrate and substrate analog becomes possible. This would explain the very substantial differences observable for resistant pneumococci in the degrees of increase in MICs of structurally different β-lactam antibiotics (14). The surprising genetic plasticity of pneumococcal PBPs is also suggested by data indicating that in three different resistant isolates, three distinct amino acid replacements may be detected in the vicinity of the active center of PBP 1 (3). Elucidation of the molecular details of remodeled PBP structure in penicillin-resistant pneumococci should also provide interesting insights into the structural basis of antimicrobial effectiveness and specificity in the β-lactam family of antibiotics.

LITERATURE CITED

1. Devitt, L., I. Riley, and D. Hansman. 1977. Human infection caused by penicillin-insensitive pneumococci. Med. J. Aust. 1:586–588.
2. Ghuysen, J. M. 1976. The bacterial D-D-carboxypeptidase-transpeptidase enzyme system: a new insight into the mode of action of penicillin, p. 1–164. In W. E. Brown (ed.), E.R. Squibb lectures on chemistry and microbial products. University of Tokyo Press, Tokyo.
3. Hakenbeck, R., H. Ellerbrok, T. Briese, S. Handwerger, and A. Tomasz. 1986. Penicillin-binding proteins of penicillin-susceptible and -resistant pneumococci: immunological relatedness of altered proteins and changes in peptides carrying the β-lactam binding site. Antimicrob. Agents Chemother. 30:553–558.
4. Handwerger, S., and A. Tomasz. 1986. Alterations in penicillin binding proteins of clinical and laboratory isolates of pathogenic pneumococci with low levels of penicillin resistance. J. Infect. Dis. 153:83–89.
5. Handwerger, S., and A. Tomasz. 1986. Alterations in kinetic properties of penicillin-binding proteins of penicillin-resistant Streptococcus pneumoniae. Antimicrob. Agents Chemother. 30:57–63.
6. Hansman, D., and M. M. Bullen. 1967. A resistant pneumococcus (letter). Lancet ii:264–265.
7. Hendley, J. O., M. A. Sande, P. M. Stewart, and J. M. Gwaltney, Jr. 1975. Spread of Streptococcus pneumoniae in families. I. Carriage rates and distribution of types. J. Infect. Dis. 132:55–61.
8. Jacobs, M. R., H. J. Koornhof, R. M. Robins-Browne, C. M. Stevenson, Z. A. Vermaak, I. Freiman, G. B. Miller, M. A. Witcomb, M. Isaacson, J. I. Ward, and R. Austrian. 1978. Emergence of multiply resistant pneumococci. N. Engl. J. Med. 299:735–740.
9. Klugman, K. P., H. J. Koornhof, A. Wasas, K. Storey, and I. Gilbertson. 1986. Carriage of penicillin resistant pneumococci. Arch. Dis. Child. 61:377–381.
10. Liu, H., and A. Tomasz. 1985. Tolerance to penicillin in multiply drug-resistant natural isolates of Streptococcus pneumoniae. J. Infect. Dis. 152:365–372.
11. Masters, P. L., W. Brumfitt, and R. L. Mendez. 1958. Bacterial flora of the upper respiratory tract in Paddington families. Br. Med. J. 1:1200–1205.
12. Simberkoff, M. S., M. Lukaszewski, A. Cross, M. Al-Ibrahim, A. L. Baltch, R. P. Smith, P. J. Geiseler, J. Nadler, and A. S. Richmond. 1986. Antibiotic-resistant isolates of Streptococcus pneumoniae from clinical specimens: a cluster of serotype 19A organisms in Brooklyn, New York. J. Infect. Dis. 153:78–82.
13. Tempest, B., J. P. Carney, and B. Eberle. 1974. Distribution of sensitivities to penicillin of types of Diplococcus pneumoniae in an American Indian population. J. Infect. Dis. 130:67–69.
13a. Tomasz, A. 1986. Penicillin-binding proteins and the antibacterial effectiveness of beta-lactam antibiotics. Rev. Infect. Dis. 8(Suppl. 3):S260–S278.
14. Tomasz, A., S. Zighelboim-Daum, S. Handwerger, H. Liu, and H. Qian. 1984. Physiology and genetics of intrinsic beta-lactam resistance in pneumococci, p. 343–347. In L. Leive and D. Schlessinger (ed.), Microbiology—1984. American Society for Microbiology, Washington, D.C.
15. Williamson, R., R. Hakenbeck, and A. Tomasz. 1980. The in vivo interaction of beta-lactam antibiotics with the penicillin-binding proteins of Streptococcus pneumoniae. Antimicrob. Agents Chemother. 18:629–637.
16. Williamson, R., and A. Tomasz. 1985. Inhibition of cell wall synthesis and acylation of the penicillin binding proteins during prolonged exposure of growing S. pneumoniae to benzylpenicillin. Eur. J. Biochem. 151:475–483.
17. Zighelboim, S., and A. Tomasz. 1980. Penicillin-binding proteins of the multiply antibiotic-resistant South African strains of pneumococci. Antimicrob. Agents Chemother. 17:434–442.
18. Zighelboim, S., and A. Tomasz. 1981. Multiple antibiotic resistance in South African strains of Streptococcus pneumoniae: mechanism of resistance to beta lactam antibiotics. Rev. Infect. Dis. 3:267–276.

III. Pathogenic Streptococci

A. Surface Components

Structural and Genetic Relationships of the Family of M-Protein Molecules of Group A Streptococci

JUNE R. SCOTT,[1] SUSAN K. HOLLINGSHEAD,[1] KEVIN F. JONES,[2] AND VINCENT A. FISCHETTI[2]

Department of Microbiology and Immunology, Emory University School of Medicine, Atlanta, Georgia 30322[1] and The Rockefeller University, New York, New York 10021[2]

The M protein of the group A streptococci appears to play a critical role in determining the virulence of these organisms by preventing their phagocytosis in the infected host (12). This protein is found on the streptococcal surface as a dimeric molecule composed of two alpha-helical chains wound around each other to form a coiled-coil fibrillar structure (16). The molecule is attached to the streptococcal cell through an apparent membrane anchor and wall stabilization region located at the carboxy-terminal end of the protein (9). The amino terminus is about 600 Å (60 nm) distal to the cell surface (Fig. 1) (16).

During infection by a group A streptococcus, the host mounts an immune response and generates type-specific antibodies to the M protein (11). Some of these anti-M antibodies are opsonic and protect the host against the invading streptococcus by facilitating phagocytosis. However, because there are over 70 different serotypes of M protein thus far identified and protection is predominantly type specific, an individual may be infected by more than one group A streptococcal serotype. In addition to the antigenic diversity of the M molecules, we have found that the proteins of strains of different serotypes may differ in apparent molecular weight on sodium dodecyl sulfate-polyacrylamide gel electrophoresis by as much as 40,000 (7). Even within one M serotype, the molecular weight range may be almost this great. Evolutionarily, for molecules having the same function in different organisms, there is usually strong conservation of size and even of amino acid sequence. In spite of the serological and size diversity among M molecules from different streptococcal strains, both the coiled-coil conformation and the antiphagocytic function appear to be conserved (7). Therefore, it seems likely that the specific amino acid sequence and molecular size are less important for function of this molecule than is its conformation.

In this regard, it has been found that human opsonic immunoglobulin G to the M6 protein is basic in charge, suggesting that it is directed to acidic determinants on the M protein (5). The PepM5 protein sequence shows a concentration of acidic amino acids within the N-terminal 25% of the molecule (14). Furthermore, several opsonic monoclonal antibodies to the M5 protein recognize epitopes in the N-terminal half (3). We suggest, therefore, that the N-terminal region is of primary functional importance for the molecule and that the coiled-coil conformation may be the vehicle by which this functional region is presented to the invading phagocyte.

emm6 Gene Cloned into *Escherichia coli*

To learn what part of the M-protein structure is required for its function and to learn about the evolutionary relationship of the different M-protein molecules, we cloned the gene for type 6 M protein (*emm6*) from group A streptococcal strain D471 into *E. coli* (18). As expected for a protein that is transported through the membrane to the surface of the streptococcus, the M protein is found in the periplasm in *E. coli* (6). Ouchterlony double immunodiffusion and reaction with monoclonal anti-M6 antibodies (18) indicated that the molecule (Coli M6) produced in the *E. coli* strain containing *emm6* is antigenically identical to the streptococcal M6 molecule. In addition, the Coli M molecule seems to be the same as the streptococcal M6 functionally, since it absorbs opsonic antistreptococcal M6 antibody and since antibody to Coli M6 is opsonic for M6 streptococci (6). In addition, we recently succeeded in transferring the cloned gene to an M⁻ streptococcal strain; this converted the strain to a functional M⁺ organism (see below) (18a).

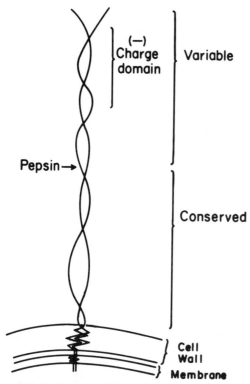

FIG. 1. Concept of the coiled-coil (16) M-protein molecule and its association with the streptococcal cell wall. A negatively charged domain identified within the N-terminal region (14) may play a role in biological activity. By use of monoclonal antibodies (10) and DNA probes (19), variable and conserved regions of the M molecule have been located (see also Fig. 2). DNA sequence analysis has identified a membrane anchor domain and a proline-rich cell wall stabilization segment within the C-terminal region of the protein (9).

Homology Among M Genes in Streptococci

To determine whether there is a region of the M6 protein that is common to other M molecules, we used DNA hybridization (20). Our labeled probe was internal to the *emm6* gene and contained all of the DNA encoding the mature protein except the 5' 32 bases and the 3' 35 bases. DNA of representative strains from 56 different M types and 4 nontypable strains hybridized with this *emm6* probe, although no DNA from unrelated gram-positive organisms hybridized. Of the other Lancefield groups of streptococci tested, only DNA from strains of groups C and G showed homology. This homology was not unexpected since some strains of these streptococcal groups have been reported to have M-like molecules on their surfaces (2, 15, 23, 24). Of the M⁻ group A strains tested, DNA from one (T28/51/4) did not hybridize with

emm6. Because this was derived from the same clinical isolate as a T28 strain that did hybridize with *emm6*, we concluded that T28/51/4 has a deletion of the M-protein gene (20). A derivative of this strain was used as the recipient for the *emm6* gene cloned in *E. coli* (see below).

The DNA sequence of *emm6* (see below) indicates the presence at the carboxy terminus of a membrane anchor region, a region involved in attaching or stabilizing the M protein in the streptococcal cell wall, and three regions of repetitive sequences (9) (Fig. 2). To localize the regions of homology with *emm6*, we subdivided our probe. For these experiments, we utilized the M13 clones constructed for DNA sequence analysis (Fig. 2). DNA hybridization with strains from representative serotypes showed that there is greater conservation within the carboxy-terminal region of the *emm* gene than the amino-terminal region (19). In agreement with this, monoclonal antibodies that bind near the middle of the M6 protein (10B6 and 10F5) each recognize epitopes on 30 of 58 different types of M proteins, while one monoclonal antibody that binds to a more amino-terminal region (10A11) recognizes epitopes on M proteins of fewer types (9 of 58) (10). Because the carboxy-terminal region is responsible for attachment of the M protein to the streptococcus, it is not surprising that it would be conserved. Also, because the amino-terminal region is more exposed to the immune system of the host (Fig. 1), it should be subject to the greatest selective pressure for change.

Only One Copy of *emm6* in Streptococci

Because antibody to M protein protects the host against infection, there is strong selective pressure for antigenic diversity of this protein. This appears similar to the antigenic variation seen for the surface proteins of some other pathogens: the pili of *Neisseria gonorrhoeae* (M. So, *in* I. Inouye, ed., *Bacterial Outer Membranes as Model Systems*, in press), the surface protein of *Borrelia hermsii* (17), and the surface glycoprotein of the trypanosomes (1). In these organisms, the antigenic variation results from recombination of the expressed gene copy with regions elsewhere in the genome of the pathogen. Using the *emm6* DNA probe described above, we found only a single region of homology in the genome of the D471 streptococcal strain from which this gene was cloned (20). Thus, diversity of M proteins must arise by a different mechanism.

Sequence of *emm6*

The DNA sequence of *emm6* was obtained by the Sanger dideoxy technique (9). The carboxy-terminal region of the M protein is typical of a

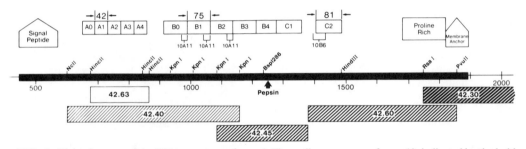

FIG. 2. Major features of the DNA sequence of *emm6*. The coding sequence of *emm6* is indicated by the bold black box. Relevant restriction enzyme sites are indicated above this line, and the scale below it is in kilobases. The predominant site at which pepsin cleaves the M6 protein is indicated by the bold arrow. The boxes in the top row indicate the locations of the signal peptide, proline-rich region, and membrane anchor as deduced from the DNA sequence (9). The three regions of reiterated DNA sequence are delimited as boxes A, B, and C, and the number of base pairs in each is indicated above a box in each region. The locations of the epitopes to which monoclonal antibodies 10A11 and 10B6 bind (10) are shown below the repeat boxes. Below the coding sequence, the boxes indicate the probes used in DNA hybridization experiments, and the degree of shading in these boxes represents the proportion of tested group A streptococcal strains that show homology with each probe (19). The darkest shading (42.30) indicates homology with all strains tested, and the unshaded box (42.63) indicates no homology with any strains except the one from which the probe was derived.

membrane anchor sequence (Fig. 2). Immediately adjacent, there is a region composed of 22% prolines and 11% glycines that are fairly regularly spaced. Because these amino acids cause bends and turns in the protein chain, this region of the molecule may intercalate in the cross-linked portion of the peptidoglycan to stabilize the M protein in the streptococcal cell wall. The membrane anchor and proline-rich regions are highly homologous to the carboxy terminus of protein A, a surface-associated protein of staphylococci. It has also been suggested that these regions are responsible for surface attachment of protein A in staphylococci (8).

An unusual feature of *emm6* is the presence of three extensive reiterated sequences (9) (Fig. 2). Region 1 is composed of five repeats of sequence A (42 bases); region 2, of five B repeats (75 bases); and region 3, of two C repeats of about 81 bases. The repeats of regions 1 and 2 are tandem and direct, and those of 3 are direct but separated by 45 bases. We have suggested that the size heterogeneity of M proteins might result from homologous recombination within these repeated sequences (9). Such recombination can generate deletions or duplications of these regions, and for region 3, recombination would also cause a deletion of the intervening nonreiterated sequence.

Recently, we succeeded in isolating spontaneous deletion mutants in *emm6* from strain D471 (5a). These occur at a frequency of about 3 to 5 in 10^4 CFU. Presumably, such mutants have not been observed previously in the laboratory because, to be detectable in Western blots, they must constitute at least 5% of the CFU in the streptococcal culture tested. Such size variants also appear to occur in nature since M proteins

of a single M type from different strains isolated either serially from one individual or from an epidemic show size differences (8; S. K. Hollingshead, V. A. Fischetti, and J. R. Scott, in press). Recent sequence analysis of mutants with shorter M proteins isolated from D471 indicates that our hypothesis about their mechanism of generation is correct: their sequence is consistent with their derivation from D471 by homologous recombination between intragenic direct repeats (Hollingshead et al., in press).

Although sequence differences have been observed among strains of the same M serotype (21; Hollingshead et al., in press), there is no information regarding the rate at which antigenic diversity arises in the M molecule. We believe that, like extragenic recombination in the genes for the gonococcal pilus subunit, the borrelia surface protein, and the trypanosome surface glycoprotein, intragenic recombination in the streptococcal M-protein gene may play a critical role in increasing the rate of generation of antigenic diversity. When new determinants arise by mutation or by recombination between imprecise repeats, they can be amplified by homologous recombination. In addition, old immunological determinants may be deleted by intragenic homologous recombination. Thus, new determinants may become more immunologically important in the molecule because they come to constitute a high percentage of the total sequence.

An M⁻ Streptococcal Strain Becomes Functionally M⁺ When *emm6* is Transferred to It

The functional integrity of the cloned *emm6* gene and the feasibility of our molecular approach to the study of the M protein were further demonstrated by transfer of the *emm6*-

containing plasmid to JRS1, a derivative of the T28/51/4 M-deletion group A streptococcal strain identified in the hybridization experiments (see above; 18a). Because there is no direct system for transformation of group A streptococci, an indirect method had to be used. We constructed a cointegrate shuttle vector composed of the *emm6* gene in the pUC9 plasmid (pJRS42.50: 9) and pVA838 (13), an *E. coli-Streptococcus* shuttle constructed by F. Macrina's group. The ampicillin resistance determinant was removed from the pUC plasmid portion and the cointegrate was designated pJRS50. This plasmid was used to transform *S. sanguis* Challis. The *S. sanguis* strain containing pJRS50 produces M protein, but this is largely secreted into the medium instead of remaining attached to the cell surface. The conjugative plasmid pVA797 (4), which is partially homologous with pVA838, was then transferred to this *S. sanguis* transformant. From results obtained by Smith and Clewell (22) using *S. sanguis* and *S. faecalis*, we anticipated that the conjugative plasmid would mobilize pJRS50, probably by forming a cointegrate at the region of homology. The resulting *S. sanguis* strain was used as a donor in mating with *S. pyogenes* JRS1, and a marker on pJRS50 was selected. The *S. pyogenes* transconjugants contain pJRS50 unlinked to any other DNA; thus, if it had been transferred by forming a cointegrate with the homologous conjugative plasmid pVA797, this cointegrate resolved in the recipient cell.

When the *S. pyogenes* transconjugants were subjected to rotation in human blood to assay for their resistance to phagocytosis, some colonies survived (18a). In the control, no cells of the recipient strain JRS1 survived the blood rotation. A transconjugant colony that survived was subject to three more rotation cycles, and the resulting strain was designated JRS2. JRS2 is resistant to phagocytosis and is opsonized by anti-M6, but not by anti-M28 antibodies. Thus, JRS2 resists phagocytosis because of the presence of type 6 M protein.

Immunofluorescence microscopy using antibody to M6 (Fig. 3) and enzyme-linked immunosorbent assay demonstrated that M6 is present on the surface of these bacteria (18a). The immunofluorescent pictures indicate that only about two-thirds of the cells in the JRS2 population express M6 on their surface. This correlates with the fact that only about two-thirds of the cells contain the plasmid (as determined by replica plating onto erythromycin). We were unable to increase the proportion of plasmid-containing cells even by continuous selection in 50 µg of erythromycin per ml. This may be because this plasmid lacks a partition determinant or, since its replicon is derived from a

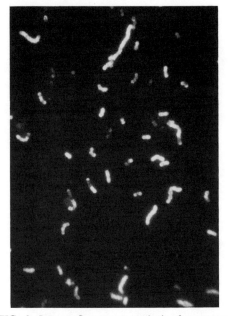

FIG. 3. Immunoflorescence analysis of streptococcal strain JRS2 with florescein-labeled anti-M6 antibodies. A sample of an overnight culture of strain JRS2, an *emm* deletion strain carrying a shuttle plasmid with *emm6*, was dried on a slide and allowed to react with anti-M6 antibodies tagged with florescein isothiocyanate. After being washed, the cells were observed with an oil immersion lens under a florescence microscope.

plasmid isolated from *S. ferus*, because it cannot replicate at the rate required to be stably maintained in *S. pyogenes*.

In addition to surviving phagocytosis, JRS2 absorbs opsonic antibodies from antiserum generated by immunization either with purified M6 protein or with a synthetic N-terminal peptide derived from M6. Furthermore, rabbits immunized with JRS2 produce antibodies that are able to opsonize M type 6 streptococci. Therefore, the plasmid containing *emm6* appears to determine all the traits associated with M protein.

Quantitative immunoblot assays indicated that JRS2 has about 0.1% as much M6 protein on its surface as does the D471 strain from which *emm6* was cloned (18a). There are two obvious explanations for this: either the *emm6* gene is not transcribed efficiently because other promoters 5' to the streptococcal DNA in pJRS50 occlude the natural *emm6* promoter, or the *emm6* promoter is regulated by *trans*-acting factors in the streptococcus which are deficient in strain JRS1. We favor the latter explanation and are currently testing this.

The ability to transfer the cloned M gene into a group A streptococcus deleted for the M protein determinant will allow us to address the issues of the relation of the structure of the M protein to its function. We are currently altering the *emm6* gene in vitro to test our ideas about the relative importance of different regions of the molecule. We can then transfer the mutated *emm6* gene into group A streptococci and determine whether it is still capable of attaching properly to the streptococcal cell surface and conferring resistance to phagocytosis on the streptococcus.

We thank W. Pulliam, P. Guenthner, J. Rhatigan, L. Malone, and M. Jarymowicz for expert technical assistance.

This work was supported by Public Health Service grants AI20723 to J.R.S. and V.A.F. and AI11822 to V.A.F. from the National Institute of Allergy and Infectious Diseases.

LITERATURE CITED

1. **Borst, P.** 1983. Antigenic variation in trypanosomes, p. 621–659. *In* J. A. Shapiro (ed.), Mobile genetic elements. Academic Press, Inc., New York.
2. **Bryans, J. T., and B. O. Moore.** 1972. Group C streptococcal infections of the horse, p. 327–338. *In* L. W. Wannamaker and J. M. Matsen (ed.), Streptococci and streptococcal diseases. Academic Press, Inc., New York.
3. **Dale, J. B., and E. H. Beachey.** 1984. Unique and common protective epitopes among different serotypes of group A streptococcal M proteins defined with hybridoma antibodies. Infect. Immun. **46**:267–269.
4. **Evans, R. P., Jr., and F. L. Macrina.** 1983. Streptococcal R plasmid pIP501: endonuclease site map, resistance determinant location, and construction of novel derivatives. J. Bacteriol. **154**:1347–1355.
5. **Fischetti, V. A.** 1983. Requirements for the opsonic activity of human IgG directed to type 6 group A streptococci: net basic charge and intact Fc region. J. Immunol. **130**:896–901.
5a. **Fischetti, V. A., M. Jarymowycz, K. F. Jones, and J. R. Scott.** 1986. Streptococcal M protein size mutants occur at high frequency within a single strain. J. Exp. Med. **164**:971–980.
6. **Fischetti, V. A., K. F. Jones, B. N. Manjula, and J. R. Scott.** 1984. Streptococcal M6 protein expressed in *Escherichia coli*. Localization, purification and comparison with streptococcal-derived M. protein. J. Exp. Med. **159**:1083–1095.
7. **Fischetti, V. A., K. F. Jones, and J. R. Scott.** 1985. Size variation of the M protein in group A streptococci. J. Exp. Med. **161**:1384–1401.
8. **Guss, B., M. Uhlen, B. Nilsson, M. Lindberg, J. Sjoquist, and J. Sjodaht.** 1984. Region X, the cell-wall-attachment part of staphylococcal protein A. Eur. J. Biochem. **138**:413–420.
9. **Hollingshead, S. K., V. A. Fischetti, and J. R. Scott.** 1986. Complete nucleotide sequence of type 6 M protein of the group A streptococcus. Repetitive structure and membrane anchor. J. Biol. Chem. **261**:1677–1686.
10. **Jones, K. F., B. N. Manjula, K. H. Johnston, S. K. Hollingshead, J. R. Scott, and V. A. Fischetti.** 1985. The location of variable and conserved epitopes among the multiple serotypes of streptococcal M protein. J. Exp. Med. **161**:623–628.
11. **Lancefield, R. C.** 1959. Persistence of type specific antibodies in man following infection with group A streptococci. J. Exp. Med. **110**:271–292.
12. **Lancefield, R. C.** 1962. Current knowledge of type-specific M antigens of group A streptococci. J. Immunol. **89**:307–313.
13. **Macrina, F. L., J. A. Tobian, K. R. Jones, R. P. Evans, and D. B. Clewell.** 1982. A cloning vector able to replicate in *Escherichia coli* and *Streptococcus sanguis*. Gene **19**:345–353.
14. **Manjula, B. N., B. L. Trus, and V. A. Fischetti.** 1985. Presence of two distinct regions in the coiled-coil structure of the streptococcal PepM5 protein: relationship to mammalian coiled-coil proteins and implications to its biological properties. Proc. Natl. Acad. Sci. USA **82**:1064–1068.
15. **Maxted, W. R., and E. V. Potter.** 1967. The presence of type 12 M-protein antigen in group G Streptococci. J. Gen. Microbiol. **49**:119–125.
16. **Phillips, G. N., Jr., P. F. Flicker, C. Cohen, B. N. Manjula, and V. A. Fischetti.** 1981. Streptococcal M protein: alpha-helical coiled-coil structure and arrangement on the cell surface. Proc. Natl. Acad. Sci. USA **78**:4689–4693.
17. **Plasterk, R. H. A., M. I. Simon, and A. G. Barbour.** 1985. Transposition of structural genes to an expression sequence on a linear plasmid causes antigenic variation in the bacterium *Borrelia hermsii*. Nature (London) **318**:257–263.
18. **Scott, J. R., and V. A. Fischetti.** 1983. Expression of streptococcal M protein in *Escherichia coli*. Science **221**:758–760.
18a. **Scott, J. R., P. C. Guenthner, L. M. Malone, and V. A. Fischetti.** 1986. Conversion of an M⁻ group A streptococcus to M⁺ by transfer of a plasmid containing an M6 gene. J. Exp. Med. **164**:1641–1651.
19. **Scott, J. R., S. K. Hollingshead, and V. A. Fischetti.** 1986. Homologous regions within M protein genes in group A streptococci of different serotypes. Infect. Immun. **52**:609–612.
20. **Scott, J. R., W. M. Pulliam, S. K. Hollingshead, and V. A. Fischetti.** 1985. Relationship of M protein genes in group A streptococci. Proc. Natl. Acad. Sci. USA **82**:1822–1826.
21. **Seyer, J. M., A. H. Kang, and E. H. Beachey.** 1980. Primary structural similarities between types 5 and 24 M proteins. Biochem. Biophys. Res. Commun. **92**:546–555.
22. **Smith, M. D., and D. B. Clewell.** 1984. Return of *Streptococcus faecalis* DNA cloned in *Escherichia coli* to its original host via transformation of *Streptococcus sanguis* followed by conjugative mobilization. J. Bacteriol. **160**:1109–1114.
23. **Woolcock, J. B.** 1974. Purification and antigenicity of an M-like protein of *Streptococcus equi*. Infect. Immun. **10**:116–122.
24. **Woolcock, J. B.** 1974. The capsule of *Streptococcus equi*. J. Gen. Microbiol. **85**:372–375.

Nucleotide Sequences That Signal the Initiation of Transcription for the Gene Encoding Type 6 M Protein in *Streptococcus pyogenes*

SUSAN K. HOLLINGSHEAD

Department of Microbiology and Immunology, Emory University School of Medicine, Atlanta, Georgia 30322

Escherichia coli is permissive in its recognition of transcription and translation signals from a wide variety of microorganisms. *Bacillus subtilis*, on the other hand, is relatively nonpermissive in its expression of heterologous genes, with expression usually being limited to genes from other gram-positive bacteria (4, 5, 11). Part of the restriction on heterologous gene expression is at the level of translation; gram-positive organisms apparently require an extensive complementarity to the 16S rRNA in the region preceding the translation initiation start site (1, 15, 17). A second part of the difference in heterologous gene expression is believed to be at the level of transcription because *B. subtilis* RNA polymerase cannot use several promoters which are transcribed efficiently by *E. coli* RNA polymerase (12, 24).

Only a single study defining a streptococcal promoter has been performed in the streptococcal host. Further studies of this kind are necessary to differentiate features common to expression of all gram-positive genes from those specific to a particular gene locus or host organism. In this communication, I report the start point of transcription in *Streptococcus pyogenes* for *emm-6*, the gene encoding the serotype 6 M protein of the group A streptococci; I then compare the *emm-6* promoter with other gram-positive promoters.

Results

Three regions preceding the *emm-6* gene that are homologous with the *E. coli* consensus sequence for promoters (8) were previously identified by examination of the DNA sequence for this gene (9). One of these, P1, was expected to function most efficiently on the basis of homology to consensus and spacing between the −35 and −10 sequences. Dideoxy sequence analysis (primer extension) was performed on mRNA from the streptococcal strain D471, which encodes the *emm-6* gene (Fig. 1).

The major start point for the mRNA in *S. pyogenes* as determined by primer extension is at nucleotide 382 of the published sequence (9; see Appendix B, this volume). Preceding the initiation nucleotide A by 7 base pairs is the

sequence TACAAT, which is in the proper position to serve as a −10 sequence for RNA polymerase. The sequence TTTACC precedes the −10 sequence by 17 base pairs, which is the optimal spacing between the −35 and −10 sequences for efficient promoter function in *E. coli* (14). This analysis shows that, as predicted from the DNA sequence, P1 is the promoter utilized by the streptococcal RNA polymerase.

A second region of the primer extension analysis which has minor stop points in all four lanes (A, G, C, and T) can be seen at nucleotide 409 (Fig. 1). It is not clear whether this sequence represents a secondary start site for mRNA. The nucleotide sequence preceding this location has very little complementarity to any known promoter sequences.

RNA was extracted from group A strain D471 and run on Northern gels by methods of Thomas (27), Meinkoth and Wahl (16), and Feinberg and Vogelstein (6). The probe was a purified *Nci*I/*Pvu*II fragment which contained most of the *emm-6* gene. The Northern gel analysis revealed a single transcript for this gene which was estimated to be 1.57 kilobase pairs in length (data not shown). The size of the transcript is consistent with the possibility of a stem and loop terminator centered about base pair 1,938 of the published sequences (see Appendix B, this volume). The predicted transcript length, with this stem and loop structure used as a terminator, was estimated at 1,552 base pairs (9).

Discussion

Moran et al. (17) have shown that promoter sequences recognized by the σ^{43} RNA polymerase of vegetative cells in *B. subtilis* display a striking conformity with the *E. coli* −35 recognition site (TTGACA) and −10 Pribnow box (TATAAT) consensus sequences of *E. coli* (22, 25, 28). (Sporulating cells of *B. subtilis* contain additional forms of the holoenzyme that differ with respect to promoter specificity.)

The above complementarity to consensus sequences, however, is not sufficient for expression in *B. subtilis* because σ^{43} RNA polymerase transcribes with low efficiency the *B. subtilis tms* and the *E. coli tac* promoters, each of which

A G C T

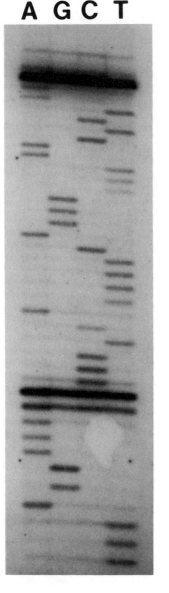

FIG. 1. Dideoxy sequence analysis of streptococcal mRNA. RNA was extracted from streptococcal strain D471 (serotype 6 M protein from the Rockefeller University collection) by incubation with lysozyme after growth in glycine (3). RNA was then purified by the method of MacDonald et al. (13) and pelleted by sedimentation through 5.7 M CsCl (7). Primer extension sequencing was by the method of Inoue and Cech (10). Lanes A, G, C, and T indicate the addition of a single dideoxy nucleotide in each reaction.

ciency of expression for heterologous genes is explained as either a requirement for greater fidelity to consensus sequences on the part of gram-positive promoters or a requirement for additional signals within the promoter.

The *Bacillus veg* promoter is utilized very efficiently by σ^{43} RNA polymerase. This promoter and several others in *Bacillus* species have an AT-rich region located 5' to the −35 site (17). In vitro studies of promoter utilization have shown that AT-rich regions are associated with efficient promoter utilization by σ^{43} RNA polymerase (17, 20, 21). Genetic evidence obtained for the maltose regulon of *S. pneumoniae* supports the hypothesis that the AT-rich regions are associated with increased promoter efficiency (26).

On the basis of the promoter identified here for the *emm-6* gene, it appears that the RNA polymerase of *S. pyogenes* recognizes sequences similar to that of σ^{43} RNA polymerase of *Bacillus* species. The promoter has excellent homology with −35 and −10 consensus sequences and has the AT-rich region previously associated with efficient utilization of gram-positive promoters.

Definitive information about transcription start points has been reported for only a few streptococcal or staphylococcal genes (Table 1; 18, 19, 23, 27, 28). The conservation of nucleotides in the −35 and −10 sequences and spacing between these two regions is a general feature of each of these five streptococcal and staphylococcal promoters (Table 1). The AT-rich region has not been found in every promoter. Most of these studies, however, have been done in *Ba-*

has excellent complementarity to the two consensus hexanucleotides (17). The lower effi-

TABLE 1. Streptococcal and staphylococcal promoter sequences (2, 18, 19, 23, 26, 28)

Gene	−35	−10	Space	Method
spc	TTCAAA	TATAAT	17	S1, *S. aureus* RNA
*erm*A	TTCTAA	CATATT	19	S1, *S. aureus* RNA
sak	TTGATT	TAAAAT	18	In vitro, *E. coli* RNA polymerase
aphA	TTGACA	TATCTT	17	S1, *B. subtilis* and *E. coli* RNA
pC221-92-base-pair replication transcript	TTGAAG	TTTAAT	NA[a]	In vitro, *B. subtilis* RNA polymerase
malM	TTGCAA	TATACT	18	Genetic evidence
malX	TTGCAA	TATACT	17	Genetic evidence

[a] Not available.

cillus species, the exceptions being studies by Stassi et al. (26) in *S. pneumoniae* and Murphy (18) in *Staphylococcus aureus*. To compare transcription systems effectively, it is important to determine the sequences recognized as promoters in the host organism as well as in heterologous organisms.

LITERATURE CITED

1. **Band, L., and D. J. Henner.** 1984. *Bacillus subtilis* requires a "stringent" Shine-Dalgarno region for gene expression. DNA **3:**17–21.
2. **Brenner, D. G., and W. V. Shaw.** 1985. The use of synthetic oligonucleotides with universal templates for rapid DNA sequencing: results with staphylococcal replicon pC221. EMBO J. **4:**561–568.
3. **Coleman, S. E., I. Van de Rijn, and A. S. Bleiweis.** 1970. Lysis of grouped and ungrouped streptococci by lysozyme. Infect. Immun. **2:**563–569.
4. **Courvalin, P., B. Weisblum, and J. Davies.** 1977. Aminoglycoside-modifying enzyme of an antibiotic-producing bacterium acts as a determinant of antibiotic resistance in *Escherichia coli*. Proc. Natl. Acad. Sci. USA **74:**999–1003.
5. **Ehrlich, S. D.** 1978. DNA cloning in Bacillus subtilis. Proc. Natl. Acad. Sci. USA **75:**1433–1436.
6. **Feinberg, A. P., and B. Vogelstein.** 1983. A technique for radiolabelling DNA restriction endonuclease fragments to high specific activity. Anal. Biochem. **132:**6–13.
7. **Glisin, V., R. Crkvenjakov, and C. Byus.** 1974. Ribonucleic acid isolated by cesium chloride centrifugation. Biochemistry **13:**2633–2637.
8. **Hawley, D. K., and W. R. McClure.** 1983. Compilation and analysis of *Escherichia coli* promoter DNA sequences. Nucleic Acids Res. **11:**2237–2254.
9. **Hollingshead, S. K., V. A. Fischetti, and J. R. Scott.** 1986. Complete nucleotide sequence of type 6 M protein of the group A streptococcus: repetitive structure and membrane anchor. J. Biol. Chem. **261:**1677–1686.
10. **Inoue, T., and T. R. Cech.** 1985. Secondary structure of the circular form of the *Tetrahymena* rRNA intervening sequence: a technique for RNA structure analysis using chemical probes and reverse transcriptase. Proc. Natl. Acad. Sci. USA **82:**648–652.
11. **Kreft, J., K. Bernhard, and W. Goebel.** 1978. Recombinant plasmids capable of replication in *Bacillus subtilis* and *Escherichia coli*. Mol. Gen. Genet. **162:**59–67.
12. **Lee, G., C. Talkington, and J. Pero.** 1980. Nucleotide sequences of a promoter recognized by *Bacillus subtilis* RNA polymerase. Mol. Gen. Genet. **180:**57–65.
13. **MacDonald, P. M., E. Kutter, and G. Mosig.** 1984. Regulation of a bacteriophage T4 late gene, *SOC*, which maps in an early region. Genetics **106:**17–27.
14. **Mandecki, W., and W. S. Reznikoff.** 1982. A *lac* promoter with a changed distance between −10 and −35 regions. Nucleic Acids Res. **10:**903–911.
15. **McLaughlin, J. R., C. L. Murray, and J. C. Rabinowitz.** 1981. Unique features in the ribosome binding site sequence of the gram positive Staphylococcus aureus betalactamase gene. J. Biol. Chem. **256:**11283–11291.
16. **Meinkoth, S., and G. Wahl.** 1984. Hybridization of nucleic acids immobilized on solid supports. Anal. Biochem. **138:**267–287.
17. **Moran, C. P., N. Lang, S. F. J. LeGrice, G. Lee, M. Stephens, P. Sonenshein, J. Pero, and R. Losick.** 1982. Nucleotide sequences that signal the initiation of transcription and translation in *Bacillus subtilis*. Mol. Gen. Genet. **186:**339–346.
18. **Murphy, E.** 1984. Nucleotide sequence of *ermA*, a macrolide-lincosamide-streptogramin B (MLS) determinant in *Staphylococcus aureus*. J. Bacteriol. **162:**633–640.
19. **Murphy, E.** 1985. Nucleotide sequence of a spectinomycin adenyltransferase AAD(9) determinant from *Staphylococcus aureus* and its relationship to AAD(3″) (9). Mol. Gen. Genet. **200:**33–39.
20. **Murray, C. L., and J. C. Rabinowitz.** 1982. Nucleotide sequences of transcription and translation initiation regions in *Bacillus* phage Phi29 early genes. J. Biol. Chem. **257:**1053–1062.
21. **Palva I., R. F. Pettersson, N. Kalkkinen, P., Lehtovaara, M. Sarvas, H. Soderlund, K. Takkinen, and L. Kaariainen.** 1981. Nucleotide sequence of the promoter and NH₂-terminal signal peptide region of the alpha-amylase gene from *Bacillus amyloliquefaciens*. Gene **15:**43–51.
22. **Rosenberg, M., and D. Court.** 1979. Regulatory sequences involved in the promotion and termination of RNA transcription. Annu. Rev. Genet. **13:**319–353.
23. **Sako, T., and N. Tsuchida.** 1983. Nucleotide sequence of the staphylokinase gene from *Staphylococcus aureus*. Nucleic Acids Res. **11:**7679–7693.
24. **Shorenstein, R. G., and R. Losick.** 1973. Comparative size and properties of the sigma subunits of RNA polymerase from *Bacillus subtilis* and *Escherichia coli*. J. Biol. Chem. **248:**6170–6173.
25. **Siebenlist, U., R. B. Simpson, and W. Gilbert.** 1980. *E. coli* RNA polymerase interacts homologously with two different promoters. Cell **20:**629–281.
26. **Stassi, D. L., J. J. Dunn, and S. A. Lacks.** 1982. Nucleotide sequence of DNA controlling expression of genes for maltosaccharide utilization in Streptococcus pneumoniae. Gene **20:**359–366.
27. **Thomas, P. S.** 1980. Hybridization of denatured RNA and small DNA fragments transferred to nitrocellulose. Proc. Natl. Acad. Sci. USA **9:**5201–5205.
28. **Trieu-Cuot, P., A. Klier, and P. Courvalin.** 1985. DNA sequences specifying the transcription of the streptococcal kanamycin resistance gene in *Escherichia coli* and *Bacillus subtilis*. Mol. Gen. Genet. **198:**348–352.

Phase Variation and M-Protein Expression in Group A Streptococci

P. CLEARY, S. JONES, J. ROBBINS, AND W. SIMPSON

Department of Microbiology, University of Minnesota, Minneapolis, Minnesota 55455

Numerous bacterial pathogens exhibit both qualitative and quantitative variation of surface antigens and other cellular products known to affect virulence. Animal parasites, from more simple viruses to sophisticated protozoa, have evolved genetic mechanisms to alter their surfaces, either to acquire access to specific tissue or to evade the selective pressure of the host's immunological defenses. Group A streptococci have also been observed to undergo such variation. Two types of variation have been recognized: alterations in the ability to resist phagocytosis, which is dependent on a fibrillar layer of M protein on the cell surface (M^+), and shifts from the synthesis of one M protein to a second, antigenically distinct molecule. Strains with new M proteins have arisen in populations experiencing frequent infections, in which initially epidemic strains were replaced by other related strains that had antigenically distinct M proteins (2, 6). At present nearly 90 different M serotypes are recognized. The phenotypic loss of M protein has been known since glossy colony morphology variants were first isolated and shown to be M^- and avirulent for mice (15). Moreover, the throats of healthy carriers gradually over time yield M^- streptococci (8). How such M^- cells can persist in the throat has yet to be explained. Concurrent with loss of M-protein expression is a change in other phenotypic markers. M^- cells are Cap^- (lack a hyaluronic acid capsule; 3), $SOR^{+/-}$ (have reduced levels of an extracellular lipoproteinase; 3), Fim^- (lack surface fimbriae; 16), Tr (lack of colony opacity; 9), and $SCFI^{+/-}$ (show diminished production of an inhibitor of chemotaxis; 1). The high frequency and pleiotropic nature of this variation suggests a nonrandom mechanism and has prompted us to search for a genetically programmed switch.

Phenotypic variation among other microorganisms has been studied in depth and is known to involve a variety of genetic mechanisms. Salmonellae invert a promoter with specific transposases to change flagellar proteins (18); influenza viruses undergo recombination and accumulate mutations which alter the antigenic specificity of the hemagglutin protein (17); and trypanosomes (5), borrelia (7), and *Neisseria gonorrhoeae* (14) enlist multiple silent genes which encode new surface antigens to alter their surface markers. Antigenic switching by *N. gonorrhoeae* is particularly relevant to group A streptococci because a negative phenotype often precedes a change in pilin antigen (10). Superimposed on these genetically programmed events is the fact that many pathogens acquire or lose virulence markers by gaining or losing extrachromosomal elements such as temperate bacteriophages or plasmids. All or some combination of the above genetic mechanisms could account for the genetic diversity of group A streptococci.

Role of Bacteriophage and M-Protein Expression

Although early studies demonstrated that "curing agents" such as ethidium bromide stimulated the segregation of M^- cells by M^+ cultures (3), an association of plasmid DNA with M-protein expression has not been made. Limiting our studies to a few matched pairs of M^+ and M^- laboratory strains, we discovered a temperate bacteriophage, SP24, that could activate an M^- variant of an M76 strain, strain CS112, to become resistant to phagocytosis (11). This phage, which is derived from an M12 culture, strain CS24, activates expression of M76 antigen upon lysogenization of strain CS112. Phage SP24 contains a unique phage *att* site and is integrated into a common chromosomal locus in both the parent M12 strain and M76$^+$ lysogens, but the relationship of integration to expression of the M76 structural gene remains unclear (12). Phage integration into the chromosome of strain CS112 initially occurs in only a minor fraction of cells. Activation of M-protein synthesis accompanies only the formation of stable lysogens, and when laboratory-constructed lysogens are cured of the prophage before integration, the cell loses the potential to become M^+. At present we are unable to ascribe a precise function to the bacteriophage and the integration event. Experiments have not determined whether stable lysogeny is sufficient to activate M expression and whether the physical location or the process of integrative recombination creates the M^+ state. To complicate matters further, infection of strain CS112 with phage

SP24 results in the substitution of a unique segment of DNA from an endogenous prophage carried by these cells (11). This substitution occurs with every infection and provides the recombinant phage with the potential to adsorb more efficiently to CS112 cells. It is not known whether the substituted DNA functions in the activation process. Although numerous M$^-$ cultures, including M$^-$ variants from strain CS24 from which phage SP24 originated, have been tested as recipients for lysogenic activation, only the M76$^-$ strain, CS112, can be activated to produce M protein. To further confuse the issue, spontaneous loss of the M$^+$ phenotype by wild-type CS24 cultures does not result in detectable changes in the SP24 prophage. The above inconsistencies question the role of phage SP24 in the transition process, i.e., changes between M$^+$ and M$^-$ states observed in wild-type isolates of group A streptococci, but does not eliminate the possibility that prophage-encoded products participate directly or indirectly in the biosynthesis of antiphagocytic M proteins. This phenomenon may not be associated with all strains of group A streptococci; strain CS112 may have acquired a unique genetic change to create an M$^-$ phenotype which can be complemented by phage SP24.

Phase Variation

The above results were difficult to interpret and suggested that transition from the M$^+$ to the M$^-$ state may proceed by more than one genetic route. Only rarely is the consequence an M$^-$ cell that can be activated by a prophage. Therefore, a more systematic approach was taken to answer the following questions. Are M$^-$ segregants the result of a unique and common genetic event, and does this organism undergo true phase variation; i.e., is transition from the M$^+$ to the M$^-$ state readily reversible? Two colonial characteristics, a glossy surface and transparent colonies, previously reported to correspond with the M$^-$ phenotype were employed to isolate 22 independent M$^-$ variants from strain CS24 (Fig. 1). Care was taken to ensure the independence of each lineage, and all variants were subcultured by single-colony passage at least seven times. To definitively identify alterations in the M-protein gene or adjacent DNA sequence, we cloned the M12 gene to expression in *Escherichia coli* for use as DNA probes. From the DNA sequence an *Hae*III restriction map was developed (Fig. 2) (13).

The first M$^-$ variants to be examined, strains CS64 and CS46, were identified by their glossy colonial morphology. The M$^-$ phenotype of these strains was highly stable and was also associated with loss of hyaluronic acid capsule and diminished expression of SCFI activity (1).

Hybridization studies of genomic DNA revealed small deletions, approximately 50 base pairs (bp), which mapped to identical *Hae*III and *Rsa*I fragments adjacent to the M-protein coding sequence (Fig. 2) (13; unpublished data). The pleiotropic nature of these deletions and their similar location suggested that they defined a regulatory switch and prompted us to further test the generality of deletion formation in high-frequency transition between M$^+$ and M$^-$ states.

M$^-$ variants of strain CS24 could be more easily identified when colonies were grown anaerobically on the clear agar medium described by Skjold and Wannamaker (9). With a stereomicroscope and obliquely transmitted light, opaque M$^+$ colonies could be distinguished from more transparent M$^-$ variants (Fig. 3). Twenty independent lines of less opaque cells were isolated and studied in detail. Opacity variants of strain CS24 could be placed into one of three categories: opaque (Op), intermediate opaque (OpI), and transparent (Tr) (Fig. 3). Op colonies (wild-type M$^+$ cultures) segregate less opaque (OpI or Tr) colonies at a frequency between 10^{-3} and 10^{-4}. Reversion to a more opaque phenotype occurred at frequencies ranging from 10^{-1} to 10^{-3} depending on the specific lineage, while some less opaque variants could not be observed to revert to a more opaque phenotype.

After extensive purification by single-colony isolations and experiments to discern sensitivity to phagocytosis, less opaque M$^-$ variants and more opaque M$^+$ revertants, from those able to revert, were analyzed for DNA rearrangements by use of Southern blot DNA-DNA hybridization with ^{32}P-labeled probe DNA, plasmid pPC101 (13). This plasmid contains an insert of 4.8 kilobases of streptococcal DNA from strain CS24 known to include the M12 coding sequence and upstream regulatory sequence (13; unpublished data). Genomic DNAs were digested with three different endonucleases, *Hae*III, *Alu*I, and *Rsa*I, to increase chances of detecting small deletions, inversions, or additions. Three groups of phase variants were found. (i) Cells readily reverted between M$^+$ opaque and M$^-$ less opaque phenotypes and had no identifiable DNA rearrangements. (ii) Two Tr M$^-$ cultures, strains A1 and A7, carried deletions less than 50 bp and 3 kilobases in size, respectively, and were phase locked in the M$^-$ state (unable to revert to the M$^+$ state). Strain A1 could revert to a more opaque colony type, but cells remained M$^-$. Strain A7 was phase locked for both phenotypes. (iii) Others with no detectable DNA rearrangement were stable with respect to their M$^-$ phenotype, but continued to revert to more opaque colony forms. Some of these had deletions.

These results establish that M-protein expres-

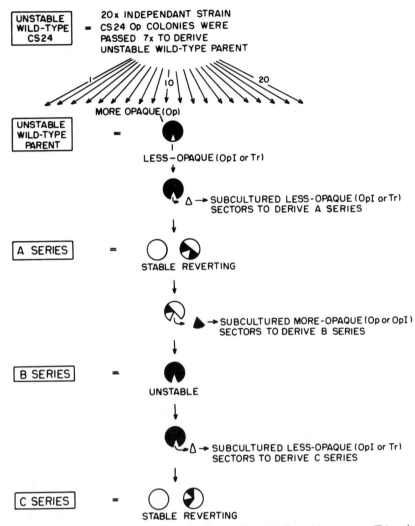

FIG. 1. Scheme employed to isolate opaque (Op), intermediate (OpI), and transparent (Tr) variants of strain CS24. Initially opaque, M+ colonies were passed through seven single-colony isolations. Likewise, each variant was subcultured at least seven times by single-colony isolation.

sion and colony morphology undergo true phase variation in group A streptococci but suggest that phase shifts need not involve a single common genetic event. M− variants carrying DNA deletions were more common than would be expected of random mutational events; however, if deletion formation were integral to the switching mechanism, then they would be expected to map outside the M12 coding sequence, possibly upstream in promoter or other regulatory sequences. Therefore, it was important to determine the location of deletions carried by strains CS64, CS46, A1, and A7 relative to the M12 coding sequence. To accomplish this, 2,300 bp of DNA in the vicinity of the M12 gene was sequenced by the Maxam-Gilbert and Sanger

methods (unpublished data), and a fine structure restriction map was derived from this sequence (Fig. 2). The position of open reading frames and primer extension studies located two transcription starts, and the longer transcript was preceded by a typical promoter consensus sequence and had a ribosome binding site (J. Robbins and P. Cleary, manuscript in preparation). In addition to these results, an 80-bp *Rsa*I fragment beginning at position 777 bp was found to activate chloramphenicol resistance (promoter activity) when cloned into the CAT vector pCM1 of Close and Rodriguez (4). This fragment was oriented in the streptococcal genome such that transcription from it would be in a direction opposite to the M12 structural gene. The rela-

Fine Structure Map Of The M12 Gene and Adjacent DNA Sequence

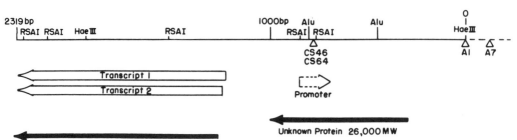

FIG. 2. Fine structure map of the M12 gene and adjacent DNA sequence. The location of restriction sites was determined by a series of double digests and confirmed from the nucleotide sequence. Open arrows indicate the beginning of two M12 transcripts. The transcript identified by the dashed arrow has not been measured, but is presumed to exist because this 80-bp fragment has promoter activity. The solid arrows represent protein starts indicated by two open reading frame codons; △ indicates the location of DNA deletions.

tionship of this promoter to M-protein expression is suggested by the fact that deletions carried by M⁻ strains CS46 and CS64 mapped in this *Rsa*I fragment. Furthermore, deletions contained in other M⁻ strains, A1 and A7, also mapped downstream from this promoter, even further from the 5′ end of the M12 structural gene, 1.2 and 1.4 kilobases, respectively.

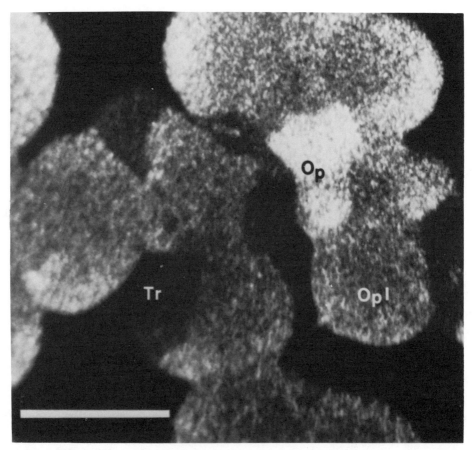

FIG. 3. Colony morphology. Strain CS24 produces Op, OpI, and Tr colony types.

The pleiotropic nature of these deletions and their location outside the M12 structural gene or its immediate promoter suggest that they define a second gene, a regulatory element, which is required for expression of the M12 antigen as well as other phenotypic markers. This conclusion is also supported by two other unpublished results: subcloned fragments of DNA must contain the M12 coding sequence and the upstream region of DNA defined by these deletions to express M12 antigen in *E. coli*; and all M$^-$ variants, including those with these deletions, either completely lack or have very reduced levels of M12-specific mRNA (J. Robbins, W. Simpson, and P. Cleary, manuscript in preparation).

Conclusions

Group A streptococci have long been known to be genetically unstable, a property that has plagued many who have attempted to investigate the pathogenesis of this organism. Our efforts to understand the basis of this instability revealed a complex interaction between bacteriophage and chromosomal genes which may be the driving force of antigenic diversity in this species.

M-protein expression and opaque colony morphology have been demonstrated to undergo coordinate phase variation at a high frequency. Genetic determinants of the Op and M$^+$ phenotypes can, however, be shown to be separable and distinct. Coordinate expression of these and other markers may be controlled by common regulatory elements defined by deletion mutations which map 5' to the M12 coding sequence and its promoter. Whether this regulatory element or DNA rearrangements or both are components of the phase switching mechanism is still a mystery and awaits further experimentation.

A portion of the DNA sequence from which the restriction map was determined was contributed by J. Spanier.

This work was supported by Public Health Service grant AI16722 from the National Institute of Allergy and Infectious Diseases. J.R. was supported by Public Health Service Training grant 5T32HL107114 as a predoctoral trainee from the National Heart and Lung Institute.

LITERATURE CITED

1. **Cleary, P., E. Chenoweth, and D. Wexler.** 1984. A streptococcal inactivator of chemotaxis: a new virulence factor specific to group A streptococci, p. 179–180. *In* Y. Kimura, S. Katami, and Y. Shiokawa (ed.), Recent advances in streptococci and streptococcal diseases. Reedbooks, Ltd., Berkshire, England.
2. **Cleary, P. P., D. Johnson, and L. W. Wannamaker.** 1979. Genetic variation in the M antigen of group A streptococci: reassortment of type specific markers and possible antigenic drift. J. Infect. Dis. **140:**747–757.
3. **Cleary, P. P., Z. Johnson, and L. Wannamaker.** 1975. Genetic instability of M protein and serum opacity factor of group A streptococci: evidence suggesting extrachromosomal control. Infect. Immun. **12:**109–118.
4. **Close, T. J., and R. L. Rodriguez.** 1982. Construction and characterization of the chloramphenicol-resistance gene cartridge: a new approach to the transcriptional mapping of extrachromosomal elements. Gene **20:**305–316.
5. **Laurent, M., E. Pays, E. Magnus, N. Van Meirvenne, G. Matthyssens, R. O. Williams, and M. Steinert.** 1983. DNA rearrangements linked to expression of a predominant surface antigen gene of trypanosomes. Nature (London) **302:**263–266.
6. **Maxted, W. R., and H. A. Valkenburg.** 1969. Variation in the M antigen of Group A streptococci. J. Med. Microbiol. **2:**199–210.
7. **Meier, J., M. Simon, and A. Barbour.** 1985. Antigenic variation is associated with DNA rearrangements in a relapsing fever borrelia. Cell **41:**403–409.
8. **Rothbard, S., and R. F. Watson.** 1948. Variation occurring in group A streptococci during human infection. Progressive loss of M substance correlated with increasing susceptibility to bacteriostasis. J. Exp. Med. **87:**521–533.
9. **Skjold, S. A., and L. W. Wannamaker.** 1986. Surface proteins in the transduction of groups A and G streptococci. J. Med. Microbiol. **21:**69–74.
10. **Spanier, J., and Cleary, P.** 1980. Bacteriophage control of antiphagocytic determinants in group A streptococci. J. Exp. Med. **152:**1393–1406.
11. **Spanier, J. G., and P. Cleary.** 1983. A DNA substitution in the group A streptococcal bacteriophage SP24. Virology **130:**514–522.
12. **Spanier, J., and P. Cleary.** 1985. Integration of bacteriophage SP24 into the chromosome of group A streptococci. J. Bacteriol. **164:**600–604.
13. **Spanier, J., S. Jones, and P. Cleary.** 1984. Small DNA deletions creating virulence in *Streptococcus pyogenes*. Science **225:**935–938.
14. **Sparling, P. F., J. G. Cannon, and M. So.** 1986. Phase and antigenic variation of pili and outer membrane protein II of *Neisseria gonorrhoeae*. J. Infect. Dis. **153:**196–201.
15. **Todd, E. W., and R. C. Lancefield.** 1928. Variants of hemolytic streptococci: their relation to the type specific substance, virulence and toxin. J. Exp. Med. **48:**751–767.
16. **Swanson, J., K. C. Hsu, and E. C. Gotschlich.** 1969. Electron microscopic studies on streptococci. I. M antigen. J. Exp. Med. **130:**1063–1091.
17. **Webster, R. G., and W. G. Laver.** 1975. Antigenic variation of influenza viruses, p. 269–314. *In* E. D. Kilbourne (ed.), The influenza viruses and influenza. Academic Press, Inc., New York.
18. **Zieg, J., M. Hilman, and M. Simon.** 1978. Regulation of gene expression by site specific inversion. Cell **15:**237–244.

Comparison of Genes Encoding Group A Streptococcal M Protein Types 1 and 12: Conservation of Upstream Sequences

ELIZABETH HAANES-FRITZ, JOHN C. ROBBINS, AND PATRICK CLEARY

Department of Microbiology, University of Minnesota, Minneapolis, Minnesota 55455

Many pathogens are capable of varying surface antigens to evade host immune defenses. In most systems, including pilus antigenic variation in gonococci (4), major surface protein variation in relapsing fever *Borrelia* spp. (8), and major surface glycoprotein variation in trypanosomes (1), switching of antigenic types is well documented. Switching between the 80 or more functionally related types (6) of group A streptococcal M proteins, however, is at best extremely rare. Although in vivo observations of antigenic switching in M proteins are documented in the literature (7) and an undetected molecular switching mechanism may indeed exist, a more likely hypothesis is that the antigenic diversity of M proteins is a product of accelerated evolution. Such an evolutionary scheme would probably involve events in addition to the normal accumulation of point mutations in a common progenitor gene. Some possibilities might include the utilization of different expression loci by different M-protein genes, the presence of multiple M-protein genes in a single strain (2, 11), or the occasional transfer of M-protein genes from one strain to another followed by recombination. To begin to understand whether events such as these might be responsible for the antigenic diversity of M proteins, we are comparing in detail the genes encoding M1 and M12 and their respective flanking regions. Here we report evidence suggesting that the M1 and M12 genes are situated in a common, highly conserved expression locus, but this same expression locus may not be common to all M types.

Cloning of DNA Encoding M1-Specific Antigens

DNA from the M1 strain CS130 was partially digested with the restriction enzyme *Sau*3A1, and the resulting fragments were then ligated into a bacteriophage lambda vector. Recombinant plaques were screened by hybridization using a radioactively labeled DNA probe containing a portion of the M12 gene cloned from strain CS24 (10). Portions of bacteriophage recombinants which hybridized to the M12 probe were then subcloned into appropriate plasmid vectors. Concentrated phage lysates and sonicated extracts of plasmid subclones were examined in double-diffusion analyses against polyclonal M1 antiserum and gave lines of identity

with a standard M1 streptococcal acid extract. Additionally, antigenic material from the clones effectively absorbed opsonic antibody from the M1 antiserum. The boundaries of the M1 gene were roughly determined by deletion mapping, and the appropriate region was sequenced (E. Haanes-Fritz and P. Cleary, manuscript in preparation). We tentatively assigned transcription and translation initiation signals to the M1 gene by analogy to those determined for the M12 gene by transcriptional studies (J. C. Robbins, J. G. Spanier, S. Jones, and P. Cleary, submitted for publication). Although additional verification of the M1 gene initiation signals is required, the assigned sites represent the only logical choices in the region of DNA eliciting M1 expression.

Conservation of Sequences Upstream of the M1 and M12 Genes

Restriction and hybridization analyses demonstrated that at least 3 kilobases of DNA flanking the 5' ends of the M1 and M12 genes were homologous. Restriction mapping of the flanking regions with enzymes recognizing six base sequences showed that most of the sites were conserved (Fig. 1). Upstream homology was further verified by restriction mapping with enzymes recognizing four base sequences and by DNA hybridizations under stringent conditions. DNA sequence analysis revealed that the regions extending from the proposed translation initiation points to 1,032 base pairs (bp) upstream of the M1 gene and 1,036 bp upstream of the M12 gene were 97% homologous (Haanes-Fritz and Cleary, in preparation). The differences between the two sequences included 25 base substitutions and four insertions or deletions of up to 3 bp each.

Conservation of the sequences upstream of different M protein genes may be highly significant. Analysis of M protein-deficient variants of the M12 strain CS24 revealed a relationship between the loss of M12 expression and the presence of small deletions (10; P. Cleary, S. Jones, J. Robbins, and W. Simpson, this volume). The deletions are upstream of the structural gene and normal transcription and translation signals of that gene. Cleary et al. propose, then, that additional upstream sequences, such as a regulatory gene, are required for optimal

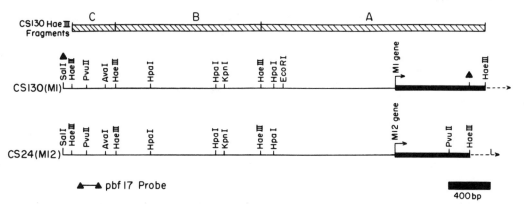

FIG. 1. Restriction maps of 5' and upstream regions of the M1 and M12 genes. The M1 and M12 structural genes are denoted by heavy lines. The *Hae*III fragments of CS130 (discussed in the text) are diagrammed by the hatched line above the M1 map. Probe pbf17 is marked with triangles.

M12 expression. The high degree of conservation between the regions flanking the M1 and M12 genes implies that similar or identical upstream sequences may be important to M1 expression as well. Furthermore, the extent of upstream homology indicates that the M1 and M12 genes are situated in a common expression locus.

Conservation of Upstream Regions in Other M Types

To determine whether other M types also had homology to the upstream regions flanking the M1 and M12 genes, *Hae*III chromosomal digests of various group A streptococcal DNAs were subjected to hybridization analysis (Fig. 2). The plasmid pbf17, a pUC9 derivative carrying a 3.7-kilobase insert corresponding to the region of the M1 clone denoted by triangles in Fig. 1, was used as the radioactively labeled probe. Hybridization conditions were chosen to allow up to 20% base-pair mismatch. In all the genomic digests, the probe hybridized to fragments identical in size to the *Hae*III B and C fragments of the M1 strain, CS130. These fragments, as diagrammed in Fig. 1, defined the region between 1.2 and 2.9 kilobases upstream of the M1 gene. In addition, the probe hybridized strongly to other heterogeneously sized fragments in some, but not all, of the genomic digests. Further hybridizations utilizing smaller probes from both the M1 and M12 clones showed that these heterogeneous fragments contained regions homologous to the portion of the *Hae*III A fragment of CS130 extending from the *Hae*III B fragment to the initiation point of the M1 gene. Furthermore, the subset of streptococcal genomic digests exhibiting homology to the upstream portion of the *Hae*III A fragment remained unchanged even when stringency levels were adjusted to allow less than 3% base-pair mismatch (data not shown).

Assuming that some other M protein genes are also expressed in the locus used by the M1 and

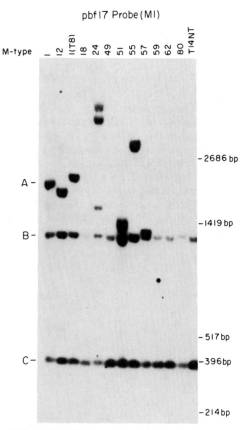

FIG. 2. *Hae*III chromosomal digests of group A streptococcal DNAs of various M types hybridized with the pbf17 probe under conditions allowing 20% base-pair mismatch. The letters along the left-hand side of the blot denote *Hae*III fragments in the M1 strain CS130 as diagrammed in Fig. 1.

M12 genes, we suspect that the heterogeneous sizes of the fragments homologous to *Hae*III fragment A are the result of restriction site polymorphisms present in the 5' ends of the respective M-protein structural genes. Variably sized restriction fragments were also detected by Scott et al. when chromosomal digests of different streptococcal strains were hybridized with DNA probes containing portions of the M6 gene (9). These authors attributed the restriction site heterogeneity to size variation among the different M-protein genes, since the molecular weights of the respective M proteins vary considerably (3). Besides size variation, though, sequence variations in the 5' ends of M-protein genes might cause the addition or deletion of particular restriction sites.

Although we do not yet understand the conspicuous absence of homology to the *Hae*III A fragment in some M types, we offer two suggestions. First, all M-protein genes may be expressed from the same location on the chromosome, but the analogous expression loci may have diverged considerably in some serotypes. Second, some M-protein genes may be situated in expression loci positioned elsewhere on the chromosome. The presence of multiple copies of the M5 gene in a single strain (5) supports the idea that M-protein genes are not limited to a single location in the streptococcal genome. Implicit in either of these explanations is the possibility that the regulatory mechanisms of M-protein gene expression may vary between different serotypes, depending on the regions flanking the genes. It is possible that the utilization of particular M-protein expression loci may represent an important event in the evolution of M-protein antigenic diversity. Hence, a comprehensive survey of hybridization patterns shown by DNAs of various M types may provide useful information in defining M-protein lineages.

We thank Steven Skjold for supplying genomic DNA from a number of streptococcal serotypes.

The project was funded by Public Health Service grant AI16722 from the National Institute of Allergy and Infectious Diseases.

LITERATURE CITED

1. **Borst, P., and G. A. M. Cross.** 1982. The molecular basis of trypanosome antigenic variation. Cell **29**:291–303.
2. **El Kholy, A., N. I. Guirguis, L. W. Wannamaker, and R. M. Krause.** 1975. Newly recognized distinctive M protein associated with certain type 1 group A streptococci. Infect. Immun. **11**:551–555.
3. **Fischetti, V. A., K. F. Jones, and J. R. Scott.** 1985. Size variation of the M protein in group A streptococci. J. Exp. Med. **161**:1384–1401.
4. **Hagblom, P., E. Segal, E. Billyard, and M. So.** 1985. Intragenic recombination leads to pilus antigenic variation in *Neisseria gonorrhoeae*. Nature (London) **315**:156–158.
5. **Kehoe, M. A., T. P. Poirier, E. H. Beachey, and K. N. Timmis.** 1985. Cloning and genetic analysis of serotype 5 M protein determinant of group A streptococci: evidence for multiple copies of the M5 determinant in the *Streptococcus pyogenes* genome. Infect. Immun. **48**:190–197.
6. **Lancefield, R. C.** 1962. Current knowledge of the type specific M antigens of group A streptococci. J. Immunol. **89**:307–313.
7. **Maxted, W. R., and H. A. Valkenburg.** 1969. Variation in the M antigen of group A streptococci. J. Med. Microbiol. **2**:199–210.
8. **Plasterk, R. H. A., M. I. Simon, and A. G. Barbour.** 1985. Transposition of structural genes to an expression sequence on a linear plasmid causes antigenic variation in the bacterium *Borrelia hermsii*. Nature (London) **318**:257–263.
9. **Scott, J. R., S. K. Hollingshead, and V. A. Fischetti.** 1986. Homologous regions within M protein genes in group A streptococci of different serotypes. Infect. Immun. **52**:609–612.
10. **Spanier, J. G., S. J. C. Jones, and P. P. Cleary.** 1984. Small deletions creating avirulence in *Streptococcus pyogenes*. Science **225**:935–938.
11. **Wiley, G. G., and A. T. Wilson.** 1960. The occurence of two M antigens in certain group A streptococci related to type 14. J. Exp. Med. **113**:451–465.

Transcription Studies of Type 12 M Protein Phase Variants

JOHN C. ROBBINS AND PATRICK CLEARY

Department of Microbiology, University of Minnesota, Minneapolis, Minnesota 55455

Phase variation of the group A streptococcal surface protein and virulence factor M protein is a biological phenomenon for which the molecular basis is unknown (P. Cleary, S. Jones, J. Robbins, and W. Simpson, this volume). Accompanying the transition from opaque to transparent colony morphology is the diminished production of M protein (8) and the creation of genomic rearrangements upstream of the type 12 M-protein gene (6; W. Simpson and P. Cleary, manuscript in preparation). In addition, M-protein expression is coordinately expressed with other phenotypic markers (1, 3).

Group A streptococcal strains CS24 and CS44 (2) undergo phase variation of type 12 M protein, changing from high levels of expression of the protein in CS24 and CS44 to low levels in CS46 and CS64, respectively; concomitant with this phenotypic change is the deletion of approximately 50 base pairs near the 5′ end of the M-protein gene (6). However, unlike true phase variants, CS46 and CS64 do not revert to high M-protein levels at a detectable rate; i.e., they are phase locked.

As part of a comprehensive study of M-protein phase variation and gene regulation, we have undertaken the task of evaluating the impact of the deletions in strains CS46 and CS64 at the level of transcription of M-protein-specific mRNA. Our studies required the purification of RNA from streptococci, which, in turn, allowed us to determine, by primer extension, the transcription start point of the M12 gene, promoter sequences, and the physical relationship between the transcription start of the M12 gene and the deletions mapping near the M12 gene.

Measurement and Characterization of M-Protein mRNA

Streptococcal RNA was prepared from mid-log-phase cells for use in Northern blot and dot-blot analyses (7) to qualitatively and quantitatively measure M-protein mRNA. Previous restriction enzyme analysis of pPC106 indicated that the 5′ end of the type 12 M protein gene exists within an *Hae*III fragment of approximately 1.9 kilobases (kb); expression assays showed the clone to encode the amino-terminal portion of the M12 protein and the direction of the gene's transcription (6). Both Northern blot (Fig. 1) and dot-blot (data not shown) analyses

were done with this restriction fragment serving as the probe, and the results show that CS46 and CS64 produce M-protein-specific mRNA but that it is diminished in these strains, relative to the parent strains, by more than 100-fold. This finding is consistent with the demonstration by double-diffusion analysis that the transparent colony morphology of strains CS46 and CS64 represents a cell type that produces little M protein (data not shown).

It is also clear from the Northern blot studies that the primary transcription unit is about 2 kb. Previous reports have placed the size of the M12 protein at approximately 58,000 daltons (9), a size requiring a gene about 2 kb in length; therefore, it may be concluded that the M12 gene is monocistronic.

However, there are at least two additional minor bands representing transcripts of approximately 3 and 5 kb detected by the nick-translated restriction fragment probe. These bands are not believed to represent minor species of M-protein transcripts because Northern blot analysis with a polynucleotide kinase-labeled synthetic oligomer probe specific for the M12 gene (Fig. 3) hybridizes only with the 2-kb transcript (J. Robbins et al., manuscript in preparation). The minor bands may represent mRNA species encoded by a region of the *Hae*III fragment upstream of the M12 gene.

Characterization of the M12 Gene Transcription Start Sites

DNA sequence analysis of the M12 gene and an undefined length of upstream sequence from pPC106 suggested the location of the M-protein amino terminus to exist more than 400 bases downstream of the deletions associated with the switch from a high to a low expression mode. This notion was tested by experimentally mapping the precise 5′ end of the M12 gene via primer extension. A synthetic 20-base oligomer complementary to the mRNA region, predicted by DNA sequence analysis to encode the amino terminus of the M12 gene, was hybridized with RNA from both CS24 and CS64 and used as a primer for reverse transcriptase. The reaction stop points (Fig. 2) suggest that two transcription start points, separated by approximately 30 base pairs, are used at nearly equal frequencies in CS24; a third start site, inactive in CS24, is

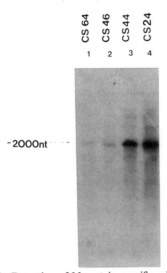

FIG. 1. Detection of M-protein-specific mRNA by Northern blot analysis of streptococcal RNA. Total cellular RNA (12 μg) was electrophoresed through a 1.4% formaldehyde gel and transferred to nitrocellulose (7). The filter was then probed with the nick-translated 1.9-kb *Hae*III restriction fragment described in the text. Prehybridization and hybridization were performed in 6× SSC (1× SSC is 0.15 M NaCl plus 0.01 5 M sodium citrate) and heparin at 42°C.

active in CS64, albeit to a lesser extent. Sequencing the M12 mRNA from CS24 (J. Robbins and P. Cleary, manuscript in preparation) precisely indicates the transcription start points (Fig. 3).

Identification of the start of the M12 gene allowed deduction of the DNA sequence of the untranslated region of the M12 mRNA, a ribosome binding site, and overlapping promoters (Fig. 3). The ribosome binding site sequence of GGAGC is in virtual agreement with the consensus sequence of GGAGG (5). The −10 and −35 regions of the most distal promoter are in various degrees of agreement with the consensus sequences of TATAAT and TTGACA (4), respectively; while the −10 region lies within a long stretch of low guanine-plus-cytosine content that might contain more than one potential recognition sequence, the −35 sequence has, at most, 50% homology. The promoter sequences partnering the transcription start site closest to the structural gene have even less homology with the −10 and −35 consensus sequences than the most distal promoter. It also appears that the −10 region of the upstream promoter overlaps the −35 region of the downstream promoter. It should be noted that the putative downstream transcription start site should be classified as requiring confirmation by S1 nuclease analysis, as avian myeloblastosis virus reverse transcrip-

tase may have stalled at this point, perhaps as a result of reading through six adenine bases.

The exact start point of the M-protein transcript in CS64 is not known; however, its corresponding promoter must overlap the promoters that are active in CS24 as the start point is interstitial to the start points in CS24.

Conclusions

The phase-locked streptococci studied here differ both quantitatively and qualitatively in their expression of M-protein-specific mRNA relative to that produced by their parent strains. The diminution of M-protein levels in CS46 and CS64 is primarily a result of lower levels of transcription of M-protein-specific mRNA and is not the result of posttranslational modification of M protein. The use of a different transcription start site and a weaker promoter at the M12 locus in CS64 suggests that the nature of the

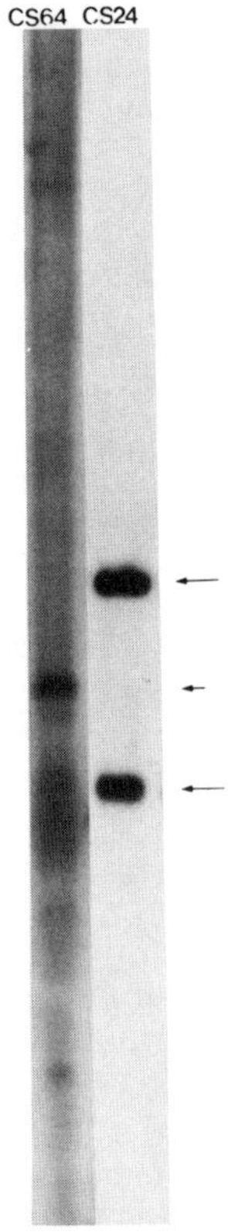

FIG. 2. M12 gene transcription start site mapping by primer extension. RNAs from CS24 (10 μg) and CS64 (40 μg) were each hybridized with a synthetic oligomer (Fig. 3). The template-primer hybrids were incubated with avian myeloblastosis virus reverse transcriptase in the presence of KCl and dATP, dGTP, dTTP, and [α-^{32}P]dCTP at 37°C. The reaction products were resolved by electrophoresis through an 8% denaturing acrylamide gel.

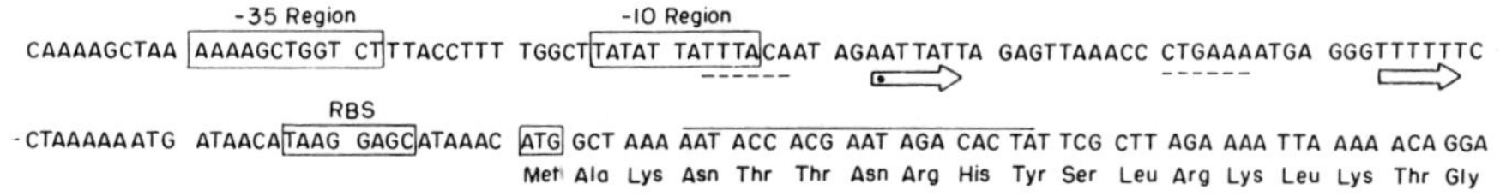

FIG. 3. DNA sequence of the M12 gene promoter region and 5' end of the M12 structural gene in CS24. The transcription start points are indicated by arrows. The promoter sequences at −10 and −35 are either boxed or indicated by a dashed underscore. The sequence complementary to the oligomer used in primer extension reactions is represented by a solid line. RBS, Ribosome binding site.

"down mutation" in this strain is the direct result of altered RNA polymerase specificity of binding to promoter sequences or a decrease in transcription initiation rate or both. Multiple neighboring and overlapping promoter sequences differing in their RNA polymerase recognition abilities could provide a means by which M-protein expression may be regulated under a variety of conditions, i.e., genomic rearrangements that otherwise threaten to fully eliminate the antiphagocytic protein's expression.

Changes of M12 protein expression in CS46 and CS64 cannot yet be directly attributed to upstream deletions; however, the deletions, located within a large open reading frame (Cleary et al., this volume), may act by disrupting expression of a *cis*-acting factor required for high levels of M12 gene transcription. Site-specific mutagenesis and deletion analysis of the cloned M12 gene, in conjunction with genomic DNA sequence analysis of the region upstream of the M12 structural gene, should help define how genomic rearrangements modify M-protein expression.

The finding that the M-protein gene is monocistronic is interesting in the light of reports that M protein is coordinately expressed with other phenotypic markers. A polycistronic transcript encoding M protein and other proteins could allow one mutation to directly affect the expression of multiple proteins. However, the possibility still exists that the mutation causing the decrease in M-protein transcription also affects expression of neighboring genes, should an operon exist, or of other genes throughout the genome. A deletion that causes the abrogation of expression of a positive regulatory element is one mechanism that could result in this effect.

We thank Jonathan G. Spanier, William Phelps, Michael Williams, and Elizabeth Haanes-Fritz for helpful discussions.

This work was supported by Public Health Service grant AI16722 from the National Institute of Allergy and Infectious Diseases. J. C. Robbins was supported by Public Health Service training grant 5T32HL107114 as a predoctoral trainee from the National Heart and Lung Institute.

LITERATURE CITED

1. **Cleary, P., E. Chenoweth, and D. Wexler.** 1984. A streptococcal inactivator of chemotaxis: a new virulence factor specific to group A streptococci, p. 179–180. *In* Y. Kimura, S. Katami, and Y. Shiokawa (ed.), Recent advances in streptococci and streptococcal diseases. Reedbooks, Ltd., Berkshire, England.
2. **Cleary, P., and Z. Johnson.** 1977. Possible dual function of M protein: resistance to bacteriophage A25 and resistance of phagocytosis by human leukocytes. Infect. Immun. **16:** 280–292.
3. **Cleary, P., Z. Johnson, and L. Wannamaker.** 1975. Genetic instability of M protein and serum opacity factor of group A streptococci: evidence suggesting extrachromosomal control. Infect. Immun. **12:**109–118.
4. **Rosenberg, M., and D. Court.** 1979. Regulatory sequences involved in the promotion and termination of RNA transcription. Annu. Rev. Genet. **13:**319–353.
5. **Shine, T., and L. Dalgarno.** 1974. The 3'-terminal sequence of *Escherichia coli* 16S ribosomal RNA: complementarity to nonsense triplets and ribosome binding sites. Proc. Natl. Acad. Sci. USA **71:**1342–1346.
6. **Spanier, J., S. Jones, and P. Cleary.** 1984. Small DNA deletions creating avirulence in *Streptococcus pyogenes*. Science **225:**935–938.
7. **Thomas, P.** 1980. Hybridization of denatured RNA and small DNA fragments transferred to nitrocellulose. Proc. Natl. Acad. Sci. USA **77:**5201–5205.
8. **Todd, E., and R. Lancefield.** 1928. Variants of hemolytic streptococci: their relation to the type specific substance, virulence and toxin. J. Exp. Med. **48:**751–767.
9. **van de Rijn, I., and V. Fischetti.** 1981. Immunochemical analysis of intact M protein secreted from cell wall-less streptococci. Infect. Immun. **32:**86–91.

Genetics of Type 5 M Protein of *Streptococcus pyogenes*

MICHAEL A. KEHOE,[1] LORNA MILLER,[1] THOMAS P. POIRIER,[2] EDWIN H. BEACHEY,[2] MAUREEN LEE,[1] AND DEAN HARRINGTON[1]

Department of Microbiology, Medical School, University of Newcastle-upon-Tyne, Newcastle-upon-Tyne, NE2 4HH, United Kingdom,[1] and Veterans Administration Medical Center and the University of Tennessee Center for the Health Sciences, Memphis, Tennessee 38104[2]

Group A streptococci (*Streptococcus pyogenes*) are important human host-specific pathogens. There are two principal sites of infection, the pharynx and the skin, from which the organism can spread and give rise to a variety of inflammatory diseases. In addition, there are two remote and serious nonsuppurative sequelae which can occur as a consequence of group A streptococcal infection at these sites, namely, poststreptococcal glomerulonephritis and acute rheumatic fever (ARF). In poststreptococcal glomerulonephritis, deposition of circulating streptococcal antigen-antibody complexes on the glomerular capillary basement membrane is thought to play an important role, though the precise mechanisms are still unclear. The pathogenesis of ARF is even less clearly understood. For over 20 years, it has been known that group A streptococci possess cell surface antigens which elicit antibodies that can cross-react with human host tissues, in particular human heart tissue (6, 7). It has been suggested that these host tissue cross-reactive antibodies initiate damaging inflammatory reactions on these host tissues.

Role of M Protein in Virulence

S. pyogenes cultures express a variety of extracellular products which have in vitro properties that implicate them as potential virulence factors. The large number of potential virulence factors and the absence of suitable mutants and good animal model systems make it difficult to assess the contribution of individual products to virulence. Nevertheless, a large body of evidence indicates that the cell surface M protein is a major virulence factor in the pathogenesis of *S. pyogenes* infections (6). The M protein forms a fibrillar structure on the streptococcal cell surface, possibly in association with other cell surface proteins and lipoteichoic acid (10). The carboxy-terminal end of M protein appears to be closely associated with the cell surface, whereas the amino-terminal end appears to extend outwards from the cell. However, the mechanism by which M protein fibrils are assembled on and attached to the cell surface remains to be elucidated.

Expression of M protein on the cell surface renders the cell resistant to phagocytosis (11).

The resistance to phagocytosis may be due, in part, to the ability of M protein to bind fibrinogen (15). In the absence of fibrinogen, M protein-positive (M^+) cells are opsonized by complement, though not as effectively as are M protein-negative cells. In the presence of fibrinogen, M^+ cells are completely resistant to opsonization by complement (15). The binding of fibrinogen to the M protein appears to mask complement receptors on the cell surface, preventing efficient complement binding.

Over 80 serotypes of M protein have been identified, and only one serotype is expressed by each strain (6). Epidemiological studies have shown that different M serotypes of group A streptococci are not equally capable of causing particular diseases. The M types that infect the pharynx tend to be different from those that infect the skin. In poststreptococcal glomerulonephritis about a dozen particular M types predominate (6). Although the epidemiology of ARF is less clear, an increasing body of evidence suggests that some M types are more rheumatogenic than others. For example, serotype M5 strains have been strongly associated with ARF in a number of distinct geographical regions throughout the world, whereas M4 strains have never been associated with ARF (1, 2). It is not known whether structural features unique to particular serotypes of M protein are important in determining the ability to cause a particular disease or whether this is determined solely by other (unidentified) properties associated with particular strains. However, the recent observations that at least some rheumatogenic M serotypes, including M5, contain epitopes within their covalent structure that are immunologically cross-reactive with human heart sarcolemmal membranes suggest that, at least in the case of ARF, the M protein molecule may contribute directly to the pathogenesis of the disease (3, 5).

Role of M Protein in Immunity

Immunity to group A streptococcal infections is due to anti-M protein antibodies which opsonize the cell, rendering it susceptible to phagocytosis (6). Thus, the M protein has been the target toward which efforts to develop a vaccine to group A streptococci have been focused. However, two significant problems have frustrated attempts over the past 50 years to develop a safe and effective vaccine. One is the presence of host tissue cross-reactive epitopes within this covalent structure of M protein (see above). Although the pathogenesis of ARF is poorly understood, the possibility that M protein epitopes which cross-react with human heart tissue might play a role in the disease requires that any vaccine intended for human use be free from such epitopes. In principle, this problem can be overcome by identifying and separating heart cross-reactive and heart non-cross-reactive M protein epitopes, to produce a well-defined vaccine. This could be achieved by peptide synthesis or by cloning limited sequences of M protein genes to produce defined M protein epitopes.

The second problem, which may prove more difficult to overcome, is that immunity is M-serotype specific and over 80 distinct M serotypes have been identified. Immunological cross-reactions between M proteins have been detected in the laboratory, and some cross-reacting antibodies have been shown to be opsonic. For example, some monoclonal antibodies raised to pepM5 (the amino-terminal half of type 5 M protein released from the cell surface by pepsin digestion) have been shown to opsonize M type 6, 19, and 24 strains (3, 4). Immunization of mice with a trivalent vaccine consisting of acid-extracted type 1, 3, and 12 M proteins has been reported to elicit protection against M type 6 and 14 streptococci, in addition to the vaccine strains (18). In these studies the immunogen eliciting opsonic cross-reactive antibodies probably consisted of M protein fragments rather than intact M protein fibrils. The fact that naturally acquired immunity is serotype specific implies that opsonogenic cross-reactive epitopes are immunorecessive when presented to the host in intact M protein fibrils on the cell surface. However, studies on other antigenically variable surface molecules (for example, the pili of *Neisseria gonorrhoeae* and influenza virus hemagglutinin) have shown that structurally conserved, normally immunorecessive regions can elicit antibodies when presented to the host in an altered sequence context, such as a synthetic peptide, and that the antibodies produced can react with the intact molecule (13, 17). Thus, defined sequences corresponding to conserved, normally immunoreactive regions of M proteins may elicit broadly cross-reactive antibodies. However, not all cross-reactive antibodies will be protective. Most immunological cross-reactions between M proteins have been identified by tests such as immunoprecipitation, which do not measure protection, rather than by opsoni-

zation tests. Whitnack et al. (16) have shown that a large proportion of the cross-reactive antibodies in anti-pepM sera are directed against epitopes which are likely to be masked by fibrinogen in vivo, and hence would not confer good protection. Unless we are fortunate enough to identify broadly cross-reactive epitopes which elicit good protection in vivo, an M protein vaccine will need to contain defined epitopes from a number of different serotypes. A better understanding of the structural and immunochemical relationships between different serotypes of M protein is required to determine whether an M protein vaccine, containing a limited number of defined epitopes, is feasible.

Genetic Analysis of Serotype 5 M Protein

Previous studies on the structure and immunochemistry of M proteins have been limited by technical difficulties in extracting intact M protein from the cell surface in sufficient quantities for analysis. Most studies to date have utilized amino-terminal M protein fragments, termed pepM, which are released from the streptococcal cell surface by limited pepsin digestion and represent only about 50% of intact molecules. To extend structural studies to entire M protein molecules and to facilitate studies on the role of M-protein structures in pathogenesis, on the feasibility of designing a safe and effective M protein vaccine, on the mechanisms which generate antigenic variation, and on the way in which M protein fibrils are assembled on the cell surface, recent studies have focused on cloned M protein genes. In other papers in this volume, studies on cloned serotype 6 and 12 M proteins are described (J. R. Scott et al., this volume; P. Cleary et al., this volume). Below we summarize our studies on a cloned serotype 5 M protein gene.

We chose to study the genetics of serotype 5 M protein because type 5 strains have been closely associated with acute rheumatic fever (1, 2). A gene bank of serotype 5 strain Manfredo was constructed in *Escherichia coli*, using a lambda phage vector, L47.1 (8), and screened by immunoblotting with anti-pepM5 sera to detect recombinant phage expressing the M5 antigen (M5$^+$). Hybrid M5$^+$ phage were detected in the gene bank at the surprisingly high frequency of about 1 in 400 (8). Since we are interested in the structure and immunochemistry of M proteins, before proceeding it was necessary to ensure that the cloning had not resulted in any alteration in these properties. Therefore, the cloned M5 gene product was concentrated from M5$^+$ lysates and compared with the M5 antigen expressed by the parent strain Manfredo (12). Immunoblotting of the cloned gene product, separated by sodium dodecyl sulfate-polyacrylamide gel electrophoresis, showed that it was identical in size to that expressed by the group A streptococcus and that, like M5 protein extracted from streptococci, it was sensitive to proteolytic cleavage, resulting in multiple polypeptide bands. In addition, the cloned gene product reacted with six anti-pepM5 monoclonal antibodies tested. Double-immunodiffusion tests employing anti-pepM5 serum displayed a line of identity between pepM5 and the cloned gene product. The cloned gene product was shown to absorb opsonic antibodies from anti-pepM5 sera and to react with human heart cross-reactive anti-pepM5 antibodies, purified by binding to and elution from human heart tissue. Finally, when concentrated M5$^+$ lysates were used to immunize rabbits, they produced a good antibody response, eliciting both protective and human heart cross-reactive antibodies (12). Thus, no differences could be detected between the M5 protein expressed by the cloned genes in *E. coli* and that expressed by the parent streptococcus strain.

Multiple Copies of the Type 5 M Protein Gene

The cloned DNA sequences from one of the M5$^+$ hybrid phage were subcloned into a low-copy-number plasmid vector to construct the M5$^+$ hybrid plasmid pMK207 (Fig. 1). The sequence encoding the M5 gene (termed *smp5*, for streptococcus M protein type 5) was mapped by isolating a combination of subclones, in vitro-generated deletion mutants, and transposon

Tn*1000* insertion mutants and analyzing these by immunoblotting total cell polypeptides, separated by sodium dodecyl sulfate-polyacrylamide gel electrophoresis, with anti-pepM5 sera (8). The *smp5* gene was located within a 2.0-kilobase-pair region of the plasmid, and the transcriptional orientation of the gene was deduced from the expression of truncated gene products by a number of the insertion mutants. When two distinct DNA sequences from within the mapped *smp5* gene were used as probes in Southern blotting experiments, several copies of DNA sequences sharing homology with the probes were detected in the parent strain Manfredo genome (8). This suggested that more than one copy of the *smp5* gene existed in the genome of this strain. To determine whether this was true, we examined additional M5$^+$ phage isolated from the gene bank.

The *smp5* gene, cloned and mapped on plasmid pMK207, contains a 1.5-kilobase-pair *Hind*III fragment (Fig. 1). Digestion with *Hind*III was used to screen DNA from M5$^+$ phage isolated from the gene bank, and phage lacking a 1.5-kilobase-pair *Hind*III fragment were identified. The cloned DNA sequences from one such phage were subcloned into a plasmid vector to construct the M5$^+$ hybrid plasmid pMK237, and restriction endonuclease cleavage sites in this plasmid were mapped (Fig. 1). Both pMK207 and pMK237 expressed M5 antigens of the same size, and no immunological differences were detected between these anti-

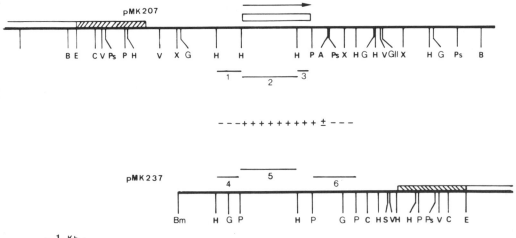

FIG. 1. Summary of sequence relationships between the M5$^+$ plasmids pMK207 and pMK237. The wide boxes at the left end of pMK207 and right end of pMK237 represent vector DNA sequences, and the thick horizontal lines represent cloned streptococcal DNA sequences. The box and arrow above pMK207 indicate the position and transcriptional orientation of the mapped *smp5* gene. The horizontal lines numbered 1 to 6 denote the sequences used as hybridization probes, and plus and minus signs denote homologous or nonhomologous regions. The letters represent the following restriction endonuclease sites: A, *Acc*I; B, *Bal*I; Bm, *Bam*HI; C, *Cla*I; E, *Eco*RI; G, *Bgl*I; GII, *Bgl*II; H, *Hind*III; P, *Pvu*II; Ps, *Pst*I; S, *Sal*I; V, *Eco*RV; and X, *Xba*I.

gens. The DNA sequences in pMK237, which shared homology with the pMK207 *smp5* gene probes, were mapped by Southern blotting experiments. These sequences were in turn used as probes in hybridizations against Southern blots of pMK207 DNA. The results of these experiments are summarized in Fig. 1. Only probes corresponding to the *smp5* gene on pMK207 hybridized. Flanking sequences did not, and they also possessed a different restriction endonuclease cleavage pattern. Together with the results of the Southern blots against chromosomal DNA, these results strongly suggest that there are at least two copies of the *smp5* gene, located at distinct sites in the strain Manfredo genome. Recently, Scott et al. (14), by using a cloned M6 gene probe which shares homology with the M5 gene, detected two bands in Southern blots against DNA from a different type 5 strain, whereas only a single band was detected with other M types tested, including M6. It is possible that multiple copies of the *smp* gene might be a characteristic feature of type 5 strains.

Sequence Homology between M Protein Genes

To determine whether M protein genes share conserved sequences, probes from different regions of the cloned *smp5* gene have been used in dot-blot hybridization experiments, performed under various stringency conditions, against DNA isolated from a large number of different group A streptococcus M types. Representative strains from related species were also included in the blots. In addition, a probe from a cloned streptolysin O gene (9) was also included as a positive hybridization control. None of the probes hybridized to DNA from group B streptococci, *S. mutans, S. faecalis,* or *S. sanguis,* even under low-stringency hybridization conditions (1.0× SSC [1× SSC is 0.15 M NaCl plus 0.015 M sodium citrate], 65°C). Under these conditions, a probe corresponding to the carboxy-terminal end of the M5 protein (probe 3 in Fig. 1) hybridized strongly to DNA from all of the group A streptococcus M types tested, including a phenotypically M⁻ strain, and also hybridized to DNA from group C and G streptococci. Even under high-stringency conditions (0.1× SSC, 65°C), this probe hybridized well to DNA from 75% of these strains. An adjacent probe, corresponding to the amino-terminal two-thirds of the M5 protein (probe 2 in Fig. 1), displayed much greater variation in its ability to hybridize to DNA from different M types. Under high stringencies it hybridized well to DNA from only a few M types and either produced weak signals or did not hybridize at all to DNA from the majority of M types tested. Under low-stringency conditions, this

probe hybridized to DNA from a wider range of M types, but still produced weak signals or did not hybridize to many of the dots. The control *slo* gene probe hybridized strongly to DNA from all the group A, C, and G streptococcus strains tested, under both high- and low-stringency conditions.

Conclusions

In summary, M protein genes do share conserved sequences, and the sequences corresponding to the carboxy-terminal end of the M protein are more strongly and widely conserved than sequences corresponding to the amino-terminal end. This is not surprising since the carboxy-terminal end of M protein is the part most closely associated with the cell surface and may require a conserved structure for its attachment to the cell. The amino-terminal end protrudes outward from the cell and is more likely to be subjected to selective immunological pressure for variation.

The fact that M protein genes share strongly conserved sequences at the 3′ end suggests that the carboxy-terminal end of M proteins may share one or more widely conserved epitopes. That this is the case has been demonstrated by the synthesis of an immunogenic peptide corresponding to a region close to the carboxy-terminal end of type 24 M protein. Antibodies raised to this synthetic peptide cross-reacted in immunoblotting experiments with a wide range of different serotypes of M protein. Unfortunately, these antibodies did not opsonize streptococci in opsonization tests performed in vitro (V. Burdette and E. H. Beachey, unpublished data), indicating that, although capable of eliciting broadly cross-reactive anti-M protein antibodies, this epitope is not accessible for protective antibody binding on the intact streptococcal cell. It seems likely that shared epitopes close to the carboxy-terminal end of M proteins may be masked by other cell wall components or M protein-bound fibrinogen in vivo. Therefore, efforts to identify epitopes which would prove useful in a vaccine will probably need to focus on the amino-terminal halves of M proteins which, as described above, are considerably more variable than the carboxy-terminal ends. Epitopes which elicit antibodies that cross-react with and opsonize a limited number of heterologous M types of streptococci have been identified in the amino-terminal ends of some M proteins, including pepM5 (4). A detailed understanding of the structural relationships between different serotypes of M proteins may allow a limited number of cross-reactive and opsonogenic epitopes to be chosen, which would provide protection against a range of epidemiologically related strains of *S. pyogenes.*

This work was supported by research grants from the United Kingdom Medical Research Council (G8426211CB), the Nuffield Foundation, and the U.S. Veterans Administration, and by Public Health Service grants AI-10085 and AI-13550 from the National Institute of Allergy and Infectious Diseases.

LITERATURE CITED

1. **Bisno, A. L.** 1980. The concept of rheumatogenic and non-rheumatogenic group A streptococci, p. 789–903. *In* S. E. Reid and J. B. Zabriskie (ed.), Streptococcal diseases and the immune response. Academic Press, Inc., New York.
2. **Bisno, A. L., X. Berrios, F. Quesney, D. M. Monroe, Jr., J. B. Dale, and E. H. Beachey.** 1982. Type-specific antibodies to structurally defined fragments of M proteins in patients with acute rheumatic fever. Infect. Immun. **38:**573–579.
3. **Dale, J. B., and E. H. Beachey.** 1982. Protective antigenic determinant of streptococcal M protein shared with sarcolemmal membrane protein of human heart. J. Exp. Med. **156:**1165–1176.
4. **Dale, J. B., and E. H. Beachey.** 1984. Unique and common protective epitopes and among different serotypes of group A streptococcal M proteins defined with hybridoma antibodies. Infect. Immun. **46:**267–269.
5. **Dale, J. B., I. Ofek, and E. H. Beachey.** 1980. Heterogeneity of type-specific and cross-reactive antigenic determinants within a single M protein of group A streptococci. J. Exp. Med. **151:**1026–1038.
6. **Fox, E. N.** 1974. M proteins of group A streptococci. Bacteriol. Rev. **38:**57–86.
7. **Kaplan, M. H., and M. Meyerserian.** 1962. An immunological cross-reaction between group A streptococcal cells and human heart tissue. Lancet **i:**706–710.
8. **Kehoe, M. A., T. P. Poirier, E. H. Beachey, and K. N. Timmis.** 1985. Cloning and genetic analysis of serotype 5 M protein determinant of group A streptococci: evidence for multiple copies of the M5 determinant in the *Streptococcus pyogenes* genome. Infect. Immun. **48:**198–203.
9. **Kehoe, M. A., and K. N. Timmis.** 1984. Cloning and expression in *Escherichia coli* of the streptolysin O determinant from *Streptococcus pyogenes*: characterization of the cloned streptolysin O determinant and demonstration of the absence of substantial homology with the determinants of other thiolactivated toxins. Infect. Immun. **43:**804–810.
10. **Ofek, I., W. A. Simpson, and E. H. Beachey.** 1982. Formation of molecular complexes between a structurally defined M protein and acylated or deacylated lipoteichoic acid of *Streptococcus pyogenes*. J. Bacteriol. **149:**426–433.
11. **Peterson, P. K., D. Schmelling, P. P. Cleary, B. J. Wilkinson, Y. Kim, and P. G. Quie.** 1979. Inhibition of alternative complement pathway opsonization by group A streptococcal M protein. J. Infect. Dis. **139:**575–585.
12. **Poirier, T. P., M. A. Kehoe, J. B. Dale, K. N. Timmis, and E. H. Beachey.** 1985. Expression of protective and cardiac cross-reactive epitopes of type 5 streptococcal M proteins in *Escherichia coli*. Infect. Immun. **48:**198–203.
13. **Rothbard, J. B., R. Fernandez, and G. K. Schoolnik.** 1984. Strain-specific and common epitopes of gonococcal pili. J. Exp. Med. **160:**208–221.
14. **Scott, J. R., S. K. Hollingshead, and V. A. Fischetti.** 1986. Homologous regions within M protein genes in group A streptococci of different serotypes. Infect. Immun. **52:**609–612.
15. **Whitnack, E., and E. H. Beachey.** 1985. Inhibition of complement-mediated opsonization and phagocytosis of *Streptococcus pyogenes* by D fragments of fibrinogen and fibrin bound to surface M protein. J. Exp. Med. **162:**1983–1997.
16. **Whitnack, E., J. B. Dale, and E. H. Beachey.** 1984. Common protective antigens of group A streptococcal M proteins masked by fibrinogen. J. Exp. Med. **159:**1201–1212.
17. **Wilson, I. A., H. L. Niman, R. A. Houghten, A. R. Cherenson, M. L. Connolly, and R. A. Lerner.** 1984. The structure of an antigenic determinant in a protein. Cell **37:**767–778.
18. **Wittner, M. K., and E. N. Fox.** 1977. Homologous and heterologous protection of mice with group A streptococcal M protein vaccines. Infect. Immun. **15:**104–108.

Surface Expression of Type 5 M Protein of *Streptococcus pyogenes* in *Streptococcus sanguis*

THOMAS P. POIRIER,[1] MICHAEL A. KEHOE,[2] ELLEN WHITNACK,[1] AND EDWIN H. BEACHEY[1]

Veterans Administration Medical Center and Department of Medicine, University of Tennessee, Memphis, Tennessee 38104,[1] and Department of Microbiology, The Medical School, The University of Newcastle upon Tyne, Newcastle upon Tyne, NE2 4HH, England[2]

It is well established that the surface M protein is the major virulence factor of group A streptococci (*Streptococcus pyogenes*) (8, 13). This fibrillar protein renders the organisms resistant to phagocytosis. The resistance to phagocytosis is overcome only by antibodies directed toward protective epitopes on the M-protein molecules (8, 13). Efforts to prepare vaccines from the isolated M-protein molecules have been hampered by the finding that purified M-protein preparations contain antigens that evoke cross-reactive immunity to host tissue, especially cardiac tissue (2–4a, 6). Indeed, it was shown that such cross-reactive epitopes reside within the M-protein molecule itself (2). This was found to be especially true of M protein derived from M serotype 5 *S. pyogenes*, the single most frequent serotype of streptococci isolated during outbreaks of acute rheumatic fever around the world (1).

In an attempt to gain a better understanding of the physicochemical properties of M protein and the relationship of various parts of the molecule to protective immunity and the pathogenesis of acute rheumatic heart disease, several M proteins have been cloned and expressed in *Escherichia coli* (5, 7, 10–12). In our laboratories, we have achieved expression of type 5 M protein by *E. coli* LE92 using the λL47.1 replacement vector (7, 10). The entire M-protein gene (*smp-5*) was subcloned into an *E. coli* plasmid, pLG339, as demonstrated by restriction enzyme and γδ insertion analysis combined with Western immunoblot analysis (7). The plasmid containing the M-protein gene was designated pMK207. The M protein expressed by this recombinant plasmid contained both the protective as well as the cardiac tissue cross-reactive epitopes (10).

Having the entire M-protein gene in hand, we undertook studies of the expression of M protein in a gram-positive organism, *Streptococcus sanguis*. The reasons for cloning M protein into these normally avirulent organisms are threefold: (i) the cloned M protein would be more likely to be expressed on the surface of gram-positive than gram-negative bacteria, and surface expression is required to study the antiphagocytic properties of M protein in a direct way; (ii) a shuttle vector system has been developed whereby genetic material cloned into *E. coli* can be readily used to transform *S. sanguis*; and (iii) genetic manipulations of the M-protein gene (*smp-5*) expressed in these organisms may provide information as to the mechanisms of transport, anchorage, expression, and epitope presentations of the M-protein molecule.

Construction of Shuttle Plasmid Containing the M-Protein Gene *smp-5*

S. sanguis V288 and the shuttle plasmid pVA838 were generously provided by Don Clewell, University of Michigan, and Frank Macrina, Virginia Commonwealth University. The construction of a shuttle plasmid containing the *smp-5* gene is illustrated in Fig. 1. Briefly, the 6.2-kilobase pair (kbp) *Eco*RV B fragment of pMK207 containing *smp-5* was added to a *Pvu*II digest of pVA838 (9) at a 50:1 ratio. Successful blunt-end replacement of the 0.38-kbp *Pvu*II fragment of pVA838 with the 6.2-kbp *Eco*RV B fragment of pMK207 was demonstrated by the loss of chloramphenicol resistance and conservation of erythromycin resistance (Em^r Cm^s) of *E. coli* V850 transformed with the modified shuttle vector. Em^r Cm^s transformants of *E. coli* expressing type 5 M protein were identified by colony immunoblotting as previously described (7).

The M5-positive transformants of *E. coli* produced immunoreactive proteins of M_r 64,000, 58,000, 55,000, and 53,000 as demonstrated by sodium dodecyl sulfate-polyacrylamide gel electrophoresis and Western immunoblotting using anti-pepM5 sera (Fig. 2). *Hin*dIII digestion of the plasmids isolated from those transformants produced fragments of the size expected for the replacement of the 0.38-kbp *Pvu*II fragment of pVA838 by the 6.2-kbp *Eco*RV fragment B of pMK207. The newly constructed plasmid, pBK100, was then purified from the *E. coli* transformants and used to transform *S. sanguis*.

Transformation of *S. sanguis* with pBK100

S. sanguis V288 cells were transformed with pBK100 as described (9). Erythromycin-resist-

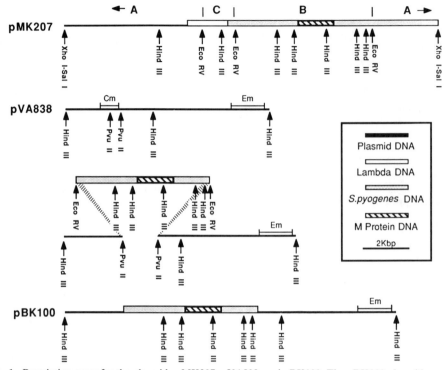

FIG. 1. Restriction maps for the plasmids pMK207, pVA838, and pBK100. The pBK100 plasmid construct is formed by blunt-end replacement of the 0.38-kbp *Pvu*II fragment of pVA838 with the 6.2-kbp *Eco*RV B fragment of pMK207.

ant transformants expressed the M_r 58,000, 55,000, and 53,000 polypeptides but not the M_r 64,000 polypeptide as demonstrated by sodium dodecyl sulfate-polyacrylamide gel electrophoresis and immunoblotting (Fig. 2). The reason for the failure to detect the M_r 64,000 protein remains unclear. Control transformants containing the shuttle vector pVA838 without M-protein genes failed to express any of the M-protein polypeptides.

Expression of Type 5 M Protein on the Surface of *S. sanguis*

To determine whether the M protein produced by the pBK100-transformed *S. sanguis* was expressed on the surface of these cells, three types of experiments were performed. First, enzyme-linked immunosorbent assay inhibition tests were performed using a pepsin extract of type 5 streptococci (pepM5) as the immobilized antigen, dilutions of rabbit anti-pepM5 as the antibody, and whole M5-producing *S. sanguis* cells as inhibitor. The intact cells inhibited the enzyme-linked immunosorbent assay in a dose-response fashion (T. P. Poirier et al., submitted for publication). It should be noted, however, that between three and four times more *S. sanguis*(pBK100) cells than control *S. pyogenes*

cells were required to achieve complete inhibition.

In the second experiment, immunofluorescence tests were performed by allowing the intact M-protein-producing *S. sanguis* cells to react with rabbit anti-pepM5 antibodies followed by fluorescein-labeled goat anti-rabbit immunoglobulin G. As can be seen in Fig. 3, approximately one-third of the transformed cells fluoresced brightly whereas the remaining organisms fluoresced weakly or not at all. Immunoelectron microscopy of anti-pepM5-antibody-treated *S. sanguis* transformants followed by ferritin-labeled goat anti-rabbit gamma-globulin antibody confirmed these findings (Poirier et al., submitted).

In the third experiment, tests of the inhibition of opsonization (2) of type 5 group A streptococci were performed by using phagocytosis-resistant M type 5 cells of *S. pyogenes* as test organisms, anti-pepM5 antibody as opsonin, and M5-transformed pBK100 cells of *S. sanguis* as inhibitor. The *S. sanguis*(pBK100) cells completely inhibited opsonization and phagocytosis of M type 5 *S. pyogenes* cells, although it was again necessary to use three to four times more of the transformed *S. sanguis* cells than control *S. pyogenes* cells to achieve maximum inhibition.

These results indicated that protective epitopes of type 5 M protein are accessible for opsonic antibody binding on the surface of the M-protein-producing *S. sanguis* cells.

In none of the above experiments did the control *S. sanguis* cells transformed with the control shuttle plasmid lacking the M-protein genes react with the anti-M protein antibodies. These results clearly demonstrate that the M protein is expressed on the surface of the transformed *S. sanguis* cells. It must be pointed out, however, that the surface expression of M proteins by these cells is heterogeneous, with only about one-third of the cells fully expressing surface M proteins. This has made it difficult to test the antiphagocytic properties of the M protein expressed on the surface of *S. sanguis* cells in a direct way because of a high background of cells that lack M protein and, therefore, do not resist phagocytosis. This problem has been partially solved in preliminary experiments by treating *S. sanguis*(pBK100) cells with fluorescein-labeled fibrinogen. The M-positive cells that bind the fibrinogen (14) are then enriched by fluorescence-activated cell sorting. These sorted, fibrinogen-binding cells are fully resistant to phagocytosis by polymorphonuclear leukocytes in fresh human blood (Poirier et al., submitted).

Conclusions

We have shown that the M protein of *S. pyogenes* can be cloned and expressed in another species of gram-positive bacteria, *S. sanguis*. The M-protein polypeptides produced

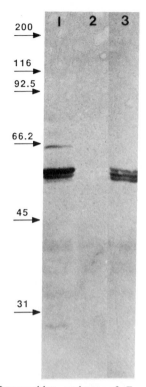

FIG. 2. Immunoblot analyses of *E. coli* V850-(pBK100) (lane 1), *S. sanguis* V288(pVA838) (lane 2), and *S. sanguis* V288(pBK100) (lane 3) allowed to react with rabbit anti-pepM5 antisera. Arrows indicate position of molecular weight standards; numbers represent thousands.

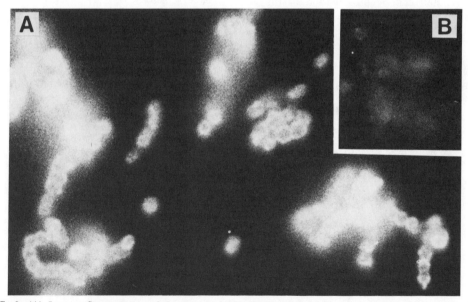

FIG. 3. (A) Immunofluorescence of *S. sanguis* V288(pBK100) allowed to react with rabbit anti-pepM5 antisera. Note the heterogeneity of fluorescence of the cells. (B) *S. sanguis* V288(pVA838) treated as above except that the film exposure was three times longer so that cells could be photographed.

by the transformed *S. sanguis* cells are similar to those produced by the same clone in *E. coli*, except that the M_r 64,000 M protein immunoreactive polypeptide produced by the latter organisms is not detected. The M protein produced by *S. sanguis* is expressed on the surface of the bacterial cells as demonstrated by the ability of the intact cells to inhibit M-protein antibodies in enzyme-linked immunosorbent assays and opsonization inhibition tests. Further evidence for a surface location for *S. sanguis*(pBK100) M5 protein is demonstrated by immunofluorescence and immunoelectron microscopy of organisms treated with anti-M protein antibody. Surface expression of the M protein, however, was heterogeneous; only about one-third of the transformed cells were reactive with anti-M protein antibodies in immunofluorescence and immunoferritin microscopy. Advantage has been taken of the ability of M protein to bind plasma fibrinogen. In preliminary studies, a population of *smp-5*-transformed *S. sanguis* cells capable of binding fluorescein-labeled fibrinogen has been selected for study by fluorescence-activated cell sorting. These fibrinogen-binding (M-positive) cells were shown to be fully resistant to ingestion by phagocytic cells in fresh human blood, a functional property primarily attributed to M protein on the surface of *S. pyogenes* cells. Our observations suggest that the *E. coli-S. sanguis* shuttle vector system may be a suitable model for the study of the effects of genetic manipulations of *smp-5* upon the mechanisms of expression, transport, anchorage, and epitope presentation of the M-protein molecule on the surface of gram-positive cocci.

We thank Edna Chiang and Jeff Goode for expert technical assistance, Michael Dockter and Denise Wall for assistance with flow cytometry, Jim Dale for discussions, and Johnnie Smith for expert secretarial assistance. Carolyn Shaw's participation in some of the early studies during a graduate student rotation through our laboratories is gratefully acknowledged.

These studies were supported by research funds from the U.S. Veterans Administration and Public Health Service research grants AI-10085 and AI-13550 from the National Institute of Allergy and Infectious Diseases.

LITERATURE CITED

1. **Bisno, A. L.** 1980. The concept of rheumatogenic and non-rheumatogenic group A streptococci, p. 789–803. *In* S. E. Read and J. B. Zabriskie (ed.), Streptococcal diseases and the immune response. Academic Press, Inc., New York.
2. **Dale, J. B., and E. H. Beachey.** 1982. Protective antigenic determinant of streptococcal M protein shared with sarcolemmal membrane protein of human heart. J. Exp. Med. **156:**1165–1176.
3. **Dale, J. B., and E. H. Beachey.** 1985. Multiple, heart cross-reactive epitopes of streptococcal M proteins. J. Exp. Med. **161:**113–122.
4. **Dale, J. B., and E. H. Beachey.** 1985. Epitopes of streptococcal M protein shared with cardiac myosin. J. Exp. Med. **162:**583–591.
4a. **Dale, J. B., and E. H. Beachey.** 1986. Sequence of myosin-crossreactive epitopes of streptococcal M protein. J. Exp. Med. **164:**1785–1790.
5. **Fischetti, V. A., K. F. Jones, B. N. Manjula, and J. R. Scott.** 1983. Streptococcal M6 protein expressed in *Escherichia coli*: localization, purification and comparison with streptococcal derived M protein. J. Exp. Med. **159:**1083–1095.
6. **Kaplan, M. H.** 1963. Immunologic relation of streptococcal and tissue antigens. I. Properties of an antigen in certain strains of group A streptococci exhibiting an immunologic cross-reaction with human heart tissue. J. Immunol. **90:**595–606.
7. **Kehoe, M. A., T. P. Poirier, E. H. Beachey, and K. N. Timmis.** 1985. Cloning and genetic analysis of serotype 5 M protein determinant of group A streptococci: evidence for multiple copies of the M5 determinant in the *Streptococcus pyogenes* genome. Infect. Immun. **48:**190–197.
8. **Lancefield, R. C.** 1962. Current knowledge of type-specific M antigens of group A streptococci. J. Immunol. **89:**307–313.
9. **Macrina, F. L., J. A. Tobian, K. R. Jones, R. P. Evans, and D. B. Clewell.** 1982. A cloning vector able to replicate in *Escherichia coli* and *Streptococcus sanguis*. Gene **19:**345–353.
10. **Poirier, T. P., M. A. Kehoe, J. B. Dale, K. N. Timmis, and E. H. Beachey.** 1985. Expression of protective and cardiac tissue cross-reactive epitopes of type 5 streptococcal M protein in *Escherichia coli*. Infect. Immun. **48:**198–203.
11. **Scott, J. R., and V. A. Fischetti.** 1983. Expression of streptococcal M protein in *Escherichia coli*. Science **221:**758–760.
12. **Spanier, J. G., S. J. C. Jones, and P. Cleary.** 1984. Small DNA deletions creating avirulence in *Streptococcus pyogenes*. Science **225:**935–938.
13. **Stollerman, G. H.** 1975. Rheumatic fever and streptococcal infection. Grune and Stratton, New York.
14. **Whitnack, E., J. B. Dale, and E. H. Beachey.** 1984. Common protective antigens of group A streptococcal M proteins masked by fibrinogen. J. Exp. Med. **159:**1201–1212.

Cloning and Partial Characterization in *Escherichia coli* of the Immunoglobulin G-Receptor Gene from Group A Streptococci

DAVID G. HEATH AND P. PATRICK CLEARY

Department of Microbiology, University of Minnesota, Minneapolis, Minnesota 55455

Several studies show that certain strains of group A streptococci bind immunoglobulins to their surface in a nonimmune manner (1, 2, 7, 10). Group A streptococci possess immunoglobulin receptors not only for immunoglobulin G (IgG) but also for human IgD and IgA (4, 12). As with protein A of *Staphylococcus aureus*, the binding of IgG to the surface of group A streptococci occurs between the surface receptor and the Fc portion of the IgG molecule (2). Electron microscopy, using ferritin-conjugated IgG, has further revealed that IgG receptors are "localized on the tips of the filamentous protrusions forming the outermost layer of the *Streptococcus* wall" (11).

Currently, there exist five known types of IgG receptors among gram-positive bacteria, based on their specificity for mammalian IgG subclasses (8). The type I receptor, represented by staphylococcal protein A, is able to bind IgG from various mammals but is unable to bind, with few exceptions, human IgG3 subclass antibodies (6, 8, 9). In contrast, the type II receptor, represented by the IgG receptor of group A streptococci, binds IgG from relatively few mammals but is able to bind human IgG3 subclass antibodies (8; unpublished data).

Cloning of the IgG-Receptor Gene

Our initial studies were aimed at cloning the M-protein gene from the group A streptococcal strain CS110 (T12, M76). To accomplish this, we radiolabeled a segment of DNA known to contain a portion of the M-protein gene from strain CS24 (T12, M12). Southern blotting, using *Hin*dIII-digested chromosomal DNA from strain CS110, revealed that a 6.0-kilobase fragment hybridized to the radiolabeled probe. We then purified chromosomal DNA from 4 to 8 kilobases and ligated this DNA into the plasmid expression vector pUC9. *Escherichia coli* JM83 was then transformed with the ligated DNA, and colony hybridizations were performed. A single positive colony was detected, and plasmid DNA isolated from this colony was shown to contain a 6.2-kilobase insert. A restriction map of this insert was then obtained by using the restriction enzymes *Bam*HI, *Sal*I, *Pvu*II, *Nci*I, and *Hae*III (Fig. 1).

Armed with a restriction map of the insert DNA cloned from strain CS110, we next subcloned various fragments of the 6.2-kilobase insert into pUC9 and again transformed *E. coli* JM83. We then prepared sonicates of the various subclones and checked for expression of the M76 protein by immunodiffusion. The rabbit antiserum used in the immunodiffusion tests was prepared against heat-killed strain CS110 and shown to be specific for the M76 protein (13). One subclone, JM83(pDH56) (Fig. 1), reacted strongly not only against the M76 antiserum but also against the preimmune rabbit serum. We then tested the sonicate of JM83(pDH56) against horse, pig, goat, rabbit, and chicken sera and against purified human IgG. The results show that the subclone JM83(pDH56) expresses a protein that binds pig, horse, rabbit, and swine sera, as well as purified human IgG (Fig. 2). A control, in which a sonicate of JM83(pUC9) was allowed to react against the sera, proved negative (data not shown).

We next grew a 4-liter culture of the subclone JM83(pDH56) and prepared protoplasts by the method of Fischetti et al. (3). After centrifugation, the supernatant was saturated with 65% ammonium sulfate to precipitate protein released from the periplasmic space. The precipitate was dialyzed against ammonium bicarbonate, lyophilized, and resuspended in phosphate-buffered saline. We then performed immunodiffusion analyses to determine whether the antigen expressed from JM83(pDH56) could bind IgG from the four human IgG subclasses, as well as the human IgG Fc fragment. As shown (Fig. 3), the protein expressed from JM83-(pDH56) binds the human IgG Fc fragment and human myeloma IgG representing all four subclasses. We therefore consider the protein expressed from JM83(pDH56) to be an IgG receptor with characteristics similar to those reported previously for IgG receptors expressed on the surface of group A streptococci.

We have found, in addition to the above data, that the IgG-receptor protein expressed from JM83(pDH56) has an apparent molecular weight of 30,000 as shown by immunoblot analysis using purified human IgG (data not shown). This

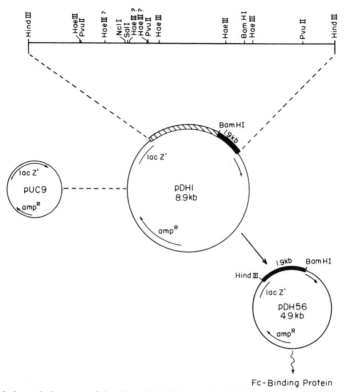

FIG. 1. Expanded restriction map of the cloned 6.2-kilobase *Hin*dIII fragment of group A streptococcal DNA from strain CS110 (T12, M76) and construction of recombinant plasmids pDH1 and pDH56. Plasmid pDH1 contains the 6.2-kilobase *Hin*dIII fragment cloned from strain CS110 into pUC9 while pDH56 contains the small 1.9-kilobase *Bam*HI-*Hin*dIII fragment which expresses the IgG-receptor gene.

molecular weight is in close agreement with that shown for the purified IgG receptor from type 15 group A streptococci (5).

We are currently working on the purification and further characterization of our cloned IgG-receptor protein. In addition, we believe that the IgG-receptor protein gene may be closely linked

FIG. 2. Immunodiffusion gel in which a sonicate of pDH56 was allowed to react against the sera shown and against purified human IgG.

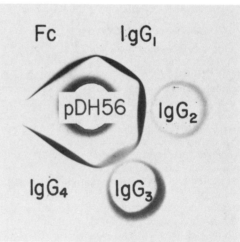

FIG. 3. Immunodiffusion gel in which a periplasmic extract of JM83(pDH56) was allowed to react against human myeloma IgG representing the four subclasses and against purified human Fc fragment.

to the M76 protein gene from strain CS110. Initial Southern hybridization data indicate that the plasmid pDH1 (Fig. 1) contains insert DNA which shares homology with the M12 gene and that this homologous region is closely linked to the gene for the IgG receptor.

LITERATURE CITED

1. **Christensen, P., A. Grubb, R. Grubb, G. Samuelsson, C. Schalen, and M. Svensson.** 1979. Demonstration of non-identity between the Fc-receptor for human IgG from group A streptococci type 15 and M protein, peptidoglycan and the group specific carbohydrate. Acta Pathol. Microbiol. Scand. Sect. C **87**:257–261.
2. **Christensen, P., B. G. Johansson, and G. Kronvall.** 1976. Interaction of streptococci with the Fc-fragment of IgG. Acta Pathol. Microbiol. Scand. Sect. C **84**:73–76.
3. **Fischetti, V. A., K. F. Jones, B. N. Manjula, and J. R. Scott.** 1984. Streptococcal M6 protein expressed in *Escherichia coli*: location, purification and comparison with streptococcal-derived M protein. J. Exp. Med. **159**:1083–1095.
4. **Forsgren, A., and A. O. Grubb.** 1979. Many bacterial species bind human IgD. J. Immunol. **122**:1468–1472.
5. **Grubb, A., R. Grubb, P. Christensen, and C. Schalen.** 1982. Isolation and some properties of an IgG Fc-binding protein from group A streptococci type 15. Int. Arch. Allergy Appl. Immunol. **67**:369–376.
6. **Ito, S., T. Miyazak, and H. Matsumoto.** 1980. Interaction between normal human IgG3 carrying Gm (b1,b3), (g), (s,t) and protein A-Sepharose CL-4B. Proc. Jpn. Acad. **56B**:226–229.
7. **Kronvall, G.** 1973. A surface component in Group A, C, and G streptococci with non-immune reactivity for immunoglobulin G. **111**:1401–1406.
8. **Myre, E. B., and G. Kronvall.** 1982. Immunoglobulin specificities of defined types of streptococcal Ig receptors, p. 209–210. *In* S. E. Holm and P. Christensen (ed.), Basic concepts of streptococci and streptococcal diseases. Reed Books Ltd., Chertsey, England.
9. **Recht, B., B. Fragione, E. Franklin, and E. van Logham.** 1981. Structural studies of a human 3 myeloma protein (GOE) that binds staph protein A. J. Immunol. **127**:917–923.
10. **Reis, K. J., M. D. P. Boyle, and E. M. Ayoub.** 1984. Identification of distinct Fc-receptor molecules on streptococci and staphylococci. J. Clin. Lab. Immunol. **13**:75–80.
11. **Ryc, M., M. Wagner, B. Wagner, and J. Havlicek.** 1982. Electron microscopic study of the location of the Fc-reacting factor on group A streptococci. Microbios **34**:7–16.
12. **Schalen, C., P. Christensen, A. Grubb, G. Samuelsson, and M.-L. Svensson.** 1980. Demonstration of separate receptors from human IgA and IgG in group A streptococci type 4. Acta Pathol. Microbiol. Scand. Sect. C **88**:77–82.
13. **Spanier, J. G., and P. P. Cleary.** 1980. Bacteriophage control of antiphagocytic determinants in group A streptococci. J. Exp. Med. **152**:1393–1406.

Human Isolates of Group A and G Streptococci Share Homologous Sequences Not Found in Animal Group G Isolates: Possible Role in Host Specificity

WARREN J. SIMPSON AND P. PATRICK CLEARY

Department of Microbiology, University of Minnesota, Minneapolis, Minnesota 55455

Group G streptococci, although rarely referred to as human pathogens, are frequently associated with human infections including streptococcal sepsis, pharyngitis, pyoderma, and necrotizing fasciitis (1). Many of these infections are associated with underlying malignancies, and often they are a direct result of group G streptococci existing as normal flora in the throat and skin of the patient (12). The taxonomy of group G streptococci is not clearly defined, but they have been divided into two groups depending on whether they form large or minute colonies (1). The latter type are commonly grouped with viridans streptococci as the species *Streptococcus anginosus* and are thought to be related to group F streptococci (3). Human isolates of group G streptococci studied in this laboratory are normally large colony formers (unpublished data), but as most clinical studies in the past made no distinction with regard to colony type, it is not known whether human isolates can be reliably distinguished by colony size. Large colony types were traditionally referred to as *Streptococcus canis* or group C strains of canine origin, although later biochemical analysis indicated that they are more correctly defined as group G streptococci (12).

M protein is a surface appendage on group A streptococci with over 80 serotypes currently recognized (4) and is primarily responsible for the antiphagocytic characteristic of this organism (7). Similarly, M-protein-like antigens are expressed by phagocyte-resistant human group G streptococci (8, 9), including several which are serologically indistinguishable from type 12 (M12) group A streptococcal M protein (10).

Mapping of M12 Sequences in Group G and A Streptococci

With the emergence of large-colony-forming group G streptococci as important opportunistic and nosocomial human pathogens, genetic studies comparing strains from animal and human sources are needed to determine whether the human isolates are host specific. Thus, the presence of an immunologically identical M12 antigen in human group A and G streptococci offers a unique opportunity to determine whether the M12 gene and adjacent sequences define an expression locus unique to human streptococcal isolates.

To assess this possibility, we first determined the extent of homology between the plasmid pPC101 and genomic DNA from the group G strain CS140. Plasmid pPC101 contains a portion of the M12 protein coding sequence and adjacent 5' sequences from the group A streptococcus strain CS24 (11) (Fig. 1). Strain CS140 is one of three human isolates that were reported by Maxted and Potter (10) to express an M12 protein antigen. As expected, plasmid pPC101 hybridized strongly to some fragments from *Hae*III digests of strain CS140 DNA. Hybridization analyses under high stringency of *Hae*III-, *Rsa*I-, and *Pvu*II-digested genomic DNA from both strains CS24 and CS140, using a variety of DNA probes (Fig. 2A), showed that homology was limited to two segments separated by a minimum length of 2.4 kilobases (Fig. 2B). The left or downstream segment appears homologous to all M12 coding sequences from strain CS24 downstream of the *Hae*III A fragment. However, approximately the first 300 base pairs from the amino-terminal end of the group A M12 gene are absent from strain CS140. The absence of this region implies that it does not specify a critical M12 antigenic determinant in group G streptococci. Other studies have indicated that the amino termini of M proteins protrude from the surface and are antigenically variable segments of M proteins (2, 6). Thus, the observation that the coding sequence of antigenically identical M12 proteins can differ by as much as 300 base pairs at the amino terminus is unexpected. Furthermore, less than 50% of the *Hae*III A fragment and most of the *Hae*III B fragment evident in strain CS24 are not duplicated in strain CS140 (Fig. 2B). Spontaneous deletions in either of these two fragments in strain CS24 reduce M-protein expression (unpublished data), suggesting that a second gene may be involved in M-protein expression in group A streptococci. Expression of the M12 antigen in group G streptococci does not appear to require this sequence.

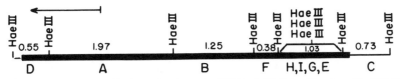

FIG. 1. *Hae*III restriction map of DNA region in strain CS24 hybridizing to the plasmid pPC101. The region corresponding to the cloned streptococcal DNA (approximately 4.8 kilobases) in pPC101 is shown schematically as the thick line. The arrowed line represents the location and direction of the M12 protein gene.

Mapping of Homologous Sequences Unique to Human Group G and A Streptococci

Genomic DNA from an additional unrelated six human and three dog strains of group G streptococci did not hybridize to the M12 probes (Fig. 2C), indicating that homology to M12 sequences was limited to those strains expressing an M12 antigen. However, despite extensive hetero-geneity of the *Hae*III-digested DNA profiles, some *Hae*III fragments similar in size were homologous to the M12 probe (Fig. 2B). These fragments, corresponding to approximately 2.4 kilobases upstream from the group A M12 structural gene (Fig. 2B and C), appear to be specific for the human group G and A streptococcal isolates. Surprisingly, the conserved region of DNA was not present in group G strains isolated from dogs.

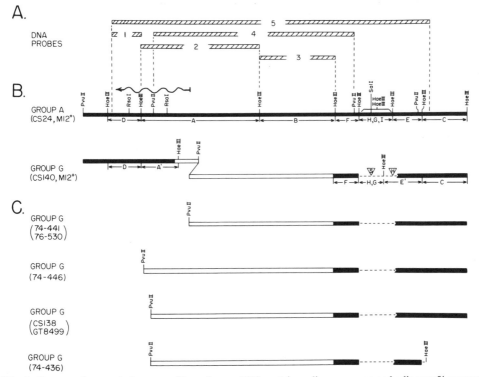

FIG. 2. Comparative restriction map of homologous M12 protein coding sequence and adjacent 5′ sequences in group A and G streptococci. (A) Streptococcal DNA probes 1, 2, 3, 4, and 5 used for mapping analysis. These probes were derived from the cloned 4.8-kilobase streptococcal DNA (probe 5) in pPC101. All hybridization experiments were performed under high stringency conditions on restriction endonuclease-digested DNA Southern blotted onto nitrocellulose. Probes 2, 3, 4, and 5 were prepared by radiolabeling the DNA in vitro with [α-^{32}P]dCTP by nick translation using a nick-translation kit (Bethesda Research Laboratories, Inc.). Probe 1 was a strand-specific M13 probe prepared by the method of Hu and Messing (5). (B) Restriction map of M12 group A strain CS24 and M12 group G strain CS140 homologous sequences. (C) Restriction map of homologous sequences in human group G streptococci. Hatched bars, Streptococcal probes; solid bars, group A and G homologous sequences; open bars, DNA sequences having no homology with probe 5; dashed lines, undefined regions of group A and G homology. The locations of *Hae*III and *Sal*I sites present in strain CS24 DNA but absent from all undefined homologous group G DNA are shown by triangles containing H and S, respectively.

Although extensive homology is conserved in this region, significant differences were evident among the various isolates. The group G strains lacked the single *Sal*I and *Hae*III sites present in strain CS24 (Fig. 2B and C). Furthermore, *Hae*III fragments forming the boundary of the upstream portion of the homologous region varied in size for the various group G isolates. These differences illustrate that, despite extensive homology in this region, rearrangements have occurred, suggesting that this region could be a hot spot for recombination. Additionally, as a result of its proximity to the M12 gene, it may define a virulence locus such as resistance to phagocytosis in both pathogenic group G and A streptococci.

Absence of these sequences in dog strains offers a means to distinguish between dog and human isolates despite their similar colony characteristics. Should such sequences be absent in group G strains from all other nonhuman sources, human isolates may require a taxonomic distinction from animal strains. An exciting prospect for further study is to determine whether the apparently human strain-specific sequences identified in this study encode determinants of host specificity.

This work was supported by Public Health Service grant AI 16722 from the National Institute of Allergy and Infectious Diseases.

We thank Steven Skjold for supplying group G isolates and Jan Smith for assistance in the preparation of this manuscript.

LITERATURE CITED

1. **Cudney, N. J. C., and A. C. Albers.** 1982. Group G streptococci: a review of the literature. Am. J. Med. Technol. **48:**37–42.
2. **Dale, J. B., and E. H. Beachey.** 1986. Localization of protective epitopes of the amino terminus of type 5 streptococcal M protein. J. Exp. Med. **163:**1191–1202.
3. **Facklam, R. R.** 1977. Physiologic differentiation of viridans streptococci. J. Clin. Microbiol. **5:**184–189.
4. **Facklam, R. R., and L. R. Edwards.** 1979. A reference laboratory's investigations of proposed M-type strains of *Streptococcus pyogenes*, capsular types of *S. agalactiae*, and new group antigens of streptococci, p. 251–253. *In* M. T. Parker (ed.), Pathogenic streptococci. Reedbooks Ltd., Chertsey, England.
5. **Hu, N.-T., and J. Messing.** 1982. The making of strand-specific M13 probe. Gene **17:**271–277.
6. **Jones, K. F., B. N. Manjula, K. H. Johnston, S. K. Hollingshead, J. R. Scott, and V. A. Fischetti.** 1985. Location of variable and conserved epitopes among the multiple serotypes of streptococcal M protein. J. Exp. Med. **161:**623–628.
7. **Lancefield, R. C.** 1962. Current knowledge of type-specific M antigens of group A streptococci. J. Immunol. **89:**307.
8. **Lawal, S. F., A. O. Coker, E. D. Salanke, and O. Ogunbi.** 1982. Serotypes among Lancefield-group G streptococci isolated in Nigeria. J. Med. Microbiol. **15:**123–125.
9. **Lawal, S. F., and O. Dosunmu-Ogunbi.** 1984. A new scheme for serotyping of group G streptococci, p. 69–70. *In* Y. Kimura, S. Kotami, and Y. Shiokawa (ed.), Recent advances in streptococci and streptococcal diseases. Reedbooks Ltd., Chertsey, England.
10. **Maxted, W. R., and E. V. Potter.** 1967. The presence of type 12 M-protein antigen in group G streptococci. J. Gen. Microbiol. **49:**119–125.
11. **Spanier, J. G., S. J. C. Jones, and P. Cleary.** 1984. Small DNA deletions creating avirulence in *Streptococcus pyogenes*. Science **225:**935–938.
12. **Vartian, C. P., I. Lerner, D. M. Shales, and K. V. Gopalakrishna.** 1985. Infections due to Lancefield group G streptococci. Medicine (Baltimore) **64:**75–88.

DNA Fingerprints of Group A, Serotype M49 Streptococci

STEPHEN A. SKJOLD[1] AND P. PATRICK CLEARY[2]

Departments of Pediatrics[1] and Microbiology,[2] University of Minnesota Medical School, Minneapolis, Minnesota 55455

Accurate grouping and typing of pathogenic bacteria are important tasks in the epidemiology of infectious disease. The identification of specific markers which differentiate bacteria into outbreak-specific strains is required. Immune specific sera (9), bacteriocins (17), bacteriophages (13, 14), and sensitivity to bacitracin (11) have all been useful for the classification of Lancefield group A streptococci. Although these methods have provided useful information in the past, they are cumbersome to perform and occasional isolates cannot be typed.

The recently developed technique of DNA fingerprinting provides an alternative approach for the characterization of streptococcal strains. Chromosomal DNA is released from bacterial cells by mutanolysin and then purified by phenol, phenol/chloroform, and ether, digested with restriction endonucleases, and electrophoresed in agarose gels (10, 15, 16). The band patterns from different isolates are then compared. This technique has been successfully used in epidemiological studies of *Vibrio cholerae* (6), *Neisseria gonorrhoeae* (2–4), *N. meningitidis* (1, 7, 8), and other organisms.

In this study we examined a collection of group A, serotype M49 streptococci (13, 14) to assess the applicability of the DNA fingerprinting technique for identifying outbreak-specific strains of this serotype and other M serotypes.

To evaluate the DNA fingerprinting method as a useful tool for the identification of outbreak-specific strains within a serotype of group A streptococci, we first tested a collection of group A, M49 streptococci. This collection comprised four pairs of M49 streptococci representing geographically different outbreaks (Fig. 1). Overall, DNA profiles were very similar; however, each pair could be clearly and reproducibly distinguished by at least one or more DNA fragments. Skin infection strains from Oxford, England (14), were the most different because of unique bands at positions A and B (Fig. 1, lanes 5 and 6). DNA from Alaska M49 nephritis strains in lane 7 and 8 had unique bands at position C, and DNA from M49 New Zealand scarlet fever/nephritis strains (12) (lanes 3 and 4) had unique bands at position D (Fig. 1). The M49 strains from an Israel food-poisoning outbreak (5) had no distinct bands in

the high-molecular-weight range but did have four characteristic, evenly spaced bands in the area marked E (Fig. 1). These data suggest DNA fingerprints are able to distinguish between strains of the same serological M type when they are epidemiologically unrelated. The consistency of these fingerprints was shown by examining DNA profiles of eight strains from New Zealand (12) which were associated with nephritis and scarlet fever (Fig. 2). DNAs digested with *Pvu*II had identical profiles.

Another question to consider is whether the DNA fingerprints reflect the specificity of the various M type. Therefore, fingerprints of four pairs of group A streptococcal strains representing serotypes M1, M12, M55, and M62 were compared (Fig. 3). Each M type proved to

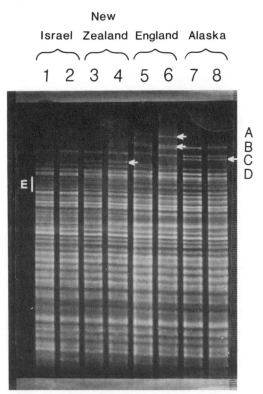

FIG. 1. *Hin*dIII fingerprints of four pairs of human isolates of group A, M49 *S. pyogenes.*

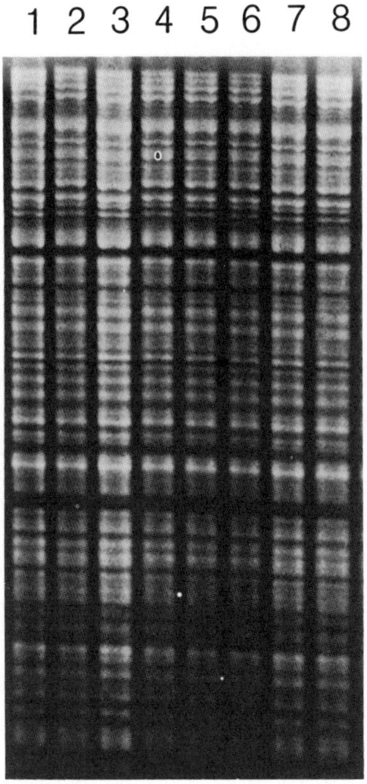

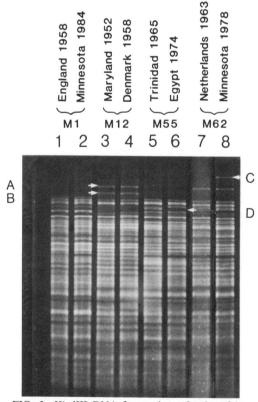

FIG. 2. *Pvu*II DNA fingerprints of group A, M49 *S. pyogenes* from New Zealand isolated from patients and contacts with acute glomerulonephritis and scarlet fever.

single-source outbreak or to differentiate between unrelated strains. The remarkable similarities observed between strains in group A serotypes M1 and M12 suggest a phylogenetic relationship between these serotypes. In support of this relationship M1 and M12 strains have been reported to coexist and to cause serial infection in populations with endemic streptococcal disease, and cloned DNA which encodes the M1 and M12 antigens has been discovered to partially overlap in nucleotide sequence (E. Haanes-Fritz, J. Robbins, and P. Cleary, this volume). Many more strains of these and other serotypes must be examined, however, before DNA fingerprints can be reliably used to predict serotype. It is clear that DNA fingerprints are outbreak specific; of course, a single strain could be involved in multiple outbreaks. Our long-range objective is to develop an extensive data bank of DNA profiles which with the aid of a computer could be employed to define the species, serotype, or epidemiological history of unknown streptococcal isolates.

have a characteristic fingerprint. The pairs of M1 and M12 strains were surprisingly identical even though they represented outbreaks occurring in different continents and, in the case of the M1 pair, were isolated over a 26-year span of time. The M55 strain associated with glomerulonephritis in a patient from Trinidad was very similar to the M55 strain isolated from the throat of a patient from Egypt, but they differed by fragment D. The M62 Minnesota strain had at least one band, fragment C, which was absent from the M62 strain from The Netherlands.

Chromosomal DNA fingerprinting is emerging as a powerful new approach for the analysis of epidemiological clusters of bacterial isolates. Similarities in banding patterns may be assessed at a glance, providing an easily interpreted result to the inexperienced eye. The investigation of group A, M49 streptococci from a worldwide range of infectious outbreaks shows the ability of this technique to display in fine detail the similarity of organisms from a

FIG. 3. *Hin*dIII DNA fingerprints of pairs of human isolates of group A *S. pyogenes* of four different M types.

This work was performed in the World Health Organization Collaborating Center for Reference and Research on Streptococci, University of Minnesota, and was supported by Public Health Service grants R01-AI20321 and R01-AI16722 from the National Institute of Allergy and Infectious Diseases.

LITERATURE CITED

1. **Bjorvatn, B., V. Lund, B.-E. Kristiansen, L. Korsnes, O. Spanne, and B. Lindqvist.** 1984. Applications of restriction endonuclease fingerprinting of chromosomal DNA of *Neisseria meningitidis*. J. Clin. Microbiol. **19**:763–765.
2. **Falk, E. S., B. Bjorvatn, D. Danielsson, B.-E. Kristiansen, K. Melby, and B. Sorensen.** 1984. Restriction endonuclease fingerprinting of chromosomal DNA of *Neisseria gonorrhoeae*. Acta Pathol. Microbiol. Immunol. Scand. Sect. B **92**:271–278.
3. **Falk, E. S., D. Danielsson, B. Bjorvatn, K. Melby, B. Sorensen, B.-E. Kristiansen, S. Lund, and E. Sandstrom.** 1985. Phenotypic and genotypic characterization of penicillinase-producing strains of *Neisseria gonorrhoeae*. Acta Pathol. Microbiol. Immunol. Scand. Sect. B **93**:91–97.
4. **Falk, E. S., D. Danielsson, B. Bjorvatn, K. Melby, B. Sorensen, and B.-E. Kristiansen.** 1985. Genomic fingerprinting in the epidemiology of gonorrhoea. Acta Derm. Venerol. **65**:235–239.
5. **Ferne, M., D. J. Cohen, T. M. Rouach, G. Dinari, and S. Bergner-Rabinowitz.** 1985. Food borne epidemics of group A, C and G streptococcal pharyngitis in Israeli military camps (1980–83), p. 22. *In* Y. Kimura, S. Kotami, and Y. Shiokawa (ed.), Recent advances in streptococci and streptococcal diseases. Reedbooks, Ltd., Berkshire, United Kingdom.
6. **Kaper, J. B., H. B. Bradford, N. C. Roberts, and S. Falkow.** 1982. Molecular epidemiology of *Vibrio cholerae* in the U.S. Gulf Coast. J. Clin. Microbiol. **16**:129–134.
7. **Kristiansen, B. E., B. Sorensen, and B. Bjorvatn.** 1984. Restriction endonuclease fingerprinting of meningococcal DNA. NIPH Ann. **7**:21–28.
8. **Kristiansen, B. E., B. Sorensen, O. Spanne, and B. Bjorvatn.** 1985. Restriction fingerprinting and serology in a small outbreak of B15 meningococcal disease among Norwegian soldiers. Scand. J. Infect. Dis. **17**:19–24.
9. **Lancefield, R. C.** 1933. A serological differentiation of human and other groups of hemolytic streptococci. J. Exp. Med. **57**:571–595.
10. **Maniatis, T., E. F. Fritsch, and J. Sambrook.** 1982. Molecular cloning: a laboratory manual. Cold Spring Harbor Laboratory, Cold Spring Harbor, New York.
11. **Maxted, W. R.** 1953. The use of bacitracin for identifying group A haemolytic streptococci. J. Clin. Pathol. **6**:224–226.
12. **Meekin, G. E., and D. R. Martin.** 1984. Autumn—the season for post-streptococcal acute glomerulonephritis in New Zealand. N.Z. Med. J. **97**:226–229.
13. **Skjold, S. A., and L. W. Wannamaker.** 1976. Method for phage typing group A type 49 streptococci. J. Clin. Microbiol. **4**:232–238.
14. **Skjold, S. A., L. W. Wannamaker, D. R. Johnson, and H. S. Margolis.** 1983. Type 49 *Streptococcus pyogenes*: phage subtypes as epidemiological markers in isolates from skin sepsis and acute glomerulonephritis. J. Hyg. **91**:71–76.
15. **Spanier, J. G., and P. P. Cleary.** 1983. A restriction map and analysis of the terminal redundancy in the group A streptococcal bacteriophage SP24. Virology **130**:502–513.
16. **Spanier, J. G., and P. P. Cleary.** 1983. A DNA substitution in the group A streptococcal bacteriophage SP24. Virology **130**:514–522.
17. **Tagg, J. R., and L. V. Bannister.** 1979. "Fingerprinting" β-haemolytic streptococci by their production of and sensitivity to bacteriocine-like inhibitors. J. Med. Microbiol. **12**:397–411.

B. Extracellular Products

Molecular Characterization of the Group A Streptococcal Exotoxin Type A (Erythrogenic Toxin) Gene and Product

JOSEPH J. FERRETTI, CHANG EN YU, WAYNE L. HYNES, AND CLAUDIA R. WEEKS

Department of Microbiology and Immunology, University of Oklahoma Health Sciences Center, Oklahoma City, Oklahoma 73190

The group A streptococci (*Streptococcus pyogenes*) secrete numerous proteins into the extracellular environment, some of which have direct toxic effects and have been implicated in human disease. In 1924, Dick and Dick (11) first described the presence of a rash-producing substance in the culture filtrates of certain pathogenic hemolytic streptococci which was to be variously termed erythrogenic toxin (41, 42), scarlatinal toxin, Dick toxin (2), scarlet fever toxin, erysipelas toxin (1), streptococcal pyrogenic exotoxin (44), and streptococcal exotoxin (28). Ten years later, Hooker and Follensby (19) reported the presence of two immunologically distinct toxins, termed type A and type B toxins, and possibly a third type. Subsequently, Watson (44) characterized the three exotoxins, types A, B, and C.

The biological activities ascribed to the streptococcal exotoxins include: erythematous skin reactions (11, 12, 19), which may be due to enhancement of delayed hypersensitivity (38), pyrogenicity (5, 27, 44), enhanced susceptibility to endotoxin shock (27, 37, 40, 41), alteration of the blood-brain barrier (37), cardiotoxicity (27, 40, 44), T-cell mitogenicity (3, 21, 22, 29, 32, 34), depression of the clearance function of the reticuloendothelial system (8, 16), and alteration of antibody response to sheep erythrocytes (9, 15, 17, 18).

The possibility that a bacteriophage was involved in the production of streptococcal exotoxin was suggested by Cantacuzene and Boncieu (6) in 1926 and by Frobisher and Brown (13) in 1927, who reported that a filterable agent associated with scarlet fever could induce exotoxin production when introduced into streptococci not previously associated with scarlet fever. These results were later confirmed by Bingel (4), and in 1964, Zabriskie (47) reported that infection of nontoxigenic *S. pyogenes* T253 with bacteriophage T12 resulted in the formation of a lysogen which secreted type A streptococcal exotoxin (erythrogenic toxin).

The molecular mechanism to account for this phage-mediated toxigenic conversion has not been elucidated over the years, although several possible explanations have been proposed; i.e., the bacteriophage contains the gene, the bacterial chromosome contains the gene and is either activated or derepressed by a phage product, or the phage is integrated in the host DNA and splits a gene coding for a nontoxic product, resulting in a truncated protein which is the type A exotoxin, and finally, a transposon contains the type A exotoxin gene.

The possibility that a structural component of phage T12, such as a capsid protein, was the exotoxin was excluded because of the lack of immunological reactivity between the bacteriophage components and antibody to type A exotoxin and also the lack of similarity between phage proteins and toxin proteins when compared by polyacrylamide gels (31). Similarly, the possibility that the bacterial chromosome contained the gene for type A exotoxin and that integration of the phage DNA into the host chromosome split a gene coding for a nontoxic product, resulting in the type A exotoxin, was dismissed when it was determined that integration of the phage genome into the host chromosome was not essential for expression of the toxin gene (31).

In addition to the difficulty in determining the molecular basis of phage T12 conversion of *S. pyogenes*, attempts to physically characterize the type A streptococcal exotoxin have been problematic because of the difficulty in isolating the protein in a pure and undegraded form. Krejci et al. (30) first reported an electrophoretically purified toxin with a molecular weight of 26,700. Nauciel et al. (33) purified the type A streptococcal exotoxin by column chromatography and performed an amino acid analysis on the 30,500-molecular-weight toxin. Further isolation procedures utilized by Kim and Watson (27) and Gerlach et al. (14) resulted in type A exotoxin with molecular weights of 29,400 and 28,000, respectively. Cunningham et al. (7) purified type A exotoxin from *S. pyogenes* NY5-10 culture

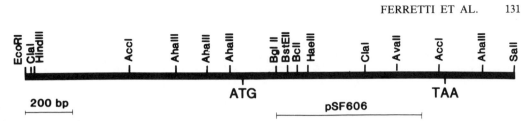

FIG. 1. Partial restriction map of the *speA* gene contained in a 1,837-base-pair (bp) fragment isolated from plasmid p1179 (46). The positions of the initiation (ATG) and termination (TAA) codons of *speA* are shown. Plasmid pSF606 containing the fragment indicated below the map was used in molecular epidemiology experiments.

supernatants by differential solubility in ethanol and acetate-buffered saline followed by ion-exchange chromatography. Two distinct fractions isolated by this method had molecular weights of 8,000 and 5,500 and were found to be pyrogenic and to alter the antibody response. Houston and Ferretti (20) isolated an 8,500-molecular-weight type A streptococcal exotoxin which induced the erythematous skin reaction. The differences observed in the molecular weight and amino acid composition of the purified type A streptococcal exotoxin are in part due to isolation of degraded forms, but are also indicative of the difficulties encountered in isolating pure type A exotoxin.

These studies were undertaken to better understand the genetics of phage conversion of *S. pyogenes* T253 by bacteriophage T12 and to allow further investigation into the type A streptococcal exotoxin and its effects in streptococcal disease. We present here data on the cloning of the streptococcal exotoxin type A gene (*speA*) from bacteriophage T12, its complete nucleotide sequence, homology studies with the *speA* gene product and other exotoxins, and a molecular epidemiology analysis of clinical isolates.

Cloning of the *speA* Gene

The molecular cloning of the *speA* gene from bacteriophage T12 has been recently described in detail (25, 45). Briefly, bacteriophage T12 was obtained from *S. pyogenes* T253(T12) after mitomycin C induction, and the DNA was obtained by phenol extraction. The phage T12 DNA was digested with *Sal*I and *Eco*RI and ligated to similarly cut vector pHP34 (35), a derivative of pBR322. After transformation into *Escherichia coli* HB101, the transformants were isolated, and cell-free extracts of the recombinant clones containing inserts were tested for the ability to produce the type A streptococcal exotoxin. One of the recombinant clones contained plasmid pA2 with a 4.75-kilobase phage T12 fragment and specified the type A exotoxin as confirmed by Ouchterlony double-diffusion assays with specific antibody to the type A exotoxin. Further subcloning of this fragment with *Sal*I and *Hind*III to similarly cut pBR322 resulted in

plasmid p1179, which contained a 1.7-kilobase fragment. A final subcloning into the *E. coli*-streptococcus shuttle vector pSA3 (10) was accomplished by cleavage with *Sal*I and *Eco*RI to form the chimeric plasmid pSA32 containing a 1.7-kilobase insert and the *speA* gene.

Southern hybridization experiments with the cloned *speA* gene confirmed its presence in phage T12 and the toxigenic T253(T12) strain and its absence in the nontoxigenic T253 strain. Thus, the *speA* gene originated in phage T12 and was not present in the nonlysogenic strain.

When the pSA32 shuttle vector containing *speA* was transferred to *Streptococcus sanguis*, the type A exotoxin was found to be secreted into the extracellular medium. This type A exotoxin possessed full biological activity and was shown to be chemically identical to exotoxin produced by *S. pyogenes* (D. Gerlach et al., submitted for publication).

Nucleotide Sequence of *speA*

The complete nucleotide sequence of the *speA* gene was determined by the dideoxy method of Sanger et al. (36) and is presented in Appendix B of this volume and elsewhere (46). A partial restriction map of the 1.8-kilobase fragment containing the *speA* gene is shown in Fig. 1. A single open reading frame of 753 base pairs is present which specifies a protein of 251 amino acids (M_r, 29,244). N-terminal amino acid analysis of the purified type A streptococcal exotoxin showed that the first nine residues were Gln-Gln-Asp-Pro-Asp-Pro-Ser-Gln-Leu and confirmed the presence of a 30-amino acid signal peptide in the unprocessed protein. Thus, the mature secreted type A streptococcal exotoxin, consisting of 221 amino acids, has an M_r of 25,787. The deduced amino acid composition of the secreted type A exotoxin is in agreement with data previously reported by Gerlach et al. (14).

The deduced amino acid sequence of the type A streptococcal exotoxin has considerable homology with staphylococcal enterotoxins B (23) and C1 (39), as shown in Fig. 2. There is a 33% overall amino acid similarity between the type A streptococcal exotoxin and the staphylococcal enterotoxin B, and in the last 100 amino acids of

```
SEB    ESQPDPKPDELHKSSKFTGLMENMK-VLYNNDH--------------------
SEC1   ESQPDPTPDELHKASKFTGLMENMK-VLYDDHY--------------------
speA                       MENNKKVLKKMVFFVLVTFLGLTISQEVFAQQDPDP
                             10        20        30

SEB    ----------------------VSAINVKSINEFFDLIYLYSIKDTKLGNYDNVR
SEC1   ----------------------VSATKVKSVDKFLAHDILYNISDKKLKNYDKVK
speA   SQLHRSSLVKNLQNIYFLYEGDPVTHENVKSVDQLLSHDLIYNVSGPNYDKLKTEL
          40        50        60        70        80        90

SEB    VEFKNKDLADKYKDKYVDVFGANYY-QCYFSKKTNNIDSHENTKRKTCMYGGVTEH
SEC1   TELLNEGLAKKYKDEVVDVYGSNYYVNCYFSSK-DNVGK--VTGGKTCMYGGITKH
speA   KNQEMATLF---KDKNVDIYGVEYYHLCYLCEN--------AERSACIYGGVTNH
          100       110       120              130

SEB    GNNQLDKYYRSITV-RVFEDGKNLLSFDVQTNKKKVTAEQLDYLTRHYLVKNKKLY
SEC1   EGNHFDNGNLONVLIRVYENKRNTISFEVQTNKKSVTAQELDIKARNFLINKKNLY
speA   EGNHLEIPKK-IVV-KVSIDGIQSLSFDIETNKKMVTAQELDYKVRKYLTDNKQLY
          140       150       160       170       180       190

SEB    EFNNSPYETGYIKFIENE-NSFWYDMMPAPGNKFDQSKYLMMYNNDKMVDSKDVKI
SEC1   EFNSSPYETGYIKFIENNGNTFWYDMMPAPGDKFDQSKYLMMYNKNKTVDSKSVKI
speA   TNGPSKYETGYIKFIPKNKESFWFDFFPEP-E-FTQSKYLMIYKDNEILDSNTSQI
          200       210       220       230       240

SEB    EVYLTTKKK
SEC1   EVHLTTKNG
speA   EVYLTTK
         250
```

FIG. 2. Amino acid sequence homology of the type A streptococcal exotoxin (*speA*), staphylococcal enterotoxin B, and staphylococcal enterotoxin C1 (SEC1). Boxes indicate identical residues; underlined residues are common in one of the two enterotoxins.

the type A streptococcal exotoxin, there are 56 amino acids identical in sequence to the staphylococcal enterotoxin B. The gene specifying the *Staphylococcus aureus* enteroxin B (*entB*) has been recently sequenced (25a) and a comparison of sequences shows a 51% overall similarity, with a striking 71% sequence homology in the last 171 nucleotides. As expected, the *speA* gene hybridizes to the *entB* gene, even under high-stringency conditions.

Cross-Reactivity of Type A Streptococcal Exotoxin and Staphylococcal Enterotoxins B and C1

The extensive amino acid homology between the exotoxins from different organisms suggested the presence of common domains which might cross-react immunologically. Ouchterlony double-diffusion analysis showed that antibody to type A streptococcal exotoxin reacted strongly with staphylococcal enterotoxin B and to a lesser extent with enterotoxin C1 (Fig. 3). These results were confirmed by Western immunoblot and immunodot experiments (Table 1). How-

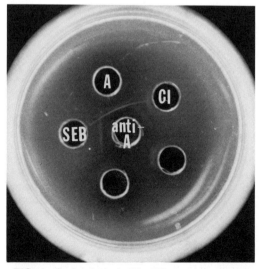

FIG. 3. Ouchterlony double diffusion of antisera to type A streptococcal exotoxin (anti-A, center well), enterotoxin B and enterotoxin C1, and type A streptococcal exotoxin.

TABLE 1. Results of dot-immunobinding of staphylococcal enterotoxins B and C1 (SEB and SEC1) and streptococcal exotoxin type A (SPEA) with homologous and heterologous antisera

Prepn	Results with antisera against:		
	SPEA	SEB	SEC1
SPEA	++++	+++	+
SEB	+++	++++	++
SEC1	++	++	++

ever, reactivity of antisera to staphylococcal enterotoxins B and C1 was much weaker with the type A streptococcal exotoxin. The overall results suggest that a conserved domain is present in the three exotoxins, which most likely originated from a common evolutionary host and which might be a common sequence present in other toxigenic gram-positive organisms. The implications of these findings are that products from other organisms containing this conserved domain may induce the hypersensitivity rash of scarlet fever, commonly ascribed to the type A streptococcal exotoxin.

Molecular Epidemiology of the *speA* Gene among Group A Clinical Isolates

The availability of the cloned *speA* gene made possible a molecular epidemiology study with clinical isolates of group A streptococci obtained from throughout the world. A specific probe consisting of 606 bases within the *speA* gene was constructed (Fig. 1). DNA-DNA hybridizations were performed with 400 strains by the colony hybridization technique (Table 2). The first group of strains analyzed contained 284 general strains obtained from patients with a variety of streptococcal diseases except scarlet fever, e.g., acute glomerulonephritis, rheumatic fever, tonsillitis, impetigo, cellulitis, pyoderma, otitis media, abcess, and septicemia. Among these general strains, 15.9% contained the *speA* gene. The

second group consisted of 116 strains obtained only from individuals described as having scarlet fever, and 40.3% contained the *speA* gene. There appeared to be an association of the *speA* gene with only certain T types, e.g., T types 1, 3/13, 6, 8/25, 12, and 14/49. When Southern hybridizations were done on nitrocellulose with the *speA* gene and samples of chromosomal DNA that had been completely digested with restriction enzymes and migrated in agarose gels, the fragment position of hybridization varied with several of the M-type strains. These results appeared to reflect a difference in chromosome structure of various T-type strains (cf. S. A. Skjold and P. P. Cleary, this volume) rather than different integration sites of the phage carrying the *speA* gene.

Discussion

The molecular cloning of the *speA* gene from bacteriophage T12 has allowed further insight into the mechanism of toxigenic conversion in *S. pyogenes*, a process similar to the phage conversion of *Corynebacterium diphtheriae* in which the structural gene for diphtheria toxin is located in phage beta (43). The questions of where the *speA* gene arose and how it was disseminated remain to be answered. The presence of a signal peptide in the unprocessed type A exotoxin suggests that the *speA* gene is not an essential phage gene and that it is of bacterial origin. Whether the phage picked up the *speA* gene in a similar manner as in the formation of a specialized transducing phage or whether the *speA* gene is part of a transposon is not clear. The high degree of homology between the streptococcus *speA* gene and the staphylococcus *entB* gene suggests a common evolutionary ancestor. Indeed, there is evidence to suggest that both genes are associated with a mobile genetic element, i.e., the *speA* gene as part of phage T12 and the *entB* gene as part of a hitchhiking

TABLE 2. Hybridization of the *speA* gene to group A clinical isolates

Country of origin	No. of general group A strains	No. *speA* positive	No. of scarlet fever strains	No. *speA* positive
Canada	45	3	17	5
Denmark	26	8	24	13
England	44	9	0	0
France	33	5	13	4
German Democratic Republic	0	0	20	2
Federal Republic of Germany	11	7	0	0
India	22	2	0	0
Japan	4	1	0	0
New Zealand	22	5	18	16
Thailand	49	4	1	1
United States	28	1	23	6
Total	284	45	116	47
Percent		15.9		40.5

transposon (26). These mechanisms of gene transfer may have allowed transfer of the *speA* gene or its progenitor to other toxigenic gram-positive organisms.

The information obtained from the deduced amino acid sequence and N-terminal sequence analysis allowed the exact determination of the molecular weight of the type A streptococcal exotoxin, which was 25,787. Recently, Johnson et al. (24) also reported the nucleotide sequence of the *speA* gene; however, their deduced amino acid sequence varies from that reported from this laboratory. Differences in their sequence include the signal peptide, which also contains some variations from classical procaryotic signal peptides, the N-terminal region, which is different from the sequence as determined by amino acid sequence analysis, and other variances found throughout the *speA* gene product.

Of special interest was the sequence homology discovered between the type A exotoxin and the staphylococcal enterotoxins. In addition to the sequence information indicating the presence of a conserved domain among these toxins, immunological cross-reactivity was also demonstrated between the toxins and heterologous antisera. The conserved domains and antigenic relatedness may account for the high degree of similarities in biological activities of these toxins, including pyrogenicity, cytotoxicity, T-cell mitogenicity, erythematous skin reaction, and alteration of the antibody response to erythrocytes. Moreover, the hypersensitivity rash of scarlet fever generally attributed to the type A exotoxin may be induced initially by a staphylococcal infection or by another organism capable of producing a protein containing the conserved domain.

The molecular epidemiology analysis of the *speA* gene in clinical isolates provides the first real evidence of its distribution among group A streptococci. Whereas the *speA* gene is present in 15% of general strains obtained from all clinical states, it was interesting that only 40% of the strains obtained from patients with scarlet fever contained the *speA* gene. The type B and C exotoxins may well be responsible for the scarlet fever produced by the remaining strains since there is evidence indicating the presence of a shared determinant among the different exotoxins that results in the enhancement of skin reactivity (28). Complicating factors in the analysis of data obtained from these strains include the variations in criteria and clinical symptoms used to evaluate whether an individual had scarlet fever. Perhaps some of these strains were obtained from patients with only a few of the clinical symptoms of scarlet fever. The host response is another variable which must be considered, with preferential reaction in some cases to other streptococcal components capable of inducing clinical symptoms similar to those of scarlet fever. A further understanding of the interactions of the streptococcal exotoxins and their role in the disease process will be possible when the genes specifying the type B and C exotoxins are available.

This work was supported by Public Health Service grant AI 19304 from the National Institutes of Health.

LITERATURE CITED

1. **Ando, K., and K. Kurachi.** 1930. The titration of scarlatinal antitoxin in white pigs. J. Immunol. **18:**341–352.
2. **Ando, K., K. Kurauchi, and H. Mishimura.** 1930. Studies on the "toxins" of hemolytic streptococci. III. On the dual nature of the Dick toxin. J. Immunol. **18:**223–255.
3. **Barsumian, E. L., P. M. Schlievert, and D. W. Watson.** 1978. Nonspecific and specific immunological mitogenicity by group A streptococcal pyrogenic exotoxins. Infect. Immun. **22:**681–688.
4. **Bingel, K. F.** 1949. Neue Untersuchungen zur Scharlachatiologie. Dtsch. Med. Wochenschr. **74:**703–706.
5. **Brunson, K. W., and D. W. Watson.** 1974. Pyrogenic specificity of streptococcal exotoxins, staphylococcal enterotoxins, and gram-negative endotoxin. Infect. Immun. **10:**347–351.
6. **Cantacuzene, J., and O. Boncieu.** 1926. Modifications subies par des streptococques d'origine non-scarlatineuse qui contact des produits scarlatineux filtrés. C.R. Acad. Sci. **182:**1185.
7. **Cunningham, C. M., E. L. Barsumian, and D. W. Watson.** 1976. Further purification of group A streptococcal pyrogenic exotoxin and characterization of the purified toxin. Infect. Immun. **14:**767–775.
8. **Cunningham, C. M., and D. W. Watson.** 1978. Alteration of clearance function by group A streptococci and its relation to suppression of the antibody response. Infect. Immun. **19:**51–57.
9. **Cunningham, C. M., and D. W. Watson.** 1978. Suppression of antibody response by group A streptococcal pyrogenic exotoxin and characterization of the cells involved. Infect. Immun. **19:**470–476.
10. **Dao, M. L., and J. J. Ferretti.** 1985. *Streptococcus-Escherichia coli* shuttle vector pSA3 and its use in the cloning of streptococcal genes. Appl. Environ. Microbiol. **49:**115–119.
11. **Dick, G. F., and G. H. Dick.** 1924. The etiology of scarlet fever. J. Am. Med. Assoc. **82:**301–302.
12. **Dochez, A. R., and F. A. Stevens.** 1927. A skin test for susceptibility to scarlet fever. J. Am. Med. Assoc. **82:**265–266.
13. **Frobisher, M., and J. H. Brown.** 1927. Transmissible toxigenicity of streptococci. Bull. Johns Hopkins Hosp. **41:**167–173.
14. **Gerlach, D., H. Knoll, and W. Kohler.** 1980. Purification and characterization of erythrogenic toxins. I. Investigation of erythrogenic toxin A produced by *Streptococcus pyogenes* NY-5. Zentralbl. Bakteriol. Mikrobiol. Hyg. **247:**177–191.
15. **Hanna, E. E., and M. Hale.** 1975. Deregulation of mouse antibody-forming cells in vivo and in cell culture by streptococcal pyrogenic exotoxin. Infect. Immun. **11:**265–272.
16. **Hanna, E. E., and D. W. Watson.** 1965. Host-parasite relationships among group A streptococci. III. Depression of reticuloendothelial function by streptococcal pyrogenic exotoxins. J. Bacteriol. **89:**154–158.
17. **Hanna, E. E., and D. W. Watson.** 1968. Host-parasite relationships among group A streptococci. IV. Suppres-

sion of antibody response by streptococcal pyrogenic exotoxin. J. Bacteriol. **95**:14–21.

18. **Hanna, E. E., and D. W. Watson.** 1973. Enhanced immune response after immunosuppression by streptococcal pyrogenic exotoxin. Infect. Immun. **7**:1009–1011.
19. **Hooker, S. B., and E. M. Follensby.** 1934. Studies of scarlet fever. I. Different toxins produced by hemolytic streptococci of scarlatinal origin. J. Immunol. **27**:177–193.
20. **Houston, C. W., and J. J. Ferretti.** 1981. Enzyme-linked immunosorbent assay for detection of type A streptococcal exotoxin: kinetics and regulation during growth of *Streptococcus pyogenes*. Infect. Immun. **33**:862–869.
21. **Hribalova, V.** 1974. Biological effects of scarlet fever toxin and the role of activation of lymphocytes. J. Hyg. Epidemiol. Microbiol. Immunol. **18**:297–301.
22. **Hribalova, V., and M. Pospisil.** 1973. Lymphocyte stimulating activity of scarlet fever toxin. Experientia **29**:704–705.
23. **Huang, I. Y., and M. S. Bergdoll.** 1970. The primary structure of staphylococcal enterotoxin B. III. The cyanogen bromide peptides of reduced and aminoethylated enterotoxin B and the complete amino acid sequence. J. Biol. Chem. **245**:3518–3525.
24. **Johnson, L. P., J. J. L'Italien, and P. M. Schlievert.** 1986. Streptococcal pyrogenic exotoxin type A (scarlet fever toxin) is related to *Staphylococcus aureus* enterotoxin B. Mol. Gen. Genet. **203**:354–356.
25. **Johnson, L. P., and P. M. Schlievert.** 1984. Group A streptococcal phage T12 carries the structural gene for pyrogenic exotoxin type A. Mol. Gen. Genet. **194**:52–56.
25a. **Jones, C. L., and S. A. Khan.** 1986. Nucleotide sequence of the enterotoxin B gene from *Staphylococcus aureus*. J. Bacteriol. **166**:34–37.
26. **Khan, S. A., and R. P. Novick.** 1982. Structural analysis of plasmid pSN2 in *Staphylococcus aureus*: no involvement in enterotoxin B production. J. Bacteriol. **149**:642–649.
27. **Kim, Y. B., and D. W. Watson.** 1970. A purified group A streptococcal pyrogenic exotoxin. Physicochemical and biological properties, including the enhancement of susceptibility to endotoxin lethal shock. J. Exp. Med. **131**:611–628.
28. **Kim, Y. B., and D. W. Watson.** 1972. Streptococcal exotoxins: biological and pathological properties, p. 33–50. *In* L. W. Wannamaker and J. M. Matson (ed.), Streptococci and streptococcal diseases. Academic Press, Inc., New York.
29. **Knoll, H., F. Petermann, and W. Kohler.** 1978. Mitogenic activity of erythrogenic toxins. II. Determination of erythrogenic toxins in culture supernatants of *Streptococcus pyogenes*. Zentralbl. Bakteriol. Parasitenkd. Infektionskr. Hyg. Abt. 1 Orig. Reihe A **240**:466–473.
30. **Krejci, L. E., A. H. Stock, E. B. Sanigar, and E. O. Kraemer.** 1942. Studies on the hemolytic streptococcus. V. The electrophoretic isolation of the erythrogenic toxin of scarlet fever and the determination of its chemical and physical properties. J. Biol. Chem. **142**:785–802.
31. **McKane, L., and J. J. Ferretti.** 1981. Phage-host interactions and the production of type A streptococcal exotoxin in group A streptococci. Infect. Immun. **34**:915–919.
32. **Nauciel, C.** 1973. Mitogenic activity of purified streptococcal erythrogenic exotoxin on lymphocytes. Ann. Immunol. (Paris) **124**:383–390.
33. **Nauciel, C., M. Raynaud, and B. Bizinni.** 1968. Purification et properties de la toxine erythrogene du streptocoque. Ann. Inst. Pasteur (Paris) **114**:796–811.
34. **Peterman, F., H. Knoll, and W. Kohler.** 1978. Mitogenic activity of erythrogenic toxins. I. Type-specific inhibition of the mitogenic activity of erythrogenic toxins by antitoxic antisera from the rabbit. Zentralbl. Bakteriol. Parasitenkd. Infektionskr. Hyg. Abt. 1 Orig. Reihe A **240**:366–379.
35. **Prentki, P., and H. M. Krisch.** 1982. A modified pBR322 vector with improved properties for the cloning, recovery, and sequencing of blunt-end fragments. Gene **17**:189–196.
36. **Sanger, F., S. Nicklen, and A. R. Coulson.** 1977. DNA sequencing with chain terminating inhibitors. Proc. Natl. Acad. Sci. USA **74**:5463–5467.
37. **Schlievert, P. M., and D. W. Watson.** 1978. Group A streptococcal pyrogenic exotoxin: pyrogenicity, alteration of blood brain barrier, and separation of sites for pyrogenicity and enhancement of lethal endotoxin shock. Infect. Immun. **21**:753–763.
38. **Schlievert, P. M., K. M. Bettin, and D. W. Watson.** 1979. Reinterpretation of the Dick test: role of group A streptococcal pyrogenic exotoxin. Infect. Immun. **26**:467–472.
39. **Schmidt, J. J., and L. Spero.** 1983. The complete amino acid sequence of staphylococcal enterotoxin C1. J. Biol. Chem. **258**:6300–6306.
40. **Schwab, J. H., D. W. Watson, and W. J. Cromartie.** 1955. Further studies of group A streptococcal factors with lethal and cardiotoxic properties. J. Infect. Dis. **96**:14–18.
41. **Stock, A. H.** 1939. Studies on the hemolytic *Streptococcus*, isolation and concentration of erythrogenic toxin of hemolytic streptococcus. J. Immunol. **36**:489–498.
42. **Stock, A. H., and R. J. Lynn.** 1961. Preparation and properties of partially purified erythrogenic toxin B of group A streptococci. J. Immunol. **86**:561–566.
43. **Uchida, T., D. M. Gill, and A. M. Pappenheimer, Jr.** 1971. Mutation in the structural gene for diphtheria toxin carried by temperate phage beta. Nature (London) New Biol. **233**:8–11.
44. **Watson, D. W.** 1960. Host-parasite factors in group A streptococcal infections. Pyrogenic and other effects of immunologic distinct exotoxins related to scarlet fever toxins. J. Exp. Med. **111**:255–284.
45. **Weeks, C. R., and J. J. Ferretti.** 1984. The gene for type A streptococcal exotoxin (erythrogenic toxin) is located in bacteriophage T12. Infect. Immun. **46**:531–536.
46. **Weeks, C. R., and J. J. Ferretti.** 1986. Nucleotide sequence of the type A streptococcal exotoxin (erythrogenic toxin) gene from *Streptococcus pyogenes* bacteriophage T12. Infect. Immun. **52**:144–150.
47. **Zabriskie, J. B.** 1964. The role of temperate bacteriophage in the production of erythrogenic toxin by group A streptococci. J. Exp. Med. **119**:761–779.

Characterization and Genetics of Group A Streptococcal Pyrogenic Exotoxins

PATRICK M. SCHLIEVERT, LANE P. JOHNSON, MARK A. TOMAI, AND JEFFREY P. HANDLEY

Department of Microbiology, Medical School, University of Minnesota, Minneapolis, Minnesota 55455

Characterization

Group A streptococcal pyrogenic exotoxins (synonyms: SPEs, streptococcal exotoxins, scarlet fever toxins, erythrogenic toxins, lymphocyte mitogens, blastogen A [type A], and keratinocyte proliferation factor) are members of a larger family of pyrogenic toxins based upon shared biological activities. Other members of the family include staphylococcal enterotoxins, staphylococcal pyrogenic exotoxins, and toxic shock syndrome toxin-1.

All of these toxins share the capacity to induce fever, enhance host susceptibility to lethal endotoxin shock, nonspecifically stimulate T-lymphocyte proliferation, enhance delayed hypersensitivity, and suppress immunoglobulin synthesis (2, 3, 5, 8, 10, 15, 22, 28, 30, 34, 36, 37, 38, 41–43). SPEs have the additional property, not shared by the other toxins, of enhancing susceptibility to specific myocardial damage by other agents such as streptolysin O and endotoxin (39, 40). The staphylococcal enterotoxins have the additional capacity to induce vomiting and diarrhea after oral administration to monkeys (3).

SPEs are true exotoxins, in that they are secreted after synthesis, rather than released upon cell lysis, and occur in four serological types designated A, B, C, and D. Biochemically, SPEs are of relatively low molecular weight and are composed of a single peptide chain that contains the biological activities. SPE type A occurs in two molecular forms, designated A_1 and A_2, which can be converted into one form after treatment with 2-mercaptoethanol (25). The latter form has biological activities comparable with those of either A_1 or A_2. It has been proposed that the additional component contained on A_1 is an unknown peptide that is bound to the active toxin through a disulfide bridge. Interestingly, cloned SPE type A also migrates as two forms (19). SPE type B occurs in three forms that differ in isoelectric point. Only type B (pI 8.3) has all of the biological activities associated with SPEs (1). SPE type C occurs also in two forms that differ in pI (6.7 and 7.0), but a given strain makes only one of the two forms (1, 31). A summary of the biochemical properties of SPEs is shown in Table 1.

Biologically, SPEs have many effects on the host, most of which result from actions on the immune system. We define the toxins by their capacities to induce fever and enhance host susceptibility to lethal endotoxin shock (43).

The most common way molecules are thought to induce fever is by way of interleukin 1 (endogenous pyrogen) release from macrophages. In addition to its role in fever production, interleukin 1 is required for T-lymphocyte proliferation in response to antigen or nonspecific mitogen stimulation. In two studies, investigators failed to show macrophage involvement in nonspecific T-lymphocyte proliferation induced by SPEs (29, 36). Because of this finding, it is possible that SPEs do not induce interleukin 1 and, thus, cause fever by a different mechanism. Schlievert and Watson (38) suggested that SPE type C causes fever after crossing the blood-brain barrier and directly stimulating the hypothalamic fever response control center.

The capacity to induce fever appears to be separate from enhancement of susceptibility to endotoxin (38), since agents that block fever do not alter the enhancement phenomenon (32, 38). The latter effect may result from the capacity of SPEs to alter reticuloendothelial system clearance function, possibly through inhibition of RNA synthesis in liver cells (35). The enhancement phenomenon is determined by pretreatment of animals with a sublethal dose of SPE for 2 to 4 h and then challenge with a sublethal dose of endotoxin, which results in death in 1 to 48 h. The enhanced susceptibility to endotoxin may be as great as 100,000-fold. If animals do not succumb to lethal shock, they show significant myocardial necrosis after 3 to 4 days. This damage, which involves all parts of the heart, occurs at toxin doses 10-fold lower than those necessary to induce lethality.

SPEs are also potent nonspecific T-lymphocyte mitogens in rabbits and humans, but not in mice (2, 10, 29, 36); the mitogenic response is greater than that seen for concanavalin A. The T-lymphocyte mitogenic effect most likely explains the capacity of SPEs to enhance delayed hypersensitivity, suppress immunoglobulin M synthesis, and enhance immunoglobulin G synthesis.

Schlievert et al. (34) showed that SPEs enhance preexistent delayed hypersensitivity de-

TABLE 1. Biochemical properties of SPEs

SPE type	Mol wt[a]	Isoelectric point[b]	Peptide chains[c]	Protease and heat stability[d]
A	26,000	5.0 (A$_1$) 5.5 (A$_2$)	1 (A$_1$) 1+ (A$_2$)[e]	+
B	17,500	8.0 8.3 9.0	1	?
C	13,000	6.7 or 7.0	1	++

[a] From references 1, 17, 31, 45.

[b] From references 1, 7, 19, 45.

[c] From references 1, 25, 45.

[d] From references 7 and 31.

[e] A$_2$ contains a low-molecular-weight peptide of unknown origin that is not required for biological activity.

veloped against heterologous as well as homologous molecules, of either streptococcal or nonstreptococcal origin. These investigators proposed that the scarlet fever rash induced by SPEs results from this effect in individuals who lack neutralizing antibodies. Thus, erythrogenic toxin may be defined as a primary toxic effect of

SPE plus the enhancement of a previously established delayed hypersensitivity to itself or another factor.

The immune suppression (of immunoglobulin M synthesis) appears to result from the nonspecific activation of T-suppressor lymphocytes (30). The enhancement of immunoglobulin G synthesis may depend on amplification of T-helper function (11–16, 24).

Together, these biological activities may result in many effects on the host and lead to illnesses such as scarlet fever. A model for the development of severe scarlet fever is shown in Fig. 1. This same model may explain induction of toxic shock syndrome by toxic shock syndrome toxin-1 and possibly enterotoxins.

Several lines of circumstantial evidence suggest that SPEs may also contribute to early events in the development of delayed sequelae such as rheumatic fever. First, these toxins are specific for group A streptococci and predispose the host to quite specific heart damage (33, 39, 40). The toxins are not made by strong hyaluronidase producers (33) which interestingly have not resulted in rheumatic fever; the toxins are expressed by rheumatogenic strains. SPEs ap-

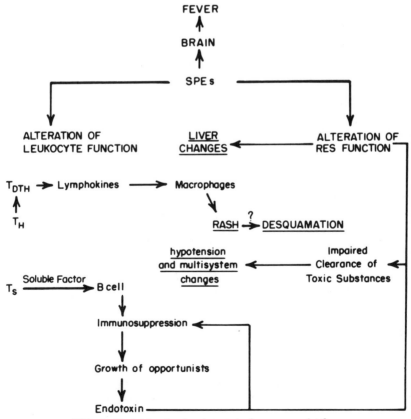

FIG. 1. Model for the development of severe scarlet fever.

pear to cross the blood-brain barrier (38) and thus may contribute to neurological effects seen in some rheumatic fever patients. Finally, the toxins have profound immune system altering effects.

Genetics

To evaluate dissemination of SPE types among streptococci and to evaluate mechanisms of control of their expression, genetic analyses of the SPEs have been done.

Frobisher and Brown in 1927 (9) reported that a filterable agent from scarlet fever strains of group A streptococci induced toxin production by nonscarlatinal strains. This was confirmed by Bingel (4) and by Zabriskie (46), who showed that bacteriophage from streptococcal strain T12g1 converted strain T25$_3$ from nonlysogenic, A toxin negative to lysogenic, A toxin positive. Subsequently, it was shown that in selected strains production of SPE types A (20, 23, 26, 27), B (26), and C (6, 20) is regulated by bacteriophage. Johnson et al. (20) proposed that A and C toxins are transferred by lysogenic conversion. Nida and Ferretti (26) showed that the B toxin could be similarly transferred. It has also been reported that a virulent mutant of a converting phage (23) and a pseudolysogenic state (26) can also lead to toxin production. Toxigenic conversion could be produced by a number of, but not all, temperate phages, and conversion could be effected in many streptococcal strains (23, 26). These findings supported the view that this may be a general phenomenon and that integration of the phage into the bacterial chromosome is not essential for the phenomenon to occur. Similar analyses of type D toxin expression have not been performed.

A major difficulty in doing experiments related to the molecular genetics of SPEs had been that of obtaining sufficient phage. Typically, phage titers of 10^7/ml had been obtained. Johnson and Schlievert (18), using a strain of T25$_3$ cured of an endogenous phage, were able to obtain nearly 10^{10} phage per ml after infection with T12 phage and induction with mitomycin C. These investigators have recently provided a restriction enzyme map of the T12 phage DNA which controls expression of A toxin (Fig. 2). The phage DNA has a molecular weight of approximately 23.5 $\times$ 10^6 and is circularly permuted. Their data suggested that DNA packaging within the phage begins at a precise site (pac) on concatemeric precursors and proceeds for a limited number of rounds.

Johnson and Schlievert (19) subsequently cloned the $speA$ gene from phage T12 into $Escherichia$ $coli$ with pBR322 as the vector plasmid. The gene was localized onto a 1.75-kilobase SalI-HindIII restriction endonuclease

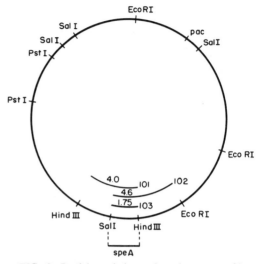

FIG. 2. Partial restriction endonuclease map of bacteriophage T12. pac, Site of initiation of packaging by the headful mechanism which proceeds in the counterclockwise direction; 101, 102, and 103, pUMN plasmids (derivatives) of pBR322 which express SPE type A in $E.$ $coli$. Sizes of plasmids indicated are in kilobases. Not all of the HindIII sites have been localized.

fragment as indicated in Fig. 2. Toxin as expressed in $E.$ $coli$ migrates as two bands, comparable with that derived from streptococci, and is biochemically, biologically, and immunologically related to streptococcal A toxin. Weeks and Ferretti (44) confirmed these findings in similar studies.

Recently, Weeks and Ferretti (45) and later Johnson et al. (17) sequenced the $speA$ structural gene and thus provided the inferred amino acid sequence. Although there are some differences in sequence obtained by the two groups, several conclusions can be drawn from the studies. First, the gene encodes a mature SPE type A protein of approximately 26,000 molecular weight. Second, the $speA$ gene encodes a 30-amino acid signal peptide that precedes the mature SPE type A protein. Third, there are two inverted repeat sequences, 3' of the $speA$ gene, that are required for toxin expression; these sequences are proposed to provide stem and loop structures necessary to stabilize the mRNA. Finally, $speA$ and SPE type A contained significant sequence homology with staphylococcal $entB$ and enterotoxin B, respectively, and SPE type A shows sequence homology with enterotoxin C; the $entC$ gene sequence is not available. Thus, it is proposed that these molecules, which share numerous biological activities, may have similar active site structure.

Johnson et al. in 1986 (21) showed that all SPE

TABLE 2. Analysis of group A streptococcal strains for production of SPE A, release of converting phage, and presence of DNA homologous to *speA* and phage T12

Strains	SPE A production[a]	Converting phage[b]	Hybridization to *speA* probe[c]	Hybridization to phage T12 probe[c]
T25₃c(T12) and 3GL16	+	+	+	+
594, C203, and NY-5	+	−	+	+
T18P and T25₃c	−	−	−	−
GT9316, GT9094, and GT8627	−	−	−	+

[a] Concentrated (200 times) culture supernatant fluids were tested with hyperimmune antisera against purified SPE A by Ouchterlony immunodiffusion.

[b] Capacity of cell-free culture fluids to induce SPE A production by indicator strains.

[c] Chromosomal DNA preparations were hybridized with ³²P-labeled *speA*-specific and phage T12-specific probes by the dot-blot hybridization technique.

A-positive streptococcal strains tested carry both *speA* and phage T12 DNA sequences adjacent to each other and thus that phages are responsible for dissemination of *speA* among streptococci. The data leading to these conclusions are summarized below.

Our laboratory maintains over 100 group A streptococcal strains; only 5 of these express SPE type A as determined by Ouchterlony immunodiffusion: T25₃c(T12), 3GL16, NY-5, C203 (both S and U), and 594 (Table 2).

When infected with phage from T25₃c(T12) or 3GL16, indicator strain T25₃c is converted to A toxin positive. DNAs obtained from T12 and 3GL16 phages yield identical restriction endonuclease digestion patterns when cut with *Eco*RI or *Sal*I (Fig. 3) or with other endonucleases, suggesting that these two phages are the same phage. Other A-toxin-positive bacterial strains did not contain phages capable of converting indicator strains T25₃c, T18P, and K56 to toxin expression (Table 2). No phages could be induced from strain 594.

The inability of SPE A-positive strains to transfer toxin production by phages does not indicate a lack of phage association with toxin. Such strains may carry defective phages or phages with different host range.

To evaluate whether strains NY-5, C203S, and 594 contain *speA* adjacent to T12 DNA sequences, Johnson et al. (21) performed Southern blot analyses of these strains, using *speA*-specific and phage T12-specific DNA probes. The *speA* probe was a 590-base-pair internal fragment. The phage probe was a 2.3-kilobase *Hin*dIII-*Sal*I fragment that maps 267 nucleotides from the 3′ end of *speA* (Fig. 4).

The *speA* and phage T12 probes hybridized to all of the SPE A-positive strains, including NY-5, C203S, and 594 (Table 2). It appears from the results that phage may be associated with toxin expression. Interestingly, DNA from some SPE A-negative (and *speA*-negative) strains hybridized to the phage probe, suggesting that phage T12 or a related phage not associated with

speA was present in those strains. Furthermore, it is noteworthy that T18P did not hybridize with the *speA* probe. This strain makes SPE type C while not making A, B, or D, and since it does not hybridize, this suggests that *speC* and *speA* are not highly homologous.

Studies were then performed to evaluate whether the strains that make SPE A carry T12 DNA sequences adjacent to *speA* in their chromosome. This hypothesis was tested by preparing *Bgl*II digests of bacterial DNA from selected strains in Table 2 and then performing Southern analysis with the *speA* and T12 probes described earlier. Restriction endonuclease *Bgl*II digestion of phage T12 generates

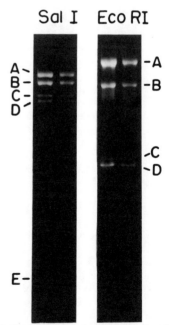

FIG. 3. Restriction endonuclease digestion patterns of phage T12 (left) and 3GL16 (right) DNA separated in agarose gels. Letters on the sides indicate the positions of DNA fragments.

FIG. 4. Physical map of phage T12 DNA. Probes used for Southern hybridization analysis are indicated by heavy lines; *att* is the attachment site for phage integration into the bacterial chromosome.

an 8-kilobase fragment of DNA that will hybridize to both probes (Fig. 4). Furthermore, because *attP* (to be discussed later) is not contained within this fragment, it should not be altered by phage integration. All SPE A-positive strains tested [T25₃c(T12), 594, NY-5, and C203S] carry *speA* in the 8-kilobase *Bgl*II fragment which also strongly hybridizes to the phage T12 probe, suggesting that the *speA* sequences are adjacent to T12 sequences.

Further analysis of *speA*-positive strains revealed some dissimilarities among their endogenous phage, however. DNA from T25₃c(T12) and 3GL16 produce hybridization patterns similar to T12 DNA when cut with *Sal*I or *Eco*RI except in the fragments where phage integration occurred. In contrast, DNAs from strains 594, NY-5, and C203S were missing fragments expected to hybridize to full-length phage T12. It is possible either that these strains suffered large deletions or that these strains were lysogenized by recombinant phages that contain only a portion of the T12 genome.

In subsequent studies by Johnson et al. (21), the position of the *attP* site for phage T12 integration into the bacterial chromosome was mapped near *speA* (Fig. 4). These studies suggest that phage T12 may have acquired *speA* from the bacterial genome by abnormal excision and that the gene has now become a permanent part of the phage. A similar mechanism has been proposed for phage acquisition of genes for diphtheria toxin and enterotoxin A.

Structure-function studies of SPEs have been hampered by inability to prepare large amounts of toxins. To overcome this, *speA* was cloned into *Bacillus subtilis* by the protocol outlined in Fig. 5. *B. subtilis* expresses 30- to 50-fold more toxin than streptococci, and the toxin is stably expressed.

SPE type A, as prepared in *B. subtilis*, migrates as proteins with pIs of 5.0 and 5.5 when tested by isoelectric focusing, and both A_1 and A_2 have molecular weights of approximately 26,000 when tested by sodium dodecyl sulfate-polyacrylamide gel electrophoresis. Further, *Bacillus*-derived SPE A is immunologically identical to streptococcal SPE A. Finally, *Bacillus*-derived SPE A is pyrogenic, enhances host susceptibility to lethal endotoxin shock, is mitogenic, and is immunosuppressive, and the activities are comparable with those of streptococcus-derived toxin. The only observed differ-

ence between *B. subtilis*- and streptococcus-obtained SPE A preparations is that those from *B. subtilis* are insoluble in water whereas those from streptococci are soluble in water. This suggests that streptococci elaborate a cofactor, which is as yet unidentified, that maintains the water solubility of streptococcal SPE A. *B. subtilis*-derived toxin is easily maintained in solution in saline.

Surprisingly, there was no proteolytic degradation of *B. subtilis* SPE A. This stability may relate to the highly resistant properties of the native toxin (7).

Since SPE A is now expressed in *B. subtilis* in active form, studies should be possible to examine structure-function relationships of this molecule and correlate those with data from struc-

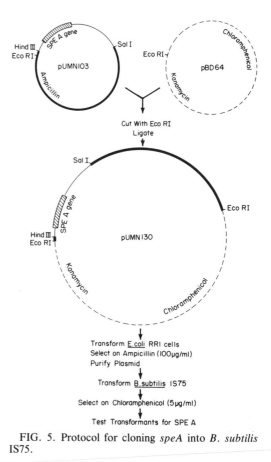

FIG. 5. Protocol for cloning *speA* into *B. subtilis* IS75.

ture-function studies of other members of the larger family.

The work related to genetics of SPEs was supported by research grant 83-697 from the American Heart Association with funds contributed in part by the Minnesota Affiliate.

Yvonne Guptill is gratefully acknowledged for typing the manuscript and Timothy Leonard is thanked for artwork and photography.

LITERATURE CITED

1. **Barsumian, E. L., C. M. Cunningham, P. M. Schlievert, and D. W. Watson.** 1978. Heterogeneity of group A streptococcal pyrogenic exotoxin type B. Infect. Immun. **20:** 512–518.
2. **Barsumian, E. L., P. M. Schlievert, and D. W. Watson.** 1978. Nonspecific and specific immunologic mitogenicity by group A streptococcal pyrogenic exotoxins. Infect. Immun. **22:**681–688.
3. **Bergdoll, M. S.** 1970. Enterotoxins, p. 265–326. *In* T. Montie, S. Kadis, and S. J. Ajl (ed.), Microbial toxins, vol. 3. Academic Press, Inc., New York.
4. **Bingel, K. F.** 1949. Neue Untersuchungen zur Scharlatachatiologie. Dtsch. Med. Wochenschr. **127:**703–706.
5. **Brunson, K. W., and D. W. Watson.** 1974. Pyrogenic specificity of streptococcal exotoxins, staphylococcal enterotoxin, and gram-negative endotoxin. Infect. Immun. **10:**347–351.
6. **Colon-Whitt, A., R. S. Whitt, and R. M. Cole.** 1979. Production of an erythrogenic toxin (streptococcal pyrogenic exotoxin) by a nonlysogenized group A streptococcus, p. 64–65. *In* M. T. Parker (ed.), Pathogenic streptococci. Reedbooks, Ltd., Chertsey, England.
7. **Cunningham, C. M., E. L. Barsumian, and D. W. Watson.** 1976. Further purification of group A streptococcal pyrogenic exotoxin and characterization of the purified toxin. Infect. Immun. **14:**767–775.
8. **Cunningham, C. W., and D. W. Watson.** 1978. Suppression of the antibody response by group A streptococcal pyrogenic exotoxin and characterization of the cells involved. Infect. Immun. **19:**470–476.
9. **Frobisher, M., Jr., and J. H. Brown.** 1927. Transmissible toxigenicity of streptococci. Bull. Johns Hopkins Hosp. **41:**167–173.
10. **Gray, E. D.** 1979. Purification and properties of an extracellular blastogen produced by group A streptococci. J. Exp. Med. **149:**1438–1449.
11. **Hale, M. L., and E. E. Hanna.** 1976. Deregulation of mouse antibody-forming cells by streptococcal pyrogenic exotoxins. II. Modification of spleen T-cell complemented nude mouse PFC responses. Cell. Immunol. **26:**168–177.
12. **Hanna, E. E., and M. Hale.** 1975. Deregulation of mouse antibody-forming cells in vivo and in cell culture by streptococcal pyrogenic exotoxin. Infect. Immun. **11:**265–272.
13. **Hanna, E. E., M. L. Hale, and M. L. Misfeldt.** 1980. Deregulation of mouse antibody-forming cells by streptococcal pyrogenic exotoxin (SPE). III. Modification of T-cell-dependent plaque-forming cell responses of mouse immunocytes is a common property of highly purified and of crude preparations of SPE. Cell. Immunol. **56:**247–257.
14. **Hanna, E. E., and D. W. Watson.** 1965. Host-parasite relationships among group A streptococci. III. Depression of reticuloendothelial function by streptococcal pyrogenic exotoxins. J. Bacteriol. **89:**154–158.
15. **Hanna, E. E., and D. W. Watson.** 1968. Host-parasite relationships among group A streptococci. IV. Suppression of antibody response by streptococcal pyrogenic exotoxin. J. Bacteriol. **95:**14–21.
16. **Hanna, E. E., and D. W. Watson.** 1973. Enhanced immune response after immunosuppression by streptococcal pyrogenic exotoxin. Infect. Immun. **7:**1009–1011.
17. **Johnson, L. P., J. J. L'Italien, and P. M. Schlievert.** 1986. Streptococcal pyrogenic exotoxin type A (scarlet fever toxin) is related to *Staphylococcus aureus* enterotoxin B. Mol. Gen. Genet. **203:**354–356.
18. **Johnson, L. P., and P. M. Schlievert.** 1983. A physical map of the group A streptococcal pyrogenic exotoxin bacteriophage T12 genome. Mol. Gen. Genet. **189:**251–255.
19. **Johnson, L. P., and P. M. Schlievert.** 1984. Group A streptococcal phage T12 carries the structural gene for pyrogenic exotoxin type A. Mol. Gen. Genet. **194:**52–56.
20. **Johnson, L. P., P. M. Schlievert, and D. W. Watson.** 1980. Transfer of group A streptococcal pyrogenic exotoxin production to nontoxigenic strains by lysogenic conversion. Infect. Immun. **28:**254–257.
21. **Johnson, L. P., M. A. Tomai, and P. M. Schlievert.** 1986. Bacteriophage involvement in group A streptococcal pyrogenic exotoxin A production. J. Bacteriol. **166:**623–627.
22. **Kim, Y. B., and D. W. Watson.** 1970. A purified group A streptococcal pyrogenic exotoxin. Physicochemical and biological properties including the enhancement of susceptibility to endotoxin lethal shock. J. Exp. Med. **131:** 611–628.
23. **McKane, L., and J. J. Ferretti.** 1981. Phage-host interactions and the production of type A streptococcal exotoxin in group A streptococci. Infect. Immun. **34:**915–919.
24. **Misfeldt, M. L., and E. E. Hanna.** 1981. Deregulation of mouse antibody-forming cells by streptococcal pyrogenic exotoxin (SPE). IV. Fractionation of a T-cell subpopulation which generates SPE-induced deregulation of anti-TNP PFC responses. Cell. Immunol. **57:**20–27.
25. **Nauciel, C., J. Blass, R. Mangalo, and M. Raynaud.** 1969. Evidence for two molecular forms of streptococcal erythrogenic toxin. Conversion to a single form by 2-mercaptoethanol. Eur. J. Biochem. **11:**160–164.
26. **Nida, S. K., and J. J. Ferretti.** 1982. Phage influence on the synthesis of extracellular toxins in group A streptococci. Infect. Immun. **36:**745–750.
27. **Nida, S. K., C. W. Houston, and J. J. Ferretti.** 1979. Erythrogenic toxin production by group A streptococci, p. 66. *In* M. T. Parker (ed.), Pathogenic streptococci. Reedbooks, Ltd., Chertsey, England.
28. **Poindexter, N. J., and P. M. Schlievert.** 1985. The biochemical and immunological properties of toxic-shock syndrome toxin-1 (TSST-1) and association with TSS. J. Toxicol. Toxin Rev. **4:**1–39.
29. **Regelmann, W. E., E. D. Gray, and L. W. Wannamaker.** 1982. Characterization of the human cellular immune response to purified group A streptococcal blastogen A. J. Immunol. **4:**1631–1636.
30. **Schlievert, P. M.** 1980. Activation of murine T-suppressor lymphocytes by group A streptococcal and staphylococcal pyrogenic exotoxins. Infect. Immun. **28:**876–880.
31. **Schlievert, P. M., K. M. Bettin, and D. W. Watson.** 1977. Purification and characterization of group A streptococcal pyrogenic exotoxin type C. Infect. Immun. **16:**673–679.
32. **Schlievert, P. M., K. M. Bettin, and D. W. Watson.** 1978. Effect of antipyretics on group A streptococcal pyrogenic exotoxin fever production and ability to enhance lethal endotoxin shock. Proc. Soc. Exp. Biol. Med. **157:**472–475.
33. **Schlievert, P. M., K. M. Bettin, and D. W. Watson.** 1979. Production of pyrogenic exotoxin by groups of streptococci: association with group A. J. Infect. Dis. **140:**676–681.
34. **Schlievert, P. M., K. M. Bettin, and D. W. Watson.** 1979. Reinterpretation of the Dick test: role of group A streptococcal pyrogenic exotoxin. Infect. Immun. **26:**467–472.
35. **Schlievert, P. M., K. M. Bettin, and D. W. Watson.** 1980. Inhibition of ribonucleic acid synthesis by group A streptococcal pyrogenic exotoxin. Infect. Immun. **27:**542–548.
36. **Schlievert, P. M., D. J. Schoettle, and D. W. Watson.** 1979. Nonspecific T-lymphocyte mitogenesis by pyrogenic exotoxins from group A streptococci and *Staphylococcus aureus*. Infect. Immun. **25:**1075–1077.
37. **Schlievert, P. M., K. N. Shands, B. B. Dan, G. P. Schmid, and R. D. Nishimura.** 1981. Identification and character-

ization of an exotoxin from *Staphylococcus aureus* associated with toxic-shock syndrome. J. Infect. Dis. **143:**509–516.

38. **Schlievert, P. M., and D. W. Watson.** 1978. Group A streptococcal pyrogenic exotoxin: pyrogenicity, alteration of blood-brain barrier, and separation of sites for pyrogenicity and enhancement of lethal endotoxin shock. Infect. Immun. **21:**753–763.

39. **Schwab, J. H., D. W. Watson, and W. J. Cromartie.** 1953. Production of generalized schwartzman reaction with group A streptococcal factors. Proc. Soc. Exp. Biol. Med. **82:**754–761.

40. **Schwab, J. H., D. W. Watson, and W. J. Cromartie.** 1955. Further studies of group A streptococcal factors with lethal and cardiotoxic properties. J. Infect. Dis. **96:**14–18.

41. **Smith, B. G., and H. M. Johnson.** 1975. The effect of staphylococcal enterotoxins on the primary in vitro immune response. J. Immunol. **115:**575–578.

42. **Sugiyama, H., E. M. McKissic, Jr., M. S. Bergdoll, and B. Heller.** 1964. Enhancement of bacterial endotoxin lethality by staphylococcal enterotoxin. J. Infect. Dis. **114:**111–118.

43. **Watson, D. W.** 1960. Host-parasite factors in group A streptococcal infections. Pyrogenic and other effects of immunologic distinct exotoxins related to scarlet fever toxins. J. Exp. Med. **111:**255–284.

44. **Weeks, C. R., and J. J. Ferretti.** 1984. The gene for type A streptococcal exotoxin (erythrogenic toxin) is located in bacteriophage T12. Infect. Immun. **46:**531–536.

45. **Weeks, C. R., and J. J. Ferretti.** 1986. Nucleotide sequence of the type A streptococcal exotoxin (erythrogenic toxin) gene from *Streptococcus pyogenes* bacteriophage T12. Infect. Immun. **52:**144–150.

46. **Zabriskie, J. B.** 1964. The role of temperate bacteriophage in the production of erythrogenic toxin by group A streptococci. J. Exp. Med. **119:**761–780.

Streptokinase: Expression of Altered Forms

HORST MALKE,[1] DAVID LORENZ,[2] AND JOSEPH J. FERRETTI[2]

Academy of Sciences of the German Democratic Republic, Central Institute of Microbiology and Experimental Therapy, DDR-69 Jena, German Democratic Republic,[1] and Department of Microbiology and Immunology, University of Oklahoma Health Sciences Center, Oklahoma City, Oklahoma 73190[2]

Streptokinase, an extracellular protein produced by a variety of pathogenic streptococci, activates in a nonproteolytic manner the blood plasma zymogen, plasminogen, to the fibrinolytic enzyme, plasmin. This reaction has been exploited in clinical medicine where streptokinase has served as a thrombolytic agent for almost three decades (14). The biochemical mechanism whereby streptokinase functions is unknown, but it has been established that the overall reaction involves the formation of a stoichiometric 1:1 streptokinase-plasminogen complex, the proteolytic activity of which is generated in the plasminogen moiety and leads to the conversion of free plasminogen to plasmin (2). The binding of streptokinase to plasminogen appears to be quite specific since there is no known streptokinase substrate other than plasminogen (or plasmin), and prototype streptokinase will activate human and cat plasminogen but not sheep, cow, or pig plasminogen. There are, however, group E streptococcal streptokinases which activate porcine plasminogen, whereas human plasminogen is refractive to activation by these proteins elaborated by streptococci pathogenic for swine (4). Prototype streptokinase also binds to human plasmin, and the resultant streptokinase-plasmin complex is capable of activating any species of mammalian plasminogen (2).

The Wild-Type Streptokinase Gene, *skc*

While streptococci did not readily lend themselves to conventional genetic analysis (9), the application of recombinant DNA technology opened up new approaches to streptococcal genetics (17) and quickly led to the cloning of the prototype streptokinase gene, *skc*, including its flanking regions involved in the control of transcription and translation (10). Primary cloning of *skc* was achieved by inserting *Streptococcus equisimilis* H46A chromosomal DNA, after partial *Sau*3A1 digestion and size fractionation, into lambda L47 vector DNA. In vitro packaging and plating on *Escherichia coli* WL95 to select for recombinant plaques yielded a total of 10 phage clones (ca. 0.1%) which produced distinct zones of caseinolysis when tested for streptokinase production by the plasminogen-casein overlay technique (10). One of the 10 *skc*-containing

primary clones was chosen to subclone *skc* onto *E. coli* plasmid vectors (pACYC184 and pBR322) to form, among others, plasmids pMF2 (10.4 kilobases [kb]) and pMF5 (6.9 kb) which contained *skc* on a 6.4-kb *Eco*RI fragment and a 2.5-kb *Pst*I fragment, respectively. The two plasmids determined streptokinase production in either orientation of *skc*, indicating that they contained the full *skc* structure, including its expression signals. Sequencing of the *Pst*I fragment (12; Appendix B of this volume) showed it to consist of 2,568 base pairs (bp), the longest open reading frame of which contained 1,320 bp coding for prestreptokinase. The protein (440 amino acid residues; M_r, 50,084) carries a 26-amino-acid-residue N-terminal extension having properties characteristic of a signal peptide. Comparison of the amino acid sequence of mature streptokinase as deduced from the nucleotide sequence of *skc* with the available amino acid sequence of a commercial streptokinase determined by Edman degradation (7) revealed minor primary structure differences. To account for these, we cannot, at present, rule out the possible existence of isostreptokinases. Also, we do not know whether the commercial streptokinase sequenced by Jackson and Tang (7) and the *skc* gene came from the same group C streptococcal strain.

Upstream from the *skc* coding region, the nucleotide sequencing revealed a canonical ribosome binding site, represented by 5'...AGGAGG(T)...3' and separated by a 5-bp spacing from the first nucleotide of the translation initiation codon, ATG. Also, the −35 region (5'...TTAAAA...3') and −10 region (5'...AATAATG...3') (spacing, 15 nucleotides) of the putative *skc* promoter are homologous to the prototypical sequences of procaryotic promoters. In addition, the 150-bp region upstream from the ribosome binding site sequence is richer in adenine plus thymine than the overall *Pst*I fragment (80% versus 60.4%). This region contains extensive upstream stretches of AT base pairs which are characteristic of certain genes from gram-positive bacteria, including *Bacillus subtilis*.

Thirty-four base pairs behind the *skc* translational stop codon, TAA, a 15-bp perfect inverted repeat begins which contains two shorter sequences both of which also exhibit intrastrand complementarity. This inverted repeat sequence, which has five GC pairs and ends with a run of six T residues, is very likely to represent the Rho-independent terminator of *skc* transcription.

Missing Homology to the Staphylokinase Gene

On the basis of their findings that the N-terminal half of the streptokinase molecule exhibits some homology to the C-terminal half, and that staphylokinase has a subunit molecular weight about half that of streptokinase and requires a dimer to activate plasminogen, Jackson and Tang (7) suggested that the streptokinase gene has evolved by duplication and fusion and also predicted close relationships between the structures of streptokinase and staphylokinase. We tested these hypotheses by generating a dot matrix to compare the *skc* sequence with the nucleotide sequence of the staphylokinase gene (16). Using a window of 20-bp segments from the DNA regions to be compared and a stringency of 14 matches, i.e., 70% homology, we failed to detect continuous matches between any regions of the two sequences. Moreover, in Southern blot hybridizations, two staphylokinase-producing clinical strains of *Staphylococcus* showed no detectable sequence homology with the 2.5-kb *Pst*I fragment used as a hybridization probe, whereas in all pathogenic group A, C, and G streptococcal strains studied, there was a perfect correlation between their ability to produce streptokinase and to hybridize with the *skc* DNA probe (5). Consistent with these results is the finding of Sako and Tsuchida (16) that there is practically no homology between the amino acid sequences of streptokinase and staphylokinase. Thus, the genes for streptokinase and staphylokinase may not be evolutionarily related and the two proteins may activate plasminogen by different mechanisms.

Using the same parameter setting for generating a dot matrix to test the *skc* gene duplication and fusion hypothesis at the nucleotide sequence level, we also failed to detect any continuous matches between the two large regions which, according to the hypothesis, might have evolved by gene fusion. Thus, the observed internal twofold homology at the amino acid sequence level may not reflect the proposed evolutionary relationships of the DNA regions encoding the two halves of the streptokinase molecule.

Expression of *skc* in Heterologous Hosts

The *skc* gene is expressed in *E. coli* when cloned in either phage lambda or plasmid vectors (10). Since expression is independent of its orientation in the plasmid insertion site, the *skc*-specific promoter must be functional in *E. coli*. However, when cloned in the *Pst*I site of pBR322, *skc* is expressed more efficiently in the sense orientation relative to the beta-lactamase (*bla*) gene promoter(s) than in the opposite orientation. This suggests that transcription from the *bla* gene promoter(s) proceeds into the *skc* gene region and contributes to its expression. Cultured *E. coli* strains transformed with plasmids carrying *skc* show streptokinase activity in all three principal locations, namely, extracellular in the medium, in the periplasm, and in the cytoplasm. Although originally *E. coli* cells were not expected to release detectable amounts of streptokinase into the medium, our data suggest that the extracellular activity is not attributable to leakage of the protein out of dead cells. However, we do not know whether the exported activity results from correct processing and active secretion. The amounts of extracellular streptokinase activity ranged from 8 to 100 U/ml, depending on the age of saturated cultures containing 3×10^9 viable cells per ml. Assuming that authentic and *E. coli* streptokinase have the same specific activity, this amount corresponds to 0.15 to 1.82 mg/liter of culture or about 600 to 7,000 streptokinase molecules released per cell. In early-stationary-phase cultures, the extracellular activity amounted to about 17% of the total streptokinase activity, the rest being found in the periplasm (30%) and in the cytoplasm (52%).

In addition to *E. coli*, heterologous *skc* gene expression was also attained in *B. subtilis* (8). When introduced into plasmid vectors capable of replication in this species, *skc* is expressed by use of its own transcription and translation signals, which appear to meet the stringent requirements of *B. subtilis* for efficient foreign gene expression. The secreted streptokinase activity began to decline toward the end of the exponential-growth phase, suggesting that *B. subtilis* exoproteases expressed at that time hydrolyzed and inactivated the foreign gene product.

To introduce *skc* into the Challis strain of *S. sanguis*, pMF5 and the streptococcal plasmid cloning vector pSM7 were fused at their unique *Eco*RI sites to form the bifunctional shuttle plasmid pSM752 (13.3 kb). This plasmid replicates and determines streptokinase production in *E. coli, B. subtilis*, and *S. sanguis*. In the latter host, pSM752 is present at the lower limit of seven to nine copies per chromosome equivalent. Streptokinase as expressed in Challis had a molecular weight of about 44,000, while the molecular weight of *S. equisimilis* streptokinase is 47,000 (11). Detailed investigation of the primary structure of the cloned Challis streptokinase revealed that it is without 31 or 32 C-terminal residues, as deduced from protein sequencing and cyanogen bromide cleavage patterns (6). Since the complete coding sequence of *skc* was present in Challis, the observed alteration of the gene product, which was still active, must have resulted from posttranslational proteolysis. Thus, it is of interest to investigate structurally altered streptokinase species with the aim of defining the functional roles of various regions of the molecule. Using another approach, we produced, sequenced, and determined the activity of deletion mutants of *skc* in an attempt to better define the role, if any, of the termini of the streptokinase molecule.

Strategy for Producing Terminal Deletions of *skc*

The starting material for producing unidirectional random deletions of *skc* was pMF5 (10). As shown in Fig. 1, *skc* was subcloned from pMF5 into pUC9 so as to allow the isolation and subsequent ligation into multimers of a *Pst*I-*Eco*RI fragment still containing the full *skc* structure. For producing serial deletions from the 5'- and 3'-coding regions, the multimers were cut with *Eco*RI and *Pst*I, respectively, to yield dimers. The open ends of the dimers were exposed to digestion with *Bal* 31, and the digestion products were monomerized by cutting with *Pst*I and *Eco*RI, respectively, and cloned into pUC8 previously cut with *Sma*I plus *Pst*I and *Sma*I plus *Eco*RI, respectively. *E. coli* JM109 clones containing increasingly smaller inserts were assayed for streptokinase activity (10). They also served as a source for inserting the various deletion products into the sequencing vectors M13mp18 and mp19 to determine the new termini of *skc* and define fusions of *skc* with *lacZ* (for a general review of *lac* fusions, see reference 18).

Properties of Mutations Removing Sequences Encoding the N Terminus of Streptokinase

Since deletions of the 5'-coding region removed the *skc* promoter and additional downstream sequences, the fragmented gene was placed under the corresponding transcriptional and translational control elements of the *lac* operon present in pUC8 (19). From a total of about 100 mutants, active streptokinase as assayed by the casein-plasminogen plate technique (10) was detected in clones having lost less than 42 codons of *skc*. Although these clones carried the signal sequence only in part or not at all, extracellular streptokinase activity was detectable without the requirement of breaking open the cells. The largest active deletion characterized, delta N10-9, removed an *skc* sequence encoding, in addition to the signal peptide, 15 N-terminal amino acids of mature streptokinase. The new N terminus of the altered protein was provided by the N-terminal hexapeptide of β-

galactosidase (Fig. 2). Of interest was the observation that the delta N10-9 clone had lost the mucoidicity characteristic of *E. coli* clones carrying the wild-type *skc* gene, indicating that the deleted sequence encodes functions conferring mucoid growth on the heterologous host. Streptokinase obtained from sonicated cells carrying delta N10-9 was serologically indistinguishable from authentic streptokinase when tested by the dot-immunobinding assay using polyclonal antistreptokinase produced in rabbits and alkaline phosphatase-conjugated anti-rabbit immunoglobulin G for visualizing the immune complexes immobilized on nitrocellulose membranes (3). In terms of comparable amounts of protein, delta N10-9 streptokinase had about the same relative activity as wild-type streptokinase produced by *E. coli* when assayed by the plate technique. The question of whether we approached the 5'-deletable maximum for activity to remain is still open since obtaining a conclusive answer would have required the additional sequencing of a great number of inactive larger deletions fused in frame to the β-galactosidase hexapeptide.

C-Terminal Deletions

The largest active deletion from the 3' end of *skc*, delta C25-4, left authentic *skc* sequences up to codon CCC-399, thus removing 41 C-terminal streptokinase-specific amino acids. The shortest inactive deletion was 10 bp longer than delta C25-4, indicating that removal of three more amino acids (K, R, P) from the C terminus causes inactivation. In pUC8, the 3' end of delta C25-4 was fused out of frame to *lacZ'*, thus determining a fusion protein, the C terminus of which consisted of 58 amino acids unrelated to both streptokinase and β-galactosidase before the translation apparatus encountered a stop codon, TGA (Fig. 3). JM109 clones carrying delta C25-4 were mucoid, indicating that the *skc*-specific sequence missing in delta C25-4 had no role in specifying mucoidicity. The truncated streptokinase cross-reacted with antibody against the authentic protein and, consistent with results obtained for posttranslationally altered streptokinase (6), had a lower activity (ca. 1/7) than wild-type streptokinase in activating plasminogen.

Double-Deletion Mutant

With delta N10-9 and delta C25-4 available, the question arose as to whether these two deletions if present together in an expression vector would allow the core *skc* sequence to determine an active protein. To construct the double deletion mutant in pUC8, the two single-deleted *skc* sequences were joined via their common unique *Bst*EII sites to form plasmid

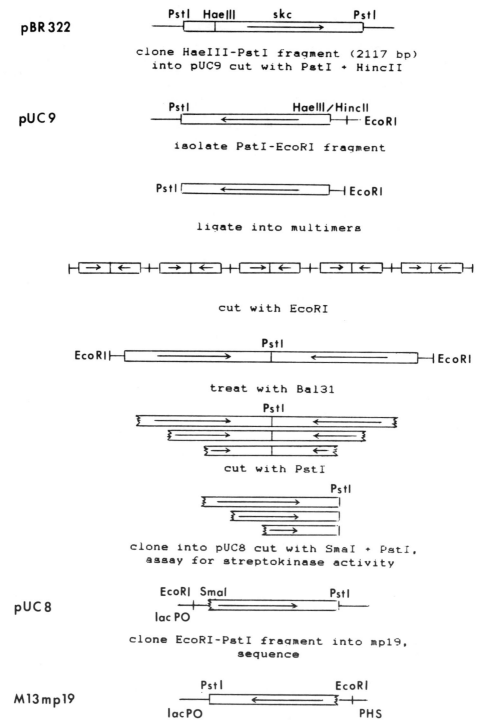

FIG. 1. Strategy for generating and characterizing unidirectional deletions of the 5′-coding region of *skc*. Recombinant DNA methods were essentially those described by Maniatis et al. (13). Lines and bars denote vector and insert DNA, respectively. Arrows indicate the orientation of *skc*. PHS, Primer hybridization site. For producing deletions from the 3′-coding region, insert multimers were cut with *Pst*I and the resultant dimers were digested with *Bal* 31. Digestion products were monomerized with *Eco*RI and assayed for activity after being cloned in pUC8 cut with *Eco*RI plus *Sma*I. *Eco*RI-*Pst*I fragments were isolated from recombinant pUC8 DNA and cloned into M13mp18 for sequencing, using the protocol published in the Amersham M13 cloning and sequencing handbook.

Wild type

```
 M   K   N   Y   L   S   F   G   M   F   A   L   L   F   A   L
ATG AAA AAT TAC TTA TCT TTT GGG ATG TTT GCA CTG CTG TTT GCA CTA
 1                               10

 T   F   G   T   V   N   S   V   Q   A   I   A   G   P   E   W
ACA TTT GGA ACA GTC AAT TCT GTC CAA GCT ATT GCT GGA CCT GAG TGG
             20                                  30

 L   L   D   R   P   S   V   N   N   S   Q   L
CTG CTA GAC CGT CCA TCT GTC AAC AAC AGC CAA TTA ...
                     40
```

Delta N10-9

```
 M   T   M   I   T   N   S ^ S   Q   L
ATG ACC ATG ATT ACG AAT TCC AGC CAA TTA ...
(1) (2) (3) (4) (5) (6)  42  43  44
```

FIG. 2. Nucleotide sequences of the 5′ ends of wild-type *skc* and deletion mutant delta N10-9 fused (∧) to *lacZ'*. Numbering proceeds from the first wild-type *skc* codon, with the signal sequence cleavage site being between amino acids 26 and 27 (12). Numbers in parentheses indicate the N-terminal hexapeptide of β-galactosidase. The single-letter amino acid code is used. Note that the *Sma*I site of pUC8 lost a CG pair in the cloning procedure.

Wild type

```
     K   R   P   E   G   E   N   A   S   Y   H   L   A   Y   D
--- AAG CGA CCC GAA GGA GAG AAT GCT AGC TAT CAT TTA GCC TAT GAT
            399                                      410

 K   D   R   Y   T   E   E   E   R   E   V   Y   S   Y   L   R
AAA GAT CGT TAT ACC GAA GAA GAA CGA GAA GTT TAC AGC TAC CTG CGT
                                420

 Y   T   G   T   P   I   P   D   N   P   N   D   K   *
TAT ACA GGG ACA CCT ATA CCT GAT AAC CCT AAC GAC AAA TAA
        430                                  440
```

Delta C25-4 (out of *lacZ* frame in pUC8)

```
     K   R   P ^ G   D   P   S   T   C   S   Q   A   W   H   W
--- AAG CGA CCC GGG GAT CCG TCG ACC TGC AGC CAA GCT TGG CAC TGG
            399                                      410

 P   S   F   Y   N   V   V   T   G   K   T   L   A   L   P   N
CCG TCG TTT TAC AAC GTC GTG ACT GGG AAA ACC CTG GCG TTA CCC AAC
                                420

 L   I   A   L   Q   H   I   P   L   S   P   A   G   V   I   A
TTA ATC GCC TTG CAG CAC ATC CCC CTT TCG CCA GCT GGC GTA ATA GCG
        430                                  440

 K   R   P   A   P   I   A   L   P   N   S   C   A   A   *
AAG AGG CCC GCA CCG ATC GCC CTT CCC AAC AGT TGC GCA GCC TGA
                    450
```

Delta C25-4 (bifunctional in-frame fusion of skc with lacZ in pUC18)

```
     K   R   P ^ G   D   P   S   T   C   R   H   A   S   L   A
--- AAG CGA CCC GGG GAT CCG TCG ACC TGC AGG CAT GCA AGC TTG GCA
            399

 L   A   V   V
CTG GCC GTC GTT ...
(7) (8) (9) (10)
```

FIG. 3. Nucleotide sequence of the 3′ ends of wild-type *skc* and deletion mutant delta C25-4 fused (∧) to *lacZ'* in pUC8 or pUC18. Numbering conventions are as in Fig. 2.

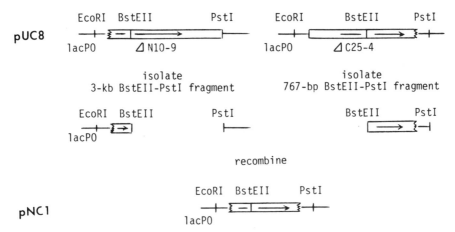

FIG. 4. Strategy for constructing plasmid pNC1 carrying the double-deleted *skc* gene. Lines and bars denote vector and insert DNA, respectively. Arrows indicate the orientation of *skc*.

pNC1 carrying the core *skc* sequence in the sense orientation relative to *lacPO* and bounded by *Eco*RI and *Pst*I sites (Fig. 4). JM109 clones carrying pNC1 produced active streptokinase, with its relative activity being determined by the C-terminal deletion. On isopropyl-β-D-thiogalactopyranoside–5-bromo-4-chloro-3-indolyl-β-D-galactopyranoside (X-Gal) indicator agar, colonies were colorless as expected from clones producing no active alpha peptide of β-galactosidase. Interestingly, when the *Eco*RI-*Pst*I fragment carrying the *skc* double deletion was inserted into pUC18 or M13mp18, the primary structure of the multiple cloning site of these vectors (19) conserved the reading frame for the β-galactosidase alpha peptide, thus leading to an in-frame fusion of *skc* with *lacZ'*. The resulting recombinant plasmid, pNC2, and recombinant phage produced blue colonies and blue plaques, respectively, on indicator agar and were surrounded by caseinolytic zones after being assayed for streptokinase activity. Thus, the fusion protein consisting of the N-terminal hexapeptide of β-galactosidase followed by the streptokinase core sequence in the center and the β-galactosidase alpha peptide at the C terminus was bifunctional, with the activity of the target protein being maintained. As determined quantitatively by using *o*-nitrophenyl-β-D-galactopyranoside as a substrate for β-galactosidase (15), the overwhelming amount of enzyme activity was cell bound (Table 1). However, cell-free supernatant fluids of isopropyl-β-D-thiogalactopyranoside-induced JM109(pNC2) cultures also showed detectable alpha complementation activity when the alpha acceptor protein was provided by an extraneous source, e.g., a lysate of JM109 (Table 1). On solid medium, there was no extracellular in situ alpha complementation, providing additional support for the

suggestion that the fusion protein did not leak out of dead cells.

Conclusions

Directed in vitro mutagenesis of the *skc* gene as reported here has shown that mature streptokinase can be deprived of at least 15 N-terminal and 41 C-terminal amino acid residues without losing activity as a plasminogen activator. The combined loss of these structures also allows the protein to maintain activity. Thus, it can be concluded that the termini of the molecule as defined above are not essential for the formation of the streptokinase-plasminogen complex and its activator capacity. In addition, the remaining core of the protein can be fused at both ends to β-galactosidase sequences to give rise to a bifunctional fusion protein. This suggests that the streptokinase core domain corresponds to an autonomous folding unit, the structure and function of which are not interfered

TABLE 1. β-Galactosidase activity of isopropyl-β-D-thiogalactopyranoside-induced JM109 cultures carrying the indicated plasmids[a]

Strain	Activity (U)[b]	
	Cell bound[c]	Supernatant[d]
JM109	<0.4	<0.2
JM109(pUC18)	666	0.5
JM109(pNC1)	1	<0.2
JM109(pNC2)	824	29

[a] Plasmids are described in the text.
[b] Miller units (15) per 10^9 late-exponential-phase cells.
[c] Cells opened by treatment with chloroform plus sodium dodecyl sulfate (15).
[d] Alpha complementation ability of cell-free supernatant fluids mixed with lysate from JM109 containing the alpha acceptor protein.

with by attached foreign sequences. Future attempts at streptokinase engineering may therefore enable the *skc* gene to be fused to nucleotide sequences coding for homing devices of streptokinase to improve its properties as a therapeutic agent.

The existence of the N-terminal deletion, delta N10-9, invalidates a current hypothesis about the mechanism of plasminogen activation by streptokinase. Bode and Huber (1) suggested that, similarly to the way trypsinogen is activated, streptokinase intrudes its specific N terminus into the isoleucine-binding pocket of plasminogen, thus structuring the specificity pocket which results in activity. In the case of streptokinase specified by delta N10-9, both the altered N-terminal sequence and the physical shortening of the molecule at the N terminus would appear to be incompatible with the N-terminal insertion hypothesis.

LITERATURE CITED

1. **Bode, W., and R. Huber.** 1976. Induction of the bovine trypsinogen-trypsin transition by peptides sequentially similar to the N-terminus of trypsin. FEBS Lett. **68:**231–236.
2. **Castellino, F. J.** 1979. A unique enzyme-protein substrate modifier reaction: plasmin-streptokinase interaction. Trends Biochem. Sci. **4:**1–5.
3. **Dao, M. L.** 1985. An improved method of antigen detection on nitrocellulose: in situ staining of alkaline phosphatase conjugated antibody. J. Immunol. Methods **82:**225–231.
4. **Ellis, R. P., and C. H. Armstrong.** 1971. Production of capsules, streptokinase, and streptodornase by *Streptococcus* group E. Am. J. Vet. Res. **32:**349–356.
5. **Huang, T. T., H. Malke, and J. J. Ferretti.** 1985. Hybridization of a cloned group C streptococcal streptokinase gene with DNA from other streptococcal species, p. 234–236. *In* Y. Kimura, S. Kotami, and Y. Shiokawa (ed.), Recent advances in streptococci and streptococcal diseases. Reedbooks Ltd., Bracknell, Berkshire, England.
6. **Jackson, K. W., H. Malke, D. Gerlach, J. J. Ferretti, and J. Tang.** 1986. Active streptokinase from the cloned gene in *Streptococcus sanguis* is without the carboxyl-terminal 32 residues. Biochemistry **25:**108–114.
7. **Jackson, K. W., and J. Tang.** 1982. Complete amino acid sequence of streptokinase and its homology with serine proteases. Biochemistry **21:**6620–6625.
8. **Klessen, C., and H. Malke.** 1986. Expression of the streptokinase gene from *Streptococcus equisimilis* in *Bacillus subtilis.* J. Basic Microbiol. **26:**75–81.
9. **Malke, H.** 1972. Transduction in group A streptococci, p. 119–133. *In* L. W. Wannamaker and J. M. Matsen (ed.), Streptococci and streptococcal diseases. Academic Press, Inc., New York.
10. **Malke, H., and J. J. Ferretti.** 1984. Streptokinase: cloning, expression, and excretion by *Escherichia coli.* Proc. Natl. Acad. Sci. USA **81:**3557–3561.
11. **Malke, H., D. Gerlach, W. Kohler, and J. J. Ferretti.** 1984. Expression of a streptokinase gene from *Streptococcus equisimilis* in *Streptococcus sanguis.* Mol. Gen. Genet. **196:**360–363.
12. **Malke, H., B. Roe, and J. J. Ferretti.** 1985. Nucleotide sequence of the streptokinase gene from *Streptococcus equisimilis* H46A. Gene **34:**357–362.
13. **Maniatis, T., E. F. Fritsch, and J. Sambrook.** 1982. Molecular cloning: a laboratory manual. Cold Spring Harbor Laboratory, Cold Spring Harbor, N.Y.
14. **Martin, M.** 1982. Streptokinase in chronic arterial diseases. CRC Press, Inc., Boca Raton, Fla.
15. **Miller, J. H.** 1972. Experiments in molecular genetics. Cold Spring Harbor Laboratory, Cold Spring Harbor, N.Y.
16. **Sako, T., and N. Tsuchida.** 1983. Nucleotide sequence of the staphylokinase gene from *Staphylococcus aureus.* Nucleic Acids Res. **11:**7679–7693.
17. **Schlessinger, D. (ed.).** 1982. Streptococcal genetics, p. 81–257. *In* Microbiology—1982. American Society for Microbiology, Washington, D.C.
18. **Silhavy, T. J., and J. R. Beckwith.** 1985. Uses of *lac* fusions for the study of biological problems. Microbiol. Rev. **49:**398–418.
19. **Yanish-Perron, C., J. Vieira, and J. Messing.** 1985. Improved M13 phage cloning vectors and host strains: nucleotide sequences of the M13mp18 and pUC19 vectors. Gene **33:**103–119.

Cloning of the Hyaluronidase Gene from *Streptococcus pyogenes* Bacteriophage H4489A

WAYNE L. HYNES AND JOSEPH J. FERRETTI

Department of Microbiology and Immunology, University of Oklahoma Health Sciences Center, Oklahoma City, Oklahoma 73190

Bacteriophages which infect group A streptococci must penetrate the mucoid hyaluronic acid capsule of the host prior to adsorption and establishment of a productive infection. Strains of streptococci possessing such a capsule are resistant to infection by virulent bacteriophages in the absence of an exogenous supply of hyaluronidase (12, 13). However, these bacterial strains are susceptible to infection by many temperate bacteriophages (12). Kjems (12) found that temperate bacteriophages have hyaluronidase activity associated with the phage particles and suggested that its presence was to provide a means for penetrating the capsule of group A streptococci, a result supported by other investigators (3, 15). Although early reports suggested that virulent phages lacked hyaluronidase activity, subsequent studies using a more sensitive assay demonstrated that the virulent phages do have associated hyaluronidase activity, albeit at a level significantly lower than the temperate bacteriophages (3).

Temperate bacteriophage infection results in the appearance of hyaluronidase in cell lysates which is antigenically distinct from any host cell enzyme (4, 12). Additionally, the hyaluronidase obtained from one bacteriophage is immunologically different from the enzyme from bacteriophages obtained from group A streptococci belonging to other serological types (4).

On the basis of the study of hyaluronic acid breakdown, Niemann et al. (15) characterized the enzyme activity associated with a bacteriophage from a type 12 strain as a hyaluronate lyase (EC 4.2.99.1). The hyaluronidase associated with a phage from a nephritogenic strain of M type 49 streptococci was suggested to be tightly bound to, or an integral element of, the phage (3). This hyaluronidase, released and solubilized by urea treatment, has been purified and characterized with respect to its physical and enzymatic properties (2).

Kjems (12) suggested that the genetic information for production of the hyaluronidase was associated with the phage DNA rather than the phage controlling a bacterial product, as has been reported to occur with production of M protein in group A streptococci (17). The current study was undertaken to examine the genetic makeup of a temperate bacteriophage and in particular to obtain and clone the phage gene specifying hyaluronidase production (*hylP*).

Cloning of the Phage Hyaluronidase Gene (*hylP*)

Bacteriophage H4489A, obtained from an M type 49 strain of *Streptococcus pyogenes* associated with an outbreak of post-streptococcal glomerulonephritis, was propagated on strain K56 without addition of hyaluronidase (19). The phage particles were collected by centrifugation, and the DNA was isolated with phenol used to disrupt the phage protein capsid. The isolated bacteriophage DNA was approximately 32 kilobases in size, and preliminary data indicated that it mapped linearly rather than being circularly permuted as has been reported for the restriction maps of streptococcal bacteriophages T12 (11) and SP24 (18).

Digestion of the phage DNA with the restriction endonuclease *Tha*I, ligation into the *Escherichia coli* plasmid vector pUC8 cleaved with *Sma*I, and subsequent transformation were performed by standard procedures. Transformed clones containing chimeric plasmids were obtained and assayed for the production of hyaluronidase. Hyaluronidase activity from bacterial colonies was detected on 2YT agar plates with an agarose overlay containing hyaluronic acid. After incubation (18 to 24 h), undigested hyaluronic acid was precipitated by using cetylpyridinium chloride (16). Colonies producing hyaluronidase showed a clear halo in a cloudy background. Liquid preparations were assayed by filling wells cut into the hyaluronic acid-agarose with the test preparation. Plates prepared for such assays did not contain a nutrient agar base.

The chimeric plasmid capable of producing hyaluronidase contained a 3-kilobase *Tha*I fragment and was referred to as pSF49; a partial restriction map of the insert is shown in Fig. 1. This *hylP*-containing fragment was located in the central portion of the streptococcal bacteriophage H4489A genome. Benchetrit et al. (2) found the hyaluronidase from an M type 49 group A streptococcus (strain GT8760) to be a 73,000-dalton glycoprotein. Such a protein would require a piece of DNA approximately 2

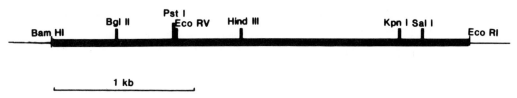

FIG. 1. Restriction map of the *hylP*-containing fragment.

kilobases in size, well within the limits of the cloned fragment.

Subcloning of the *hylP*-containing fragment, using *Bam*HI and *Eco*RI, into pUC9, such that the fragment was in the opposite orientation with regard to the *lacOP* region, still gave clones with hyaluronidase activity. The enzymatic activity observed with the fragment in this orientation was greater than that in the other direction, suggesting that the cloned fragment contained its own promoter region, presumably located at the *Bam*HI end. Similar results were obtained when the *hylP*-containing fragment was subcloned into the *E. coli* M13 phage vectors mp18 and mp19, with clones of the latter showing greater enzymatic activity.

Herd et al. (10) found that neither of the hyaluronidases obtained from two streptococcal strains possessed enzymatic activity to chondroitin sulfate. This lack of activity is in contrast to mammalian and some other bacterial enzymes which are active against both hyaluronic acid and chondroitin sulfate (14). The inability to degrade chondroitin sulfate has been suggested as a possible reason for the lack of toxicity of streptococcal hyaluronidase after injection into rabbits (7). The activity of both the bacteriophage (H4489A) and the *E. coli* clone containing *hylP* was directed at hyaluronic acid, with no activity detected against chondroitin sulfate.

Greiling et al. (8) found that *N*-tosyl-L-phenylalanine chloromethyl ketone (TPCK) and *N*-α-*p*-tosly-L-lysine chloromethyl ketone (TLCK), site-specific reagents which react with histidine, partially inhibit the activity of hyaluronidase from a group A streptococcus. The reason the treated enzyme was not completely inhibited was not explained. Culture supernatant of the *E. coli* clone containing *hylP* was mixed with TPCK as described (8) and incubated at 37°C for 4 h prior to assay on hyaluronic acid-agarose plates. TPCK reduced the enzymatic activity of the cloned enzyme by approximately 50%, as determined by the zone diameter.

Preliminary immunological analysis of the hyaluronidase produced by *E. coli*(pSF49) was carried out with antisera prepared against bacteriophage H4489A. Niemann et al. (15) suggested that the active center of the hyaluronidase from the type 12 phage was part of the

knoblike or starlike structures at the end of the tail. Using the hyaluronidase purified from the M type 49 bacteriophage, Benchetrit et al. (4) found it to be available as an antigen in the intact bacteriophage, more specifically as part of the tail.

Using serum prepared against the whole phage, we were able to show inhibition of the enzymatic activity of both the phage preparation and the cloned enzyme (Fig. 2). This serum had no effect on the activity of testicular hyaluronidase, while serum from a nonimmune animal showed no enzymatic inhibition, indicating that the effect observed with the immune sera was a specific interaction.

Cloning of *hylP* from the streptococcal bacteriophage H4489A opens numerous avenues of investigation, some of which are being pursued in our laboratory. Determination of the nucleotide sequence of *hylP* would allow construction of a specific probe for use as a tool in the examination of clinical isolates of *S. pyogenes* for the presence of bacteriophage-determined hyaluronidase. Such a probe may also allow detection not only of hyaluronidase genes from other streptococcal bacteriophages but also of chromosomally determined hyaluronidase. This may allow clarification of the number and serological types of group A streptococcal strains

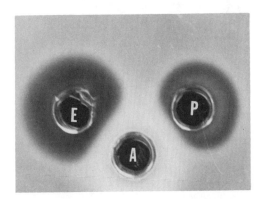

FIG. 2. Inhibition of hyaluronidase activity by antiserum prepared against bacteriophage H4489A. P, Phage H4489A hyaluronidase; E, streptococcal phage hyaluronidase produced by *E. coli*; A, antiserum prepared against phage H4489A.

capable of producing hyaluronidase, as reports have varied from as few as 21% of strains tested and an association with only a few M types (5) to more recent reports which suggest that all M types are capable of producing hyaluronidase (1).

Some insight may also be gained into an in vivo function of hyaluronidase, often considered to be a virulence factor, probably identical with the so called "spreading factor" (7), as well as providing access to a possible reserve of energy for the invading bacterium (6). Determination of any role in the initial establishment of an infection or in the delayed sequelae, poststreptococcal glomerulonephritis, is also important in view of the immunological specificity, similar to that seen with M protein, of the phage enzyme and the possibility that patients with streptococcal infections are being immunized with bacteriophage antigens. Data presented by Halperin et al. (9) indicated that patients with group A streptococcal infections do indeed mount a response to bacteriophage hyaluronidase.

This work was supported by Public Health Service grant AI 19304 from the National Institutes of Health.

LITERATURE CITED

1. Benchetrit, L. C., C. C. Avelino, L. Barrucand, A. S. Figueiredo, and C. M. De Oliveira. 1984. Hyaluronidase production by groups A, B, C, and G streptococci: a statistical analysis. Zentralbl. Bakteriol. Parasitenkd. Infektionskr. Hyg. Abt. 1 Orig. Reihe A 257:27–37.
2. Benchetrit, L. C., E. D. Gray, R. D. Edstrom, and L. W. Wannamaker. 1978. Purification and characterization of a hyaluronidase associated with a temperate bacteriophage of group A, type 49 streptococci. J. Bacteriol. 134:221–228.
3. Benchetrit, L. C., E. D. Gray, and L. W. Wannamaker. 1977. Hyaluronidase activity of bacteriophages of group A streptococci. Infect. Immun. 15:527–532.
4. Benchetrit, L. C., L. W. Wannamaker, and E. D. Gray. 1979. Immunological properties of hyaluronidases associated with temperate bacteriophages of group A streptococci. J. Exp. Med. 149:73–83.
5. Crowley, N. 1944. Hyaluronidase production of haemolyt-

ic streptococci of human origin. J. Pathol. Bacteriol. 56: 27–35.
6. Faber, V. 1952. Streptotoccal hyaluronidase: review and the turbidimetric method for its determination. Acta Pathol. Microbiol. Scand. 31:345–354.
7. Ginsburg, I. 1972. Mechanisms of cell and tissue injury induced by group A streptococci: relation to poststreptococcal sequalae. J. Infect. Dis. 126:294–340.
8. Greiling, H., H. W. Stuhlsatz, T. Eberhard, and A. Eberhard. 1975. Studies on the mechanism of hyaluronate lyase action. Connect. Tissue Res. 3:135–139.
9. Halperin, S. A., E. D. Gray, P. Ferrieri, and L. W. Wannamaker. 1975. Enzyme-linked immunosorbent assay for identification and measurement of antibodies to group A streptococcal bacteriophage. J. Lab. Clin. Med. 106:505–511.
10. Herd, J. K., J. Tschida, and L. Motycka. 1974. The detection of hyaluronidase on electrophoresis membranes. Anal. Biochem. 61:133–143.
11. Johnson, L. P., and P. M. Schlievert. 1983. A physical map of the group A streptococcal pyrogenic exotoxin bacteriophage T12 genome. Mol. Gen. Genet. 189:251–255.
12. Kjems, E. 1958. Studies on streptococcal bacteriophages. 3. Hyaluronidase produced by the streptococcal phage-host cell system. Acta Pathol. Microbiol. Scand. 44:429–439.
13. Maxted, W. R. 1952. Enhancement of streptococcal bacteriophage lysis by hyaluronidase. Nature (London) 170: 1020–1021.
14. Meyer, K. 1971. Hyaluronidases, p. 307–320. In P. D. Boyer (ed.), The enzymes, 3rd ed., vol. 5. Academic Press, Inc., New York.
15. Niemann, H., A. Birch-Andersen, E. Kjems, B. Mansa, and S. Stirm. 1976. Streptococcal bacteriophage 12/12-borne hyaluronidase and its characterization as a lyase (EC 4.2.99.1) by means of streptococcal hyaluronic acid and purified bacteriophage suspensions. Acta Pathol. Microbiol. Scand. Sect. B 84:145–153.
16. Richman, P. G., and H. Baer. 1980. A convenient plate assay for the quantitation of hyaluronidase in hymenoptera venoms. Anal. Biochem. 109:376–381.
17. Spanier, J. G., and P. P. Cleary. 1980. Bacteriophage control of antiphagocytic determinants in group A streptococci. J. Exp. Med. 152:1393–1406.
18. Spanier, J. G., and P. P. Cleary. 1983. A restriction map and analysis of the terminal redundancy in the group A streptococcal bacteriophage SP24. Virology 130:503–513.
19. Wannamaker, L. W., S. Skjold, and W. R. Maxted. 1970. Characterization of bacteriophages from nephritogenic group A streptococci. J. Infect. Dis. 121:407–418.

Molecular-Level Characterization of the *Streptococcus faecalis* Plasmid pAD1 Hemolysin/Bacteriocin Determinant

MICHAEL S. GILMORE,[1] DETLEV BEHNKE,[2] BRUCE A. ROE,[3] AND DON B. CLEWELL[4]

Department of Microbiology and Immunology, University of Oklahoma College of Medicine, Oklahoma City, Oklahoma 73190[1]; Zentral Institut für Mikrobiologie und Experimentelle Therapie, Jena, German Democratic Republic[2]; Department of Chemistry, University of Oklahoma, Norman, Oklahoma 73069[3]; and University of Michigan Dental Research Institute, Ann Arbor, Michigan 48109[4]

Bacterial hemolysins can be classified into two types depending upon the mechanism by which target cell lysis is achieved. Simple hemolysins, like the thiol-activated streptolysin O, effect target cell lysis by binding the target membrane as monomers. Monomeric membrane binding, in the case of thiol-activated hemolysins, is followed by oligomerization and presumably transmembrane channel formation (15). In contrast to the simple hemolysins in which a single protein species can effect target cell lysis (albeit by oligomerization), complex hemolysins require the activities of two dissimilar proteins to cause hemolysis.

A model for hemolysis resulting from the activities of two distinct protein functions was originally proposed by Bernheimer et al. (1) to describe the "CAMP" reaction. In the CAMP reaction, observed at the junction of a cross-streak of *Staphylococcus aureus* and *Streptococcus agalactiae* strains, sphingomyelinase secreted by *S. aureus* sensitizes the erythrocytes by enzymatically converting membrane sphingomyelin to relatively nonpolar *N*-acylsphingosine (ceramide). The CAMP factor, secreted by strains of *S. agalactiae*, binds the newly exposed *N*-acylsphingosine and nonenzymatically effects lysis of the sensitized erythrocytes at the junction of the cross-streak of the two species (1).

A complex hemolysin system which appears to be mechanistically identical to the CAMP factor/sphingomyelinase hemolysin system has recently been found to exist in strains of *Bacillus cereus* (M. S. Gilmore, M. Wachter, J. Kreft, and W. Goebel, manuscript in preparation). In the *B. cereus* system (cereolysin AB or CerAB), both the sphingomyelinase (CerB) and the CAMP-like membrane disrupting factor (CerA) are secreted by the same bacterium. Nucleotide sequence analysis of the CerAB-encoding genes revealed that the genes are tandemly arranged on the bacterial chromosome, separated by a 70-base-pair (bp) spacer (M. S. Gilmore and W. Goebel, manuscript in preparation). Preliminary S1 mapping experiments suggest that CerA and CerB are encoded by a polycistronic message.

Other complex hemolysin systems have been found in strains of *Listeria monocytogenes* (12) and *S. aureus* (11). Although the genomic organization of these complex hemolysins has yet to be determined, two-component hemolysin systems appear to be widely distributed among gram-positive genera. Figure 1 illustrates the ability of isolated components from gram-positive complex hemolysins to complement activities derived from evolutionarily divergent gram-positive bacterial genera, demonstrating their functional relatedness.

A complex hemolysin system has been described in strains of *Streptococcus faecalis* (formerly *zymogenes*) (2, 6). The *S. faecalis* hemolysin has been shown to contribute to the virulence of the bacterium in an animal model (9) and also in epidemiological studies (Y. Ike and D. B. Clewell, this volume). The *S. faecalis* hemolysin is unique among gram-positive hemolysins for two reasons: (i) the *S. faecalis* hemolysin is plasmid encoded (10), and (ii) the hemolysin also possesses bacteriocin activity (2). Because of the uniqueness of the *S. faecalis* hemolysin and its medical importance, it was of interest to investigate the organization and regulation of the *S. faecalis* hemolysin genes.

Localization of Hemolysin-Encoding Genes on *S. faecalis* Plasmid pAD1

Nonhemolytic derivatives of *S. faecalis* plasmid pAD1 were generated by transposon mutagenesis as previously described (3). In earlier studies (3), nonhemolytic mutants of pAD1 were shown to possess transposon Tn*916* and Tn*917* insertions in the contiguous *Eco*RI F, H, and D restriction fragments (see Fig. 2). Recently, nonhemolytic mutants of pAD1 have been obtained which possess transposon inserts in the fragment F proximal region of *Eco*RI fragment C (E. Ehrenfeld, personal communication). Fully hemolytic transposon mutants of pAD1 have been obtained which possess insertions in the central and distal portions of *Eco*RI fragments C and D. The hemolysin determinant on plasmid pAD1 then spans *Eco*RI restriction fragments H and F and extends into fragments C and D.

Complementation Analysis of pAD1 Transposon Mutants

Since an *S. faecalis* hemolysin was previously shown to have two components, termed A and L (6), it was of interest to establish the two-component nature of the pAD1 hemolysin and to determine whether the nonhemolytic phenotype demonstrated by *S. faecalis* harboring Tn*916* and Tn*917* mutants of pAD1 resulted from lack of production of one or both components. To determine whether expression of one or both components was blocked, cross-streaks of independently obtained mutants were made on brain heart infusion agar plates containing 5% human erythrocytes. As illustrated in Fig. 3, nonhemolytic transposon mutants of pAD1 fall into two complementation classes. All nonhemolytic mutants thus far studied express one of the components of the hemolysin; no insertion mutants have yet been found to be defective in both loci. If the hemolysin components were transcribed as a polycistronic message, as appears to be the case for CerAB, a polar effect of insertions on the expression of downstream genes would be expected. Since this polar effect has not been observed among transposon mutants studied, the two components of the hemolysin/bacteriocin appear to be independently transcribed.

The kinetics of hemolysis resulting from the action of *S. faecalis* hemolysin components on erythrocyte membranes have been studied (6).

Granato and Jackson observed that increasing the concentration of one component, L substance, while holding the concentration of the second component constant resulted in an increasing rate of hemolysis. In contrast, when L substance was held constant and the second component, A substance, was increased, an initial increase in the rate of hemolysis at low A substance concentrations, followed by a decrease in rate at higher A substance concentrations, was observed (6). Such an augmentation of hemolysis by an increase in the concentration of one component and such an inhibition of hemolysis by an increase in the concentration of the second component can be observed on blood agar plates and has been widely used to identify the genotype of the respective mutants. As shown in Fig. 3, streaking an $A^- L^+$ mutant through a streak of wild-type hemolytic *S. faecalis* results in an enlarged area of hemolysis at the junction of the two strains on blood agar. In contrast, streaking an $A^+ L^-$ mutant through the wild-type hemolytic strain results in near total inhibition of hemolysis at the junction of the two strains, presumably as a consequence of excess component A production.

Molecular sizes of components A and L of an *S. faecalis* hemolysin have previously been determined to be 27,000 and 11,000, respectively (7, 8). Assuming that signal peptides are required for secretion, a minimum coding capacity of approximately 1,200 nucleotides total would be required to encode the hemolysin.

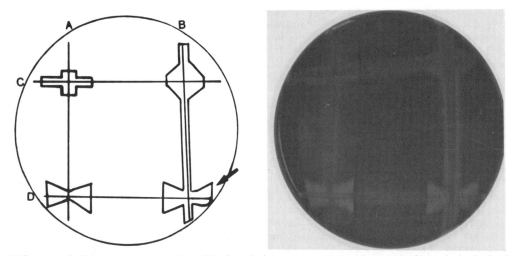

FIG. 1. (Left) Schematic representation of the inoculation pattern and resulting zones of hemolysis obtained when various complex hemolysin-producing strains (A, B, C, and D) were cross-streaked on blood agar. The bacterial strains pictured are (A) *B. subtilis* BR151 expressing recombinant CerB cloned from a cereolysin AB-producing strain of *B. cereus*, (B) a locally obtained clinical isolate of *S. aureus*, (C) *B. subtilis* BR151 expressing CerA, (D) a locally obtained clinical isolate of *S. agalactiae*. The classical CAMP reaction occurring between the *S. aureus* and the *S. agalactiae* isolates is indicated by an arrow. (Right) A photograph of the actual blood agar plate.

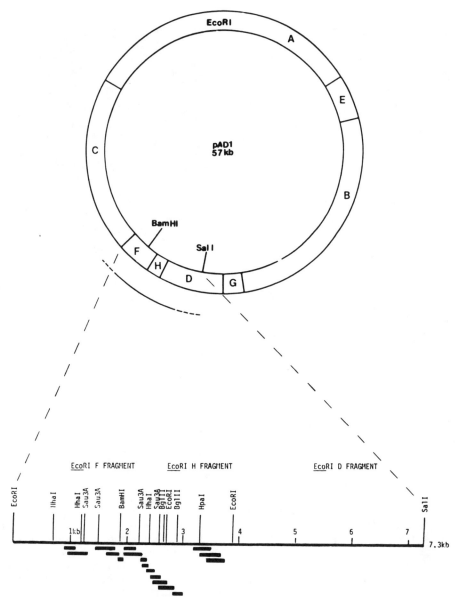

FIG. 2. Physical map of the *Eco*RI restriction pattern of the 57-kb hemolysin/bacteriocin-encoding *S. faecalis* plasmid pAD1. Transposon mapping has allowed the hemolysin genes to be localized to the region surrounding *Eco*RI fragments F and H. The portion of the plasmid illustrated in the enlargement has been cloned, and nucleotide sequence determination for this section of the plasmid is in progress. Regions where extended contiguous nucleotide sequence information is complete are indicated by bold lines.

Transposon analysis of pAD1, however, has revealed that the hemolysin genes are distributed over a minimum of 3.9 kilobases (kb) (the sum of the sizes of *Eco*RI fragments F and H). This discrepancy between the calculated minimum coding capacity and the observed size of the plasmid region required for hemolysin expression may be explained by a large intergenic space. Since both components appear to be independently transcribed, there is no requirement for the components to be closely linked. However, since a narrow optimum in the ratio of component A to component L has been observed to be required for hemolysis, coordinate regulation of the two genes would be expected. An alternative explanation for the large discrepancy between the calculated minimum coding capacity and the observed DNA

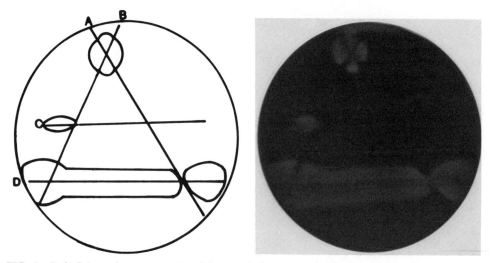

FIG. 3. (Left) Schematic representation of the inoculation pattern and resulting zones of hemolysis obtained when various transposon insertion mutants of pAD1 were cross-streaked on blood agar. The strains pictured are (A) FA2-2(pAM307); (B) JH2-2(pOU107) pAD1::Tn916, Tn916 being located in EcoRI fragment D; (C) JH2-2(pOU109) pAD1::Tn916, Tn916 being located in EcoRI fragment F; and (D) JH2-2(pAD1). (Right) A photograph of the actual blood agar plate after 36 h of incubation at 37°C. Note the inhibition of pAD1-mediated hemolysis at the junction of A and D but the occurrence of hemolysis at the junction of A and B. Conversely, hemolysis is enhanced at the junction of B and D.

requirement is that one or both components are synthesized as much larger precursors, or that auxiliary functions are required for component maturation.

Nucleotide Sequence Determination for the pAD1 Hemolysin Genes

Fragments of pAD1 deduced from transposon analyses to be required for the expression of the hemolysin/bacteriocin have been cloned. The region spanning from the BamHI site within EcoRI fragment F to the SalI site within EcoRI fragment D, and the entire EcoRI fragment F (see Fig. 2), have been separately cloned into the vector pUC8 in Escherichia coli. This was performed to facilitate isolation of subfragments and physical mapping of the hemolysin-encoding region. Reconstitution of the continuity of these two fragments does not result in the expression of detectable levels of hemolysis by E. coli, even when induced with isopropyl-β-D-thiogalactopyranoside. Sau3A, AluI, and RsaI fragments of these two clones were sequenced by the chain termination method of Sanger et al (13).

A total of 6,090 unique nucleotides have been obtained from the cloned 7.3-kb region. Approximately 2,500 nucleotides can be mapped precisely to a 3-kb section of the hemolysin encoding region within EcoRI fragments F and H (see Fig. 2). The 470-bp contiguous sequence obtained between Fig. 2 map positions 3 and 4 initiates the only uninterrupted open reading

frame thus far detected. The open reading frame, initiated with an ATG methionine codon, encodes a hydrophobic region devoid of charged amino acids 26 residues in length (Fig. 4). This is typical for a signal peptide which would be expected to occur on precursors of both secreted components A and L. Moreover, the precise location of Tn917 integration into this region has recently been reported (14). In this nonhemolytic transposon mutant, pAM307, Tn917 was observed to integrate within the series of seven thymine residues immediately upstream of the putative ribosome binding site utilized in translation of the open reading frame (see Fig. 4). No other open reading frames in either orientation can be identified in the vicinity of this transposon insertion site. The adjacent 1,090-bp contiguous nucleotide sequence illustrated in Fig. 2 between map positions 2 and 3 does not appear to encode an open reading frame. The Tn917-harboring mutant plasmid pAM307 exhibits an $A^+ L^-$ phenotype when streaked across a wild-type hemolytic S. faecalis strain. That is, hemolysis at the junction of the intersecting strains is inhibited as illustrated in Fig. 3. The open reading frame initiating immediately downstream from the Tn917 insertion site would therefore be expected to encode component L of the hemolysin.

The assignment of the open reading frame as that which encodes component L is tenuous despite the supporting evidence from transposon integration site determination and phenotype identification. Component L has a reported molecular weight of 11,000 (8), requiring approximately 300-bp coding capacity in addition to sequences for a secretory signal peptide, assuming no other processing steps are involved in component L maturation. If the open reading frame encoded component L, it would be expected to continue only about 150 bp further than shown in Fig. 2, and not into *Eco*RI fragment D. This possibility is diminished by the observation that strains harboring pAM307 complement nonhemolytic mutants possessing transposon inserts in *Eco*RI fragment D and not those with insertions in fragment F. Additionally, Tn*916* mutants of pAD1 with insertions in fragment D have also been found which hyperexpress the hemolytic phenotype.

Conclusion

The genetic organization and mechanism of action of the *S. faecalis* hemolysin appear to differ from those described for other gram-positive complex hemolysins. Cloned component CerA from the *B. cereus* complex hemolysin CerAB does show a limited ability to complement the defect in *S. faecalis* pAD1 mutants of the A⁻ L⁺ genotype. However, it has not been determined whether this complementation results from a simple additive effect or is a truly synergistic effect of these heterologous hemolysin components on the erythrocyte membrane.

Components A and L of the *S. faecalis* hemolysin are encoded by genes spaced at least 1,090 bp apart and appear to be independently transcribed. In contrast, the genes for the *B. cereus* CerAB hemolysin are spaced 70 bp apart and are transcribed as a polycistronic mRNA. The *S. faecalis* hemolysin is encoded by a large, highly transmissible plasmid (5, 10) and possesses bacteriocin activity (2). This class of large *S. faecalis* plasmids also encodes resistance to the hemolysin/bacteriocin (4). Resistance to the

```
                 RSAI                              AHAIII
AATACTTAAAAGCTCTGTAGTACAGAATTTCAGTATGAATTTGATTTAAATTAATGTTTC


                                              SAU3A
TTATATTGATAAATTTTTAGTATGCCACTAATGTATTTTAGTAATAGATCAACTGGAGAA


TTAGTGTTTAGAGCGAACGATTAAATATTTATATTAGGCAAATATTGTCTCAAAAGGTAA


         Tn917 INSERTION
                SD                         HPAI/HINCII
TAACAACGATTTAATAGATAGTCTTT/TTTTAGGATATATTTGTTTTTAATGGTTAACTAT
                                          MetValAsnTyr
               (LeuPheLeu)-----------

TCTATTTTACTGACAATAATAGCTCTTgTcCTAATCTCTTTAATAGCTTTTTTAAGTATT
SerIleLeuLeuThrIleIleAlaLeuValLeuIleSerLeuIleAlaPheLeuSerIle
-----------PUTATIVE HYDROPHOBIC SIGNAL PEPTIDE-------------

ATAAATTCGCATACAATAAAAAGATTTGTAGATAAAGAAATAATGGAACAAGGAAACGTC
IleAsnSerHisThrIleLysArgPheValAspLysGluIleMetGluGlnGlyAsnVal
---------

CAGAGAATTATTACAGAAGCAATTGAAGGAATTGAAACCATTAAATCTGCTAATGCAGAA
GlnArgIleIleThrGluAlaIleGluGlyIleGluThrIleLysSerAlaAsnAlaGlu


AAGAGTTTTTTGTTAAATTGGAAAAACATGTTTACGTCTCAACTATTAATTACA...
LysSerPheLeuLeuAsnTrpLysAsnMetPheThrSerGlnLeuLeuIleThr...
```

FIG. 4. Nucleotide sequence of the open reading frame identified between map positions 3 and 4 of Figure 2. Plasmid pAD1 derivative pAM307 possesses a Tn*917* insertion where indicated and exhibits an A⁺ L⁻ genotype. The amino-terminal 26 amino acids are highly hydrophobic, as would be predicted for a secreted protein possessing a signal peptide, such as hemolysin/bacteriocin component L.

S. faecalis hemolysin/bacteriocin is correlated with the appearance of a new D-alanyl lipoteichoic acid species (4). From a teleological perspective, it can be speculated that the bacteriocin resistance or immunity function may be encoded between the genes for hemolysin components A and L. Such genetic organization would prevent acquisition by a recipient of a functional hemolysin/bacteriocin determinant without the immunity-conferring gene during an interrupted mating. Such genetic organization would also explain the unexpectedly large region of the plasmid shown by transposon mapping to be involved in the expression of the *S. faecalis* hemolysin/bacteriocin. Additional nucleotide sequence determination will be necessary to determine the genetic organization and to detect evidence for the possible coordinate regulation of the *S. faecalis* plasmid pAD1 hemolysin/bacteriocin genes.

LITERATURE CITED

1. **Bernheimer, A. W., R. Linder, and L. S. Avigad.** 1979. Nature and mechanism of action of the CAMP protein of group B streptococci. Infect. Immun. **3:**838–844.
2. **Brock, T. D., and J. M. Davie.** 1963. Probable identity of a group D hemolysin with a bacteriocine. J. Bacteriol. **86:**708–712.
3. **Clewell, D. B., P. K. Tomich, M. C. Gawron-Burke, A. E. Franke, Y. Yagi, and F. Y. An.** 1982. Mapping of *Streptococcus faecalis* plasmids pAD1 and pAD2 and studies relating to transposition of Tn*917*. J. Bacteriol. **152:**1220–1230.
4. **Davie, J. M., and T. D. Brock.** 1966. Effect of teichoic acid on resistance to the membrane-lytic agent of *Streptococcus zymogenes.* J. Bacteriol. **92:**1623–1631.
5. **Dunny, G. M., and D. B. Clewell.** 1975. Transmissible toxin (hemolysin) plasmid in *Streptococcus faecalis* and its mobilization of a noninfectious drug resistance plasmid. J. Bacteriol. **124:**784–790.
6. **Granato, P. A., and R. W. Jackson.** 1969. Bicomponent nature of lysin from *Streptococcus zymogenes.* J. Bacteriol. **108:**865–868.
7. **Granato, P. A., and R. W. Jackson.** 1971. Characterization of the A component of *Streptococcus zymogenes* lysin. J. Bacteriol. **107:**551–556.
8. **Granato, P. A., and R. W. Jackson.** 1971. Purification and characterization of the L component of *Streptococcus zymogenes* lysin. J. Bacteriol. **108:**804–808.
9. **Ike, Y., H. Hashimoto, and D. B. Clewell.** 1984. Hemolysin of *Streptococcus faecalis* subspecies *zymogenes* contributes to virulence in mice. Infect. Immun. **45:**528–530.
10. **Jacob, A. E., G. J. Douglas, and S. J. Hobbs.** 1975. Self-transferable plasmids determining the hemolysin and bacteriocin of *Streptococcus faecalis* var. *zymogenes.* J. Bacteriol. **121:**863–872.
11. **Kreger, A. S., K.-S. Kim, F. Zaboretsky, and A. W. Bernheimer.** 1971. Purification and properties of staphylococcus delta hemolysin. Infect. Immun. **3:**449–465.
12. **Parrisius, J., S. Bhakdi, M. Roth, J. Tranum-Jensen, W. Goebel, and H. P. R. Seeliger.** 1986. Production of listeriolysin by beta-hemolytic strains of *Listeria monocytogenes.* Infect. Immun. **51:**314–319.
13. **Sanger, F., A. Coulsen, G. Barrett, A. Smith, and B. Roe.** 1980. Cloning in single-stranded bacteriophage as an aid to rapid DNA sequencing. J. Mol. Biol. **143:**161–178.
14. **Shaw, J., and D. B. Clewell.** 1985. Complete nucleotide sequence of macrolide-lincosamine-streptogramin B-resistance transposon Tn*917* in *Streptococcus faecalis.* J. Bacteriol. **164:**782–796.
15. **Smith, C. J., and J. L. Duncan.** 1978. Thiol-activated (oxygen-labile) cytolysins, p. 129–183. *In* J. Jeljaszewicz and T. Wadstrom (ed.), Bacterial toxins and cell membranes. Academic Press, London.

High Incidence of Hemolysin Production by *Streptococcus faecalis* Strains Associated with Human Parenteral Infections: Structure of Hemolysin Plasmids

YASUYOSHI IKE[1] AND DON B. CLEWELL[2]

Department of Microbiology, Gunma University School of Medicine, Maebashi, Japan,[1] and Departments of Oral Biology and Microbiology, and The Dental Research Institute, Schools of Dentistry and Medicine, The University of Michigan, Ann Arbor, Michigan 48109[2]

Hemolysin production is common among gram-positive and gram-negative bacteria; however, in most cases its role in virulence is not clear. There have been reports that urinary tract infections caused by *Escherichia coli* are more likely to involve hemolytic strains (8, 9, 20). *Streptococcus faecalis*, a member of the normal intestinal flora, is frequently the causative agent of urinary tract infections and can be involved in other disease states such as endocarditis. *S. faecalis* subsp. *zymogenes* is distinguished from other *S. faecalis* strains by its production of a cytotoxin able to lyse human, rabbit, and horse erythrocytes. Strains producing this beta-hemolysin also produce a bacteriocin, and it is believed that hemolysin and bacteriocin are mediated by the same genetic determinant (1, 2, 3, 14). The plasmid pAD1 (57 kilobases) in *S. faecalis* confers hemolysin-bacteriocin expression and resistance to UV light, as well as a conjugative mating response to the peptide sex pheromone cAD1 excreted by recipient plasmid-free cells (6, 7, 15, 22). We recently reported that the hemolysin-bacteriocin determinant of pAD1 significantly enhanced virulence in intraperitoneal infections in mice (17). The role of *S. faecalis* hemolysin in human infections has been obscure, and to our knowledge, no epidemiological data showing a correlation of the hemolytic trait with a specific type of human infection have been reported. In this communication, we report that a high proportion of *S. faecalis* clinical isolates are hemolytic and that the hemolytic traits are frequently encoded by a conjugative plasmid. The majority of the hemolysin plasmids respond to the sex pheromone cAD1 and bear structural similarities to pAD1.

Results

High frequency of *S. faecalis* among clinical isolates of Lancefield group D streptococci. Among 106 clinical isolates of group D streptococci, about 90% of the strains were *S. faecalis*; the others were *S. faecium* (Table 1). Among 100 healthy medical students at Gunma University, 23 had *S. faecalis* in their fecal specimens,

whereas 84 harbored *S. faecium*. Thus, while *S. faecium* was found to be a more common inhabitant of the gut, *S. faecalis* was much more common in parenteral infections.

The majority of *S. faecalis* clinical isolates are hemolytic. When horse blood was used for detection of hemolysin production by the clinical isolates, 31% of the *S. faecalis* strains were hemolytic (data not shown); however, when human or rabbit blood was used, about 60% were hemolytic (Table 1). Interestingly, the hemolytic strains which produced gelatinase produced a hemolytic zone on human and rabbit blood agar but not on horse blood. The various sources of the clinical isolates of *S. faecalis* and the related percentages of those that were hemolytic are shown in Table 1. For each source, half or more of the strains were hemolytic. The frequency among the 23 hemolytic strains of *S. faecalis* from feces of healthy students was very low (4, or 17%) compared with that of the clinical isolates. In the following experiments, hemolysin production was observed on human blood agar.

Apparent pheromone response by clinical isolates. In *S. faecalis*, certain conjugative plasmids transfer at a relatively high frequency (10^{-1} to 10^{-3} per donor cell) in broth (3). Cells harboring these plasmids generally respond to peptide pheromones produced by recipient cells and synthesize an adhesive surface protein that facilitates the formation of mating aggregates (5, 7, 10, 12, 16, 24). If the donor strain is mixed with a cell-free filtrate ("sex pheromone") of a recipient, a self-aggregation, or clumping, occurs within about 90 min. Among several beta-hemolytic *S. faecalis* isolates examined to date, the hemolysin-bacteriocin trait was found associated with such a conjugative plasmid (3). The clinical isolates obtained in this study were each exposed to a culture filtrate of the plasmid-free *S. faecalis* strain JH2-2. As shown in Table 2, the majority (85%) of the hemolytic strains exhibited a clumping response. In addition, the hemolytic strains were usually (90%) resistant to one or more drugs. Only about 54% of the nonhemolytic strains were also resistant.

TABLE 1. Presence of *S. faecalis* and *S. faecium* in clinical isolates containing group D streptococci[a] and incidence of hemolysin production by isolates of *S. faecalis*[b]

Source	No. of specimens	Specimens with *S. faecium*	Specimens with *S. faecalis*	Hemolytic strains of *S. faecalis* (%)
Urine	52	2	50	25 (50)
Pus	23	3	20	14 (70)
Vagina	12		12	6 (50)
Sputum	7		7	6 (85)
Bile	5	3	2	2
Blood	2	1	1	1
Unknown	5		5	4
Total (%)	106 (100)	9 (8.5)	97 (91.5)	58 (60)

[a] The clinical isolates of Lancefield group D streptococci were received from Gunma University Hospital, Maebashi City, and Isesaki City Hospital, Isesaki City, Japan. The clinical isolates were from focuses in which group D streptococci were contained at a density more than 10^5/ml of material. Bile esculin azide agar plates (Difco Laboratories) were used as primary isolation media for group D streptococci. The species of group D streptococci were identified by serum to group D streptococcus of Phadebact Strept D test (Pharmacia) and the API 20 STREP system (Analytab Products, Plainview, N.Y.).

[b] Hemolysin detection was on Todd-Hewitt agar containing 4% human blood (Gunma University Hospital).

TABLE 2. Pheromone response and drug resistance of Hly$^+$ and Hly$^-$ strains of *S. faecalis*

Response to pheromone[a] and drug resistance patterns[b]	Isolation frequency[c]		P[d]
	Hly$^+$ strains, 58 tested (%)	Hly$^-$ strains, 39 tested (%)	
Positive clumping response	49 (84.5)	19 (48.7)	<0.001
Drug sensitive	6 (10.3)	18 (46.2)	<0.001
Tc	12 ⎫	8 ⎫	
Sm	1 ⎬ 13 (22.4)	⎬ 10 (25.6)	
Em	⎭	2 ⎭	
Tc Em	3 ⎫	3 ⎫	
Tc Sm	1 ⎪ 7 (12.1)	1 ⎬ 4 (10.3)	
Tc Cm	2 ⎬	⎭	
Tc Km	1 ⎭		
Tc Cm Sm	1 ⎫		
Tc Km Gm	1 ⎪		
Tc Km Sm	1 ⎪		
Tc Em Cm	1 ⎬ 7 (12.1)		
Tc Em Sm	2 ⎪		
Tc Km Gm	1 ⎭		
Tc Em Km		2 (5.1)	
Em Cm Km Gm	5 ⎫		
Tc Em Km Gm	1 ⎬ 8 (13.8)		
Tc Em Sm Km	2 ⎭		
Tc Em Cm Km		1 ⎫ 3 (7.7)	
		2 ⎭	
Tc Em Cm Sm Km	2 ⎫	2 (5.1)	
Tc Em Sm Km Gm	4 ⎬ 17 (29.3)		
Tc Em Cm Km Gm	11 ⎭		

[a] The aggregation with exposure to pheromone was performed as previously described (10, 12). Pheromone corresponded to a culture filtrate of the plasmid-free strain JH2-2. Generally, 0.5 ml of culture filtrate from late logarithmically growing cells was mixed with 0.5 ml of fresh N2GT broth and 20 μl of overnight-cultured cells to be tested for their ability to respond. The mixtures were cultured for 4 h at 37°C with shaking and were examined for clumping.

[b] Overnight cultures of the strains in AB3 (Difco) broth were diluted by 100 times with fresh AB3 broth. One loopful of the dilutions was plated on the AB3 broth agar plates containing each drug, and the plates were incubated for 18 h at 37°C. Antibiotic concentrations used were: erythromycin (Em), 25 μg/ml; streptomycin (Sm), 500 μg/ml; kanamycin (Km), 500 μg/ml; gentamicin (Gm), 200 μg/ml; chloramphenicol (Cm), 25 μg/ml; tetracycline (Tc), 3 μg/ml; rifampin (Rif), 25 μg/ml; and fusidic acid (Fus), 25 μg/ml.

[c] Hly$^+$, Hemolytic; Hly$^-$, nonhemolytic.

[d] Significance was determined by using the Fisher exact test for 2×2 contingency tables.

TABLE 3. Conjugative transfer of hemolysin production (Hly) of *S. faecalis*[a]

Phenotype of strain harboring transferable hemolytic trait	No. of donor strains with designated phenotype	No. of hem+ transconjugants per donor cell	Phenotype of transconjugant (% of transconjugant with designated phenotype)	No. of donor strains that gave rise to the indicated phenotypes
Hly	3	$10^{-2}–10^{-3}$	Hly	3
Hly Tc	8	10^{-3}	Hly	8
Hly Tc Em	2	10^{-2}	Hly (10), Hly Em (A)[b] (90)	2
Hly Tc Em	1	10^{-6}	Hly Em (B)	1
Hly Tc Cm	2	$10^{-3}–10^{-4}$	Hly (10), Hly Cm (C) (90)	2
Hly Tc Cm Sm	1	10^{-3}	Hly	1
Hly Tc Km Gm	1	10^{-3}	Hly (50), Hly Km Gm (D) (50)	1
Hly Tc Em Sm Km	1	10^{-3}	Hly	1
Hly Em Cm Km Gm	1	10^{-4}	Hly	1
Hly Tc Em Cm Sm Km	1	10^{-3}	Hly	1
Hly Tc Em Cm Km Gm	11	10^{-3}	Hly (10–20), Hly Em Cm Km Gm (E) (90–80)	8
		10^{-3}	Hly (10), Hly Em Cm Km Gm (80), Hly Em Cm (F) (10)	1
		10^{-3}	Hly (10), Hly Km Gm (G) (90)	2

[a] Overnight cultures of donors of Hly+ strains and recipient FA2-2 (Rif, Fus) (6) were used. Broth matings were carried out as previously described (11, 18). For the transfer of hemolysin properties, the mating mixtures were diluted 10^{-1}, 10^{-2}, and 10^{-4} times with fresh N2GT broth. A 0.1-ml amount of each dilution was plated on selective Todd-Hewitt agar plates containing 4% human blood and rifampin and fusidic acid for counterselection of the donor strain. After overnight incubation of the selective plates at 37°C, the hemolytic colonies (transconjugants) were randomly selected from each plate and purified, and the drug resistance of the transconjugants was examined. Thirty-two (55%) of 58 donors transferred Hly marker and are shown in this table.

[b] The transconjugants which expressed representative phenotypes are designated A to G. Each strain of these transconjugants was used for the second transfer shown in Table 4.

Transferability of hemolysin production. To determine the transferability of the hemolytic trait, mating experiments were performed in broth. Of the 58 hemolytic strains, 32 (55%) were able to donate the determinant for hemolysin production (Table 3). All of these strains aggregated upon exposure to "pheromone," and in most cases transconjugants arose at about 10^{-3} per donor. Transconjugants expressing only the hemolytic property were found in all the matings except one, and the hemolytic trait in these derivatives could be transferred to another recipient (i.e., from FA2-2 to JH2SS) at 10^{-3} to 10^{-1} per donor (data not shown).

As shown in Table 3, resistance traits were frequently transferred (unselected) with the hemolysin trait. To determine linkage between the hemolytic property and the drug resistance of those transconjugants, one strain was selected in each case from the transconjugants which expressed the representative phenotype (the strains are designated as A to G in Table 3), and second-mating experiments were performed in broth (Table 4). When the mating mixtures were plated on nonselective media, transconjugants expressing only the hemolytic trait were found in all the matings except that involving strain B. When selected on drug plates, transconjugants expressing drug resistance without the hemolytic property were found (Table 4). In the case of strain B, the erythromycin resistance determinant was more frequently transferred than the hemolytic determinant, suggesting that the nontransferable hemolysin plasmid was mobilized by the transferable erythromycin resistance plasmid. In none of the matings was there evidence for linkage between the hemolytic property and drug resistance.

Of the 26 strains that did not transfer the hemolysin trait in broth, 17 exhibited induced clumping upon exposure to the JH2-2 culture filtrate. To determine whether transfer of the hemolytic trait would occur on a solid surface, filter mating experiments were performed as previously described (13) (data not shown). Only one strain was able to donate the hemolytic determinant, at about 10^{-5} per donor.

Structure and UV-resistance determinant of the hemolysin plasmids. *S. faecalis* recipients excrete multiple pheromes, each specific for a donor harboring a related pheromone-responding plasmid. Once a plasmid is acquired by the recip-

TABLE 4. Separation of hemolysin production and drug resistance by conjugal transfer

Strain[a]	Phenotype of donor[b] strain used for second transfer	Drug added to selective agar plate[c] (no. of transconjugants per donor cell)	Phenotype of transconjugant[d] (% of transconjugants with designated phenotype)
A	Hly Em	None (10^{-2}) Em (10^{-2})	Hly Tra$^+$ (10), Hly Em (90) Em Tra$^+$ (1), Hly Em (99)
B	Hly Em	None (10^{-6}) Em (10^{-3})	Hly Em Em Tra$^+$ (99.9), Hly Em (0.1)
C	Hly Cm	None (10^{-2}) Cm (10^{-2})	Hly Tra$^+$ (10), Hly Cm (90) Cm Tra$^-$ (10), Hly Cm (90)
D	Hly Km Gm	None (3×10^{-3}) Km or Gm (3×10^{-3})	Hly Tra$^+$ (25), Hly Km Gm (75) Km Gm Tra$^-$ (0.3), Hly Km Gm (99.7)
E	Hly Em Cm Km Gm	None (10^{-2}) Em, Cm, Km, or Gm (10^{-2})	Hly Tra$^+$ (10), Hly Em Cm Km Gm (90) Em Cm Km Gm Tra$^+$ (60), Hly Em Cm Km Gm (40)
F	Hly Em Cm	None (5×10^{-2}) Em or Cm (10^{-2})	Hly Tra$^+$ (5), Hly Em Cm (95) Em Cm Tra$^+$ (10), Hly Em Cm (90)
G	Hly Km Gm	None (4×10^{-2}) Km or Gm (10^{-2})	Hly Tra$^+$ (10), Hly Km Gm (90) Hly Gm Tra$^+$ (40), Hly Km Gm (60)

[a] The strains were derived from the transconjugants shown in Table 3.
[b] Host of donor strain is FA2-2 (Rif Fus).
[c] Recipient strain is JH2SS (Str Spc) (21). Selective agar plates were Todd-Hewitt broth agar containing 4% human blood, streptomycin (500 μg/ml), and spectinomycin (500 μg/ml) plus the drugs indicated in the table (Em, erythromycin; Cm, chloramphenicol; Km, kanamycin; Gm, gentamicin).
[d] Tra, Transferability and clumping response. Transferability was tested by third transfer from JH2SS to FA2-2 in broth mating.

ient, excretion of the related pheromone activity ceases or is greatly reduced, and a peptide that behaves as a specific competitive inhibitor is excreted (4, 16). Other pheromones (unrelated) continue to be produced. If a strain containing a pheromone-responding plasmid does not respond (aggregate) to a culture filtrate of JH2-2(pAD1), the related pheromone for that plasmid is probably cAD1, the pAD1-specific pheromone.

S. faecalis OG1S strains into which 31 conjugative hemolysin plasmids were introduced re-sponded to culture filtrate of JH2-2, but only three strains responded to culture filtrate of JH2-2(pAD1). Thus, the related pheromone for 28 (about 90%) of the hemolysin plasmids appears to be cAD1. Eleven plasmids that responded to cAD1 were chosen for structural analyses (Table 5). For these experiments, we included two hemolysin plasmids which were identified in strains isolated in a university hospital in Tokyo. Figure 1 shows the *Eco*RI restriction profile of each plasmid separately. The

TABLE 5. Phenotypes of representative strains used in structure analysis of hemolysin plasmids that responded to cAD1

Hly plasmid examined	Transfer frequency of Hly plasmid from FA2-2 to JH2SS	Phenotype of wild strains in which the Hly plasmid was identified	Source[a]
pMG701	3×10^{-2}	Hly	J
pMG702	2×10^{-2}	Hly Tc Sm	J
pMG703	10^{-1}	Hly Tc Cm Em Km Gm	G
pMG704	10^{-2}	Hly Tc Cm Em Km Gm	I
pMG705	4×10^{-2}	Hly Tc Cm Em Km Gm	G
pMG706	10^{-2}	Hly Tc Km Gm	I
pMG707	2×10^{-2}	Hly Tc Cm Sm	I
pMG708	10^{-1}	Hly Tc Cm	I
pMG709	5×10^{-2}	Hly Tc Cm	G
pMG710	10^{-1}	Hly Tc	I
pMG711	2×10^{-1}	Hly	G

[a] Initial of hospital: J, Jikei University Hospital, Tokyo; G, Gunma University Hospital, Maebashi City; I, Isesaki City Hospital, Isesaki City. Hospitals were located in geographically dispersed cities in Japan.

a b c d e f g h i j k l m n

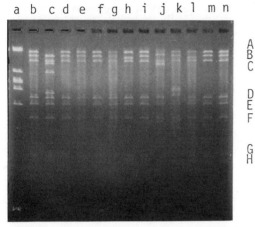

A
B
C

D
E
F

G
H

FIG. 1. Agarose gel electrophoresis of *Eco*RI restriction fragments of Hly plasmids that responded to cAD1. (a) λ DNA; (b and n) pAD1 fragments A through H with corresponding molecular masses of 12.5, 10.0, 7.8, 2.57, 2.15, 1.57, 0.7, and 0.5 megadaltons; (c through m), pMG701 through pMG711.

plasmids exhibited an *Eco*RI restriction profile similar to that of pAD1. Five of the plasmids were essentially indistinguishable from pAD1. The plasmids also have *Sal*I restriction profiles similar to that of pAD1 (three fragments each) (Fig. 2).

pAD1 bears a determinant for resistance to UV light. This trait is observed when the plasmid is located in a UV-sensitive host such as *S. faecalis* UV202 (23). Each of the 11 plasmids that responded to cAD1 was transferred to strain

a b c d e f g h i j k l m n

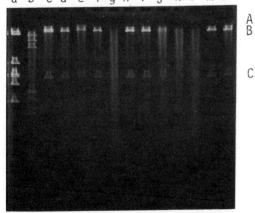

A
B

C

FIG. 2. Agarose gel electrophoresis of *Sal*I restriction fragments of Hly plasmids that responded to cAD1. (a) *Eco*RI fragments of λ DNA; (b) *Eco*RI fragments of pAD1; (n) *Sal*I fragments of pAD1 A through C with corresponding molecular masses of 17.4, 16.5, and 3.8 megadaltons; (c through m) pMG701 through pMG711.

UV202. UV resistance of these plasmids was analyzed as described elsewhere (5a). UV202 strains containing the 11 plasmids each showed a UV resistance phenotype (data not shown).

Dicussion

In this study, the majority of streptococcal group D parenteral infections were found to involve hemolytic strains of *S. faecalis*. This is in contrast to the situation in fecal specimens from healthy individuals where a much lower percentage (17%) of *S. faecalis* strains were hemolytic. (In only 23 of 100 fecal samples could *S. faecalis* strains be detected.) A similar contrast in hemolysin expression between fecal and parenteral strains has been reported for *Escherichia coli* (8, 9, 20). The majority of these hemolytic *S. faecalis* strains exhibited a clumping response when exposed to a culture filtrate of the plasmid free strain JH2-2, and in over half of the isolates, the hemolysin trait could be transferred conjugatively in broth. The clumping response data are consistent with a previous report (12) showing that hemolytic clinical isolates from patients in Ann Arbor, Michigan, were significantly more likely to exhibit a clumping response than nonhemolytic strains. Most of the isolates were resistant to one or more antibiotics, and in many cases resistance traits were transferred (unselected) with the hemolysin trait. However, direct linkage between the hemolysin determinant and resistance determinants was not evident.

Hemolysin determinants in *S. faecalis* usually exist on plasmids (3), and comparison of several determinants showed strong homology (19). The hemolysin plasmids pAD1, pAMγ1, pJH2, and pOB1 all confer a pheromone response, and hybridization studies have found them to be closely related (19). Two of the plasmids, pAD1 from strain DS16 (Ann Arbor, Michigan) and pAMγ1 from strain DS5 (Miami, Florida), are almost identical, and both respond to the peptide pheromone cAD1 (19). pAD1 encodes UV resistance, and the UV resistance determinant is located between 23.2 and 25.4 kilobases on the pAD1 (57 kilobases) map (5a). In this study, 90% (28 of 31) of hemolysin plasmids identified in clinical isolates from two geographically dispersed hospitals in Japan responded to cAD1, and 11 plasmids analyzed were all very similar to pAD1 on the basis of restriction patterns. UV resistance determinants were also associated with those 11 hemolysin plasmids. These results indicate that pAD1-like hemolysin plasmids are highly disseminated in *S. faecalis* strains in Japan.

The conjugative hemolysin plasmid pPD5, which transfers only on filter membranes, does not appear to confer a pheromone response (24).

In the present study only one of those hemolytic strains that did not transfer the hemolytic trait in broth was transferable in filter matings. In general, *S. faecalis* plasmids that require mating on solid surfaces transfer at frequencies on the order of 10^{-4} per donor. It is possible that among the strains found unable to transfer the hemolytic trait, even in filter matings, some have the determinant located on the chromosome.

Together with the previous 50% lethal dose data in mice showing a significant contribution of hemolysin to virulence (17), the epidemiological data presented here suggest that the hemolysin of *S. faecalis* may play a role in human infections. It appears that in most cases the hemolysin determinant can be transferred at relatively high frequencies and is therefore probably plasmid borne. In its natural environment, the bacteriocin activity associated with the hemolysin probably contributes to self-selection. That is, nearby nonhemolytic (i.e., bacteriocin sensitive) strains may be killed if they do not acquire the plasmid. Finally, the high incidence of multiple drug resistance in hemolytic strains raises the possibility that a conjugative hemolysin plasmid might play a significant role in the dissemination (mobilization) of the resistance determinants.

This work was supported by Japanese Ministry of Education, Science and Culture grant 60570185.

LITERATURE CITED

1. **Bassinger, S. F., and R. W. Jackson.** 1968. Bacteriocin (hemolysin) of *Streptococcus zymogenes*. J. Bacteriol. **96:** 1895–1902.
2. **Brock, T. D., and J. M. Davie.** 1963. Probable identity of a group D hemolysin with a bacteriocine. J. Bacteriol. **86:** 708–712.
3. **Clewell, D. B.** 1981. Plasmids, drug resistance, and gene transfer in the genus *Streptococcus*. Microbiol. Rev. **45:** 409–436.
4. **Clewell, D., F. An, B. White, and C. Gawron-Burke.** 1985. *Streptococcus faecalis* sex pheromone (cAM373) also produced by *Staphylococcus aureus* and identification of a conjugative transposon (Tn*918*). J. Bacteriol. **162:**1212–1220.
5. **Clewell, D. B., and B. Brown.** 1980. Sex pheromone cAD1 in *Streptotoccus faecalis*: induction of a function related to plasmid transfer. J. Bacteriol. **143:**1063–1065.
5a. **Clewell, D. B., E. Ehrenfeld, R. Kessler, Y. Ike, A. Franke, M. Madion, J. Shaw, R. Wirth, F. An, M. Mori, C. Kitada, M. Fujino, and A. Suzuki.** 1986. Sex pheromone systems in *Streptococcus faecalis*, p. 131–142. *In* S. B. Levy and R. P. Novick (ed.), Antibiotic resistance genes: ecology, transfer and expression. Banbury report vol. 24. Cold Spring Harbor Laboratory, Cold Spring Harbor, N.Y.
6. **Clewell, D. B., P. K. Tomich, M. C. Gawron-Burke, A. E. Franke, Y. Yagi, and F. Y. An.** 1982. Mapping of *Streptococcus faecalis* plasmids pAD1 and pAD2 and studies relating to transposition of Tn*917*. J. Bacteriol. **152:**1220–1230.
7. **Clewell, D. B., Y. Yagi, Y. Ike, R. A. Craig, B. L. Brown, and F. An.** 1982. Sex pheromones in *Streptococcus*

faecalis: multiple pheromone systems in strain DS5, similarities of pAD1 and pAMγ1, and mutants of pAD1 altered in conjugative properties, p. 97–100. *In* D. Schlessinger (ed.), Microbiology—1982. American Society for Microbiology, Washington, D.C.
8. **Cook, E. M., and S. P. Ewins.** 1975. Properties of strains of *Escherichia coli* isolated from a variety of sources. J. Med. Microbiol. **8:**107–111.
9. **Dudgeon, L. S., E. Wordley, and F. Bawtree.** 1921. On bacillus coli infections of the urinary tract, especially in relation to haemolytic organisms. J. Hyg. **20:**137–164.
10. **Dunny, G. M., B. L. Brown, and D. B. Clewell.** 1978. Induced cell aggregation and mating in *Streptococcus faecalis*: evidence for a bacterial sex pheromone. Proc. Natl. Acad. Sci. USA **75:**3479–3483.
11. **Dunny, G. M., and D. B. Clewell.** 1975. Transmissible toxin (hemolysin) plasmid in *Streptococcus faecalis* and its mobilization of a noninfectious drug resistance plasmid. J. Bacteriol. **124:**784–790.
12. **Dunny, G. M., R. A. Craig, R. Carron, and D. B. Clewell.** 1979. Plasmid transfer in *Streptococcus faecalis*. Production of multiple sex pheromones by recipients. Plasmid **2:** 454–465.
13. **Franke, A. E., and D. B. Clewell.** 1980. Evidence for conjugal transfer of a *Streptococcus faecalis* transposon (Tn*916*) from a chromosomal site in the absence of plasmid DNA. Cold Spring Harbor Symp. Quant. Biol. **45:**77–80.
14. **Granato, P. A., and R. W. Jackson.** 1969. Bicomponent nature of lysin from *Streptococcus zymogenes*. J. Bacteriol. **100:**865–868.
15. **Ike, Y., and D. B. Clewell.** 1984. Genetic analysis of the pAD1 pheromone response in *Streptococcus faecalis*, using transposon Tn*917* as an insertional mutagen. J. Bacteriol. **158:**777–783.
16. **Ike, Y., R. A. Craig, B. A. White, Y. Yagi, and D. B. Clewell.** 1983. Modification of *Streptococcus faecalis* sex pheromones after acquisition of plasmid DNA. Proc. Natl. Acad. Sci. USA **80:**5369–5373.
17. **Ike, Y., H. Hashimoto, and D. B. Clewell.** 1984. Hemolysin of *Streptococcus faecalis* subsp. *zymogenes* contributes to virulence in mice. Infect. Immun. **45:**528–530.
18. **Jacob, A. E., and S. J. Hobbs.** 1974. Conjugal transfer of plasmid-borne multiple antibiotic resistance in *Streptococcus faecalis* var. *zymogenes*. J. Bacteriol. **117:**360–372.
19. **LeBlanc, D. J., L. N. Lee, D. B. Clewell, and D. Behnke.** 1983. Broad geographical distribution of a cytotoxin gene mediating beta-hemolysis and bacteriocin activity among *Streptococcus faecalis* strains. Infect. Immun. **40:**1015–1022.
20. **Minshew, B. H., J. Jorgensen, G. W. Counts, and S. Falkow.** 1978. Association of hemolysin production, hemagglutination of human erythrocytes, and virulence for chicken embryos of extraintestinal *Escherichia coli* isolates. Infect. Immun. **20:**50–54.
21. **Tomich, P. K., F. Y. An, and D. B. Clewell.** 1980. Properties of erythromycin-inducible transposon Tn*917* in *Streptococcus faecalis*. J. Bacteriol. **141:**1366–1374.
22. **Tomich, P., F. An, S. Damale, and D. B. Clewell.** 1979. Plasmid-related transmissibility and multiple drug resistance in *Streptococcus faecalis* subsp. *zymogenes* strain DS16. Antimicrob. Agents Chemother. **15:**828–830.
23. **Yagi, Y., and D. B. Clewell.** 1980. Recombination-deficient mutant of *Streptococcus faecalis*. J. Bacteriol. **143:** 966–970.
24. **Yagi, Y., R. E. Kessler, J. H. Shaw, D. E. Lopatin, F. An, and D. B. Clewell.** 1983. Plasmid content of *Streptococcus faecalis* strain 39-5 and identification of a pheromone (cPD1)-induced surface antigen. J. Gen. Microbiol. **129:** 1207–1215.

C. Pneumococci

Towards a Genetic Analysis of the Virulence of *Streptococcus pneumoniae*

JOHN A. WALKER, REBECCA L. ALLEN, AND GRAHAM J. BOULNOIS

Department of Microbiology, University of Leicester, Leicester, LE1 7RH, United Kingdom

Streptococcus pneumoniae is an important human pathogen, being responsible for lower respiratory tract infections, septicemia, and meningitis. It elaborates a number of proteins, such as a sulfhydryl-activated cytotoxin termed pneumolysin, a human immunoglobulin A1 (IgA1)-specific protease, hyaluronidase, and neuraminidase. Each of these has properties suggestive of a role in virulence, although none has been shown to contribute conclusively to the pathogenic potential of the pneumococcus. Since diverse genera of gram-positive bacteria also produce one or more proteins with activities analogous to those produced by pneumococci, the pneumococcus may therefore represent a good model for studying the role of these proteins in virulence.

Pathogenesis of Pneucococcal Infections

The pneumococcus is a member of the normal microflora of the nasopharynx, with carriage rates often as high as 60%. The reasons for invasion from this to other sites in the adult are unclear and are probably multifactorial. Important features are acquisition of a pneumococcal serotype not previously encountered by that individual and to which there will be no antibody, and a change in the status of the specific or nonspecific defenses of the individual (4, 5). That the latter is probably crucial in the pathogenesis of infection is illustrated by the frequency with which pneumococcal infections are preceded by damage to the respiratory tract epithelium either by viral infection or by other agents. Meningitis and otitis media are often sequelae to respiratory tract infections which are presumably accompanied by a transient bacteremia. Alternatively, pneumococcal meningitis can result from physical injury that permits direct access of pneumococci to the meninges. Interestingly, pneumococcal meningitis has been documented when bacteria gain direct access to the circulation in the absence of identifiable infection elsewhere (4, 5). Splenectomy and conditions such as multiple myeloma that result in an impaired ability to produce antibody also predispose individuals to pneumococcal infection and underline the importance of specific antibody in protection against pneumococcal infection (4, 5).

Putative Pneumococcal Virulence Factors

The polysaccharide capsule of the pneumococcus has attracted considerable interest since it has been shown to contribute to virulence and has been used successfully for vaccination. These topics will not be discussed further here, but the interested reader is referred to a recent review (5).

The pneumococcus elaborates a number of proteins with properties such that they may be considered as putative virulence factors. In support of the notion that these indeed constitute a battery of virulence factors, it is worth noting that pathogens from diverse genera of gram-positive bacteria also elaborate proteins with biological activities similar or identical to those produced by the pneumococcus. However, it must be stressed that none of these products has been shown to contribute conclusively to virulence. Thus the pneumococcus can be considered as a model system for the analysis of these factors.

S. pneumoniae produces a sulfhydryl-activated (thiol-activated) toxin that is a member of a family of toxins produced by diverse genera of gram-positive bacteria. This family of toxins (Table 1) share common physical and biological properties and give rise to antibodies that are broadly cross-reactive and which will neutralize and precipitate other toxins of this family (14, 37). These toxins must therefore possess similar primary or secondary structures to account for these properties. It is interesting to note that there is no detectable DNA sequence homology between the cloned gene for streptolysin O and genes for a number of these related toxins, including pneumolysin (25).

The sulfhydryl-activated toxins require reduction for maximal activity (37), a requirement that indicates that cysteine residues in the proteins are crucial for activity. Initially, it was assumed

TABLE 1. Some examples of diverse bacterial genera that produce sulfhydryl-activated toxins (2, 37)

Genus	Species	Toxin
Bacillus	B. alvei	Alveolysin
	B. cereus	Cereolysin
	B. thuringiensis	Thuringiolysin
Clostridium	C. perfringens	Perfringiolysin
	C. tetani	Tetanolysin
Listeria	L. monocytogenes	Listeriolysin
Streptococcus	S. pneumoniae	Pneumolysin
	S. pyogenes	Streptolysin O

that activation by reducing agents involved the breakage of at least one disulfide bridge to yield free sulfhydryl groups (1, 37). However, this is difficult to reconcile with the observation that perfringiolysin has only one half cystine residue (39). In the case of alveolysin four cysteine residues were found, but no involvement of these residues in disulfide bridge formation could be demonstrated (15). Toxic activity of alveolysin was lost in parallel with alkylation of a single cysteine residue, an observation that led to the hypothesis that one of the cysteine residues provides a thiol group that is essential for toxin activity (12, 13, 15). In this respect these toxins were likened to thiol proteases (2). The precise role this essential cysteine residue plays in toxin activity remains unclear.

All members of this family of toxins are irreversibly inactivated by cholesterol (19, 22, 37). Since these toxins are lytic only for cells with cholesterol in their membranes and because the pattern of inhibition by cholesterol shares many features with the toxin-erythrocyte interaction, cholesterol has been suggested to serve as the membrane receptor for this class of toxin (1, 37). It seems unlikely that the essential cysteine residue (see above) plays any part in the cholesterol binding reaction, since oxidized toxin or toxin pretreated with thiol blocking reagents still binds cholesterol (22). However, such treatments do prevent binding of toxin to erythrocytes, an effect attributed to steric hindrance of the toxin-erythrocyte interaction (22).

Domains of streptolysin O involved in its fixation to membrane (the fixation site) can be separated from those involved in the lytic process by a variety of criteria (1). The so-called lytic site of the toxin remains poorly defined. One attractive possibility is that it may be involved in the oligomerization of toxin that gives rise to arc- and ring-shaped structures in streptolysin O-treated membranes. These structures are composed of about 70 to 80 toxin molecules, are devoid of cholesterol, and form transmembrane lesions analogous to those formed by staphylococcal α-toxin and the C5b-C9 comple-

ment complex (8). It will be interesting if the essential —SH groups discussed above are an integral part of this lytic site.

Sublytic concentrations of pneumolysin have pronounced effects on the activity of human polymorphonuclear leukocytes (PMNs), lymphocytes, and platelets (10, 21, 32). For example, at pneumolysin concentrations which had no effect on PMN viability, the toxin inhibited respiratory burst, chemotactic response, and random migration. These inhibitory effects were not observed when pneumolysin was pretreated with cholesterol, indicating that, as for lytic activity, toxin-cholesterol interactions on the PMN membrane are a crucial aspect of PMN inhibition (10). While this may indicate that lytic action and anti-PMN activity are two facets of the same toxic activity, it has been suggested that the lesions produced at low and high concentrations of cereolysin are different (9). Thus it is tempting to speculate that the sulfhydryl-activated toxins represent a family of multifunctional proteins.

The role of pneumolysin in the pathogenesis of pneumococcal infections is unclear. Given the in vitro effects of pneumolysin on PMNs it is tempting to assume that inhibition of PMN activities crucial to their antimicrobial function allows pneumococcal infections to progress. Purified pneumolysin has been shown to activate the classical pathway of complement and considerably reduces its opsonic activity for the pneumococcus (7, 34). This activation of complement may result in generalized inflammation and may also divert complement away from the invading bacteria (7, 34). However, the precise role pneumolysin plays (if any) in pneumococci (23) and pneumococcal infections remains an open question. As judged by the appearance of antibodies to pneumolysin in pneumococcal infections, it would appear that pneumolysin is produced in vivo during infection (38). Furthermore, mice vaccinated with purified pneumolysin were afforded limited protection from challenge with virulent pneumococci (33). Together, these observations may indicate some role for pneumolysin in virulence; however, two observations may argue against this. First, pneumolysin, unlike the other sulfhydryl-activated toxins, is not secreted from the bacterium (20), and second, serum cholesterol might be expected to inactivate any toxin released from the cell as a consequence of autolysis of the pneumococcus. Clarification of the role pneumolysin plays (if any) in pneumococcal infection awaits further experimentation.

In common with many pathogens of the mucosal surfaces, the pneumococcus secretes a protease that specifically cleaves human IgA1 but not IgA2 (35). Cleavage of the immunoglobulin occurs in the octapeptide repeat sequence in the hinge region of IgA1, a region missing from IgA2. The cleavage site for the pneumococcal enzyme is identical to that of the equivalent enzymes secreted by other streptococcal species (35). Since the Fab_α cleavage products of IgA1 appear to retain antigen binding capacity, Fab_α might coat pneumococci during an infection and help limit access of components of the host's immune system to the pneumococcal cell surface (29). Pneumococcal IgA protease is discussed in more detail by S. A. Pratt and G. J. Boulnois (this volume).

Pneumococci, like many alpha-hemolytic streptococci, produce a neuraminidase (27). This enzyme cleaves $\alpha2$-3 and $\alpha2$-8 linked N-acetyl neuraminic acid, carbohydrate chains encountered in glycoprotein of human brain tissue (11). These glycoproteins may be important in neuronal cell adhesion, and their alteration may be crucial in the pathogenesis of meningitis (11). The observations that prognosis was poor for patients with pneumococcal meningitis associated with abnormally high levels of N-acetyl neuraminic acid in the cerebrospinal fluid (31) and that injection of purified pneumococcal neuraminidase into mice gave rise to neurological signs (26) perhaps indicate a role for neuraminidase in the pathogenesis of pneumococcal meningitis.

Pneumococci also produce hyaluronidase or spreading factor, an enzyme that depolymerizes hyaluronic acid (17). This polymer is an important component of the extracellular matrix, and its destruction may facilitate spread of pneumococci through tissues. A role for this enzyme in pneumococcal infection has yet to be established conclusively.

We have recently initiated a study of the contribution each of these proteins makes to pneumococcal virulence. Our general approach is to isolate the gene for the protein in question and to study both the gene and its product to learn more about the putative virulence factor in question. In addition, the cloned gene will provide a route for the deletion of the pneumococcal version of the gene so that we might try to assess the role of that factor in virulence using animal models for pneumococcal infection. Our preliminary findings relating to the structure and function of pneumolysin are described here; those relating to the IgA1 protease are described by Pratt and Boulnois (this volume).

Molecular Cloning of the Pneumolysin Gene

To facilitate studies on both the structure-function relationships of these sulfhydryl-activated toxins and the role of pneumolysin in infection, we have recently cloned and characterized the gene for pneumolysin (J. A. Walker, R. L. Allen, and G. J. Boulnois, manuscript in preparation).

Libraries of *S. pneumoniae* DNA sequences were constructed (Walker and Boulnois, unpublished data) in the λ phage insertion vector λgt10 (18). Overlaying either recombinant phage plaques or colonies with sheep erythrocytes revealed the presence of hemolytic recombinants.

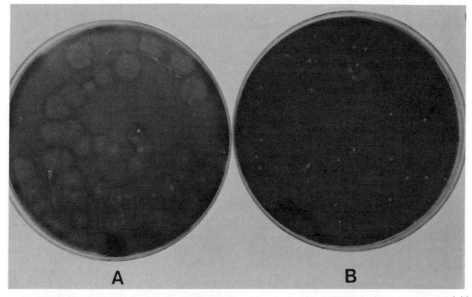

FIG. 1. Hemolysis associated with λPLY003 plaques. Plates carrying λPLY003 plaques were overlaid with soft top agar containing a 7.5% suspension of washed sheep erythrocytes (A) or with the same erythrocyte suspension containing 10 mg of cholesterol per ml (B).

TABLE 2. Properties of hemolysin produced by λPLY003

Phage	Hemolytic activity in phage lysates (HU/ml)[a]			
	Untreated[b]	+ DTT[c]	+ DTT + cholesterol[d]	+ DTT + anti-SLO[e]
λgt10	0	0	0	0
λPLY003	1.6×10^3	1.2×10^4	1.6×10^3	8×10^2

[a] Hemolytic units are expressed as the reciprocal of the dilution of lysate yielding 50% lysis of sheep erythrocytes.
[b] Lysate untreated prior to dilution.
[c] Lysate incubated with 1 mM dithiothreitol (DTT) prior to dilution.
[d] Lysate incubated with 1 mM DTT and treated with cholesterol prior to dilution.
[e] Lysate incubated with 1 mM DTT and treated with anti-streptolysin O serum prior to dilution.

A recombinant λ phage, λPLY003, that expresses a potent hemolysin has been studied in detail. When overlaid with blood agar, every plaque generated by λPLY003 was surrounded by a zone of hemolysis (Fig. 1A), and this activity could be inhibited when cholesterol was included in the agar overlay (Fig. 1B). In semiquantitative hemolytic assays (Table 2), the hemolysin produced in λPLY003 lysates showed enhanced activity when pretreated with thiol reagents such as dithiothreitol, and this enhanced activity could be inhibited by cholesterol and antisera raised against streptolysin O. Since these are properties expected of pneumolysin, it was concluded that λPLY003 carried all or part of the structural gene for pneumolysin.

Restriction endonuclease mapping revealed that the insert in λPLY003 was 5.5 kilobases in length. A 2.5-kilobase TthIII fragment from within the insert was subcloned into pUC8 to yield plasmid pJW252, which, when introduced to Escherichia coli hosts, directed the production of hemolysin.

In an in vitro transcription-translation system (36) pJW252 directs the synthesis of two polypeptides with estimated molecular weights of about 58,400 and 56,100 in addition to vector products (Fig. 2). Clearly, the smaller of the products is the major species produced. Both of these polypeptides were precipitated by serum raised against streptolysin O (data not shown), and thus both may be candidates for the pneumolysin polypeptide since the reported molecular weight of pneumolysin ranges from 52,000 to 60,000 (28, 33). These observations indicated that pJW252 probably carries the intact structural gene for pneumolysin and that this gene, at least in the in vitro system, may express two pneumolysin-related polypeptides. It is interesting to note that production of two amino acid sequence-related polypeptides has been documented for the streptolysin O gene (6, 25).

DNA Sequence of the Pneumolysin Gene

On the basis of the results described above, it seemed likely that pJW252 carried the structural gene for pneumolysin. The complete nucleotide sequence of the 2.5-kilobase TthIII fragment carried by pJW252 was therefore determined by using the shotgun cloning and "dideoxy" sequencing methods (3). Examination of the sequence of this fragment revealed two open reading frames of sufficient length to encode the products observed in the in vitro transcription-translation system. These reading frames yield

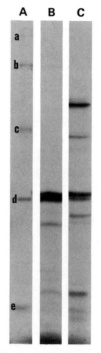

FIG. 2. In vitro transcription-translation products of pJW252. Purified pJW252 or pUC8 DNA was added to a cell-free extract of E. coli containing [^{35}S]methionine and capable of coupled transcription-translation of exogenous DNA. Radiolabeled polypeptides were analyzed by sodium dodecyl sulfate-polyacrylamide gel electrophoresis and fluorography. Track A, Size standards (a, 92,500; b, 69,000; c, 46,000; d, 30,000; e, 14,300); tracks B and C, polypeptides directed by pUC8 and pJW252, respectively.

```
        10                    30                    50

AGCCGGATCT AGCTCGTAAT CCTTTACAAG ACCAACCTTG ATTGACTTAG ATAAGGTATT
                                           ?-35

        70                    90                   110

TATGTTGGAT AATACGGTTA TTCCGACTTC TTATCTAGCC AGACGGCGAC GCAATGTCTC
?-10

       130                   150                   170

AGAAGAATTG TACGAGGAAA TTTTGGATCA CTTAGTCCAA CCACGGCTGA TTTCGCTGAA

       190                   210                   230

CAAGTCTGAG TTTATGCAAC TCAATCCAGG AACTTATTAG GAGGTAGAAG ATGGCAAATA
                                           ?rbs

                                                        M   A   N   K

       250                   270                   290

AAGCAGTAAA TGACTTTATA CTAGCTATGA ATTACGATAA AAAGAAACTC TTGACCCATC

  A   V   N   D   F   I   L   A   M   N   Y   D   K   K   K   L   L   T   H   Q
```

FIG. 3. DNA sequence around the 5' end of the pneumolysin gene. The annotations −10 and −35 refer to the postulated promoter sequence, and rbs refers to the putative ribosome binding site. The two possible ATG translation starts are underlined, and the predicted N-terminal amino acid sequence of pneumolysin is indicated.

polypeptides of 459 and 471 amino acids from two distinct, but in frame ATG initiator codons (Fig. 3). The use of both of these starts in a single transcript may account for the appearance of the two pneumolysin-related polypeptides in the in vitro expression system. The larger product had a predicted molecular weight of 52,800. That the reading frame that gave rise to this product was the likely coding sequence for pneumolysin was suggested by the finding of a ribosome binding site (Shine-Dalgarno sequence) 7 base pairs upstream of the postulated ATG start codon. The slightly smaller sequence was not associated with a convincing ribosome binding site, but it is curious that apparently this is the major product synthesized in the in vitro system. Further upstream from this reading frame are sequences that resemble E. coli (16) and Bacillus subtilis (30) promoter sequences (Fig. 3). Whether these are the signals used by both E. coli and the pneumococcus is not yet clear.

The predicted amino acid sequence of pneumolysin is shown in Fig. 4. The polypeptide is highly hydrophobic in character, as is the native protein (24). The predicted amino acid sequence of pneumolysin reveals a single cysteine residue at position 428. Since the sulfhydryl toxins are apparently single subunit proteins (37), this finding would seem to preclude the notion that thiol activation of these toxins reflects the breakage of disulfide bridges. This single cysteine residue presumably defines a crucial region of the toxin and is the target for further analysis.

```
  1   MANKAVNDFI  LAMNYDKKKL  LTHQGESIEN  RFIKEGNQLP  DEFVVIERKK

 51   RSLSTNTSDI  SVTATNDSRL  YPGALLVVDE  TLLENNPTLL  AVDRAPMTYS

101   IDLPGLASSD  SFLQVEDPSN  SSVRGAVNDL  LAKWHQDYGQ  VNNVPARMQY

151   EKITAHSMEQ  LKVKFGSDFE  KTGNSLDIDF  NSVHSGEKQI  QIVNFKQIYY

201   TVSVDAVKNP  GDVFQDTVTV  EDLKQRGISA  ERPLVYISSV  AYGRQVYLKL

251   ETTSKSDEVE  AAFEALIKGV  KVAPQTEWKQ  ILDNTEVKAV  ILGGDPSSGA

301   RVVTGKVDMV  EDLIQEGSRF  TADHPGLPIS  YTTSFLRDNV  VATFQNSTDY

351   VETKVTAYRN  GDLLLDHSGA  YVAQYYITWD  ELSYDHQGKE  VLTPKAWDRN

401   GQDLTAHFTT  SIPLKGNVRN  LSVKIRECTG  LAWEWWRTVY  EKTDLPLVRK

451   RTISIWGTTL  YPQVEDKVEN  D*
```

FIG. 4. Predicted amino acid sequence of pneumolysin.

Conclusions and Prospects

The genes for pneumolysin and pneumococcal IgA1 protease (see Pratt and Boulnois, this volume) have been cloned and a preliminary analysis has been completed. The DNA sequencing studies have identified a single cysteine residue in pneumolysin that presumably locates an important region of the protein worthy of further study. Further regions of interest in terms of the structure and function of these toxins should be revealed by the comparison of the primary amino acid sequence of several of these toxins. To this end, we (J. A. Walker, C. Geoffrey, J. Alouf, and G. J. Boulnois, unpublished data) have almost completed the sequence of the alveolysin gene that has recently been cloned (C. Geoffrey and J. Alouf, personal communication). The gene for streptolysin O has also been cloned (25), and its DNA sequence is currently being determined (M. K. Kehoe, personal communication). The three-way comparison of the amino acid sequence of these sulfhydryl toxins should prove fascinating and ought to highlight domains of these proteins worthy of further study.

We have constructed in vitro a sequence-defined deletion of the pneumolysin gene and are currently in the process of recombining this deletion into the pneumococcal version of the gene. This we hope will allow us to begin a systematic study of the contribution pneumolysin makes to pneumococcal virulence.

We are currently screening our gene libraries for the genes for hyaluronidase, neuraminidase, and capsule production. As in our studies on pneumolysin, we hope not only to obtain considerable information on the structure and function of these important factors but also to be able to shed some light on their role (if any) in the pathogenesis of pneumococcal infection.

This work was supported by grants from the Medical Research Council of the United Kingdom to G.J.B. R.L.A. was supported by a studentship from the Science and Engineering Research Council of the United Kingdom.

We thank J. Alouf, Institut Pasteur, Paris, for kindly providing antiserum against streptolysin O.

LITERATURE CITED

1. **Alouf, J.** 1980. Streptococcal toxins (streptolysin O, streptolysin S, erythrogenic toxin). Pharm. Ther. **11**:661–717.
2. **Alouf, J. E., and C. Geoffrey.** 1984. Structure activity relationships in sulfhydryl-activated toxins, p. 165–171. In J. E. Alouf, F. J. Fehrenbach, J. H. Freer, and J. Jeljaszewicz (ed.), Bacterial protein toxins. Academic Press, Inc., New York.
3. **Andersen, S.** 1981. Shotgun DNA sequencing using DNase I-generated fragments. Nucleic Acid Res. **9**:3015–3027.
4. **Austrian, R.** 1979. The pneumococcus at Hopkins: early portents of future developments. Johns Hopkins Med. J. **144**:192–201.
5. **Austrian, R.** 1984. Pneumococcal infections, p. 257–288. In R. Germanier (ed.), Bacterial vaccines. Academic Press, Inc., New York.
6. **Bhakdi, S., M. Roth, A. Szeigoleit, and J. Tranum-Jensen.** 1984. Isolation and identification of two hemolytic forms of streptolysin-O. Infect. Immun. **46**:394–400.
7. **Bhakdi, S., and J. Tranum-Jensen.** 1985. Complement activation and attack on autologous cell membranes induced by streptolysin-O. Infect. Immun. **48**:713–719.
8. **Bhakdi, S., J. Tranum-Jensen, and S. Szeigoleit.** 1985. Mechanism of membrane damage by streptolysin-O. Infect. Immun. **47**:52–60.
9. **Cowell, J. L., K.-S. Kim, and A. W. Bernheimer.** 1978. Alteration by cereolysin of the structure of cholesterol-

containing membranes. Biochim. Biophys. Acta **507**:230–241.

10. **Ferrante, A., B. Rowan-Kelly, and J. C. Paton.** 1984. Inhibition of in vitro human lymphocyte response by the pneumococcal toxin pneumolysin. Infect. Immun. **46**:585–589.

11. **Finne, J.** 1985. Polysialic acid—a glycoprotein carbohydrate involved in neural adhesion and bacterial meningitis. Trends Biochem. Sci. **10**:129–132.

12. **Geoffrey, C., and J. E. Alouf.** 1982. Interaction of alveolysin, a sulfhydryl-activated bacterial cytolytic toxin with thiol group reagents and cholesterol. Toxicon **1**:239–241.

13. **Geoffrey, C., and J. E. Alouf.** 1983. Selective purification by thiol-disulfide interchange chromatography of alveolysin, a sulfhydryl-activated toxin of Bacillus alvei. J. Biol. Chem. **256**:9968–9972.

14. **Geoffrey, C., and J. E. Alouf.** 1984. Antigenic relationship between sulfhydryl-activated toxins, p. 241–243. In J. E. Alouf, F. J. Fehrenbach, J. H. Freer and J. Jeljaszewicz (ed.), Bacterial protein toxins. Academic Press, Inc., New York.

15. **Geoffrey, C., A.-M. Gilles, and J. E. Alouf.** 1981. The sulfhydryl groups of the thiol-dependent cytolytic toxin from Bacillus alvei. Evidence for one essential sulfhydryl group. Biochem. Biophys. Res. Commun. **3**:781–788.

16. **Hawley, D. K., and R. McClure.** 1983. Compilation and analysis of Escherichia coli promoter DNA sequences. Nucleic Acids Res. **11**:2237–2255.

17. **Humphrey, J. H.** 1944. Hyaluronidase production by pneumococci. J. Pathol. Bacteriol. **56**:273–275.

18. **Huynh, T., R. A. Young, and R. W. Davis.** 1985. Construction and screening of cDNA libraries in λgt10 and λgt11, p 49–78. In D. M. Glover (ed.), DNA cloning, vol. 1. IRL Press, Oxford.

19. **Johnson, M. K.** 1972. Properties of purified pneumococcal hemolysin. Infect. Immun. **6**:755–760.

20. **Johnson, M. K.** 1977. Cellular location of pneumolysin. FEMS Microbiol. Lett. **2**:243–245.

21. **Johnson, M. K., D. Boese-Marrazzo, and W. A. Pierce.** 1981. Effects of pneumolysin on human polymorphonuclear leukocytes and platelets. Infect. Immun. **34**:171–176.

22. **Johnson, M. K., C. Geoffrey, and J. E. Alouf.** 1980. Binding of cholesterol by sulfhydryl-activated cytolysins. Infect. Immun. **27**:97–101.

23. **Johnson, M. K., D. Hamon, and G. K. Drew.** 1982. Isolation and characterization of pneumolysin-negative mutants of Streptococcus pneumoniae. Infect. Immun. **37**:837–839.

24. **Johnson, M. K., R. Knight, and G. K. Drew.** 1982. The hydrophobic character of thiol-activated cytolysins. Biochem. J. **207**:557–560.

25. **Kehoe, M. K., and K. N. Timmis.** 1984. Cloning and expression in Escherichia coli of the streptolysin O determinant from Streptococcus pyogenes: characterization of the cloned streptolysin O determinant and demonstration of the absence of substantial homology with determinants of other thiol-activated toxins. Infect. Immun. **43**:804–810.

26. **Kelly, T. T., and D. Greiff.** 1970. Toxicity of pneumococcal neuraminidase. Infect. Immun. **2**:115–117.

27. **Kelly, T. T., D. Greiff, and S. Forman.** 1966. Neuraminidase activity of Diplococcus pneumoniae. J. Bacteriol. **91**:601–603.

28. **Kreger, A. S., and A. W. Bernheimer.** 1969. Physical behavior of pneumolysin. J. Bacteriol. **98**:263–275.

29. **Mansa, B., and M. Kilian.** Retained antigen-binding activity for Fab$_\alpha$ fragments of human monoclonal immunoglobulin A1 (IgA1) cleaved by IgA1 protease. Infect. Immun. **52**:171–174.

30. **Moran, C. P. J., N. Lang, S. F. J. LeGrice, G. Lee, M. Stephens, A. L. Sonenshein, J. Pero, and R. Losick.** 1982. Nucleotide sequences that signal the initiation of transcription and translation in Bacillus subtilis. Mol. Gen. Genet. **186**:339–346.

31. **O'Toole, R. D., L. Goode, and C. Howe.** 1971. Neuraminidase activity in bacterial meningitis. J. Clin. Invest. **50**:979–985.

32. **Paton, J. C., and A. Ferrante.** 1983. Inhibition of human polymorphonuclear leukocyte respiratory burst, bactericidal activity, and migration by pneumolysin. Infect. Immun. **41**:1212–1216.

33. **Paton, J. C., R. A. Lock, and D. J. Hansman.** 1983. Effect of immunization with pneumolysin on survival time of mice challenged with Streptococcus pneumoniae. Infect. Immun. **40**:548–552.

34. **Paton, J. C., B. Rowan-Kelly, and A. Ferrante.** 1984. Activation of human complement by the pneumococcal toxin pneumolysin. Infect. Immun. **43**:1085–1087.

35. **Plaut, A. G.** 1983. The IgA1 proteases of pathogenic bacteria. Annu. Rev. Microbiol. **37**:603–622.

36. **Pratt, J. M., G. J. Boulnois, V. D. Darby, E. Orr, E. Wahle, and I. B. Holland.** 1981. Identification of gene products programmed by restriction endonuclease DNA fragments using an E. coli in vitro system. Nucleic Acid Res. **9**:4459–4474.

37. **Smyth, C. J., and J. L. Duncan.** 1978. Thiol-activated (oxygen-labile) cytolysins, p. 129–183. In J. Jeljaszewicz and T. Wadstrom (ed.), Bacterial toxin and cell membranes. Academic Press, Inc., New York.

38. **Sutliffe, W. D., and A. Zoffuto.** 1969. Pneumolysin and antipneumolysin. Antimicrob. Agents Chemother. **11**:3350.

39. **Yamakawa, Y., A. Ito, and H. Sato.** 1977. Theta toxin of Clostridium perfringens. Biochim. Biophys. Acta **494**:301–313.

Insertionally Inactivated Mutants That Lack Pneumococcal Surface Protein A

LARRY S. McDANIEL, JANET YOTHER, W. DOUGLAS WALTMAN II, AND DAVID E. BRILES

The Cellular Immunobiology Unit of the Tumor Institute, Department of Microbiology, and The Comprehensive Cancer Center, University of Alabama at Birmingham, Birmingham, Alabama 35294

It has long been established that anticapsular antibodies opsonize *Streptococcus pneumoniae* (16, 18) and rapidly promote the clearance of pneumococci from the blood (5, 16). However, there is also evidence from a number of laboratories that antibodies directed against cell wall antigens can facilitate the removal of pneumococci from the blood (7) and, as a result, protect mice against infection (2, 4, 9, 13, 14, 19).

We have been investigating antibodies against cell surface components other than the capsule which can protect against fatal pneumococcal infection (8, 8a). In our studies to characterize surface components of *S. pneumoniae* and to determine the biological role of these molecules, we have produced hybridomas that secrete monoclonal antibodies reactive with several pneumococcal antigens including antigenic determinants of pneumococcal teichoic acid and pneumococcal surface protein A (PspA). In this paper we will describe our studies involving PspA.

We recently reviewed the production and characterization of monoclonal antibodies reactive with PspA (8a). Two of these antibodies, Xi64 (immunoglobulin M) and Xi126 (immunoglobulin G2b), protect mice against infection with certain pneumococcal strains carrying PspA (9a). It has been possible to determine a relative molecular weight for PspA by using our monoclonal antibodies to develop immunoblots of cell lysates. This procedure indicates that PspA from strain R36A has a molecular weight of 84,000 (R36A is the nonencapsulated pneumococcal strain against which the monoclonal antibodies were produced). PspA appears to have multiple antigenic determinants since all four of the antibodies we have produced that react with it recognize different epitopes on the molecule. PspA from different pneumococcal isolates varies both in its expression of these antigenic determinants and in its molecular weight.

PspA has the same apparent molecular weight in Rx1, a strain derived from R36A (12), and in D39, the encapsulated type 2 isolate from which R36A was derived (3). However, we have detected a variation in the molecular weight of PspA by as much as ±20,000 among 30 pneumo-

coccal isolates representing 10 different capsular serotypes (W. D. Waltman, L. S. McDaniel, and D. E. Briles, manuscript in preparation). We are also examining isolates which do not react with our current anti-PspA antibodies to determine whether they express genetic variants of PspA.

PspA appears to be accessible to antibodies on encapsulated viable pneumococci. This conclusion is based on the observations that our antiprotein antibodies bind to a number of viable encapsulated isolates (9a), that we can immunoprecipitate surface-labeled proteins from encapsulated pneumococci, and that our antibodies promote the removal of encapsulated pneumococci from the blood of mice, resulting in protection against infection (8a). Therefore, either the capsule must have gaps that expose underlying cell wall components or the PspA molecule must protrude through the capsule. Figure 1 is a highly schematic depiction of our view of the surface of an encapsulated pneumococcus showing the possible features described above.

To aid in our studies of the PspA molecule, we have constructed mutants which fail to produce PspA (9b). We used the process of insertional inactivation (10, 15) to target the insertion of pVA891 into a Rx1 chromosomal gene necessary for the production of PspA. The pVA891 plasmid fails to replicate autonomously in streptococci (6). The scheme for our molecular manipulations is outlined in Fig. 2.

After transformation of Rx1 with pVA891 carrying the targeting chromosomal fragments, erythromycin-resistant transformants were screened using a colony blot assay (9a) to identify mutants which failed to produce PspA. We identified three mutants designated WG44.1, WG44.2, and WG44.3 from a single mutant library that were PspA⁻. Immunostaining of electroblotted lysate proteins separated on a sodium dodecyl sulfate-polyacrylamide gel demonstrated that these mutants failed to produce any detectable PspA (Table 1).

We next cloned the pVA891-associated *pspA* fragment of WG44.1 into *Escherichia coli* (9b). This was accomplished by extracting total DNA from WG44.1 and using this to transform *E. coli*. It has been proposed that, when insertion into

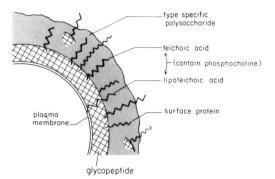

FIG. 1. Surface of an encapsulated pneumococcus showing surface proteins, such as PspA, teichoic acid, and lipoteichoic acid. These components could extend beyond the capsule polysaccharide coat, or there may be gaps in the capsule which would allow antibodies to react with wall-associated components. It is also possible that in the growth of the cell wall some wall components might become cleaved but not reattached, resulting in their release along with any PspA or teichoic acids attached to them into the capsular layer.

the chromosome occurs via a target fragment, there is a duplication of the target fragment (10, 11). Therefore, it is possible for the duplicated fragment to undergo a recombination event that would regenerate the plasmid carrying the target fragment. This plasmid would then be capable of autonomous replication in *E. coli* and could be detected by transformation of *E. coli* and plating on chloramphenicol-containing plates. We have isolated such a derivative of pVA891 carrying a pneumococcal fragment of about 550 base pairs and designated this plasmid pKSD300. This fragment has homology with a gene necessary for the production of PspA because when a pneumococcal strain expressing PspA is transformed with pKSD300 all resulting transformants fail to produce PspA.

To determine whether our cloned fragment specifically hybridized with pneumococcal DNA, we carried out Southern blot analysis of restriction enzyme digests of Rx1 chromosomal DNA. We isolated the cloned fragment from pKSD300, labeled it with ^{32}P, and found that the probe hybridized to unique fragments of the Rx1 digests (9b).

This specificity was confirmed by colony hybridization experiments (Fig. 3). Of 102 pneumococcal isolates tested, 18 of 25 which produced PspA reacted with the probe while only 3 of 77 which lacked PspA reacted with the probe. Additionally, neither our antibodies nor our probe reacted with five group A or five group B streptococcal isolates tested. We have interpreted the correlation between the reactions of the anti-PspA antibodies and the probe as evidence that we have cloned a fragment of the

structural gene that encodes PspA. Since our probe reacts with some isolates that fail to react with the antibodies and fails to react with some isolates that do react with the antibodies, it is possible that we have cloned a fragment of *pspA*

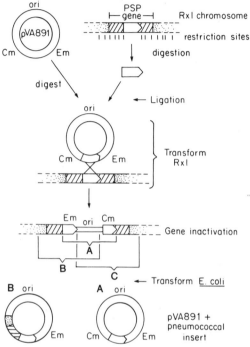

FIG. 2. Diagram showing insertional inactivation of a gene encoding a pneumococcal surface protein (PSP) in Rx1, and the generation of a derivative of pVA891 carrying a fragment of the inactivated gene into *E. coli*. Rx1 chromosomal DNA is cut with restriction enzymes, and the fragments produced are ligated into the plasmid pVA891. These derivatives of pVA891 are then transformed into Rx1. Because the derivative plasmids lack a pneumococcal origin of replication, all erythromycin-resistance transformants obtained are those in which the plasmid has integrated into the chromosome via the homology of its particular chromosomal fragment. By use of a colony blot assay, it is possible to identify those erythromycin-resistant colonies in which a gene required for the production of a PSP has been inactivated. Once the Psp⁻ isolates are identified, the pVA891 plasmid with its targeting fragment of the gene required for PSP production can be cloned into *E. coli* by two different procedures. (A) A plasmid containing pVA891 with its pneumococcal fragment can spontaneously form because of the direct repeat of the fragment in the pneumococcal chromosome. This spontaneous plasmid can be detected by transformation of *E. coli*. This is how we obtained pKSD300 (see text). (B or C) Restriction enzymes that cut pVA891 once, leaving intact replication functions and a selectable marker, would generate fragments which could be circularized by ligation and transformed into *E. coli*. This figure has been adapted from Morrison et al. (10).

TABLE 1. Reactivity of mutants with monoclonal antibodies[a]

Strain	Anti-PspA		Anti-phosphorylcholine, 22.1A4
	Xi64	Xi126	
Rx1	+	+	+
WG44.1	−	−	+
WG44.2	−	−	+
WG44.3	−	−	+

[a] Total-cell lysates of each strain were prepared. The anti-PspA antibodies were used to immunostain (Western blot) electroblotted proteins after separation by sodium dodecyl sulfate-polyacrylamide gel electrophoresis. The anti-phosphorylcholine antibody was used as a control when the lysates were spotted onto a nitrocellulose membrane (dot blot) to test reactivity.

that is different from the region of the gene that encodes the antigenic determinants recognized by our antibodies. This interpretation is consistent with our observation of the presence or absence of individual PspA determinants as detected by antibodies on different pneumococcal isolates (9a). We will be better able to analyze the immunogenic determinants of PspA once we have cloned the entire gene that encodes PspA.

It is important to point out that the 102 pneumococcal isolates we screened represented 21 different capsular serotypes. We found that the reactivity of the antibodies and the probe with the isolates was totally independent of the capsular serotype (McDaniel et al., submitted).

To gain insight into the possible function of PspA, we examined transformation frequencies and the rate of growth of our mutants. We found that PspA was not required for transformation since the PspA⁻ mutants were readily transformed by a chromosomal streptomycin resistance determinant. However, there was a consistent 4- to 10-fold decrease in the frequency of

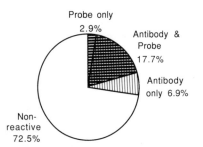

FIG. 3. Ability of 102 pneumococcal isolates (representing 21 serotypes) to react with our two antibodies (Xi64 and Xi126) in a colony blot assay or to bind the isolated pneumococcal insert from pKSD300 in a colony hybridization procedure. The pie chart shows the percentage of isolates reacting with the indicated marker.

transformation of the mutants when compared with the PspA⁺ parental strain Rx1. We also observed a slight decrease of twofold or less in transformation frequency for strains carrying random insertions of pVA891 in their chromosomes at sites not required for PspA production. Therefore, while PspA is not required for transformation, its absence from the surface of pneumococci may affect the frequency of transformation. We also found that the absence of PspA had no effect on the growth rate of one of our mutants in a complex medium (9b).

We have also examined the potential of PspA as an immunogen by testing its ability to induce protective antibodies in mice. For these studies we used X-linked immunodeficient *xid* mice, which fail to make an antibody response to carbohydrate-associated antigens including pneumococcal capsular polysaccharide and phosphorylcholine of the cell wall teichoic acids (17). Protective anti-phosphorylcholine antibodies are readily produced when normal (non-*xid*) mice are injected with heat-killed pneumococci.

Groups of mice were given weekly intravenous injections of heat-killed Rx1 (PspA⁺) or WG44.1 (PspA⁻). The mice were then challenged intravenously with the encapsulated PspA⁺ type 3 strain WU2 and observed for survival for 10 days (Table 2). The documented failure of *xid* mice to make anti-type 3 responses (1) made it unlikely that any protection seen was due to an anticapsular response. This possibility was even further minimized by the fact that immunizing strains were nonencapsulated variants of D39, a type 2 strain.

Immunization with the PspA⁺ strain resulted in the protection of *xid* mice whereas immunization with the PspA⁻ strain did not result in protection. This finding indicated not only that PspA is able to elicit a protective immune response but that it can do so in the partially immunodeficient *xid* mouse that cannot respond to polysaccharide antigens.

The pneumococcal capsular polysaccharide is the major determinant of virulence for this bac-

TABLE 2. Immunization of CBA/N (*xid*) mice with PspA⁺ and PspA⁻ pneumococci[a]

Immunizing strain	PspA	% of mice alive at:	
		2 days	10 days
Rx1	+	100	100
WG44.1	−	50	0
None	−	10	0

[a] Mice were given two weekly intravenous injections of 2×10^8 heat-killed pneumococci in Ringer's lactate. Six days after the last injection, the mice were challenged intravenously with 10 times the 50% lethal dose of the type 3 strain WU2.

TABLE 3. Survival of mice intravenously challenged with PspA[+] versus PspA[−] pneumococci[a]

Mice	Challenge dose[b]	No. of mice alive at 10 days/no. of mice infected
(CBA/N × DBA/2)F$_1$ males (*xid*)	300 PspA[+]	1/10 (10)[c]
	300 PspA[−]	2/17 (11.8)
(CBA/N × DBA/2)F$_1$ females (non-*xid*)	10^6 PspA[+]	2/8 (25)
	10^6 PspA[−]	9/16 (56)

[a] PspA[+] is the type 2 strain D39. PspA[−] pneumococci are the type 2 strains JY53 and JY54 which were obtained by transforming D39 with pKSD300. The PspA[−] data represent pooled results with both JY53 and JY54.
[b] CFU of pneumococci injected.
[c] The number in parentheses is the percent survival.

terium. While it has been suggested that a number of factors, such as hemolysin or other toxic substances, may also play a role in the ability of this bacterium to cause disease, to our knowledge, no one has directly shown that any factor other than the capsule can influence pneumococcal virulence. In our studies we analyzed the role of PspA in pneumococcal virulence. The encapsulated type 2 strain D39 was transformed with pKSD300 (20) to introduce the PspA[−] mutation. We found that the mutation resulted in a small but significant increase in the survival of challenged mice when compared with those challenged with the parental strain D39 (Table 3). We further analyzed the affect of PspA by examining the rate of removal of PspA[−] compared with PspA[+] pneumococci from the blood of injected mice. Table 4 shows that there was a 10-fold reduction in the number of mutants as compared with the parent D39 strain within 1 h after intravenous injection. The results indicate that, in this strain of pneumococcus, PspA plays a role in the early removal of pneumococci from the blood and could be a contributing factor to the virulence of pneumococci.

TABLE 4. Clearance of PspA[+] versus PspA[−] pneumococci from the blood of infected CBA/N mice

Time	Postinoculation CFU/mouse		P[a]
	D39	JY53 and JY54	
1 min	13,700	9,400	NS
4 h	6,000	300	<0.05
8 h	65,000	8,300	<0.5
24 h	5.3×10^7	1.1×10^7	NS

[a] P values were calculated by using the two-sample rank test to compare the CFU per mouse data of the individual mice inoculated with D39 versus those inoculated with the mutant pneumococci. NS, Not significant.

We acknowledge the generosity and enthusiasm of Walter R. Guild and Moses Vijayakumar. We are also thankful to Colynn Forman, Nora Liu, and Lynn McGarry for their excellent technical assistance, Maxine Aycock for preparation of the illustrations, and Ann Brookshire for preparation of the manuscript.

This work was supported by Public Health Service grants CA 16673, CA 13148, AI 21548, and AI 00498 from the National Institutes of Health.

LITERATURE CITED

1. **Amsbaugh, D. F., C. T. Hansen, B. Prescott, P. W. Stashak, D. R. Barthold, and P. J. Baker.** 1972. Genetic control of the antibody response to type III pneumococcal polysaccharide in mice. I. Evidence that an X-linked gene plays a decisive role in determining responsiveness. J. Exp. Med. **136**:931–949.
2. **Au, C. C., and T. K. Eisenstein.** 1981. Nature of the cross-reactive antigen in subcellular vaccines of *Streptococcus pneumoniae*. Infect. Immun. **31**:160–168.
3. **Avery, O. T., C. M. MacLeod, and M. McCarty.** 1944. Studies on the chemical nature of the substance inducing transformation of pneumococcal types. Induction of transformation by a desoxyribonucleic acid fraction isolated from pneumococcus Type III. J. Exp. Med. **79**:137–158.
4. **Briles, D. E., M. Nahm, K. Schroer, J. Davie, P. Baker, J. Kearney, and R. Barletta.** 1981. Antiphosphocholine antibodies found in normal mouse serum are protective against intravenous infection with type 3 *Streptococcus pneumoniae*. J. Exp. Med. **153**:694–705.
5. **Brown, E. J., S. W. Hosea, and M. M. Frank.** 1983. The role of antibody and complement in the reticuloendothelial clearance of pneumococci from the bloodstream. Rev. Infect. Dis. **5**(Suppl.):S797–S805.
6. **Macrina, F. L., R. P. Evans, J. A. Tobian, D. L. Hartley, D. B. Clewell, and K. R. Jones.** 1983. Novel shuttle vehicles for *Escherichia-Streptococcus* transgeneric cloning. Gene **25**:145–150.
7. **McDaniel, L. S., W. H. Benjamin, Jr., C. Forman, and D. E. Briles.** 1984. Blood clearance by anti-phosphocholine antibodies as a mechanism of protection in experimental pneumococcal bacteremia. J. Immunol. **133**:3308–3312.
8. **McDaniel, L. S., and D. E. Briles.** 1985. Protective effects of antibodies to pneumococcal cell wall proteins, p. 103–105. *In* L. Leive (ed.), Microbiology—1985. American Society for Microbiology, Washington, D.C.
8a. **McDaniel, L. S., and D. E. Briles.** 1986. Monoclonal antibodies against surface components of *Streptococcus pneumoniae*, p. 143–164. *In* A. J. L. Macario and E. C. deMacario (ed.), Monoclonal antibodies against bacteria, vol. 3. Academic Press, Inc., Orlando, Fla.
9. **McDaniel, L. S., G. Scott, J. F. Kearney, and D. E. Briles.** 1984. Monoclonal antibodies against protease-sensitive pneumococcal antigens can protect mice from fatal infection with *Streptococcus pneumoniae*. J. Exp. Med. **160**:386–397.
9a. **McDaniel, L. S., G. Scott, K. Widenhofer, J. M. Carroll, and D. E. Briles.** 1986. Analysis of a surface protein of *Streptococcus pneumoniae* recognised by protective monoclonal antibodies. Microb. Pathogen. **1**:519–531.
9b. **McDaniel, L. S., J. Yother, M. N. Vijayakumar, L. McGarry, W. R. Guild, and D. E. Briles.** 1987. Use of insertional inactivation to facilitate studies of biological properties of pneumococcal surface protein A (PspA). J. Exp. Med. **165**:381–394.
10. **Morrison, D. A., M. C. Trombe, M. K. Hayden, G. A. Wasazk, and J. Chen.** 1984. Isolation of transformation-deficient *Streptococcus pneumoniae* mutants defective in control of competence, using insertion-duplication mutagenesis with the erythromycin resistance determinant in pAMB1. J. Bacteriol. **159**:870–876.
11. **Pozzi, G., and W. R. Guild.** 1985. Modes of intergration of heterologous plasmid DNA into the chromosome of *Strep-*

tococcus pneumoniae. J. Bacteriol. **161:**909–912.

12. **Shoemaker, N. B., and W. R. Guild.** 1974. Destruction of low efficiency markers is a slow process occuring at a heteroduplex stage of transformation. Mol. Gen. Genet. **128:**283–290.

13. **Szu, S. C., S. Clarke, and J. B. Robbins.** 1983. Protection against pneumococcal infection in mice conferred by phosphocholine-binding antibodies: specificity of the phosphocholine binding and relation to several types. Infect. Immun. **39:**993–999.

14. **Thompson, H. C. W., and I. S. Snyder.** 1971. Protection against pneumococcal infection with a ribosomal preparation. Infect. Immun. **3:**16–23.

15. **Vijayakumar, M. N., S. D. Priebe, G. Pozzi, J. M. Hageman, and W. R. Guild.** 1986. Cloning and physical characterization of chromosomal conjugative elements in streptococci. J. Bacteriol. **166:**972–977.

16. **White, B.** 1938. The biology of the pneumococcus. Oxford University Press, New York.

17. **Wicker, L. S., and I. Scher.** 1986. X-linked immune deficiency (*xid*) of CBA/N mice. Curr. Top. Microbiol. Immunol. **124:**87–102.

18. **Wright, H. D.** 1927. Experimental pneumococcal septicemia and antipneumococcal immunity. J. Pathol. Bacteriol. **30:**185–252.

19. **Yother, J., C. Forman, B. M. Gray, and D. E. Briles.** 1982. Protection of mice from infection with *Streptococcus pneumoniae* by antiphosphocholine antibody. Infect. Immun. **36:**184–188.

20. **Yother, J., L. S. McDaniel, and D. E. Briles.** 1986. Transformation of encapsulated *Streptococcus pneumoniae.* J. Bacteriol. **168:**1463–1465.

Molecular Cloning of the Immunoglobulin A1 Protease Gene from *Streptococcus pneumoniae*

S. A. PRATT AND G. J. BOULNOIS

Department of Microbiology, University of Leicester, Leicester, LE1 7RH, United Kingdom

Secretory immunoglobulin A (sIgA) is the major protective antibody of the mucosal surfaces (22). Formed by the mucosal plasma cells, sIgA consists of two monomers of IgA attached to a J chain and a secretory component (12). Two subclasses of IgA have been defined and differ primarily in the amino acid sequence of the α chain (4). IgA1 has a 13-amino acid sequence in the hinge region between the $C_\alpha 1$ and $C_\alpha 2$ domains of the heavy chain which is absent from IgA2 (3, 20, 22).

A group of opportunistic pathogens, which include *Streptococcus pneumoniae*, *Neisseria meningitidis*, *Neisseria gonorrhoeae*, and *Haemophilus influenzae*, although taxonomically diverse, have one property in common which may aid the invasive process: secretion of a human IgA1-specific protease (10, 14–16, 18). Hydrolysis of IgA1 by the IgA1 proteases of a variety of pathogens occurs in the hinge region of the immunoglobulin. The hinge consists of a duplicated octapeptide, and a given protease always cuts at the same point, in one half of the duplication, even though a nearly identical bond exists a few amino acids away (3, 9, 20).

N. gonorrhoeae and *N. meningitidis* may each produce two distinct enzymes, with different cleavage specificities (Fig. 1), although a single isolate usually produces only one protease (15, 16, 18). Four IgA1 proteases have been detected among *H. influenzae* isolates. Types 1 and 2 cleave different peptide bonds (15) (Fig. 1). The target for the type 3 enzyme has not been elucidated, and type 4 is unusual in that it apparently cleaves IgA1 at two distinct sites (6, 13). Three streptococcal species, *S. pneumoniae*, *S. sanguis*, and *S. mitior*, produce IgA1 proteases (7, 10, 17), but in contrast to the IgA1 proteases of gram-negative bacteria, which cleave in one octapeptide repeat, the equivalent enzymes produced by gram-positive bacteria cleave in the other repeated peptide (Fig. 1).

Given the properties of IgA1 proteases and their distribution among bacteria that are pathogens of mucosal surfaces, it is tempting to conclude that these enzymes may contribute to bacterial virulence. Although their role in disease has yet to be conclusively established, two functions for the proteases have been postulated.

First, they may act as an aid to bacterial attachment to the correct tissue. IgA1 protease on the outer membrane of the gonococcus has been suggested to combine with IgA1 on the luminal surfaces of host mucosal cells as an integral part of the attachment process (21). Second, they may help the infecting agent avoid the immune system in two ways. It has been proposed for the gonococcus that IgA1 protease activity protects the infecting organism by degrading IgA antibodies directed against gonococcal antigens. Alternatively, the Fab_α cleavage fragments of IgA1 directed against bacterial antigens may bind nonproductively to these antigens and restrict access of other components of the immune system. As with the Fab fragments of IgG, it seems likely that Fab_α fragments generated by IgA1 proteases retain antigen binding activity. While this was initially thought not to be the case (19), recently, free Fab_α fragments, generated from IgA1 myeloma proteins with defined antigen specificity by the protease of *H. influenzae*, were shown to be capable of neutralizing their corresponding antigen (11).

In an attempt to demonstrate a role in virulence for IgA1 proteases, organ cultures of human fallopian tube mucosa which produce IgA1 have been used to compare the properties of an IgA1 protease-deficient mutant of the gonococcus with its wild-type parent. Histopathological studies revealed no difference between the mutant and the wild-type organism with respect to attachment and invasion. It was therefore concluded that the gonococcal IgA1 protease is not essential for infection of previously uninfected human genital mucosa (2). To what extent this in vitro assay reflects the in vivo situation remains to be determined. The role of IgA1 proteases, if any, in the infective process remains elusive.

The genetic analysis of the IgA1 proteases of gram-negative bacteria has provided considerable information about these interesting enzymes (1, 5, 8). We report here the first steps in a similar analysis of the IgA1 proteases of gram-positive bacteria.

Molecular Cloning of the Pneumococcal IgA1 Protease Gene

To facilitate studies of the pneumococcal IgA1 protease and its relationship to the other strep-

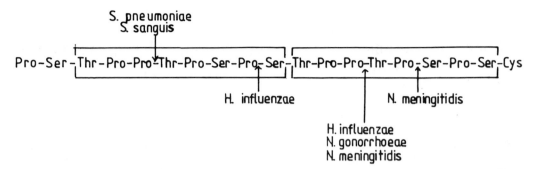

FIG. 1. Sites of cleavage in the human IgA1 heavy chain of IgA1 proteases elaborated by a variety of pathogenic bacteria.

tococcal IgA1 proteases, the gene for the pneumococcal enzyme has been cloned. A clinical isolate of *S. pneumoniae* known to produce IgA1 protease was chosen as a source for chromosomal DNA. Culture supernatant fluids of this strain were incubated with human myeloma IgA1, and the products were separated by electrophoresis in nondenaturing polyacrylamide gels. IgA1 and its cleavage products were identified by Western blotting (Fig. 2). Proteins were electrophoretically transferred to a nitrocellulose filter, and the filter was incubated with rabbit antibodies specific for human α and λ chains. Bound antibody was detected after addition of a goat anti-rabbit immunoglobulin conjugated to horseradish peroxidase and the appropriate substrate. The anti-α and -λ antibodies recognize both intact IgA1 (Fig. 2, tracks 1 and 4) and the cleavage product (Fig. 2, tracks 3 and 6). Thus, the myeloma IgA1 has a λ light chain specificity.

High-molecular-weight chromosomal DNA was purified from the above strain and cleaved with *Sau*3A; fragments of 45 kilobases were selected. These fragments were cloned in the cosmid vector *cos*4. Vector *cos*4 was linearized by cleavage with *Pvu*II, and the resulting blunt ends were dephosphorylated. After purification of the treated DNA, these linear molecules are cleaved with *Bam*HI. This procedure generates two arms of the vector, each carrying *cos* sites, and eliminates the formation of concatemeric vector molecules in the subsequent ligation reaction. The *cos*4 arms were ligated to the *Sau*3A-cleaved chromosomal DNA. Recombinant molecules were packaged into phage heads in vitro and introduced into *Escherichia coli* LE392 by infection. The cosmid vector *cos*4 carries an ampicillin resistance gene as a selectable marker. In this way a gene library representative of pneumococcal DNA was produced.

Screening of the cosmid library involved pooling clones in groups of six and assaying for IgA1

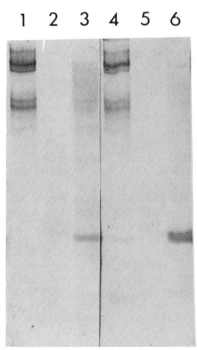

FIG. 2. IgA1 protease activity in *S. pneumoniae* culture supernatants. IgA1 was incubated with pneumococcal culture supernatants. The cleavage products were separated in a nondenaturing polyacrylamide gel and transferred electrophoretically to nitrocellulose paper. The filter was probed with rabbit anti-human immunoglobulin antibodies specific for α and λ chains, and the bound antibodies were detected with goat anti-rabbit immunoglobulin conjugated to horseradish peroxidase in conjunction with enzyme substrate. Tracks: (1) IgA1 alone; (2) *S. pneumoniae* culture supernatant alone; (3) *S. pneumoniae* culture supernatant incubated with IgA1 (filters 1, 2, and 3 were probed with rabbit anti-human α chains); (4) IgA1 alone; (5) *S. pneumoniae* culture supernatant alone; (6) *S. pneumoniae* culture supernatant incubated with IgA1 (filters 4, 5, and 6 were probed with rabbit anti-human λ chains).

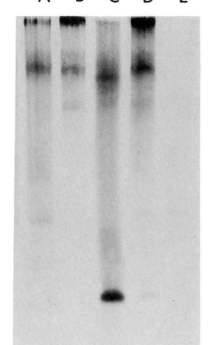

A B C D E

FIG. 3. IgA1 protease expressed in *E. coli.* Cell-free extracts of LE392(pGB1001) generated by sonication were incubated with myeloma IgA1, and cleavage products were detected as described in the legend to Fig. 2. Tracks: (A) Sonicates of *E. coli* LE392 incubated with IgA1; (B) IgA1 alone; (C) *S. pneumoniae* culture supernatant incubated with IgA1; (D) *E. coli* LE392(pGB1001) sonicate incubated with IgA1; (E) *E. coli* LE392(pGB1001) sonicate alone.

protease activity after cell disruption by sonication. One pool of six clones had IgA1 protease activity, and each member of this pool was then tested individually to identify the IgA1 protease-positive clone. IgA1 protease activity of LE392-(pGB1001) yielded the same size cleavage product (Fig. 3, track D) as the enzyme produced by the original pneumococcal parent (Fig. 3, track C). LE392 had no such activity (Fig. 3, track A). Immunoelectrophoretic analysis of the IgA1 cleavage products generated by LE392(pGB-1001) gave two precipitin arcs that correspond to Fab$_\alpha$ and Fc$_\alpha$ fragments of IgA1. The precipitin arcs match precisely those produced from IgA1 by the native enzyme from the pneumococcus (data not shown).

After one subculture IgA1 protease activity of LE392(pGB1001) was lost. This might be explained as a consequence of deletion of the cosmid or a gross rearrangement of the cloned streptococcal DNA. The first explanation is apparently not correct, since the size of pGB1001 remains constant at about 45 kilobases during

several subcultures. We are currently examining the second hypothesis.

S.A.P. was supported by a studentship from the Medical Research Council of the United Kingdom.

We thank J. A. Walker and N. J. High for guidance in using cosmid cloning procedures.

LITERATURE CITED

1. **Bricker, J., M. Mulks, E. R. Moxon, A. G. Plaut, and A. Wright.** 1985. Physical and genetic analysis of DNA regions encoding the immunoglobulin A proteases of different specificities produced by *Haemophilus influenzae.* Infect. Immun. **47:**370–374.
2. **Cooper, M. D., Z. A. McGee, M. H. Mulks, J. M. Koomey, and T. L. Hindman.** 1984. Attachment to and invasion of human fallopian tube mucosa by an IgA1 protease-deficient mutant of *Neisseria gonorrhoeae* and its wild-type parent. J. Infect. Dis. **150:**737–744.
3. **Frangione, B., and C. Wolfenstein-Todel.** 1972. Partial duplication in the "hinge" region of IgA1 myeloma proteins. Proc. Natl. Acad. Sci. USA **69:**3673–3676.
4. **Grey, H. M., C. A. Abel, W. J. Yount, and H. G. Kunkel.** 1968. A subclass of human γA-globulins (γA2) which lacks the disulphide bonds linking heavy and light chains. J. Exp. Med. **128:**1223–1236.
5. **Halter, R., J. Pohlner, and T. F. Meyer.** 1984. IgA protease of *Neisseria gonorrhoeae*: isolation and characterization of the gene and its extracellular product. EMBO J. **3:**1595–1601.
6. **Insel, R. A., P. Z. Allen, and I. D. Berkowitz.** 1982. Types and frequency of *Haemophilus influenzae* IgA1 proteases. Semin. Infect. Dis. **4:**225–231.
7. **Kilian, M., and K. Holmgren.** 1981. Ecology and nature of immunoglobulin A1 protease-producing streptococci in the human oral cavity and pharynx. Infect. Immun. **31:**868–873.
8. **Koomey, J. M., and S. Falkow.** 1984. Nucleotide sequence homology between the immunoglobulin A1 protease genes of *Neisseria gonorrhoeae, Neisseria meningitidis,* and *Haemophilus influenzae.* Infect. Immun. **43:**101–107.
9. **Liu, Y. S. V., T. L. K. Low, A. Infante, and F. W. Putnam.** 1976. Complete covalent structure of a human IgA1 immunoglobulin. Science **193:**1017-1020.
10. **Male, C. J.** 1979. Immunoglobulin A1 protease production by *Haemophilus influenzae* and *Streptococcus pneumoniae.* Infect. Immun. **26:**254–261.
11. **Mansa, B., and M. Kilian.** 1986. Retained antigen-binding activity for Fabα fragments of human monoclonal immunoglobulin A1 (IgA1) cleaved by IgA1 protease. Infect. Immun. **52:**171–174.
12. **McNabb, P. C., and T. B. Tomasi.** 1981. Host defense mechanisms at mucosal surfaces. Annu. Rev. Microbiol. **35:**477–496.
13. **Mulks, M. H., S. J. Kornfeld, B. Frangione, and A. G. Plaut.** 1982. IgA protease specificity is related to serotype in *Haemophilus influenzae.* J. Infect. Dis. **146:**266–274.
14. **Mulks, M. H., S. J. Kornfeld, and A. G. Plaut.** 1980. Specific proteolysis of human IgA by *Streptococcus pneumoniae* and *Haemophilus influenzae.* J. Infect. Dis. **141:**450–456.
15. **Mulks, M. H., and A. G. Plaut.** 1978. IgA protease production as a characteristic distinguishing pathogenic from harmless *Neisseriaceae.* N. Engl. J. Med. **299:**973–976.
16. **Mulks, M. H., A. G. Plaut, H. A. Feldman, and B. Frangione.** 1980. IgA proteases of two distinct specificities are released by *Neisseria meningitidis.* J. Exp. Med. **152:**1442–1447.
17. **Plaut, A. G., R. J. Genco, and T. B. Tomasi, Jr.** 1974. Isolation of an enzyme from *Streptococcus sanguis* which specifically cleaves IgA. J. Immunol. **113:**289–291.
18. **Plaut, A. G., J. V. Gilbert, M. S. Artenstein, and J. D.**

Capra. 1975. *Neisseria gonorrhoeae* and *Neisseria meningitidis*: extracellular enzyme cleaves human immunoglobulin A. Science **190**:1103–1105.

19. **Plaut, A., J. V. Gilbert, and R. Wistar.** 1977. Loss of antibody activity in human immunoglobulin A exposed to extracellular immunoglobulin A protease of *Neisseria gonorrhoeae* and *Streptococcus sanguis*. Infect. Immun. **17**: 130–135.

20. **Plaut, A. G., R. Wistar, Jr., and J. D. Capra.** 1974. Differential susceptibility of human IgA1 immunoglobulins to streptococcal IgA protease. J. Clin. Invest. **54**: 1295–1300.

21. **Swanson, J.** 1980. Adhesion and entry into cells: a model of the pathogenesis of gonorrhoea, p. 17–40. *In* H. Smith, J. J. Skehel, and M. J. Turner (ed.), The molecular basis of microbial pathogenicity. Verlag Chemie.

22. **Tomasi, T. B., Jr.** 1972. Secretory immunoglobulins. N. Engl. J. Med. **287**:500–506.

Molecular and Genetic Analysis of the M Protein of
Streptococcus equi

JORGE E. GALAN AND JOHN F. TIMONEY

Department of Veterinary Microbiology, New York State College of Veterinary Medicine, Cornell University,
Ithaca, New York 14853

Streptococcus equi causes strangles, a highly contagious disease of *Equidae* characterized by inflammation of the upper respiratory tract and abscessation of the regional lymph nodes. At least two virulence factors have been identified on the surface of *S. equi*: the hyaluronic acid capsule and the M-like protein, a surface structure with antiphagocytic properties (1, 21). Thin sections of *S. equi* reveal the presence of hair-like projections on its surface (J. E. Galan, Ph.D. thesis, Cornell University, Ithaca, N.Y., 1986) that are very similar to those described for the M protein of *Streptococcus pyogenes* (17). Until recently, the M-like protein of *S. equi* has been the subject of only limited study. The presence of a trypsin-labile protein on *S. equi* was first reported by Bazeley et al., who suggested that it was a protective antigen (2). Later, Moore and Bryans (14) and Erickson and Norcross (4) described a similar antigen. Woolcock (21) extracted an antigen by acid treatment that induced opsonic antibodies in rabbits and horses. In all these studies, no effort was made to characterize the protein in terms of molecular mass and other physical properties.

More recently, the immunologically reactive proteins in culture supernatants and acid extracts of *S. equi* have been studied in our laboratory (19). We have observed that several hydrolytic fragments of the M protein are present in acid extracts of *S. equi* and that the bactericidal epitope is carried on a fragment with a molecular weight (M_r) of 29,000. Further studies (10) have implicated the M-like protein in the pathogenesis of purpura hemorrhagica, an immune-mediated disease of horses often seen as a sequela of strangles. The sera of animals in the acute stage of purpura hemorrhagica contain immune complexes of immunoglobulin A (IgA) and M antigen of *S. equi*. Analysis of the specificities of the immunoglobulins found in complexes dissociated in 6 M urea revealed reactivity with several polypeptide fragments in acid extracts of *S. equi* including the 29-kilodalton fragment that carries the bactericidal epitope of the M protein. Similar results were observed with sera obtained by immunizing guinea pigs with polyethylene glycol precipitates of serum

taken from a horse with acute purpura hemorrhagica (10).

The clinical and pathological features of strangles resemble closely those of *S. pyogenes* infections in humans. In fact, purpura hemorrhagica is similar to Henoch-Schonlein purpura and Berger's nephrosis, diseases of humans sometimes seen after upper respiratory tract infections caused by group A streptococci or other microorganisms (3, 5).

Unlike group A streptococci, for which at least 74 M types have been described (7), precipitin and passive protection studies suggest that there is only one serotype of *S. equi* M protein (1, 14). In addition to the antigenic variation among the different types of group A M protein (8), there is a great variation in size of the M molecule within strains of the same M type and among strains of different M types (7). Using immunoblotting, DNA fingerprinting, and Southern hybridization analysis, we have compared a number of isolates of *S. equi* obtained in different parts of the world and have been unable to detect variations in the size of protoplast M protein or its acid hydrolytic products, nor have we been able to detect differences in the electrophoretic pattern of total cell DNA digests (J. E. Galan and J. F. Timoney, submitted for publication).

Serum antibodies against the M protein neutralize its antiphagocytic properties so that the organism can be successfully phagocytized (13). These antibodies are commonly but incorrectly termed bactericidal antibodies and have traditionally been associated with protection against group A streptococcal infections in humans (8). In horses, the presence of serum bactericidal antibodies correlates rather poorly with protection (18). In fact, we have demonstrated that protection against strangles is correlated with the presence of mucosal nasopharyngeal antibodies to acid-extracted polypeptide fragments with M_r of 41,000 and 46,000 that are different from the fragment carrying the bactericidal determinant (9). In the same study (9) we found that serum IgG and IgA antibodies had specificities different from those of mucosal nasopharyngeal IgG and IgA, and we concluded that the local and systemic immune responses of the

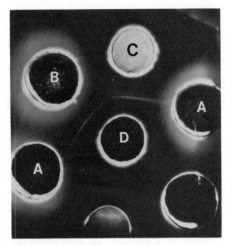

FIG. 1. Immunodiffusion reactions of lysates of *E. coli* Y1089(λ gt11) (A) and Y1089(λ gt11/SEM7) (B) and acid extract of *S. equi* (C) with *S. equi* M-protein antiserum (D).

horse to protein antigens of *S. equi* were independent.

In the studies described below, we used recombinant DNA techniques to clone and express in *Escherichia coli* the M-protein gene of *S. equi*. The availability of an antiserum to the *S. equi* M-protein gene product expressed in *E. coli* should greatly simplify the analysis of the relationship of the different polypeptide fragments in acid extracts of *S. equi*. In addition, the genetic dissection of the *S. equi* M molecule should help to better define the epitope(s) involved in the development of purpura hemorrhagica.

Cloning the M-Protein Gene of *S. equi* in *E. coli*

One- to 10-kilobase fragments of *Eco*RI-digested total cell DNA from *S. equi* CF32 were ligated into the *Eco*RI site of the λ gt11 cloning vector (22). After in vitro packaging of the DNA molecules (11), the phage particles were plated on *E. coli* Y1090 and screened by plaque immunoblotting for the production of *S. equi* M protein. Approximately 10,000 plaques were screened using M-protein-specific antiserum, and three potentially positive clones were detected. One clone that clearly expressed the *S. equi* M-protein determinant upon rescreening was selected and lysogenized into *E. coli* Y1089. This recombinant phage was termed λ gt11/SEM7.

Immunochemical Characterization of λ gt11/SEM7

E. coli Y1089(λ gt11/SEM7) lysates formed a line of identity with *S. equi* acid extract when allowed to react in immunodiffusion with *S. equi* M-protein antiserum (Fig. 1B and C). *E. coli* Y1089(λ gt11) lysates formed no precipitin lines

when allowed to react with *S. equi* M-protein antiserum under the same conditions (Fig. 1A). Immunoblot analysis of isopropyl-β-D-thiogalactopyranoside-induced and noninduced λ gt11/SEM7 lysates showed that the cloned M protein appeared as three polypeptides with M_r of 58,000, 53,000 and 50,000 and some lower-molecular-weight bands (Fig. 2B and C, panel I). No difference in the level of expression of induced (Fig. 2B, panel I) and noninduced (Fig. 2C, panel I) lysogens was detected, indicating that transcription of the protein was independent of the β-galactosidase promoter present in λ gt11. *E. coli* Y1089(λ gt11) lysates did not react with *S. equi* M-protein antiserum (Fig. 2A, panel I). The multiple molecular weights of the M protein are as yet unexplained. The top three bands have mobilities similar to the protoplast M protein of *S. equi*, which also exhibits three closely spaced bands of 58,000, 53,000, and 50,000 (Galan and Timoney, submitted). This strongly suggests that the complete M-protein gene has been cloned because the protoplast M molecule can be considered to be the intact form of the M protein (20). The lower bands probably represent degradation products since they are much less apparent in fresh preparations. Of interest is the fact that very similar molecular weight distributions have been observed for the

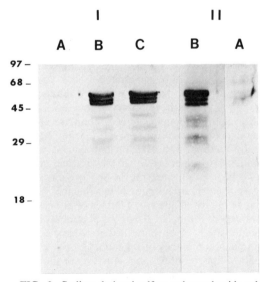

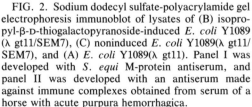

FIG. 2. Sodium dodecyl sulfate-polyacrylamide gel electrophoresis immunoblot of lysates of (B) isopropyl-β-D-thiogalactopyranoside-induced *E. coli* Y1089 (λ gt11/SEM7), (C) noninduced *E. coli* Y1089(λ gt11/SEM7), and (A) *E. coli* Y1089(λ gt11). Panel I was developed with *S. equi* M-protein antiserum, and panel II was developed with an antiserum made against immune complexes obtained from serum of a horse with acute purpura hemorrhagica.

TABLE 1. Inhibition of opsonic antibodies by
lysates of *E. coli* Y1089(λ gt11) and
Y1089(λ gt11/SEM7)

Lysate	CFU[a]
None	6
E. coli Y1089(λ gt11)	23
E. coli Y1089(λ gt11/SEM7)	320

[a] CFU obtained in a bactericidal assay performed as described in reference 9.

group A streptococcal M-protein types 5 and 6 expressed in *E. coli* (6, 12, 15).

Absorption studies with λ gt11/SEM7 lysates showed that the cloned M protein effectively removed opsonic antibodies from *S. equi* M-protein antiserum (Table 1). *E. coli* Y1089(λ gt11) lysate did not absorb opsonic antibodies from the same antiserum (Table 1).

Purpuragenic Determinants in λ gt11/SEM7

To determine whether λ gt11/SEM7 contained epitopes involved in the pathogenesis of purpura hemorrhagica, lysates were separated by sodium dodecyl sulfate-polyacrylamide gel electrophoresis, transferred to nitrocellulose sheets, and allowed to react with an antiserum made against immune complexes obtained from serum from a horse with acute purpura hemorrhagica. The antiserum recognized the same polypeptides as those recognized by the *S. equi* M-protein antiserum (Fig. 2B, panel II). No reactivities were detected in the *E. coli*(λ gt11) lysate (Fig. 2A, panel II).

Southern Blot Analysis

Phage λ gt11/SEM7 was digested with *Eco*RI and separated on 0.7% agarose. The size of the insert was estimated as 5.1 kilobases. Insert DNA was isolated from low-temperature gel agarose, labeled with [^{32}P]dCTP, and used as a probe in high-stringency Southern hybridization studies (16). Total cell DNAs from a series of *S. equi* strains were digested with *Eco*RI and hybridized with the probe. Only one fragment of 5.1 kilobases from each strain hybridized. DNA homology with the cloned fragment was also detected in *S. zooepidemicus* and *S. pyogenes* M types 5, 6, and 12.

Reactivities of the Antiserum to λ gt11/SEM7

Antiserum to λ gt11/SEM7 recognized several polypeptide fragments in an acid extract of *S. equi* including the 46-, 41-, and 29-kilodalton polypeptides. Because the entire M-protein gene has been cloned and expressed in *E. coli*, the obvious explanation for the varying molecular sizes is that acid hydrolysis generates different-sized fragments of the same molecule. It follows therefore that the 41- and 46-kilodalton polypep-

tide fragments in acid extracts of *S. equi* are indeed regions of the M molecule that carry epitopes active in stimulating mucosal antibodies, while the 29-kilodalton fragment carries the bactericidal determinant.

It appears, then, that the M protein of *S. equi* is of some complexity since regions of the molecule seem to be functionally heterogeneous. Genetic dissection of the M molecule in combination with immunological techniques using the purpura immune complex antiserum should allow us to locate the regions of the M protein implicated in the pathogenesis of purpura hemorrhagica. We would also like to examine the sequence of the *S. equi* M-protein gene since it may offer clues as to why *S. equi* M protein is antigenically stable and shows no size variation compared with its group A counterpart.

LITERATURE CITED

1. **Bazeley, P. L.** 1943. Studies with streptococci. V. Some relations between virulence of *S. equi* and immune response in the host. Aust. Vet. J. **19:**62–85.
2. **Bazeley, P. L., S. Baldwin, H. Dickson, and J. R. Thayer.** 1949. The keeping qualities of strangles vaccine. Aust. Vet. J. **25:**130–133.
3. **Borges, W. H.** 1972. Anaphylactoid purpura. Med. Clin. North Am. **56:**201–203.
4. **Erickson, E. D., and N. L. Norcross.** 1975. The cell surface antigens of *S. equi.* Can. J. Comp. Med. **39:**110–115.
5. **Evans, D. J., D. Gwyn Williams, and D. K. Peters.** 1973. Glomerular deposition of properdin in Henoch-Schonlein syndrome and idiopathic focal nephritis. Br. Med J. **3:**326–328.
6. **Fischetti, V. A., K. F. Jones, B. N. Manjula, and J. R. Scott.** 1984. Streptococcal M6 protein expressed in *E. coli*: localization, purification and comparison with streptococcal-derived M protein. J. Exp. Med. **159:**1083–1095.
7. **Fischetti, V. A., K. F. Jones, and J. R. Scott.** 1985. Size variation of the M protein in group A streptococci. J. Exp. Med. **160:**1384–1401.
8. **Fox, E.** 1974. M proteins of group A streptococci. Bacteriol. Rev. **38:**57–86.
9. **Galan, J. E., and J. F. Timoney.** 1985. Mucosal nasopharyngeal immune response of horses to protein antigens of *Streptococcus equi.* Infect. Immun. **47:**623–628.
10. **Galan, J. E., and J. F. Timoney.** 1985. Immune complexes in purpura hemorrhagica of the horse contain IgA and M antigen of *S. equi.* J. Immunol. **135:**3134–3137.
11. **Hohn, B.** 1979. In vitro packaging of lambda and cosmid DNA. Methods Enzymol. **68:**229–309.
12. **Kehoe, M. A., T. P. Poirier, E. H. Beachey, and K. N. Timmis.** 1985. Cloning and genetic analysis of serotype 5 M protein determinant of group A streptococci: evidence for multiple copies of the M5 determinant in *Streptococcus pyogenes* genome. Infect. Immun. **48:**190–197.
13. **Lancefield, R. C.** 1962. Current knowledge of the type-specific M antigen of group A streptococci. J. Immunol. **89:**307–313.
14. **Moore, B. O., and J. T. Bryans.** 1970. Type specific antigenicity of group C streptococci from diseases of the horse, p. 231–238. *In* J. T. Bryans and H. Gerber (ed.), Proceedings of the Second International Conference on Equine Infectious Diseases, Paris, 1969. S. Karger, Basel.
15. **Scott, J. R., and V. A. Fischetti.** 1983. Expression of streptococcal M protein in *E. coli.* Science **221:**758–760.
16. **Southern, E. M.** 1975. Detection of specific sequences among DNA fragments separated by gel electrophoresis. J. Mol. Biol. **89:**503–517.

17. **Swanson, J., K. C. Hsu, and E. C. Gotschlich.** 1969. Electron microscopic studies on streptococci. I. M antigen. J. Exp. Med. **130:**1063–1091.

18. **Timoney, J. F., and D. Eggers.** 1985. Serum bactericidal responses to *S. equi* of horses following infection or vaccination. Equine Vet. J. **17:**306–310.

19. **Timoney, J. F., and J. Trachman.** 1985. Immunologically reactive proteins of *Streptococcus equi.* Infect. Immun. **48:**29–34.

20. **Van De Rijn, I., and V. A. Fischetti.** 1981. Immunochemical analysis of intact M protein secreted from cell wall-less streptococci. Infect. Immun. **32:**86–91.

21. **Woolcock, J. B.** 1974. Purification and antigenicity of an M-like protein of *Streptococcus equi.* Infect. Immun. **10:**116–122.

22. **Young, R. A., and R. W. Davis.** 1983. Efficient isolation of genes by using antibody probes. Proc. Natl. Acad. Sci. USA **80:**1194–1198.

Biological Role(s) of the Pneumococcal *N*-Acetylmuramic Acid-L-Alanine Amidase

RUBENS LÓPEZ, JOSÉ M. SÁNCHEZ-PUELLES, CONCEPCIÓN RONDA, PEDRO GARCÍA, JOSÉ L. GARCÍA, AND ERNESTO GARCÍA

Centro de Investigaciones Biológicas, Consejo Superior de Investigaciones Científicas, Veláquez, 144, 28006 Madrid, Spain

The physiological roles of the bacterial autolysins have been a matter of continuous speculations for a long time. These enzymes have been thought to play a central role in several basic biological functions in bacteria (13, 15). The exact assignment of a precise biological function to the bacterial autolysins has been impossible so far because of the difficulties of obtaining mutations in the genes encoding these enzymes. *Streptococcus pneumoniae* contains a powerful autolytic enzyme, an *N*-acetylmuramic acid-L-alanine amidase (7, 11), that has been suggested to be involved in processes such as daughter cell separation and genetic transformation, and in the irreversible effects caused by treatment of the bacterial cells with β-lactam antibiotics (15). We have recently developed a technique to rapidly detect autolysin-deficient (Lyt⁻) mutants of *S. pneumoniae* (5) that has allowed us to clone the pneumococcal autolysin gene (*lytA*) in *Escherichia coli* (4). This approach has provided an important tool to develop the genetics of the *lytA* gene: the complete nucleotide sequence of the gene has been determined (3) and several Lyt⁻ mutants have been characterized at the nucleotide level (E. García, J. L. García, P. García, C. Ronda, J. M. Sánchez-Puelles, and R. López, this volume).

The isolation and characterization of a mutant showing a complete deletion of the gene encoding the pneumococcal amidase provide important advantages to the study of the biological consequences of the absence of this autolysin. In addition, this mutant provides the ideal background to introduce the *lytA* gene in a plasmid adapted to *S. pneumoniae*. The present paper illustrates some of the ongoing studies using this approach.

Isolation and Genetic Characterization of the Lyt⁻ Mutant M31 of *S. pneumoniae*

S. pneumoniae was heavily mutagenized with *N*-methyl-*N*'-nitro-*N*-nitrosoguanidine (8), and a mutant showing the Lyt⁻ phenotype was isolated by using the filter technique described recently (5). The DNA of this mutant was used to transform strain M11 (Hex⁻ Lyt⁺), and one of the Lyt⁻ transformants (strain M31) was se-lected for further studies in this unmutagenized background. We have cloned and expressed the E form of the pneumococcal autolysin in *E. coli*. These cells contain a plasmid (pGL30) of 12 kilobases (Fig. 1). Several derivatives of this plasmid have been prepared; one of them, pGL80, containing a 1,213-base-pair *Hin*dIII fragment of pneumococcal DNA, possesses the complete genetic information for the pneumococcal amidase. All these plasmids are unable to establish in *S. pneumoniae* as independent replicons when introduced by genetic transformation although they can be integrated in the bacterial chromosome provided that homologous regions are present both in the chromosome of the receptor cell and in the donor recombinant plasmid (2, 10). The first striking feature of strain M31 was its inability to be transformed to the Lyt⁺ phenotype using the recombinant plasmids (i.e., pGL30 or pGL80) containing the *lytA* gene, although it was fully competent as demonstrated when other markers were tested. In contrast, M12 (4), a pneumococcal mutant that presents a transitional change (García et al., this volume) in the nucleotide sequence of the *lytA* gene, was fully transformable to the Lyt⁺ phenotype when the recombinant plasmids used as donor DNAs contained an intact *lytA* gene (i.e., pGL30 and pGL80). These results were the first clues suggesting the possibility that M31 completely lacked the *lytA* gene. The results of Southern blots of M31 DNA hybridized to pGL80 are shown in Fig. 2. No hybridization bands are visible, demonstrating that M31 has completely lost the *lytA* gene. In contrast, the DNA of the parental strain Rst7 (*lytA*⁺) hybridized with pGL80.

Biological Consequences of the Deletion of the *lytA* Gene

The primary biological consequences of deletion of the *lytA* gene were as follows. (i) Small chains (six to eight cells) formed and lysis did not occur in the stationary phase of growth when M31 was incubated at 37°C (Fig. 3), in contrast to the normal "diplo" cell formation and to the stationary-phase lysis shown by the wild-type strain. (ii) When M31 cells were treated with

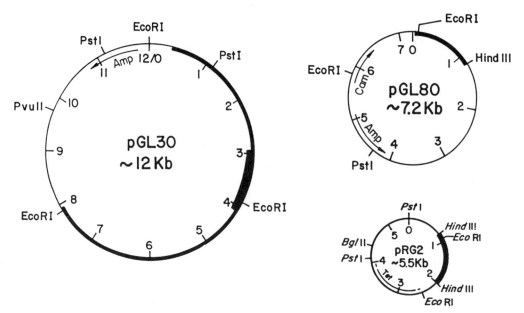

FIG. 1. Maps of the *lytA* recombinant plasmids. Light line, Vector DNA; heavy line, DNA of chromosomal origin. The box in pGL30 indicates the localization of the *Hin*dIII insert cloned in pGL80 and pRG2. The *Hin*dIII insert containing the *lytA* gene has been sequenced (3).

penicillin, a tolerant (i.e., bacteriostatic) response was found rather than the lytic one shown by the strain M11 (not shown). (iii) Whereas it had been thought that the amidase was the only peptidoglycan hydrolase present in this microorganism, cultures of strain M31 did autolyze upon incubation at 30°C but not at 37°C. Crude sonicated extracts of M31 grown at 30°C showed a lytic activity against choline-containing pneumococcal walls but were unable to degrade ethanolamine-containing ones. Moreover, the degradation of the walls was higher at 30°C than at 37°C. Analyses of the cell wall degradation products suggested that the new lytic activity corresponds to a glycosidase(s).

Biological Consequences of the Presence of the *lytA* Gene in a Recombinant Plasmid Adapted to *S. pneumoniae*

The recombinant plasmid pGL80 (Fig. 1) contains a 1,213-base-pair *Hin*dIII insert. This insert has been completely sequenced (3), showing that it contains only one open reading frame encoding the pneumococcal amidase. We excised the *lytA* gene (i.e., the *Hin*dIII insert) from pGL80 and ligated this gene to the pneumococcal plasmid pLS1 (14) previously digested with *Hin*dIII. Afterwards, the plasmid mixture was used to transform competent cells of the M31 strain. The use of this recipient strain prevents the integration of the *lytA* gene present in the plasmid into the bacterial chromosome since the host chromosome does not contain any homologous region with the recombinant plasmid (2, 10). Pneumococcal cells that exhibit a tetracycline-resistant phenotype were selected. Ten of 500 tetracycline-resistant transformants also showed a Lyt$^+$ phenotype. The plasmid present in one of these transformants, named pRG2 (Fig. 1), was isolated and its physical map was confirmed by electrophoretic analyses of the intact and digested plasmid. The transformed cells containing pRG2 (M51) grew as normal "diplo" cells and lysed at the end of the exponential phase of growth, in remarkable contrast to the physiological behavior shown by the parental strain M31 (Fig. 3). M51 lysed immediately at the end of the exponential phase of growth, whereas a noticeable stationary phase is found in the wild-type strain R6. In addition, the tolerant response of M31 against penicillin was completely changed to a bacteriolytic one when M51 was treated with 0.03 U of antibiotic per ml, and the lytic response of M51 was remarkably faster than that of the wild-type strain R6. These results suggested an increase in the autolytic activity present in M51 with respect to the wild-type strain. Analyses of crude sonicated extracts showed that M51 contains about fivefold more autolytic activity than the wild-type strain whereas M31 and M32 (a frameshift mutant of the *lytA* gene) (García et al., this volume) completely lacked amidase activity.

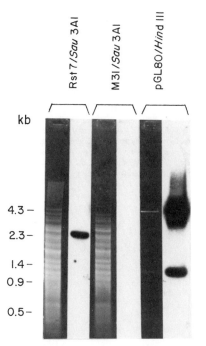

FIG. 2. Ethidium bromide-stained agarose gels and Southern blot analyses of restriction fragments obtained from Rst7, M31, and pGL80 DNAs. Ethidium bromide-stained gels show the restriction patterns obtained from the digestion of the different DNAs employed (e.g., M31/*Sau*3A1 indicates that M31 DNA was digested with *Sau*3A1). The adjacent panel corresponds to the same gel blotted to Hybond membranes (Amersham International) and hybridized at 65°C for 5 h with ^{32}P-labeled pGL80 used as probe. The sizes of the restriction fragments (kilobases) are indicated at the side.

Discussion

M31 is the first mutant of *S. pneumoniae* that has a complete deletion of the *lytA* gene encoding the pneumococcal amidase. This mutant shows a normal growth rate in test-tube experiments and can undergo genetic transformation. The absence of the amidase does not affect any

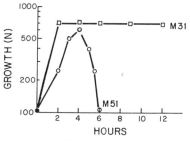

FIG. 3. Curves of growth of pneumococcal strains M31 and M51. Growth (and lysis), in C+Y medium (16), at 37°C was followed by nephelometry.

essential physiological function in test-tube experiments since the cells can grow at a normal rate. Nevertheless, the autolysis of the cells at the end of the stationary phase of growth ("suicidal tendencies") and the formation of small chains may represent important biological functions in the natural environment (9, 12). The presence of a recombinant plasmid containing the *lytA* gene in transformed cells of M31 (strain M51) restores the synthesis of the amidase. M51 recovers the capacity to form single cells or doublets and to lyse at the end of the exponential phase of growth. It has been suggested that regulation in vivo is achieved by a restricted transport of the amidase through the membrane rather than by lipoteichoic acid, as a specific inhibitor (6), preventing the access of the enzyme to the cell wall (1). Whatever the mechanism of control, it seems to be sufficient to regulate the amidase in M51, which contains a remarkably higher level of amidase activity than the wild-type strain R6.

The formation of long chains in strains containing high levels of amidase activity (M51) or completely lacking enzyme (M31) when the cells were grown in media containing ethanolamine or a high concentration of choline (unpublished data) is a response that cannot be ascribed to the amidase as previously suggested (1, 16). We are currently investigating the possible relationship between the formation of long chains and the new lytic activity, corresponding to a glycosidase, that has been found in M31. Finally, our results confirm the basic role of the pneumococcal amidase in the bactericidal nature of β-lactam antibiotics.

This work was supported by Comision Asesora de Investigacion Cientifica y Tecnica (grant 144) and by the Spain-USA Joint Committee for Scientific and Technological Cooperation (grant CDB 840 20, 35).

LITERATURE CITED

1. **Briese, T., and R. Hakenbeck.** 1985. Interaction of the pneumococcal amidase with lipoteichoic acid and choline. Eur. J. Biochem. **146:**417–427.
2. **Claverys, J. P., J. M. Louarn, and A. M. Sicard.** 1981. Cloning of *Streptococcus pneumoniae* DNA: its use in pneumococcal transformation and in studies of mismatch repair. Gene **13:**65–73.
3. **García, P., J. L. García, E. García, and R. López.** 1986. Nucleotide sequence and expression of the pneumococcal autolysin gene from its own promoter in *Escherichia coli*. Gene **43:**265–272.
4. **García, E., J. L. García, C. Ronda, P. García, and R. López.** 1985. Cloning and expression of the pneumococcal autolysin gene in *Escherichia coli*. Mol. Gen. Genet. **201:** 225–230.
5. **García, E., C. Ronda, J. L. García, and R. López.** 1985. A rapid procedure to detect the autolysin phenotype in *Streptococcus pneumoniae*. FEMS Microbiol. Lett. **29:**77–81.
6. **Höltje, J. V., and A. Tomasz.** 1975. Lipoteichoic acid: a specific inhibitor of autolysin activity in pneumococcus. Proc. Natl. Acad. Sci. USA **72:**1690–1694.

7. **Howard, L. V., and H. Gooder.** 1974. Specificity of the autolysin of *Streptococcus (Diplococcus) pneumoniae*. J. Bacteriol. **117:**796–804.

8. **Lacks, S.** 1970. Mutants of *Diplococcus pneumoniae* that lack deoxyribonucleases and other activities possibly pertinent to genetic transformation. J. Bacteriol. **101:**373–383.

9. **McCarty, M.** 1985. The transforming principle, p. 57–58. W. W. Norton and Co., New York.

10. **Morrison, D. A., M. C. Trombe, M. K. Hayden, G. A. Waszak, and J. Chen.** 1984. Isolation of transformation-deficient *Streptococcus pneumoniae* mutants defective in control of competence, using insertion-duplication mutagenesis with the erythromycin resistance determinant of pAMB1. J. Bacteriol. **159:**870–876.

11. **Mosser, J. L., and A. Tomasz.** 1970. Choline-containing teichoic acid as a structural component of pneumococcal cell wall and its role in sensitivity of lysis by an autolytic enzyme. J. Biol. Chem. **245:**287–298.

12. **Ottolenghi, E., and R. D. Hotchkiss.** 1962. Release of genetic transforming DNA from pneumococcal cultures during growth and disintegration. J. Exp. Med. **116:**491–519.

13. **Rogers, H. J., H. R. Perkins, and J. B. Ward.** 1980. Microbial cell walls and membranes. Chapman and Hall, London.

14. **Stassi, D. L., P. López, M. Espinosa, and S. A. Lacks.** 1981. Cloning of chromosomal genes in *Streptococcus pneumoniae*. Proc. Natl. Acad. Sci. USA **78:**7028–7032.

15. **Tomasz, A.** 1984. Building and breaking of bonds in the cell wall of bacteria—the role for autolysins, p, 3–12. *In* C. Nombela (ed.), Microbial cell wall synthesis and autolysis. Elsevier Science Publishers, Amsterdam.

16. **Tomasz, A., M. Westphal, E. B. Briles, and P. Fletcher.** 1975. On the physiological functions of teichoic acids. J. Supramol. Struct. **3:**1–16.

Molecular Genetics of the Pneumococcal Amidase: Characterization of *lytA* Mutants

ERNESTO GARCÍA, JOSÉ L. GARCÍA, PEDRO GARCÍA, CONCEPCIÓN RONDA, JOSÉ M. SÁNCHEZ-PUELLES, AND RUBENS LÓPEZ

Centro de Investigaciones Biológicas, Consejo Superior de Investigaciones Científicas, Velázquez, 144, 28006 Madrid, Spain

Autolysins, also called murein hydrolases, are enzymes that can hydrolyze covalent bonds in bacterial cell walls. Most bacterial species contain one or more autolytic enzymes of different enzymatic specificity, which suggests that these enzymes fulfill essential physiological functions. The autolysins seem to play important roles in several fundamental biological phenomena (e.g., cell wall enlargement, cell division, lysis induced by β-lactam antibiotics, and genetic transformation) (11). However, the genetics of autolytic enzymes are largely unknown since mutations in the structural gene of a bacterial autolysin have not been characterized and, therefore, the real role of these enzymes remains to be clarified. *Streptococcus pneumoniae* represents an appropriate model system for these studies since it contains a powerful autolysin, an *N*-acetylmuramic acid-L-alanine amidase (7) previously considered the only autolysin of this species. Furthermore, we have recently cloned, in *Escherichia coli*, a 7.5-kilobase *Bcl*I fragment of pneumococcal DNA containing the *lytA* gene encoding the amidase (3) (Fig. 1), taking advantage of a rapid and reliable filter technique (4) to distinguish amidase-contaning (Lyt$^+$) and amidase-defective (Lyt$^-$) pneumococcal strains. The complete nucleotide sequence of the *lytA* gene has been determined, showing that it codes for a 36,532-dalton polypeptide (5). The genetic characterization of several mutants affected in the *lytA* gene is presented here.

Molecular Characterization of an Autolytic-Defective Mutant (cwl-1) of *S. pneumoniae*

The first Lyt$^-$ mutant of *S. pneumoniae* (strain cwl-1) was isolated by Lacks (9). Strain R6ly4-4 was constructed by transformation of the wild-type strain R6 with cwl-1 DNA (2). More recently, the isolation of other amidase-defective mutants of pneumococcus, namely, DOC 3, DOC 4, and 6-7, has been described (S. Zighelboim, Ph.D. thesis, Rockefeller University, New York, N.Y., 1980). All these mutants share the physiological properties shown by R6ly4-4: (i) growth in small chains, (ii) absence of lysis in the stationary phase of growth, and

(iii) tolerance when treated with β-lactam antibiotics. In addition, we recently reported that R6ly4-4, as well as DOC 3, synthesizes a temperature-sensitive autolytic enzyme (6). The mutation present in strain cwl-1, designated as *lytA4* (4), was complemented by the cloned wild-type *lytA* gene, indicating that this particular mutation is located in the pneumococcal amidase gene (5) and, therefore, represents the first case of a mutation in the structural gene of a bacterial autolysin. For this reason, it was important to characterize the *lytA4* mutation at the nucleotide level. The *lytA* gene is located between two *Sau*3A1 sites in the *S. pneumoniae* chromosome (Fig. 1). Chromosomal DNA obtained from strain R6ly4-4 (*lytA4*) was digested with *Sau*3A1, and after electrophoresis in low-melting-temperature agarose, the region corresponding to the appropriate length (about 2.3 kilobases) was excised. The DNA fragments were purified and ligated to *Bam*HI-digested pBR322. The ligation mixture was used to transform competent cells of *E. coli* HB101. Plasmid minipreparations from the Amp^r Tet^s *E. coli* transformants were used to transform pneumococcal strain M11 (Lyt$^+$) to the Lyt$^-$ phenotype. Following this procedure, a recombinant plasmid, named pGL70, was isolated. As expected, this plasmid was unable to transform pneumococcal strain M12 (*lytA4*) to the Lyt$^+$ phenotype, indicating that it contains the *lytA4* marker. In addition, pGL70 encodes a 36-kilodalton polypeptide (i.e., E-form amidase) in *E. coli* maxicells (Fig. 2).

As an earlier step to determining the sequence alteration present in the *lytA4* mutation, it was necessary to ascertain the precise localization of this marker in the *lytA* gene. This could be accomplished by transforming pneumococcal strain M12 (*lytA4*) with different restriction fragments of the *lytA*$^+$ allele cloned into the replicative form of M13 DNA (5). It was found that recombination of the *lytA4* mutation was achieved only when the cloned pneumococcal inserts contained a *Hin*dIII-*Taq*I fragment of 517 base pairs (Fig. 3), indicating that the mutation is located in this region. After mapping of the *lytA4* mutation, plasmid pGL70 was digested with

FIG. 1. Physical map of the 7.5-kilobase *Bcl*I fragment of chromosomal DNA of *S. pneumoniae* cloned in pGL30 (3). The black box corresponds to the 1,213-base-pair *Hin*dIII fragment containing the *lytA* gene (5). This fragment was cloned into *Hin*dIII-cut pBR325. The restriction map of the recombinant plasmid, pGL80, is shown in the accompanying paper (R. López, J. M. Sánchez-Puelles, C. Ronda, P. García, J. L. García, and E. García, this volume).

both *Hin*dIII and *Taq*I restriction endonucleases, and the resulting fragments were ligated either to M13mp10 or M13mp11 previously digested with both *Hin*dIII and *Acc*I. The sequence determination showed that the only alteration present in the *lytA4* mutation was a CG to TA transition at position 370 from the left *Hin*dIII site (Fig. 3). This transition causes the change of a glycine to a glutamic acid in the predicted amino acid sequence of the mutant autolysin (Fig. 3). This change occurs in one of the hydrophobic parts of the enzyme (5), and since glutamic acid is a very hydrophilic amino acid whereas glycine is almost neutral in this

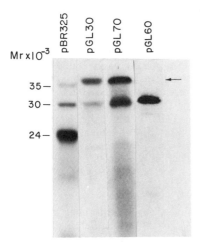

FIG. 2. Fluorogram of labeled plasmid-coded proteins in UV-irradiated maxicells. *E. coli* CSR603 maxicells containing the plasmids indicated at the top of the figure were labeled with [³⁵S]methionine and subjected to sodium dodecyl sulfate-polyacrylamide gel electrophoresis as previously described (12). The arrow indicates the position of the E-form pneumococcal amidase.

respect, the *lytA4* mutation causes a noticeable alteration in the hydropathic profile of this region that might be responsible for the thermosensitivity of the amidase encoded by this gene (6).

When strains cwl-1, DOC 3, DOC 4, and 6-7 were intercrossed, no Lyt⁺ recombinants were detected, suggesting that their mutations might be located close together. The availability of plasmids pGL70 (*lytA4*) and pGL32 (Δ*lytA32*) (see below) allowed a more precise mapping of these mutations since higher transformation frequencies can be obtained when the cloned *lytA* gene, instead of chromosomal DNA, is used as donor DNA (3). We found a 4% recombination frequency between *lytA4* and Δ*lytA32*, indicating that both markers were separated by about 100 base pairs (1). This value was later confirmed by sequencing (see below). The same recombination frequency was obtained when the *lytA* mutants DOC 3, DOC 4, and 6-7 were transformed with pGL32 (Δ*lytA32*). In contrast, no Lyt⁺ recombinants were found when pGL70, instead of pGL32, was used as donor DNA. These results, taken together, strongly suggest that the *lytA* mutations of strains DOC 3, DOC 4, and 6-7 are identical to that present in cwl-1 (*lytA4*).

Construction of a Genetic Map of the *lytA* Gene

To determine the physiological significance of the *lytA* gene, we isolated other different, spontaneous or otherwise, Lyt⁻ mutants. During the construction of a genomic library of pneumococcal DNA in *E. coli* plasmids (3), we observed that some preparations of recombinant plasmids gave rise, in transformation of the pneumococcal strain M11 (*LytA*⁺), to the appearance of Lyt⁻ transformants. Subsequently, we isolated a clone of *E. coli* which contained the recombinant plasmid pGL32 able to transform competent cells of *S. pneumoniae* to the Lyt⁻ phenotype. Restriction enzyme analyses indicated that pGL32 had the same restriction pattern and size as pGL30 (Fig. 1), suggesting that pGL32 probably arose by spontaneous mutation during cloning manipulations. The mutation present in pGL32 was introduced into *S. pneumoniae* by genetic transformation, and the nucleotide defect of one of the Lyt⁻ transformants (strain M32) was determined as described above. Figure 3 shows that the Δ*lytA32* mutation corresponds to deletion of a single GC base pair at position 478. This deletion generates a termination codon (TAA) in the reading frame encoding the pneumococcal amidase. The frame shift leads also to another termination codon (TGA) six codons downstream from the TAA stop. In agreement with this finding, sodium dodecyl sulfate-polyacrylamide gel electrophoresis of [³⁵S]methionine-labeled *E. coli* maxicells harboring pGL60 (a derivative of pGL32 constructed by subcloning the 2.3-kilobase *Sau*3A1 fragment including the *lytA* gene [see Fig. 1] into the *Bam*HI site of pBR322) did not show any band corresponding to the E-form amidase (Fig. 2).

Using a combined procedure of chemical mutagenesis (8, 14) of pGL80 (Fig. 1) and subsequent transformation of *S. pneumoniae* M11 (*lytA⁺*) with the mutagenized plasmid mixture,

we have also isolated several new *lytA* mutants. These mutants were mapped by recombination with the replicative form of M13 DNAs containing different restriction fragments of the *lytA* gene (see above) as well as by genetic recombination (13) between various *lytA* markers. The resulting genetic and physical maps are shown in Fig. 3. In addition to the *lytA4* and Δ*lytA32* markers, three new mutations have been sequenced (Fig. 3). The mutation *lytA63* corresponds to a GC to AT transition in position 270, causing the replacement of a histidine residue by a tyrosine residue. The markers *lytA27* and *lytA41* also correspond to GC to AT transitions at nucleotides 890 and 922, respectively, although these two mutations cause the appearance of termination codons in the reading frame of the amidase (Fig. 3).

Concluding Remarks

The role of autolytic enzymes in bacteria is controversial (11). Most of the conclusions obtained until now came from the study of the so-called Lyt mutants, but none of these mutants has been characterized as having a defect in the structural gene of a particular enzyme. It has been proposed recently (10) that the Lyt mutants of *Bacillus subtilis* should be renamed *fla* (for flagella) on the basis of their genetic defect. We describe here, for the first time, the characterization of several mutations located in

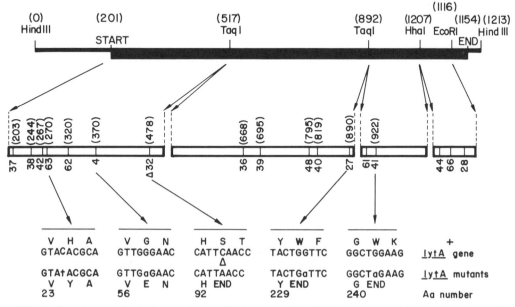

FIG. 3. Genetic and physical map of the 1,213-base-pair *Hin*dIII fragment containing the *lytA* gene. The genetic map is indicated by numbers. Numbers in parentheses indicate the nucleotide location of the genetic sites. The initiation and termination codons of the *lytA* gene are indicated by START and END, respectively. The partial nucleotide sequence is displayed for the noncoding strand. Mutations are indicated in lowercase lettering. The partial amino acid sequence (in one-letter code) is also shown. Δ, Deletion.

the structural gene of the pneumococcal amidase. The availability of these mutants provides a useful tool to develop our ongoing studies on the mechanisms of regulation of this important autolysin (e.g., transport of the amidase through the membrane, localization of the active center of the enzyme, etc.). The development of a genetic system for the analysis of amidase-deficient strains of *S. pneumoniae* described in this study has provided a procedure for the isolation and characterization of a mutant completely lacking the *lytA* gene (López et al., this volume). The study of the biological properties of this mutant has allowed us to draw definite conclusions on the role(s) of a bacterial autolytic enzyme.

This work was supported by Comisión Asesora de Investigación Científica y Técnica (grant 144) and by the Spain-USA Joint Committee for Scientific and Technological Cooperation (grant CDB 840 20, 35).

LITERATURE CITED

1. **Claverys, J. P., J. C. Lefevre, and A. M. Sicard.** 1979. Molecular studies on the marker effect in pneumococcus, p. 135–150. *In* S. W. Glover and L. O. Butler (ed.), Transformation 1978. Cotswold Press Ltd., Oxford.
2. **Fisher, H., and A. Tomasz.** 1984. Production and release of peptidoglycan of wall teichoic acid polymers in pneumococci treated with beta-lactam antibiotics. J. Bacteriol. **157:**507–513.
3. **García, E., J. L. García, C. Ronda, P. García, and R. López.** 1985. Cloning and expression of the pneumococcal autolysin gene in *Escherichia coli.* Mol. Gen. Genet. **201:**225–230.
4. **García, E., C. Ronda, J. L. García, and R. López.** 1985. A rapid procedure to detect the autolysin phenotype in *Streptococcus pneumoniae.* FEMS Microbiol. Lett. **29:**77–81.
5. **García, P., J. L. García, E. García, and R. López.** 1986. Nucleotide sequence and expression of the pneumococcal autolysin gene from its own promoter in *Escherichia coli.* Gene **43:**265–272.
6. **García, P., E. García, C. Ronda, R. López, R. Z. Jiang, and A. Tomasz.** 1986. Mutants of *Streptococcus pneumoniae* that contain a temperature-sensitive autolysin. J. Gen. Microbiol. **132:**1401–1405.
7. **Howard, L. V., and H. Gooder.** 1974. Specificity of the autolysin of *Streptococcus (Diplococcus) pneumoniae.* J. Bacteriol. **117:**796–804.
8. **Lacks, S.** 1966. Integration efficiency and genetic recombination in pneumococcal transformation. Genetics **53:**207–253.
9. **Lacks, S.** 1970. Mutants of *Diplococcus pneumoniae* that lack dexyribonucleases and other activities possibly pertinent to genetic transformation. J. Bacteriol. **101:**373–383.
10. **Pooley, H. M., and D. Karamata.** 1984. Genetic analysis of autolysin-deficient and flagellaless mutants of *Bacillus subtilis.* J. Bacteriol. **160:**1123–1129.
11. **Rogers, H. J., H. R. Perkins, and J. B. Ward.** 1980. Microbial cell walls and membranes. Chapman and Hall, London.
12. **Sancar, A., A. M. Hack, and N. D. Rupp.** 1979. Simple method for identification of plasmid-coded proteins. J. Bacteriol. **137:**692–693.
13. **Sicard, A. M., and H. Ephrussi-Taylor.** 1965. Genetic recombination in DNA-induced transformation of pneumococcus. II. Mapping the *amiA* region. Genetics **52:**1207–1227.
14. **Tiraby, J. G., and M. Fox.** 1974. Marker discrimination and mutagen-induced alterations in pneumococcal transformation. Genetics **77:**440–458.

Molecular Cloning of a Competence Control Region from *Streptococcus pneumoniae* by Use of Transcription Terminator Vectors in *Escherichia coli*

MARK S. CHANDLER[1] AND DONALD A. MORRISON[2]

Department of Microbiology and Immunology[1] and Laboratory for Cell, Molecular, and Developmental Biology,[2] University of Illinois at Chicago, Chicago, Illinois 60680

Competence for genetic transformation in *Streptococcus pneumoniae* (pneumococcus) is a specialized cellular state characterized by the binding and nicking of double-stranded DNA at the cell surface, the uptake of a single-strand fragment of DNA accompanied by the simultaneous degradation of the complementary strand, and finally integration of that DNA at homologous regions of the chromosome (4, 6, 8, 9, 14–16). Competence is a temporary state, induced in all the cells of a culture by an extracellular protein called competence factor (7, 11, 20, 21, 26–28). Addition of a small amount of competence factor to physiologically noncompetent cells causes the appearance of a high level of competence factor in the culture medium, followed soon after by the development of competence (12). Induction of competence causes a dramatic shift in protein synthesis to a few, mainly new proteins (13). Two-dimensional gel electrophoresis of proteins pulse-labeled during competence shows at least 14 competence-specific proteins, not made at all in noncompetent cells (12). Following competence, cells resume producing the same profile of proteins as before competence (12, 13). It is believed that these competence-specific proteins are important in transformation. The major competence-induced protein, eclipse complex protein, is found associated with incoming single-stranded DNA, protecting that DNA from nucleases within the cell during transformation (17). The other competence-induced proteins have various cellular locations, some in the cytoplasm, some in the membrane, and one extracellular (29); the role of these proteins in transformation has yet to be determined.

Morrison et al. described a transformation-defective mutant of pneumococcus, CP1415, in which induction of competence for genetic transformation is blocked, apparently as a result of failure to release competence factor (18). In growing cultures this strain is never spontaneously competent. Normal levels of competence can be induced only by the addition of high levels of competence factor. However, competence factor is not released during in-duced competence. In addition, DNA processing and recombination are normal during induced competence of this strain. The mutation, designated *com-15*, was obtained by insertion-duplication mutagenesis. Random mutagenesis was accomplished by ligating the *ermB* determinant of pAMβ1 to *Taq*I fragments of pneumococcal DNA and transforming these chimeric DNA molecules into the pneumococcal chromosome, causing insertional inactivation of chromosomal genes. The chromosomal fragments provided a homologous sequence for targeting the chimeric DNA into the recipient chromosome. Erythromycin-resistant transformants were then screened for their transformation phenotype.

To identify and map genes involved in the control of competence, we sought to recover DNA from the neighborhood of this *com::ermB* insertion-duplication mutation.

Terminator Vector

Earlier studies by Stassi and Lacks showed that another pneumococcal locus, *mal*, caused plasmid instability in pBR322 (24). They suggested that excessive transcription of the pBR322 vector interferes with the maintenance of the plasmid (24). With this in mind, our cloning experiments included a vector that contains transcription terminators. Such a vector is pKK232-8, one of a family of vectors developed by Brosius from pBR322 (1). It carries a promoterless CAT gene, preceded by the M13mp8 polylinker cloning site and translational stop codons in three reading frames. A transcription terminator suppresses readthrough into the CAT gene from upstream plasmid sites, while downstream of the CAT gene is placed a tandem duplication of the terminal portion of the *Escherichia coli rrnB* rRNA operon containing transcription terminators (T1, T2) and the 5S RNA gene. Insertion of a DNA fragment containing an active promoter can be selected by its activation of the CAT gene, and highly active promoters can be tolerated because of the protection afforded by the distal terminators (2).

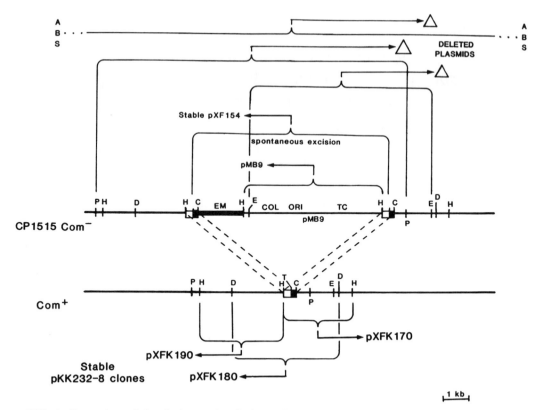

FIG. 1. Comparison of the cloning results obtained when vectors with and without efficient transcription termination signals were used (see text). In the lower diagram the original targeting *Taq*I fragment used to form the *com* mutation is indicated by the solid bar; the 500-bp segment cloned in pMB9 is indicated by the extended bar. Restriction enzyme site abbreviations: A, *Apa*I; B, *Bam*HI; C, *Cla*I; D, *Dra*I; E, *Eco*RI; H, *Hin*dIII; P, *Pst*I; S, *Sal*I; T, *Taq*I.

Strong Promoters

We found that large pieces of DNA from the *com* region were unstable in the standard *E. coli* cloning vectors pMB9 and pBR325. However, large pieces from this region were stable in the *E. coli* vector pKK232-8. In this vector, we have obtained clones of overlapping, independently isolated fragments from the *com* locus of wild-type pneumococci, providing a restriction map covering 5.9 kilobases at this competence control locus. We infer that this region contains one or more sequences which act as strong promoters in *E. coli* and destabilize unprotected cloning vectors.

The original *ermB* insert and approximately 500 base pairs (bp) of neighboring pneumococcal DNA were cloned in the *E. coli* vector pMB9. This composite plasmid, pXF154 (pMB9::*ermB*::*com*) was reintroduced into the pneumococcus chromosome to form CP1515, again causing a Com⁻ mutant phenotype. The *ermB*::*com* part of pXF154 was also subcloned into the streptococcal vector pMV158 (3) to provide a conve-

nient labeled probe for *com* region DNA in subsequent cloning experiments. The source of pneumococcal DNA for cloning was either a wild-type strain or the Com⁻ mutant CP1515. Cloning experiments used the embedded pMB9 as a prospective vector, or the plasmid pKK232-8 (Fig. 1). Recombinant plasmids were detected, respectively, by selection for tetracycline and erythromycin resistance or by activation of the promoterless CAT gene. Recombinant plasmids were analyzed by digestion with appropriate restriction enzymes to determine the size of inserted DNA and to determine whether intact insert::vector junctions were present. Loss of a junction or loss of vector sequences was taken to indicate rearrangement or deletion accompanying cloning. Plasmids bearing *com* DNA were identified by colony hybridization with the 500-bp probe described above. Attempts to clone DNA segments from the *com* region into pMB9 were generally unsuccessful. Specific segments tried are shown by the brackets in Fig. 1. For all the larger segments indicated, yields were low and recombinant plasmids obtained were deleted or rearranged (see below). In contrast, the same strategy readily yielded clones of pMB9 itself from this strain, as well as pMB9 plasmids carrying the duplicated 500-bp segment (Fig. 1).

Three overlapping segments of the *com* region were successfully cloned from wild-type cells in the vector pKK232-8 (Fig. 1). The fragments, identified by probing digests of chromosomal DNA with the 500-bp probe, were enriched by electrophoresis and inserted into the vector. The clones hybridizing with the 500-bp probe were stable and possessed the expected junction sites. Furthermore, this cloned DNA displayed the expected Com+ transforming actvitiy.

To test whether one of these cloned segments required a terminator for stability, an attempt was made to subclone pneumococcal DNA from pXFK180 (Fig. 1) into pBR325. Recombinant plasmids were detected by the insertional inactivation of the CAT gene. Among the recombinant (Cmˢ) plasmids obtained, 80% were rearranged, and 20% had multiple inserts, including terminator fragments from the donor plasmid. Typically, these rearranged plasmids contained only one of the expected ligation junction sites (sometimes neither), and the entire recombinant plasmid was often smaller than the original vector itself. Thus, transcription terminators appear to be required for the stability of this region in *E. coli* plasmids, indicating that this region contains sequences that act as strong promoters in *E. coli*.

Discussion

S. pneumoniae appears to contain frequent sites in its chromosome which are difficult to clone in *E. coli*. The *com* locus is not stable in pBR325; the same is true for several *rec* loci examined in this laboratory. As mentioned above, strong promoters on the pneumococcal *mal* fragment prevent it from being cloned in the *E. coli*(pBR322) system (24). Attempts to clone in *E. coli* the *hexA* and *hexB* genes of pneumococcus in vectors without transcription terminators have also failed (10, 22). The authors suggest that transcription from these *hex* loci into the vector may interfere with control of plasmid replication (10, 22). Unpublished results from this laboratory have shown that, for many insertion-duplication mutants, significant lengths of neighboring DNA cannot be recovered in standard *E. coli* vectors not containing strong transcription terminators. Attempts to make complete libraries of pneumococcal DNA have also encountered a large fraction of fragments unstable in plasmid vectors, including the cosmid vector pHC79, pBR325, pACYC184, and pUC19.

One class of DNA site known to cause instability in *E. coli* plasmid vectors, strong promot-

ers, can be stably maintained when paired with efficient transcription terminators (2, 5, 25). As promoter sequences are known to be A-T rich, pneumococcal DNA (61% A-T [23]) may be especially likely to contain sequences with promoter activity (19). Using pKK232-8, we have cloned three overlapping fragments from the competence control locus and obtained a restriction map of this region (Fig. 1). The *malX* promoter has been shown to be an extremely efficient promoter when inserted upstream of a promoterless CAT gene in pKK232-8 (unpublished data) and can be cloned in *E. coli* if transcription terminators are incorporated in the cloning vector (unpublished data). Finally, using such terminator vectors we have also found that the yield of fragments in a pneumococcal library increases significantly.

Using vectors with transcription terminators, we were able to clone, in *E. coli*, pneumococcal DNA which could not be recovered in standard cloning vectors lacking transcription terminators. These examples suggest that using vectors with efficient transcription terminators should enable the cloning of many more streptococcal loci.

We thank J. Brosius for generously providing pKK232-8 and W. R. Guild for pMV158.

This work was supported by Public Health Service grant AI19875 from the National Institutes of Health.

LITERATURE CITED

1. **Brosius, J.** 1984. Plasmid vectors for the selection of promoters. Gene 27:151–160.
2. **Brosius, J.** 1984. Toxicity of an overproduced foreign gene product in *Escherichia coli* and its use in plasmid vectors for the selection of transcription terminators. Gene 27:161–172.
3. **Burdett, V.** 1980. Identification of tetracycline-resistant R-plasmids in *Streptococcus agalactiae* (group B). Antimicrob. Agents Chemother. 18:753–760.
4. **Fox, M. S., and M. K. Allen.** 1964. On the mechanism of deoxyribonuclease integration in pneumococcal transformation. Proc. Natl. Acad. Sci. USA 52:412–419.
5. **Gentz, R., A. Langner, A. C. Y. Chang, S. N. Cohen, and H. Bujard.** 1981. Cloning and analysis of strong promoters is made possible by the downstream placement of a RNA termination signal. Proc. Natl. Acad. Sci. USA 78:4936–4940.
6. **Gurney, T., Jr., and M. S. Fox.** 1968. Physical and genetic hybrids formed in bacterial transformation. J. Mol. Biol. 32:83–100.
7. **Javor, G., and A. Tomasz.** 1968. An autoradiographic study of genetic transformation. Proc. Natl. Acad. Sci. USA 60:1216–1222.
8. **Lacks, S. A.** 1977. Binding and entry of DNA in bacterial transformation, p. 177–232. *In* J. L. Reissig (ed.), Microbial interactions, series B, vol. 3, Chapman & Hall, Ltd., London.
9. **Lacks, S., B. Greenberg, and K. Carlson.** 1967. Fate of donor DNA in pneumococcal transformation. J. Mol. Biol. 29:327–347.
10. **Martin, B., H. Prats, and J.-P. Claverys.** 1985. Cloning of the *hexA* mismatch-repair gene of *Streptococcus pneumoniae* and identification of the product. Gene 34:293–303.
11. **McCarty, M., H. E. Taylor, and O. T. Avery.** 1947. Bio-

chemical studies of environmental factors essential in transformation of pneumococcal types. Cold Spring Harbor Symp. Quant. Biol. **11**:177–183.

12. **Morrison, D. A.** 1981. Competence-specific protein synthesis in *Streptococcus pneumoniae*, p. 39–54. *In* M. Polsinelli and G. Mazza (ed.), Transformation—1980. Cotswold Press, Oxford.

13. **Morrison, D. A., and M. F. Baker.** 1979. Competence for genetic transformation in pneumococcus depends on synthesis of a small set of proteins. Nature (London) **282**:215–217.

14. **Morrison, D. A., and W. R. Guild.** 1972. Transformation and deoxyribonucleic acid size: extent of degradation on entry varies with size of donor. J. Bacteriol. **112**:1157–1168.

15. **Morrison, D. A., and W. R. Guild.** 1973. Structure of deoxyribonucleic acid on the cell surface during uptake by pneumococcus. J. Bacteriol. **115**:1055–1062.

16. **Morrison, D. A., and W. R. Guild.** 1973. Breakage prior to entry of donor DNA in pneumococcus transformation. Biochim. Biophys. Acta **299**:545–556.

17. **Morrison, D. A., and B. Mannarelli.** 1979. Transformation in pneumococcus: nuclease resistance of deoxyriboncleic acid in the eclipse complex. J. Bacteriol. **140**:655–665.

18. **Morrison, D. A., M.-C. Trombe, M. K. Hayden, G. A. Waszak, and J.-D. Chen.** 1984. Isolation of transformation-deficient *Streptococcus pneumoniae* mutants defective in control of competence, using insertion-duplication mutagenesis with the erythromycin resistance determinant of pAMβ1. J. Bacteriol. **159**:870–876.

19. **Mulligan, M. E., and W. R. McClure.** 1986. Analysis of the occurrence of promoter-sites in DNA. Nucleic Acids Res. **14**:109–126.

20. **Pakula, R., and W. Walczak.** 1963. On the nature of competence of transformable streptococci. J. Gen. Microbiol. **31**:125–133.

21. **Porter, R. D., and W. R. Guild.** 1969. Number of transformable units per cell in *Diplococcus pneumoniae*. J. Bacteriol. **97**:1033–1035.

22. **Prats, H., B. Martin, and J.-P. Claverys.** 1985. The *hexB* mismatch repair gene of *Streptococcus pneumoniae*: characterisation, cloning and identification of the product. Mol. Gen. Genet. **200**:482–489.

23. **Schildkraut, C. L., J. Marmur, and P. Doty.** 1961. The formation of hybrid DNA molecules and their use in studies of DNA homologies. J. Mol. Biol. **3**:595–617.

24. **Stassi, D. L., and S. A. Lacks.** 1982. Effect of strong promoters on the cloning in *Escherichia coli* of DNA fragments from *Streptococcus pneumoniae*. Gene **18**:319–328.

25. **Stueber, D., and H. Bujard.** 1982. Transcription from efficient promoters can interfere with plasmid replication and diminish expression of plasmid specified genes. EMBO J. **1**:1399–1404.

26. **Tomasz, A.** 1965. Control of the competent state in pneumococcus by a hormone-like cell product: an example for a new type of regulatory mechanism in bacteria. Nature (London) **208**:155–159.

27. **Tomasz, A.** 1966. Model for the mechanism controlling the expression of competent state in pneumococcus cultures. J. Bacteriol. **91**:1050–1061.

28. **Tomasz, A., and J. L. Mosser.** 1966. On the nature of the pneumococcal activator substance. Proc. Natl. Acad. Sci. USA **55**:58–66.

29. **Vijayakumar, M. N., and D. A. Morrison.** 1986. Localization of competence-induced proteins in *Streptococcus pneumoniae*. J. Bacteriol. **165**:689–695.

IV. Oral Streptococci

Cloning and Expression of a Structural Gene for Adhesion (Type 1) Fimbriae in *Streptococcus sanguis*

PAULA M. FIVES-TAYLOR AND TODD J. PRITCHARD

Department of Medical Microbiology, The University of Vermont, Burlington, Vermont 05405

Adherence to and colonization of host tissue are primary steps leading to the disease state. Dental caries and periodontal disease are two of the most common microbial infections in humans. The causal relationship between these diseases and dental plaque formation was demonstrated many years ago (5, 10, 11, 13, 15). Early plaque development involves aerobic, predominantly gram-positive organisms succeeded by a complex, filamentous, predominantly anaerobic population (16). It is probable that the first organism to adhere to the tooth initiates plaque development by altering the environment to favor the attachment of the later flora. *Streptococcus sanguis*, a gram-positive coccus, is one of the first organisms to adhere to new or cleaned teeth (9), is found in high numbers in dental plaque, and has been shown to have a high affinity for the saliva-coated tooth surface. For these reasons, an understanding of the molecular mechanisms involved in the adhesion of this organism is of interest.

Recent studies have implicated fimbriae in the adhesion of *S. sanguis* to the in vitro tooth model, saliva-coated hydroxyapatite (4, 6–8, 12). While immunological methods have proved useful in the study of fimbriae from *S. sanguis* (4, 12), the isolation and biochemical analyses of these fimbriae by conventional methods has been fraught with difficulties. These studies were undertaken to identify and characterize putative subunits and precursors of *S. sanguis* fimbriae by cloning the respective genes into *Escherichia coli*.

Chromosomal DNA from *S. sanguis* FW213 was partially digested with *Eco*RI and ligated into the expression vector pOP203(A$_2^+$) (7a). Plasmid pOP203(A$_2^+$) is a pMB9 Tc^r derivative which contains the lactose operator-promoter region fused to the Qβ phage A$_2$ gene. When the A$_2$ gene is expressed, it is lethal to the *E. coli* host (14, 17). The lysis gene was inactivated by inserting the chromosomal DNA of the streptococcus in the middle of the gene. This construct permits positive selection of Tc^r recombinants in the presence of an inducer of the lactose operon. The ligation mixture was used to transform *E. coli* SK1592, and 4,500 presumptive recombinant colonies were picked from the plates containing tetracycline and isopropyl-β-D-galactopyranoside and were subcultured. Analysis of 20 randomly selected colonies showed that all contained plasmids greater than the 7-kilobase (kb) size of the vector. Restriction with *Eco*RI revealed that these plasmids had distinct restriction patterns and that the average size insert was 3.2 kb (data not shown). On the basis of the Clarke and Carbon (2) formula, this library should be greater than 99.9% complete.

The recombinant clones were screened by colony immunoassays with AdAb, an *S. sanguis* polyclonal antibody made specific for fimbriae by successive adsorptions with a nonfimbriated mutant until the sera no longer reacted with the mutant in the enzyme-linked immunosorbent assay (4). Only two clones, VT616 and VT618, were positive (Fig. 1, row 3). The location of the genes for these antigens to the plasmids was confirmed by isolating the plasmids from VT616 and VT618 and transforming SK1592 and JA228 (*E. coli* hosts) in two separate experiments. Colony immunoassays showed that antigen production was transferred with the plasmid (Fig. 1, row 6). *E. coli* V871 and JA228 were applied as negative controls (Fig. 1, row 1).

Chromosomal DNA from *S. sanguis* FW213 and plasmid DNAs from VT615, VT616, and VT618 were isolated and restricted with *Eco*RI and probed with plasmid pVT618 (Fig. 2). A 6-kb *Eco*RI fragment from *S. sanguis* FW213 and *E. coli* clones VT616 and VT618 showed homology with the 6-kb insert from pVT618, indicating that the pVT618 insert is streptococ-

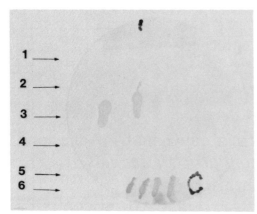

FIG. 1. Colony immunoassays. The set of PAb⁺ clones was probed with AdAb, a fimbrial specific antibody. *E. coli* SK1592 and JA228 were added as negative controls in row 1. Row 6 contains the four *E. coli* strains (JA228 and SK1592) transformed with recombinant plasmids isolated from VT616 and VT618 seen in row 3.

cal DNA and that the insert in pVT616 shares homology with the insert in pVT618. All clones

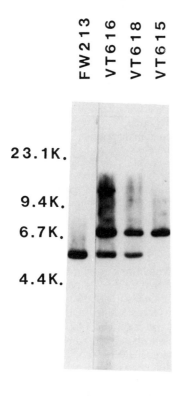

PROBE:pVT618

FIG. 2. Southern blot hybridization of pVT618 *Eco*RI fragments. Lane 1, *S. sanguis* FW213 chromosomal DNA cleaved with *Eco*RI; lanes 2–4, *E. coli* recombinant clones.

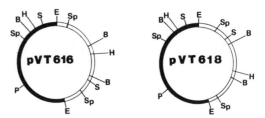

FIG. 3. Restriction endonuclease site maps of pVT616 and pVT618. Sp, *Sph*I; E, *Eco*RI; S, *Sst*I; B, *Bam*HI; P, *Pvu*II; H, *Hin*dIII.

showed homology with the 7.2-kb vector of pVT618.

Restriction endonuclease maps of pVT616 and pVT618 are shown in Fig. 3. The inserts in both plasmids are 6 kb in size and have no *Eco*RI sites. Restriction digestion with other enzymes shows that both inserts have two *Bam*HI, one *Hin*dIII, one *Sst*I, two *Sph*I, and no *Pvu*II sites, confirming the Southern analysis data which show that these two inserts are the same fragment of DNA. Restriction mapping, however, indicates that these two inserts have been cloned in reverse orientation, a useful property for further genetic analyses.

The clones, VT616 and VT618, produced an M_r 30,000 protein that, in agreement with the colony immunoassays, reacted in AdAb (Fig. 4). The protein was produced whether or not the cells were induced, suggesting that the lactose operator-promoter was not necessary for expression of the gene and that the fimbrial promoter may be present on the 6-kb insert. Equal numbers of cells of clone VT618 (three experiments) produced more protein when induced than when not induced, suggesting that transcripts were being made from the lactose operator promoter as well as the unknown promoter. There was no reaction with the host, SK1591. We lysed VT618 and immunoprecipitated the antigen(s) that reacted with AdAb. As shown in Fig. 3 (VT618 IP), the 30-kilodalton protein was immunoprecipitated. The relationship between the 30-kilodalton protein specified in the clone and *S. sanguis* fimbriae purified from cross-immunoelectrophoretic gels (4) is shown in Fig. 4. One minor component of the fimbrial preparation is indistinguishable in size from the immunoreactive products of the clones. In addition to this component, a number of larger polypeptides including a major one at 43 kilodaltons and a high-molecular-weight smear with a distinguishable band at 200 kilodaltons were observed. Reactions with monoclonal antibodies and tandem immunoelectrophoresis demonstrated that all these bands have the same epitope (data not shown). Type 2 fimbriae of *Actinomyces viscosus* have been similarly pu-

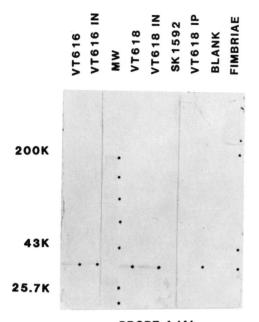

VT616 | VT616 IN | MW | VT618 | VT618 IN | SK1592 | VT618 IP | BLANK | FIMBRIAE

200K

43K

25.7K

PROBE AdAb

FIG. 4. Identification of immunoreactive *S. sanguis* fimbrial proteins expressed in *E. coli*. Preparations were subjected to sodium dodecyl sulfate-polyacrylamide gel electrophoresis, and the proteins were transferred and probed with AdAb. Lanes 1 and 2, VT616 and V616 induced with isopropyl-β-D-galactopyranoside; lane 3, molecular weight markers; lanes 4 and 5, VT618 and VT618 induced with isopropyl-β-D-galactopyranoside; lane 6, *E. coli* SK1592; lane 7, the immunoprecipitate formed when AdAb was added to a lysate of VT618; lane 9, fimbriae from *S. sanguis* FW213 cells.

rified and cloned, and a similar pattern of multiple bands also has been noted (3). This paper and that of Morris et al. (12) are the only reports on the biochemistry of gram-positive fimbriae, but the data do suggest that gram-positive fimbriae are very distinct from gram-negative fimbriae. Why the fimbrial epitope appears in multiple forms is unknown. Several reasonable suggestions are that (i) fimbriae are only partially dissociated in sodium dodecyl sulfate, β-mercaptoethanol, and urea; (ii) the fimbrial peptides may be glycosylated; (iii) some cell wall material remains attached to the peptides; and (iv) fimbriae are composed of nonidentical subunits with an epitope in common. Incomplete dissociation can be ruled out as we know native gels give the same number of bands (data not shown), suggesting that multiple peptides are associated with the cell and that the fimbriae really are resistant to dissociation. Glycosylation remains a viable alternative as all the higher-molecular-weight species are sensitive to periodate (manuscript in prepara-

tion). We are presently testing the fimbrial preparations with endoglycosidase H in an effort to resolve this alternative. Attachment of cell wall material is less likely, as treatment of the fimbrial preparation with mutanolysin produces no change in the molecular masses (in preparation). Whether or not fimbriae are composed of nonidentical subunits awaits genetic analysis and sequencing of the fimbrial gene.

These results demonstrate that at least one streptococcal fimbrial gene can be cloned and expressed in *E. coli*. Our laboratory is proceeding with the genetic analysis of the fimbrial product.

This investigation was supported by Public Health Service grant R01-DE05606 from the National Institute of Dental Research.

LITERATURE CITED

1. **Carlson, J., H. Grahnen, G. Jonsson, and S. Wikner.** 1970. Establishment of *Streptococcus sanguis* in the mouths of infants. Arch. Oral Biol. **15**:1142–1148.
2. **Clarke, L., and J. Carbon.** 1976. A colony bank containing synthetic ColE1 hybrid plasmids representative of the entire *E. coli* genome. Cell **9**:91–99.
3. **Donkersloot, J. A., J. O. Cisar, M. E. Wax, R. J. Harr, and B. M. Chassy.** 1985. Expression of *Actinomyces viscosus* antigens in *Escherichia coli*: cloning of a structural gene (*fimA*) for the 2 fimbriae. J. Bacteriol. **162**:1075–1078.
4. **Fachon-Kalweit, S., B. Elder, and P. Fives-Taylor.** 1985. Antibodies that bind to fimbriae block adhesion of *Streptococcus sanguis* to saliva-coated hydroxyapatite. Infect. Immun. **48**:617–624.
5. **Fitzgerald, R. J., and P. H. Keyes.** 1960. Demonstration of the etiologic role of streptococci in experimental caries in the hamster. J. Am. Dent. Assoc. **61**:9–19.
6. **Fives-Taylor, P.** 1982. Isolation and characterization of a *Streptococcus sanguis* FW213 mutant nonadherent to saliva-coated hydroxyapatite heads, p. 206–209. *In* D. Schlessinger (ed.), Microbiology—1982. American Society for Microbiology, Washington, D.C.
7. **Fives-Taylor, P., and D. Thompson.** 1985. Surface properties of *Streptococcus sanguis* FW213 mutants nonadherent to saliva-coated hydroxyapatite. Infect. Immun. **47**:752–759.
7a. **Fives-Taylor, P. M., F. L. Macrina, T. J. Pritchard, and S. S. Peene.** 1987. Expression of *Streptococcus sanguis* antigens in *Escherichia coli*: cloning of a structural gene for adhesion fimbriae. Infect. Immun. **55**:123–128.
8. **Gibbons, R. J., I. Etherden, and S. Skobe.** 1983. Association of fimbriae with the hydrophobicity of *Streptococcus sanguis* FC-1 and adherence to salivary pellicles. Infect. Immun. **41**:414–417.
9. **Gibbons, R. J., S. S. Socransky, S. deAranjo, and J. Van Houte.** 1964. Studies of predominant cultivated microbiota of dental plaques. Arch. Oral Biol. **9**:365–370.
10. **Gibbons, R. J., and J. Van Houte.** 1973. On the formation of dental plaques. J. Periodontol. **44**:347–360.
11. **Jenkins, G. N.** 1968. The mode of formation of dental plaque. Caries Res. **2**:130.
12. **Morris, E. J., N. Ganeshkumar, and B. C. McBride.** 1985. Cell surface components of *Streptococcus sanguis*: relationship to aggregation, adherence, and hydrophobicity. J. Bacteriol. **164**:255–262.
13. **Orland, F., J. Blaney, R. Harrison, J. Reymers, P. Trexier, M. Wagner, M. Gordon, and T. Luckey.** 1954. Use of the germ free animal technic in the study of experimental dental caries. I. Basic observations on rats reared free of

all microorganisms. J. Dent. Res. **33:**147–174.

14. **Pucci, M. J., and F. L. Macrina.** 1985. Cloned *gtfA* gene of *Streptococcus mutans* LM7 alters glucan synthesis in *Streptococcus sanguis.* Infect. Immun. **48:**704–712.

15. **Socransky, S. S.** 1970. Relationship of bacteria to the etiology of periodontal disease. J. Dent. Res. **49:**203.

16. **Socransky, S. S., A. D. Manganiello, D. Propas, V. Oram, and J. Van Houte.** 1971. The oral microbiota of man from birth to senility. J. Periodontol. **42:**485.

17. **Winter, R. B., and L. Gold.** 1983. Overproduction of bacteriophage Qβ maturation (A2) protein leads to cell lysis. Cell **33:**877–885.

Identification of Virulence Components of Mutans Streptococci

ROY R. B. RUSSELL AND MARTYN L. GILPIN

Dental Research Unit, Royal College of Surgeons of England, Downe, Kent BR6 7JJ, England

The coherent mass of bacterial growth referred to as dental plaque, which accumulates on exposed surfaces and in the pits and fissures of teeth, contains representatives of several dozen different genera. Prominent among the types regularly isolated are various streptococci, including *Streptococcus sanguis, S. mitior*, and members of the group collected under the name "mutans streptococci." It is this latter group which is believed to be closely associated with the development of dental caries.

Although it has been known for over a decade that what was once known as *S. mutans* actually consists of a number of distinct species, the various specific names have been only very slowly adopted, and much of the literature refers to the various serotypes a through h, based on wall polysaccharide antigens (Table 1). The two species of greatest importance are those still properly referred to as *S. mutans* (serotypes c, e, and f) and *S. sobrinus* (serotypes d, g, and h) because these are widespread in humans. A number of studies from a variety of countries have shown that around 90% of isolates of mutans streptococci from humans are *S. mutans* and between 8 and 40% are *S. sobrinus*.

Virulence of Mutans Streptococci

A number of different traits have been proposed to contribute to the ability of mutans streptococci to cause dental caries, and although some of these have been shown to be significant in rodent models of the disease, there is as yet no evidence to implicate any as virulence determinants in humans. The main properties suggested as being of importance in cariogenesis may be divided into those involved in plaque formation and those involved in the metabolic activities in plaque.

Plaque formation is dependent upon an array of specific binding interactions between bacteria and salivary components, between surface components of different bacteria, and between bacteria and complex macromolecules of salivary or bacterial origin. The main physiological attributes of significance are acid production, acid tolerance, and the ability to survive in plaque by competing with other bacteria by efficient utilization of intracellular and extracellular polymers. If we are to understand the processes leading to dental caries and to rationally devise methods for prevention of the disease, it is essential that we acquire a detailed knowledge at the molecular level of the components of *S. mutans* involved.

Sucrose Metabolism of Mutans Streptococci

Prominent among the proposed virulence traits are those activities concerned with metabolism of sucrose. Sucrose can not only be used as a growth substrate (and source of acid) but can also be converted to extracellular polymers by two classes of enzyme: glucosyltransferases (GTFs, EC 2.4.1.5), which split sucrose to yield free fructose and various polymeric glucans, and fructosyltransferases (FTFs, EC 2.4.1.10), which yield free glucose and polymeric fructans. Both GTF and FTF exist in a number of different forms and can produce a number of types of polymer, but there is considerable uncertainty as to how many different enzymes there are, how they interact with one another, and how they are regulated. In addition, there is little information on how the enzymes of different mutans species are related or how the present complex situation may have evolved.

Cloning of Genes for *S. sobrinus* GTF

S. sobrinus MFe28 (mutans serotype h) is unusual in that virtually all of the polymer which it makes from sucrose is a water-insoluble glucan (1). Since it is such a polymer which is believed to be of most importance in the accumulation of the bacteria on hard surfaces, we selected it for study. We have reported the cloning of two distinct GTF genes in the bacteriophage λ vector L47.1 (3). Recombinant phage plaques in which either *gftS* or *gtfI* was expressed could be recognized by the fact that on sucrose-containing medium they accumulated a heap of glucan with an appearance resembling a bacterial colony. The enzymes produced in *Escherichia coli* containing the *gtf* genes have now been characterized by us in collaboration with H. Mukasa (J. Gen. Microbiol., in press). The results, summarized in Table 2, show the close similarity of enzymes synthesized in *E. coli* to those from *S. sobrinus* and, furthermore, reveal that the enzyme forming soluble glucan is that named GTF-S2 by Mukasa (6). An enzyme

TABLE 1. Mutans group of streptococci

Species	Serotype	Guanine plus cytosine (mol%)	Host
S. cricetus	a	42–44	Hamsters
S. rattus	b	41–43	Rats
S. mutans	c, e, f	36–38	Humans, monkeys
S. sobrinus	d, g, h	44–46	Humans, monkeys
S. ferus	c	43–45	Wild rats
S. macacae	c	35–36	Monkeys

with similar properties has also been described by McCabe (5). We have not yet isolated the gene for GTF-S1 from our recombinant gene bank.

Antigenic Relationships of GTF-I

GTF-I can readily be purified from recombinant E. coli by a single affinity chromatography column (2), and antiserum against it can be raised. This antiserum was used to demonstrate antigenic identity of the cloned GTF-I with GTF-I from S. sobrinus (serotype h) by immunodiffusion. Strong reactions with S. sobrinus strains of serotypes d and g were also found, though none was observed with concentrated culture supernatants of other mutans streptococci. By immunodiffusion, no cross-reaction between GTF-I and GTF-S1 or GTF-S2 could be detected. In contrast to the results obtained with immunodiffusion, the more sensitive Western blot technique indicated the presence of antigens related to GTF-I in all mutans streptococci, serotypes a through h. (S. ferus was not tested.)

Genetic Studies of gtfI

The gtfI gene was subcloned from the bacteriophage vector into plasmid pBR322 as a single 5-kilobase HindIII insert. Since 4.3 kilobases would be required to code for a GTF of molecular weight 160,000, the insert should consist almost entirely of gtfI. The limits of the gene

TABLE 2. Comparison of properties of GTFs of S. sobrinus and of recombinant E. coli carrying cloned gtf genes

Enzyme	Glucan linkage	Mol wt (approx)	pI	Dextran primer requirement
S. sobrinus GTF				
GTF-S1	α-1,3,6	150,000	3.9	Yes
GTF-S2	α-1,6	150,000	5.5	No
GTF-I	α-1,3	160,000	4.9	Yes
Cloned GTF				
GTF-S	α-1,6	150,000	5.7	No
GTF-I	α-1,3	160,000	4.9	Yes

have been partially defined by deletions as shown in Fig. 1. Deletions extending to the BglII or AvaI sites shown at the right-hand end of the gtfI gene in Fig. 1 resulted in no expression of GTF, while the deletion into the SacI site on the left of the map resulted in formation of a truncated and enzymatically inactive molecule of 130,000 molecular weight. The results lead us to believe that the promoter of gtfI lies at the right end of the map shown.

To seek for stretches of DNA homologous to gtfI, we prepared an internal 3.3-kilobase HpaI–HpaI fragment and used it as a probe in Southern blot hybridization with HpaI-digested chromosomal DNA from other organisms. Strong hybridization with DNA from the other S. sobrinus strains (serotypes d and g) and also with that of S. cricetus (a) was seen. Weaker but still distinct reactions were seen with S. mutans (c, e, f) and also with S. rattus (b) and S. macacae (c). This wide distribution of homologous sequences is much wider than that reported for any other gene in the mutans streptococci. It should be noted, however, that we have yet to demonstrate that the homologous sequences are in gtfI genes.

Cloning of a Gene for S. mutans FTF

Several authors have reported studies of the properties of S. mutans FTF, but there is considerable disagreement about the number of electrophoretically distinct FTFs and their molecular weights. Since there is evidence that much of the heterogeneity may arise as a consequence of proteolytic degradation (6, 8), we have modified our earlier method for detecting FTF after sodium dodecyl sulfate-polyacrylamide gel electrophoresis (7) by introducing an improved periodic acid-Schiff stain (4). This allows us to detect GTF and FTF activity in unconcentrated culture supernatants and to find (Fig. 2) that all strains of S. mutans examined contain FTF activity of 95,000 and 80,000 daltons. In preparations stored for some days (e.g., the serotype f strain in Fig. 2) or in cultures which have been concentrated, we also find a series of active bands between 85,000 and 70,000 daltons, while Western blotting with anti-FTF serum detects even smaller breakdown products.

A recombinant bacteriophage vector carrying an ftf gene from S. mutans Ingbritt (c) expresses bands of activity corresponding precisely to those found in S. mutans, indicating that all forms of the enzyme are derived as products of a single gene. This result is in agreement with the recent report by Sato and Kuramitsu (10), and we have also been able to demonstrate antigenic identity of the FTF made by our recombinants and one provided by H. Kuramitsu.

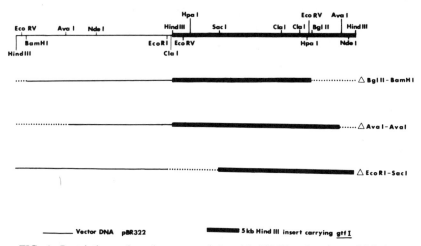

FIG. 1. Restriction endonuclease map of plasmid pMLG1 and regions of deletion.

Relationship of FTF to Glucan-Binding Protein

Antiserum raised against *S. mutans* FTF, purified from *E. coli* infected with recombinant bacteriophage, was used to investigate its relationship with another molecule with FTF activity previously described by us and first recognized as a glucan-binding protein (9). Although FTF and glucan-binding protein have closely similar sizes and isoelectric points and make the same type of β-2,1-linked fructan, they are antigenically entirely distinct and no cross-reaction could be found by immunodiffusion, Western blotting, or line immunoelectrophoresis.

Possible Relationships of Polymer-Forming and Polymer-Binding Proteins

One of the most intriguing questions to be answered concerns the origin of the multiplicity of different enzymes in the mutants streptococci which are involved with sucrose-derived polymers. As progress is rapidly being made toward the identification of the various components and their antigenic and genetic characterization, we should soon start to get an indication of the occurrence of gene duplication or gene-fusion events which could generate a family of related products. We have recently observed (by Western blotting with anti-glucan-binding protein serum) an antigenic cross-reaction between glucan-binding protein and GTF not only within *S. mutans* but also across a species barrier with the GTF of *S. sobrinus*. This unexpected link between glucan synthesis, glucan binding, and fructan synthesis leads us to propose the hypothetical scheme in Fig. 3, which illustrates a series of overlapping relationships which may be reflected at the DNA sequence or protein-structural level. These relationships could therefore be detected by DNA hybridization or immunological methods. Although there is as yet little evidence to support such a scheme, it suggests that GTFs, some of which can bind glucans, may be related to nonenzymatic glucan-binding proteins which would function as receptors at cell surfaces. These glucan-binding proteins would in turn share some properties with those of FTFs

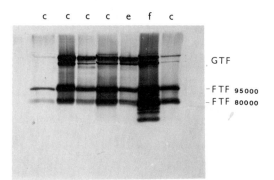

FIG. 2. Detection of bands of glucan and fructan synthesized from sucrose after sodium dodecyl sulfate-polyacrylamide gel electrophoresis. *S. mutans* cultures used were, from left: Ingbritt (c), 3209 (c), GS-5 (c), K34 (c), C40 (e), 151 (f), Ingbritt (c).

FIG. 3. Schematic representation of possible relationships between GTF, FTF, and nonenzymatic glucan-binding proteins.

which show a similar capacity to bind. Furthermore, although this paper has concentrated upon sucrose-related phenomena, it is likely that many of the other properties associated with virulence will reveal complex relationships among the collection of species making up the mutans streptococci.

We are grateful to Gordon Dougan for advice and assistance with genetic techniques and to Hidehiko Mukasa for collaboration in GTF characterization.

LITERATURE CITED

1. **Beighton, D., R. R. B. Russell, and H. Hayday.** 1981. The isolation and characterization of *Streptococcus mutans* serotype *h* from dental plaque of monkeys (*Macaca fascicularis*). J. Gen. Microbiol. **124:**414–416.
2. **Douglas, C. W. I., and R. R. B. Russell.** 1982. Effect of specific antisera on adherence properties of the oral bacterium *Streptococcus mutans*. Arch. Oral Biol. **27:** 1039–1045.
3. **Gilpin, M., R. R. B. Russell, and P. Morrissey.** 1985. Cloning and expression of two *Streptococcus mutans* glucosyltransferases in *Escherichi coli*. Infect. Immun. **49:** 414–416.
4. **Konat, G., H. Offner, and J. Mellah.** 1984. Improved sensitivity for detection and quantitation of glycoproteins on polyacrylamide gels. Experientia **40:**303–304.
5. **McCabe, M. M.** 1985. Purification and characterization of a primer-independent glucosyltransferase from *Streptococcus mutans* 6715-13 mutant 27. Infect. Immun. **50:**771– 777.
6. **Mukasa, H.** 1986. Properties of *Streptococcus mutans* glucosyltransferases, p. 121–132. *In* L. Menaker, S. Hamada, H. Kiyono, S. M. Michalek, and J. R. McGhee (ed.), Cellular, molecular and clinical aspects of *Streptococcus mutans*. Elsevier Science Publishers, Amsterdam.
7. **Russell, R. R. B.** 1979. Use of Triton X-100 to overcome the inhibition of fructosyltransferase by SDS. Anal. Biochem. **97:**173–175.
8. **Russell, R. R. B., E. Abdulla, M. L. Gilpin, and K. Smith.** 1986. Characterization of *Streptococcus mutans* surface antigens, p. 61–70. *In* S. D. Hamada, L. Menaker, H. Kiyono, S. Michalek, and J. R. McGhee (ed.), Cellular, molecular and clinical aspects of *Streptococcus mutans*. Elsevier Science Publishers, Amsterdam.
9. **Russell, R. R. B., A. C. Donald, and C. W. I. Douglas.** 1983. Fructosyltransferase activity of a glucan-binding protein from *Streptococcus mutans*. J. Gen. Microbiol. **129:**3243–3250.
10. **Sato, S., and H. K. Kuramitsu.** 1986. Isolation and characterization of a fructosyltransferase gene from *Streptococcus mutans* GS-5. Infect. Immun. **52:**166–170.

Evidence for a Duplicated DNA Sequence Associated with a Glucosyltransferase Gene in *Streptococcus mutans*

MICHAEL J. PUCCI, KEVIN R. JONES, AND FRANCIS L. MACRINA

Department of Microbiology and Immunology, Virginia Commonwealth University, Richmond, Virginia 23298

Specific oral streptococci, which include *Streptococcus mutans* and other related species, are recognized as the etiological agents of dental caries (4). Human dietary studies and animal models implicate sucrose in cariogenicity (5, 14). *S. mutans* cells first attach to the smooth tooth surface by a sucrose-independent event, possibly involving glycoproteins or lipoteichoic acids (12). Next, a sucrose-dependent step results in firmer attachment and in cell-to-cell aggregation on the tooth surface. This attachment and adherence step appears to be linked to the synthesis of glucan polymers by glucosyltransferase (GTF) enzymes which utilize sucrose as substrate (6). In particular, mutants deficient in the synthesis of water-insoluble glucans displayed reduced cariogenicity in animals (15).

The number of GTFs and the importance of each enzyme in virulence is still unclear. For example, it appears that there may be at least three GTFs in *S. mutans* Bratthall serotype c strains (3). Molecular genetic approaches to this problem should allow each enzyme to be individually cloned and characterized. We have previously described one GTF gene, *gtfA*, which appeared to play some role in overall exopolysaccharide synthesis (9). This gene, first described by Robeson et al. (11), encoded a 55,000-dalton polypeptide which synthesized a small, water-soluble glucan in *Escherichia coli*. The *gtfA* gene from *S. mutans* LM7 (serotype e) was cloned onto a shuttle plasmid vector and introduced into *Streptococcus sanguis* Challis (9). The resultant recombinant synthesized elevated levels of both water-soluble and water-insoluble glucan. The latter activity is significant, as *S. sanguis* normally does not synthesize water-insoluble glucans. We describe here the isolation and preliminary characterization of another GTF gene, *gtfC*, from *S. mutans* LM7 (Bratthall serotype e). The gene product, a 150,000-dalton polypeptide, displayed both sucrose hydrolytic and polymer synthetic activities. Restriction endonuclease mapping and Southern hybridization data suggested that at least a portion of this gene is duplicated in *S. mutans* serotypes c, e, and f.

Initial Cloning

Our cloning strategy for GTF genes from *S. mutans* was as follows. Chromosomal DNA was isolated from *S. mutans* LM7 as previously described (9) and partially digested with *Eco*RI restriction endonuclease under conditions which generated an array of high-molecular-weight (~9 to 20 kilobases [kb]) fragments (8). These fragments were purified by sucrose density gradient centrifugation and ligated with bacteriophage λ EMBL4 purified arms (7). The recombinant phage DNA then was packaged and allowed to infect *E. coli* NM538. About 1,000 of the resulting plaques constituted the equivalent of a genomic library of *S. mutans* LM7, as determined by the formula of Clarke and Carbon (2). The library next was screened immunologically using antisera raised in rabbits against *S. mutans* extracellular proteins (10). Immunopositive plaques were picked and purified, and plate lysates were assayed for sucrose hydrolytic activity as described previously (9). Three of 48 lysates were able to generate reducing sugar activity from sucrose by the Nelson assay. These three lysates were assayed for the production of exopolysaccharides from sucrose substrate (9), and one, λ-50, displayed the ability to synthesize methanol-insoluble polysaccharide from [U-^{14}C]-sucrose (Table 1).

A lysate was then prepared from 1 liter of *E. coli* culture. A sample of this lysate was saved for further studies, and the remainder was used to obtain λ-50 DNA by standard methods (13) for restriction endonuclease mapping and subcloning. This DNA was digested with several restriction enzymes to generate the map shown in Fig. 1. The *S. mutans* LM7 chromosomal insert consisted of two *Eco*RI fragments, 8.1 and 4.7 kb, totaling 12.8 kb. The large *Eco*RI fragment contained unique *Sph*I, *Sst*I, and *Pst*I sites, and several *Bam*HI and *Hind*III sites existed in both fragments. No *Cla*I sites were present in either *Eco*RI fragment but the λ arms did contain *Cla*I sites approximately 200 base pairs (bp) outside each outer *Eco*RI site.

TABLE 1. Methanol-insoluble polysaccharide synthesis

Sample	Total protein (µg)	cpm after 3 h	
		+ Primer	− Primer
Background		396 ± 18	396 ± 18
λ-50 lysate	52.0	3,685 ± 60	3,146 ± 87
GS5 extracellular	3.0	15,195 ± 1,085	ND[a]

[a] ND, Not determined.

FIG. 1. Restriction endonuclease map of λ-50 DNA. *S. mutans* LM7 chromosomal DNA insert totals approximately 12.8 kb. Restriction endonuclease abbreviations: B, *Bam*HI; C, *Cla*I; E, *Eco*RI; H, *Hin*dIII; P, *Pst*I; Sp, *Sph*I; Ss, *Sst*I. The bar indicates 1 kb.

Localization of the *gtfC* Gene

We next attempted to determine both the approximate location of the *gtf* gene and size of its polypeptide product. The large *Eco*RI fragment was subcloned intact and in two portions, from the outer *Eco*RI to the *Sst*I site and from the *Sst*I site to the inner *Eco*RI site. The positive selection vector, pOP203(A$_2^+$), was used in these subcloning experiments (9). Cultures of the *E. coli* strains containing these recombinant plasmids were disrupted by sonication, supernatants were separated from cellular debris, and samples of these supernatants were subjected to Western blot analysis along with the λ-50 lysate (Fig. 2). Using the rabbit antisera to *S. mutans* extracellular proteins, lane C (*Sst*I-*Eco*RI fragment) displayed one immunopositive polypeptide of about 90,000 daltons (Fig. 2). Lane B (*Eco*RI-*Sst*I fragment) showed no detectable immunopositive polypeptides. Lane D (λ-50 lysate) contained several immunoreactive bands ranging from approximately 150,000 to 120,000 daltons. Since lane D represented the entire *gtf* gene, we have used the largest size, 150,000 daltons, to represent the intact, functional GTF polypeptide. At this point, the gene appeared to be distinct from the previously cloned *gtfA* gene, and we designated the new gene *gtfC*. The 2.9-kb *Sst*I-*Eco*RI fragment encoded a 90,000-dalton polypeptide which we assumed to be a truncated version of the 150,000-dalton *gtfC* polypeptide. This allowed the approximate positioning of the coding region as shown by the arrow in Fig. 1. Reducing sugar assays supported the contention that *gtfC* begins in the 2.9-kb *Sst*I-*Eco*RI fragment. The sonic extract from the *E. coli* strain containing

this cloned fragment was sucrase positive while strain containing the 5.2-kb *Eco*RI-*Sst*I fragment was sucrase negative (data not shown). These data suggested that only about 60% of the *gtfC* polypeptide is needed for sucrose hydrolysis.

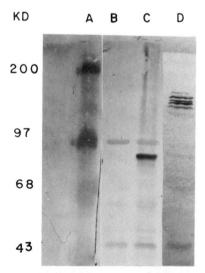

FIG. 2. Immunoblot of λ-50 and subclones. Lane A contains molecular weight markers: 200, 97, 68, and 43 kilodaltons. Lane B represents the immunoreactive protein profile of the 5.2-kb *Eco*RI-*Sst*I fragment cloned into pOP203(A$_2^+$). Lane C represents the profile of the 2.9-kb *Sst*I-*Eco*RI fragment cloned into pOP203(A$_2^+$). The immunoreactive polypeptide is approximately 90,000 daltons. Lane D represents the immunoreactive protein profile of λ-50 lysate. The immunoreactive proteins range from about 150,000 to 120,000 daltons.

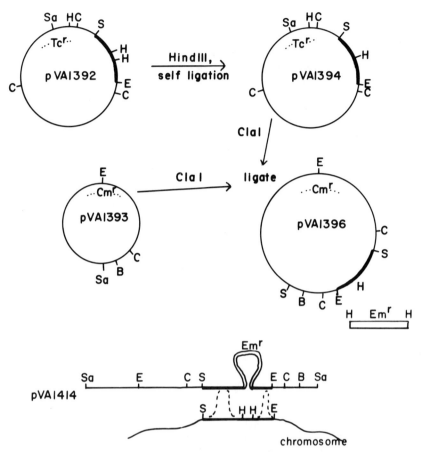

FIG. 3. Site-specific insertional mutagenesis of *gtfC*. Plasmid pVA1392 consists of *gtfC* *Sst*I-*Eco*RI cloned into pOP203(A$_2$+). This recombinant plasmid was digested with *Hin*dIII and self-ligated, which eliminated the 300-bp *Hin*dIII fragment. This plasmid, pVA1394, was digested with *Cla*I, and the fragment containing the *gtfC* DNA was subcloned into pVA1393, which is pACYC184 minus its *Hin*dIII site. This recombinant plasmid, pVA1396, was cleaved with *Hin*dIII, and an erythromycin resistance (Em^r) gene from pVA677 was cloned into the *Hin*dIII site. The recombinant plasmid, pVA1414, was then linearized with *Sal*I and transformed into *S. mutans*. Em^r transformants were selected. Abbreviations: B, *Bam*HI; C, *Cla*I; E, *Eco*RI; H, *Hin*dIII; S, *Sst*I; and Sa, *Sal*I. Dashed lines indicate crossovers at areas of homology.

Evidence for DNA Sequence Duplication

Analysis of the restriction map of the cloned *S. mutans* LM7 chromosomal DNA revealed that a sequence of five restriction endonuclease sites appeared to be repeated in both the 8.1- and 4.7-kb *Eco*RI fragments. The region containing these five sites is represented as a stippled box in Fig. 1. The five restriction sites are *Bam*HI, *Hin*dIII, *Hin*dIII, *Bam*HI, and *Eco*RI. This region appears to be contained entirely within the *gtfC* coding region. This led us to hypothesize that all or part of the *gtfC* gene was duplicated just downstream from the first copy. To further investigate this possibility, a *gtfC*-specific probe was obtained. The 300-bp fragment lying between the two *Hin*dIII sites of the repeated region (see stippled box in Fig. 1) was purified

by electroelution (8) and cloned into the unique *Hin*dIII site on *E. coli* plasmid pACYC184 (1). This recombinant plasmid, pVA1436, was nick-translated and used to probe chromosomal DNA from strains representing *S. mutans* serotypes a through g and *S. sanguis*, in a Southern blot DNA hybridization experiment. The results revealed hybridization to 8.1- and 4.7-kb fragments when the DNA was digested with *Eco*RI for *S. mutans* GS5 (serotype c), V403 (c), LM7 (e), and OMZ175 (f). No hybridization was evident with any of the other serotypes or with *S. sanguis*. These data indicated that the 300-bp *Hin*dIII fragment, presumably completely within the *gtfC* gene, was present within two *Eco*RI fragments of 8.1 and 4.7 kb in the chromosomes of *S. mutans* serotype c, e, and f strains.

Using the restriction endonuclease map of *gtfC*, a strategy was devised to create a site-specific mutation in vivo in *S. mutans* GS5, a transformable serotype c strain. This strategy is outlined in Fig. 3 and involved replacing the 300-bp *Hin*dIII fragment with an erythromycin (Em) resistance gene from pVA677. This recombinant plasmid, pVA1414, was then linearized and transformed into *S. mutans* GS5. Em^r transformants should have involved double-cross-over recombinational events which resulted in a strand exchange. *S. mutans* GS5 Em^r transformants were obtained, including a few with insertions in both repeated regions. Preliminary examination of these mutants indicated no differences in exopolysaccharide synthesis when compared with the wild type.

Conclusions

We have cloned a GTF gene, designated *gtfC*, whose gene product appeared to synthesize a water-soluble exopolysaccharide. The approximate position and direction of transcription was determined. The *gtfC* gene product was a polypeptide of 150,000 daltons. Restriction endonuclease mapping revealed a repeated region of DNA corresponding to the coding region of *gtfC*. This suggested that sequences of *gtfC* are duplicated downstream from the first copy. At present, it is not clear whether the entire gene is duplicated. A 300-bp *gtfC*-specific probe showed that the gene and the putative duplicated sequence were present in *S. mutans* serotypes c, e, and f. A strategy to create in vivo mutations in *gtfC* was detailed, and preliminary results indicated no detectable differences in polymer synthesis. Our results suggest that it is possible that at least one class of GTF genes may be present in multiple copies. This has implications for the control of glucan production and illustrates the potential complexity of exopolysaccharide synthesis mediated by *S. mutans*.

This work was supported by Public Health Service grant DE 04224 from the National Institute for Dental Research. M.J.P. was the recipient of National Research Service award DE 05371 from the National Institute of Dental Research.

We thank V. Goodall for assistance in the preparation of this manuscript.

LITERATURE CITED

1. **Chang, A. C. Y., and S. N. Cohen.** 1978. Construction and characterization of amplifiable multicopy deoxyribonucleic acid cloning vehicles derived from the P15A cryptic miniplasmid. J. Bacteriol. **134:**1144–1156.
2. **Clarke, L., and J. Carbon.** 1976. A colony bank containing synthetic ColE1 hybrid plasmids representative of the entire *E. coli* genome. Cell **9:**91–99.
3. **Curtiss, R., III.** 1985. Genetic analysis of *Streptococcus mutans* virulence. Curr. Top. Microbiol. Immunol. **118:**253–277.
4. **Gibbons, R. J., and J. van Houte.** 1975. Dental caries. Annu. Rev. Med. **26:**121–136.
5. **Gustafasson, B. E., C. F. Quensel, L. S. Lanke, C. Lundquist, H. Grahnen, B. F. Bonow, and B. Krasse.** 1954. The Vipeholm dental caries study. The effect of different levels of carbohydrate intake on caries activity in 436 individuals observed for five years. Acta. Odontol. Scand. **11:**232–364.
6. **Hamada, S., and H. D. Slade.** 1980. Mechanisms of adherence of *Streptococcus mutans* to smooth surfaces *in vitro*, p. 107–135. *In* E. H. Beachey (ed.), Bacterial adherence: receptors and recognition, ser. B, vol. 6. Chapman and Hall, London.
7. **Handrix, R., J. Roberts, F. Stahl, and R. Weisberg.** 1983. Lambda II. Cold Spring Harbor Laboratory, Cold Spring Harbor, N.Y.
8. **Maniatis, T., E. F. Fitsch, and J. Sambrook.** 1982. Molecular cloning: a laboratory manual, p. 164–165, 282–283. Cold Spring Harbor Laboratory, Cold Spring Harbor, N.Y.
9. **Pucci, M. J., and F. L. Macrina.** 1985. Cloned *gtfA* gene of *Streptococcus mutans* LM7 alters glucan synthesis in *Streptococcus sanguis.* Infect. Immun. **48:**704–712.
10. **Pucci, M. J., J. G. Tew, and F. L. Macrina.** 1986. Human serum antibody response against *Streptococcus mutans* antigens. Infect. Immun. **51:**600–606.
11. **Robeson, J. P., R. G. Barletta, and R. Curtiss III.** 1983. Expression of a *Streptococcus mutans* glucosyltransferase gene in *Escherichia coli.* J. Bacteriol. **153:**211–221.
12. **Rolla, G., O. J. Iverson, and P. Bonevoll.** 1978. Lipoteichoic acid—the key to adhesiveness of sucrose-grown *Streptococcus mutans.* Adv. Exp. Med. Biol. **107:**607–615.
13. **Silhavy, T. J., M. L. Berman, and L. W. Enquist.** 1984. Experiments with gene fusions. Cold Spring Harbor Laboratory, Cold Spring Harbor, N.Y.
14. **Tanzer, J. M.** 1979. Essential dependence of smooth surface caries on, and augmentation of, fissure caries by sucrose and *Streptococcus mutans* infection. Infect. Immun. **25:**526–531.
15. **Tanzer, J. M., M. L. Freedman, R. J. Fitzgerald, and R. H. Larson.** 1974. Diminished virulence of glucan synthesis-defective mutants of *Streptococcus mutans.* Infect. Immun. **10:**197–203.

Genetic Analysis of *Streptococcus mutans* Glucosyltransferases

H. K. KURAMITSU, T. SHIROZA, S. SATO, AND M. HAYAKAWA

Northwestern University Medical School, Chicago, Illinois 60611

A variety of approaches have suggested that one important cariogenic property of *Streptococcus mutans* strains is their ability to synthesize water-insoluble adhesive glucans (4). These polysaccharides are synthesized by the concerted action of multiple glucosyltransferases (GTFs) (2). The serotype g strain *Streptococcus sobrinus* 6715 produces three distinct GTFs: a primer-independent soluble glucan-synthesizing enzyme, another primer-dependent soluble glucan-synthesizing enzyme, and an insoluble glucan-synthesizing GTF (15). In contrast, only two GTFs have been demonstrated so far in serotype c strains of *Streptococcus mutans*. One of these, GTF-S, synthesizes soluble glucans (8) while the other enzyme, GTF-I, produces predominantly insoluble glucans (10, 12).

Mutants of *S. mutans* which are apparently defective in GTF-I activity have been isolated in several laboratories and are altered in sucrose-dependent colonization of smooth surfaces (4). However, in most cases it has not been possible to demonstrate that the phenotypic properties displayed by these mutants result from a single mutation in the GTF-I gene alone. In addition, mutants defective in GTF-S activities have not yet been described since selection procedures to identify such mutants are not available. Therefore, the ability to construct readily defined GTF mutants in *S. mutans* would greatly enhance our ability to determine the relative role of each GTF in the sucrose-dependent colonization of tooth surfaces.

In addition to the GTF-S and GTF-I enzymes possessing molecular weights ranging between 140,000 and 160,000, a lower-molecular-weight GTF produced by these organisms has been described (14). The gene coding for this later enzyme, *gtfA*, has been recently introduced into *S. sanguis* and increases the synthesis of insoluble glucans by these organisms (13). This observation has suggested that the *gtfA* product may also play a role in insoluble glucan synthesis in *S. mutans*. However, more recent results have indicated that insertional inactivation of the *gtfA* gene in strain GS-5 (serotype c) does not significantly affect glucan synthesis by these organisms (M. Pucci and F. L. Macrina, personal communication).

Despite the interest in the enzymology of the GTFs produced by *S. mutans*, little information is currently available regarding the regulation of GTF synthesis. The expression of GTF activity appears to be constitutive since sucrose is not required for production of the enzyme (6). However, when strain Ingbritt (serotype c) is grown in a chemostat, GTF levels can vary up to 10-fold depending on the growth rate of the organism (1). The molecular basis for such variation under steady-state growth conditions has not been explained up to now. In addition, it is not known whether the GTFs are coordinately regulated (constitute a single *gtf* operon).

One promising approach toward resolving the roles of each of these enzymes in colonization is that of cloning their respective genes. Several laboratories have recently reported the isolation of *gtf* genes from *S. mutans* (3, 5, 9). A clone, λDS-76, coding for GTF activity isolated from a strain GS-5 lambda phage clone bank has recently been characterized in this laboratory (H. Aoki et al., submitted for publication). The ability to isolate such a gene from transformable strain GS-5 should allow mapping of the gene relative to other cariogenic markers on the chromosome as well as the construction of defined GTF mutants following insertional inactivation. Such approaches should help to define more clearly the role of the gene products in cariogenicity.

Characterization of Homogeneous Cloned GTF Preparations

Previous results indicated that partially purified GTF preparations from the DS-76 clone synthesized both water-soluble and insoluble glucans (9). The GTF gene has been further subcloned from the original phage clone as a 7.7-kilobase *Cla*I fragment into vector pACYC-184 (Aoki et al., submitted). The enzyme has now been purified to homogeneity following DEAE BioGel-A, high-performance liquid chromatography (HPLC) gel filtration and DEAE-HPLC fractionation (Fig. 1). Two homogeneous proteins (145 and 135 kilodaltons [kDa]) expressing GTF activity have been isolated. Both proteins are somewhat smaller than the major 155-kDa GTF activity detected in culture fluids of strain GS-5 (Fig. 1). Since the GS-5 insert

A   B

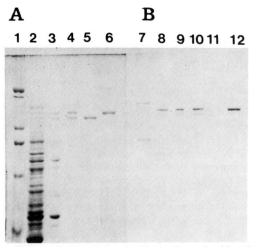

FIG. 1. Sodium dodecyl sulfate-polyacrylamide gel electrophoresis analysis of purified GTF from recombinant clones. (A) Coomassie blue staining of enzyme fractions. (B) Activity staining after periodic acid-Schiff staining. Lanes: 1, molecular weight standards (200,000, 116,000, 94,000); 2, MH76 crude extract; 3, DEAE-BioGel-A fraction; 4, HPLC gel filtration fraction; 5, DEAE-HPLC fraction A; 6, DEAE-HPLC fraction B; 7, GS-5 culture fluids; 8, MH76 crude extract; 9, DEAE-BioGel-A fraction; 10, HPLC gel filtration fraction; 11, DEAE-HPLC fraction A; 12, DEAE-HPLC fraction B.

contains only a single gtf gene (Aoki et al., submitted), it is likely that the cloned proteins represent processed products of the intact gtf gene.

Both purified enzyme fractions synthesize primarily insoluble glucans in the absence of exogenous dextran T10 (Table 1). The addition of dextran did not increase the formation of insoluble glucan but did significantly increase the levels of soluble glucan produced by each enzyme.

Expression of Cloned GTF Activity in Heterologous Oral Streptococci

Several different approaches have suggested that the DS-76 clone codes for an enzyme syn-

thesizing primarily insoluble glucan (GTF-I): the purified enzyme synthesizes 85 to 90% insoluble glucan (Table 1); the cloned enzyme reacts strongly with antibody prepared against a homogeneous serotype c GTF-I (12) provided by H. Mukasa (National Defense Medical College, Japan); and in vitro inactivation of the cloned gene followed by transformation of strain GS-5 with the altered gene results in a significant reduction in insoluble glucan synthesis (Aoki et al., submitted). To further confirm that DS-76 codes for GTF-I activity, the cloned gene was introduced into two other oral streptococci (*S. milleri* and *S. sanguis*) which do not normally synthesize insoluble glucans. Initially, the gtf gene was transferred on a 5.3-kilobase SphI fragment into the shuttle vector pVA856 (11) (Fig. 2). The resultant plasmid was then introduced into two transformable strains: *S. milleri* NCTC 10707 and *S. sanguis* V685.

After selection for erythromycin-resistant transformants, it was observed that many of the transformants synthesized insoluble glucan. Some of the transformants did not express cloned GTF activity and appeared to have undergone deletions in the entering chimeric shuttle plasmid. Neither *S. sanguis* nor *S. milleri* was capable of synthesizing significant levels of insoluble glucan before the introduction of pSS30 or after transformation with shuttle plasmid pVA856. In addition, the culture fluids of

TABLE 1. Glucan synthesis by GTF-I fractions from recombinant *E. coli*[a]

| Fraction | cpm/18 h | | | |
| | − Dextran T10 | | + Dextran T10 | |
	Insoluble	Soluble	Insoluble	Soluble
A (135 kDa)	1,227	220	1,195	1,179
B (145 kDa)	2,089	223	2,417	840

[a] Enzyme fractions A (20 μl) and B (10 μl) were assayed by utilizing the standard [^{14}C]glucose:sucrose assay system (4).

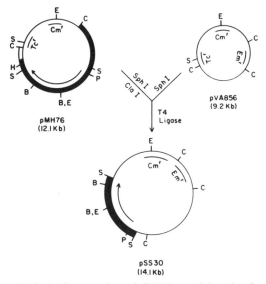

FIG. 2. Construction of GTF-I-containing shuttle plasmid pSS30. —, Vectors; □, portion of lambda 47.1 DNA; ■, GS-5 DNA insert. The arrows indicate the approximate location of the GTF-I gene and the direction of transcription. Only the relevant restriction sites are depicted: B, BamHI; C, ClaI; E, EcoRI; H, HindIII; S, SphI.

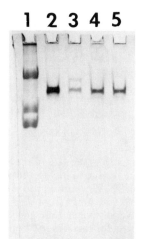

1 2 3 4 5

FIG. 3. Sodium dodecyl sulfate-polyacrylamide gel electrophoresis analysis of extracellular GTF activities expressed in transformed heterologous streptococci. The GTF activities were detected after incubation with sucrose and periodic acid-Schiff staining. Lanes: 1, Coomassie blue staining of molecular weight markers (200,000, 116,000, 92,500); 2, *S. milleri*(pSS30); 3, *S. sanguis* V685(pSS30); 4, *E. coli*(pSS30); 5, *E. coli* (pMH76).

such transformants displayed GTF activities which comigrated with the GTF of the original *Escherichia coli* subclone MH76 after sodium dodecyl sulfate-polyacrylamide gel electrophoresis (Fig. 3). These results further confirm that the cloned gene codes for GTF-I activity and additionally indicate that GS-5 extracellular proteins can be both expressed and secreted in *S. milleri* and *S. sanguis*. Since neither transformant displayed sucrose-dependent colonization of smooth surfaces, it is clear that GTF-I activity alone is not sufficient for this property. In addition, since *S. sanguis* V685 could synthesize water-soluble glucan prior to the introduction of the cloned GTF-I gene, additional *S. mutans* factors appear to be required for colonization in addition to the two types of GTF enzymes. It should now be possible to identify such factors by utilizing additional *S. mutans* cloned genes introduced into heterologous streptococci.

This investigation was supported by Public Health Service grant DE-03258 from the National Institute of Dental Research.

LITERATURE CITED

1. **Ellwood, D. C.** 1976. Chemostat studies of oral bacteria, p. 785–798. *In* H. M. Stiles, W. J. Loesche, and T. C. O'Brien (ed.), Proceedings: microbial aspects of dental caries (a special supplement to Microbiology Abstracts), vol. 3. Information Retrieval, Inc., Washington, D.C.
2. **Fukushima, K., R. Motoda, K. Takada, and T. Ikeda.** 1981. Resolution of *Streptococcus mutans* glucosyltransferase into two components essential to water-insoluble glucan synthesis. FEBS Lett. **128**:213–216.
3. **Gilpin, M. L., R. R. B. Russell, and P. Morrissey.** 1985. Cloning and expression of two *Streptococcus mutans* glucosyltransferases in *Escherichia coli* K-12. Infect. Immun. **49**:414–416.
4. **Hamada, S., and H. D. Slade.** 1980. Biology, immunology, and cariogenicity of *Streptococcus mutans*. Microbiol. Rev. **44**:331–384.
5. **Jacobs, W. R., J. F. Barrett, J. E. Clark-Curtiss, and R. Curtiss III.** 1986. In vivo repackaging of recombinant cosmid molecules for analysis of *Salmonella typhimurium*, *Streptococcus mutans*, and mycobacterial genomic libraries. Infect. Immun. **52**:101–109.
6. **Janda, W. M., and H. K. Kuramitsu.** 1978. Production of extracellular and cell-associated glucosyltransferase activity by *Streptococcus mutans* during growth on various carbon sources. Infect. Immun. **19**:116–122.
7. **Kuramitsu, H.** 1974. Characterization of cell-associated dextransucrase activity from glucose-grown cells of *Streptococcus mutans*. Infect. Immun. **10**:227–235.
8. **Kuramitsu, H. K.** 1975. Characterization of extracellular glucosyltransferase activity of *Streptococcus mutans*. Infect. Immun. **12**:738–749.
9. **Kuramitsu, H. K., and H. Aoki.** 1986. Isolation and manipulation of *Streptococcus mutans* glucosyltransferase genes, p. 199–204. *In* S. Hamada, S. Michalek, H. Kiyono, L. Menaker, and J. McGhee (ed.), Molecular microbiology and immunobiology of *Streptococcus mutans*. Elsevier Science Publishing, Inc., New York.
10. **Kuramitsu, H. K., and L. Wondrack.** 1983. Insoluble glucan synthesis by *Streptococcus mutans* serotype c strains. Infect. Immun. **42**:763–770.
11. **Macrina, F. L., R. P. Evans, J. A. Tobian, D. L. Hartley, D. B. Clewell, and K. R. Jones.** 1983. Novel shuttle plasmid vehicles for *Escherichia coli-Streptococcus* transgeneric cloning. Gene **25**:145–150.
12. **Mukasa, H., A. Tsumori, and A. Shimamura.** 1985. Isolation and characterization of an extracellular glucosyltransferase synthesizing insoluble glucan from *Streptococcus mutans* serotype c. Infect. Immun. **49**:790–796.
13. **Pucci, M. J., and F. L. Macrina.** 1985. Cloned *gtfA* gene of *Streptococcus mutans* LM-7 alters glucan synthesis by *Streptococcus sanguis*. Infect. Immun. **48**:704–712.
14. **Robeson, J. P., R. G. Barletta, and R. Curtiss III.** 1983. Expression of a *Streptococcus mutans* glucosyltransferase gene in *Escherichia coli*. J. Bacteriol. **153**:211–221.
15. **Shimamura, A., H. Tsumori, and H. Mukasa.** 1983. Three kinds of glucosyltransferases from *Streptococcus mutans* 6715 (serotype g). FEBS Lett. **157**:79–84.

Genetic Analysis of Surface Proteins Essential for Virulence of *Streptococcus sobrinus*

ROY CURTISS III, RAUL GOLDSCHMIDT, JOHN BARRETT, MARILYN THOREN-GORDON, DANIEL J. SALZBERG, HETTIE H. MURCHISON, AND SUZANNE MICHALEK

Department of Biology, Washington University, St. Louis, Missouri 63130

The *Streptococcus mutans* group of microorganisms comprises five genospecies, all of which have the capability of colonizing on the tooth surface to result in dental caries (2, 9). The initial attachment of these bacteria is sucrose independent and involves a weak interaction of proteins and lipoteichoic acid on the bacterial cell surface with salivary glycoproteins deposited on the tooth surface. This initial absorption is followed by a more tenacious attachment due to the synthesis of water-insoluble glucans by glucosyltransferases which, in conjunction with glucan-binding proteins, lead to cell-cell aggregation and the formation of dental plaque.

We have been particularly interested in the initial colonization events and have made use of derivatives of the serotype g strain 6715 of *Streptococcus sobrinus*. In these studies, we have employed a diversity of genetic, biochemical, and immunological approaches in conjunction with animal infectivity studies with mutant derivatives (6). We present here additional data pertaining to the surface protein antigen A (SpaA) and dextranase (Dex) and the nature of their relationship and likely involvement in sucrose-independent adherence to salivary glycoproteins forming the pellicle on the tooth surface.

Surface Protein Antigen A

Some years ago, we cloned the gene for the SpaA protein from *S. sobrinus* 6715, using cosmid cloning and immunological screening (10). The SpaA protein comprises approximately 35% of the protein on the cell surface of *S. sobrinus* and has an apparent molecular weight of 210,000. The pYA726 recombinant plasmid (Fig. 1) has a portion (see R. G. Holt and J. O. Ogundipe, this volume) of the SpaA gene cloned at the *Bam*HI site of the cloning vector pACYC184. If one labels *Bam*HI-digested pYA726 by nick translation (13) and uses it to hybridize against *Bam*HI-digested chromosomal DNA from representative serotypes of all members of the *S. mutans* group, one observes hybridization, even under conditions of low stringency, only to DNA from the serotype d and serotype g strains (Fig. 2). It should be noted that, in addition to hybridization to DNA

fragments larger than the size of the 8.4-kilobase insert presumably due to incomplete digestion of chromosomal DNA, we observed weak hybridization to DNA fragments from serotype d and g strains that were smaller than the 8.4-kilobase insert. The basis for this observation, which has been made by three members of our group, and the nature of the sequence hybridizing to the DNA segment encompassing the *spaA* gene are unknown. *Bam*HI-digested pACYC184 labeled by nick translation does not hybridize to DNA from any of the *S. mutans*, *S. sobrinus*, and *Streptococcus cricetus* strains (data not shown).

Escherichia coli strains with the pYA726 recombinant plasmid specify a protein of approximately 160,000 molecular weight termed SpaA1 (see Holt and Ogundipe, this volume). Antiserum raised against the SpaA1 protein purified from recombinant *E. coli* reacts with a protein antigen present in extracts of *S. sobrinus* serotype g and d strains, *S. mutans* serotype c, e, and f strains, and an *S. cricetus* serotype a strain (Fig. 3). No reaction was observed with serotype b *Streptococcus rattus* strains. It is evident that, although the DNA sequences encoding the SpaA protein in different serotypes were sufficiently divergent to preclude hybridization, some major antigenic determinant(s) was nevertheless conserved to cause a substantial degree of cross-reaction. The surface protein in the serotype c, e, and f strains that is immunologically related to the SpaA protein (4, 10) has been termed antigen I/II (17), antigen B (18) or antigen P1 (8).

Using antibodies to the *E. coli*-produced SpaA1 protein, we were able to isolate mutants of *S. sobrinus* that were deficient in SpaA protein on their surface (5). Of the original 17 mutants isolated, 16 were lacking in dextranase activity whereas one was still capable of hydrolysis of α-1,6 glucan. These mutants were used to infect gnotobiotic rats. It was observed that the mutants lacking the SpaA protein and dextranase were completely devoid of cariogenicity and were present on teeth at titers three logs lower than observed for the wild-type *S. sobrinus* parent strain (3). On the other hand, the mutant that lacked SpaA protein but still possessed dextranase had an intermediate level of

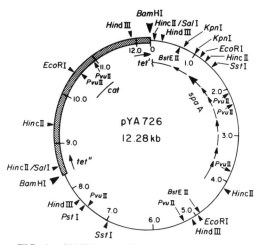

FIG. 1. pYA726 recombinant vector containing the *spaA1* segment of the *S. sobrinus spaA* gene cloned in the vector pACYC184.

cariogenicity and was more proficient at colonizing the tooth surfaces in these rats (3). These results led us to ask whether dextranase was derived from either the SpaA protein or a unique protein that was immunologically related to the SpaA protein.

Dextranase

By using agar media with an overlay of blue dextran, it was possible to isolate *S. sobrinus* mutants deficient in dextranase activity (Table 1). Most mutants were completely devoid of

dextranase activity, either extracellular or intracellular, although one mutant, UAB119, had a defect which prevented efficient translocation of dextranase from the cytoplasm to the cell surface. These Dex⁻ mutants often had reduced cariogenicity in gnotobiotic rats (Table 2), but in no case were they as avirulent as were the SpaA⁻ Dex⁻ mutants isolated by using inability to react with anti-SpaA serum. A number of the Dex⁻ mutants such as UAB108, UAB113, UAB122, UAB158, and UAB244 are defective in glucan synthesis and produce large quantities of a water-soluble glucan that inhibits plaque formation by members of the *S. mutans* group, both in vitro and in vivo (14, 19).

To further study dextranase, a method has been developed for incorporating blue dextran in sodium dodecyl sulfate-polyacrylamide gels and renaturing the proteins in the gels to visualize dextranase activity (Fig. 4) (1b). From this we learned that the highest molecular weight of dextranase was 175,000, with other forms at 160,000, 150,000, and lower. Although the detection of multiple sizes of dextranase could imply multiple genes, the ease with which Dex⁻ mutants are isolated suggests a single gene. Furthermore, we discovered proteases on the surface of *S. sobrinus* which can convert in vitro high-molecular-weight dextranase to enzymatically active lower-molecular-weight dextranases (1a). This suggested that these multiple molecular weight forms were probably due to proteolytic breakdown occurring on the *S. sobrinus* cell surface.

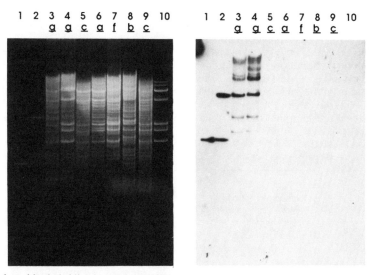

FIG. 2. Southern blot hybridization of *Bam*HI-digested nick-translated labeled pYA726 containing the *spaA1* gene to *Bam*HI-digested DNA from pACYC184 (lane 1), pYA726 (lane 2), *S. sobrinus* UAB66 (lane 3), *S. sobrinus* mutant UAB121 (lane 4), *S. mutans* serotype c strain V318 (lane 5), serotype a *S. cricetus* HS6 (lane 6), *S. mutans* serotype f strain OMZ175 (lane 7), *S. rattus* serotype b strain BHT (lane 8), serotype c *S. mutans* MT8148 (lane 9), and λ DNA (lane 10).

FIG. 3. Ouchterlony analysis placing antisera against the recombinant *E. coli*-specified SpaA1 protein in the center well and extracts of *S. sobrinus* serotypes d and g, *S. mutans* serotypes c, e, and f, and an *S. cricetus* serotype a strains in the surrounding wells.

TABLE 1. Dextranase activity of wild-type and Dex⁻ *S. sobrinus* mutants[a]

Strain	Phenotype	U/mg of protein	
		Extracellular	Intracellular
UAB66	Dex⁺	13.8	0.54
UAB107	Dex⁻	<0.01	<0.01
UAB108	Dex⁻	<0.01	<0.01
UAB113	Dex⁻	<0.01	<0.01
UAB119	Dex±	2.6	2.9
UAB122	Dex⁻	<0.01	<0.01
UAB158	Dex⁻	<0.01	<0.01
UAB244	Dex⁻	<0.01	<0.01

[a] Dextranase activity determined by measuring release of reducing sugar by hydrolysis of dextran (15).

Using sequential affinity chromatography with hemoglobin-Sepharose to remove proteases, followed by gel filtration, we were able to purify the 175,000- and 160,000-molecular-weight dextranases (Fig. 5) (1a). These purified proteins were then used to raise antisera in rabbits. These antisera, along with antisera against SpaA proteins produced by recombinant *E. coli* and *S. sobrinus* and against glucan-binding proteins, were evaluated for ability to immunoprecipitate dextranase activity in the presence of *Staphylococcus aureus* protein A (Table 3). It is evident that the antiserum against the 175,000-molecular-weight dextranase is most efficient in causing immunoprecipitation of dextranase activity (1). It is also evident that the antiserum against SpaA protein, whether made by recombinant *E. coli* or by *S. sobrinus*, is also capable of causing immunoprecipitation of dextranase activity. As shown in Fig. 6, dextranase activity can be recovered from the immunoprecipitates. These results indicate that the SpaA and dextranase proteins are immunologically relat-

ed and share at least one antigenic determinant (1).

More recently, we have successfully cloned the gene for dextranase from *S. sobrinus* (1a, 11). The important finding is that the DNA sequences encoding dextranase are unable to hybridize to a DNA probe containing the *S. sobrinus spaA* gene (Fig. 7). It is therefore evident that SpaA and dextranase are encoded by separate genes which, nevertheless, specify antigenic determinants that are related. We have recently learned that many of the recombinant clones expressing dextranase are unstable and lose certain DNA sequences so that the protein with dextranase activity in each clone has an apparent molecular weight of 130,000 to 145,000. Nevertheless, this derivative dextranase protein did not react with antiserum against the SpaA protein.

Discussion

Our results indicate that the SpaA and dextranase proteins are encoded by separate DNA sequences. It is also evident that the proteins share at least one important epitope and that only when both proteins are absent from the *S. sobrinus* cell surface does one observe total

TABLE 2. Mean caries scores of wild-type and Dex⁻ mutants[a]

Strain	Buccal score		Sulcal score		Proximal score		Mean no. of *S. sobrinus* CFU per mandible (×10⁶)
	Enamel	Dentinal slight	Dentinal slight	Dentinal extensive	Enamel	Dentinal slight	
UAB66	21.2 ± 0.5	17.8 ± 0.6	18.5 ± 0.6	10.4 ± 0.5	5.5 ± 0.5	4.9 ± 0.6	2.5
UAB107	11.6 ± 1.0	9.5 ± 0.5	12.9 ± 0.5	1.1 ± 0.3	0.8 ± 0.3	0.0	2.0
UAB108	8.8 ± 0.5	4.2 ± 0.4	12.2 ± 0.4	3.0 ± 0.3	4.4 ± 0.8	1.6 ± 0.8	1.7
UAB113	12.6 ± 0.5	8.4 ± 0.6	13.6 ± 0.2	3.6 ± 0.7	4.0 ± 0.6	0.4 ± 0.4	0.91
UAB119	9.5 ± 0.7	6.6 ± 0.6	11.2 ± 0.3	0.6 ± 0.2	4.8 ± 0.5	2.1 ± 0.4	ND[b]
UAB122	9.5 ± 0.4	7.8 ± 0.4	10.5 ± 0.5	2.9 ± 0.5	0.1 ± 0.1	0.0	0.82
UAB158	10.9 ± 0.6	8.8 ± 0.4	10.5 ± 0.3	2.2 ± 0.4	0.4 ± 0.3	0.1 ± 0.1	1.2

[a] Values are mean ± standard error of the mean as determined by the method of Keyes (12).
[b] Not determined.

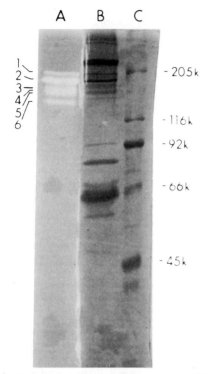

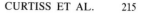

FIG. 4. Sodium dodecyl sulfate-polyacrylamide gel containing blue dextran used for electrophoresis of *S. sobrinus* cell surface and extracellular proteins stained with Coomassie blue (lane B) or renatured and allowed to hydrolyze blue dextran (lane A). Molecular weight markers are shown in lane C.

avirulence and inability to induce caries in the germfree rat. Since derivative recombinant *E. coli* clones expressing dextranase do not exhibit

TABLE 3. Immunoprecipitation of dextranase activities

Antiserum	Dextranase activity after immunoprecipitation (%)[a]
1. Preimmune	93
2. Glucan-bound protein..............	42
3. 160,000 dextranase	24
4. 175,000 dextranase	9
5. *S. sobrinus* SpaA.................	36
6. Extracellular protein..............	19
7. *E. coli* SpaA1....................	33

[a] Percent dextranase activity after immunoprecipitation with various antisera and *S. aureus* protein A was determined by assaying dextranase activity remaining in reaction mixtures compared with dextranase activity in an *S. sobrinus* extracellular protein fraction that had not been immunoprecipitated.

the cross-reactive determinants but retain dextranase activity, one can consider dextranase as a bifunctional protein. We currently favor the idea that one function of dextranase is concerned with its enzymatic activity to hydrolyze α-1,6 glucosyl linkages and that this activity is important in normal glucan synthesis (14, 19). The other domain of dextranase is cross-reactive with the SpaA protein and, in keeping with results that indicate a role for antigen I/II, B, or SpaA in sucrose-independent adherence to salivary glycoproteins (7, 16), would also serve to facilitate sucrose-independent adherence. We believe that SpaA and the antigenically similar portion of dextranase provide backup to each other in this important, initial colonization function. Experiments are in progress to evaluate these possibilities.

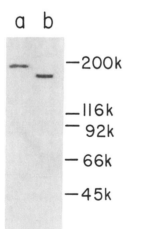

FIG. 5. Sodium dodecyl sulfate-polyacrylamide gel electrophoresis of purified 175,000 and 160,000 dextranases obtained from culture supernatant fluids of *S. sobrinus* (1a).

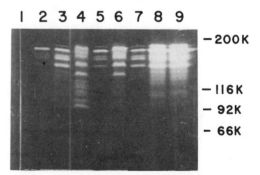

FIG. 6. Blue dextran-sodium dodecyl sulfate-polyacrylamide gel electrophoresis of the immunoprecipitates from the experiment presented in Table 3 analyzed after renaturation of proteins with the hydrolysis of blue dextran (1). Lanes 1 through 7 analyze the immunoprecipitates from the immunoprecipitation reactions 1 through 7 listed in Table 3. Lanes 8 and 9 contain nonimmunoprecipitated samples of extracellular protein from *S. sobrinus* culture supernatant fluids.

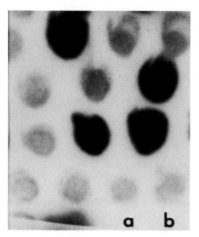

a   b

FIG. 7. Hybridization of DNA sequences to a DNA probe containing *spaA*. DNA samples from a number of cosmid recombinants expressing different *S. mutans* cell surface proteins were bound to nitrocellulose and filter hybridized with the 8.4-kilobase fragment containing the *spaA1* gene from pYA726 that had been nick translated. The spots marked a and b represent cosmid recombinants expressing dextranase, whereas the spots giving strong hybridization all specify SpaA protein.

We thank Christine Pearce and Margaret Buoncristiani for assistance in the preparation of this manuscript.

Research was supported by Public Health Service grants DE06673 and DE06801 from the National Institutes of Health.

LITERATURE CITED

1. **Barrett, J. F., T. A. Barrett, and R. Curtiss III.** 1986. Biochemistry and genetics of dextranase from *Streptococcus mutans* 6715, p. 205–215. *In* S. Hamada, S. M. Michalek, H. Kiyono, L. Menaker, and J. R. McGhee (ed.), Molecular microbiology and immunobiology of *Streptococcus mutans*. Elsevier Science Publishers, New York.

1a. **Barrett, J. F., T. A. Barrett, and R. Curtiss III.** 1987. Purification and partial characterization of the multicomponent dextranase complex of *Streptococcus sobrinus* and cloning of the dextranase gene. Infect. Immun. **55:**792–802.

1b. **Barrett, J. F., and R. Curtiss III.** 1986. Renaturation of dextranase activity from culture supernatant fluids of *Streptococcus sobrinus* after sodium dodecyl sulfate polyacrylamide gel electrophoresis. Anal. Biochem. **158:**365–370.

2. **Curtiss, R., III.** 1985. Genetic analysis of *Streptococcus mutans* virulence. Curr. Top. Microbiol. Immunol. **118:**253–277.

3. **Curtiss, R., III, R. Goldschmidt, R. Pastian, M. Lyons, S. M. Michalek, and J. Mestecky.** 1986. Cloning virulence determinants from *Streptococcus mutans* and the use of recombinant clones to construct bivalent oral vaccine strains to conver protective immunity against *S. mutans*-induced dental caries, p. 173–180. *In* S. Hamada, S. M. Michalek, H. Kiyono, L. Menaker, and J. R. McGhee (ed.), Molecular microbiology and immunobiology of *Streptococcus mutans*. Elsevier Science Publishers, New York.

4. **Curtiss, R., III, R. G. Holt, R. G. Barletta, J. P. Robeson, and S. Saito.** 1983. *Escherichia coli* strains producing *Streptococcus mutans* proteins responsible for colonization and virulence. Ann. N.Y. Acad. Sci. **409:**688–696.

5. **Curtiss, R., III, S. A. Larrimore, R. G. Holt, J. F. Barrett, R. Barletta, H. H. Murchison, S. M. Michalek, and S. Saito.** 1983. Analysis of *Streptococcus mutans* virulence attributes using recombinant DNA and immunological techniques, p. 95–104. *In* R. J. Doyle and J. E. Ciardi (ed.), Proceedings: glucosyltransferases, glucans, sucrose and dental caries (a special supplement to Chemical Senses). Information Retrieval, Inc., Washington, D.C.

6. **Curtiss, R., III, H. H. Murchison, W. E. Nesbitt, J. F. Barrett, and S. M. Michalek.** 1985. Use of mutants and gene cloning to identify and characterize colonization mechanisms of *Streptococcus mutans*, p. 187–193. *In* S. Mergenhagen and B. Rosan (ed.), Molecular basis for oral microbial adhesion. American Society for Microbiology, Washington, D.C.

7. **Douglas, C. W. I., and R. R. B. Russell.** 1985. Effect of specific antisera upon *Streptococcus mutans* adherence to saliva-coated hydroxyapatite. FEMS Microbiol. Lett. **25:**211–214.

8. **Forester, H., N. Hunter, and K. W. Knox.** 1983. Characteristics of a high molecular weight extracellular protein of *Streptococcus mutans*. J. Gen. Microbiol. **129:**2779–2788.

9. **Hamada, S., and H. D. Slade.** 1980. Biology, immunology, and cariogenicity of *Streptococcus mutans*. Microbiol. Rev. **44:**331–384.

10. **Holt, R. G., Y. Abiko, S. Saito, M. Smorawinska, J. B. Hansen, and R. Curtiss III.** 1982. *Streptococcus mutans* genes that code for extracellular proteins in *Escherichia coli* K-12. Infect. Immun. **38:**147–156.

11. **Jacobs, W. R., J. F. Barrett, J. E. Clark-Curtiss, and R. Curtiss III.** 1986. In vivo repackaging of recombinant cosmid molecules for analyses of *Salmonella typhimurium*, *Streptococcus mutans*, and mycobacterial genomic libraries. Infect. Immun. **52:**101–109.

12. **Keyes, P. H.** 1958. Dental caries in the molar teeth of rats. II. A method for diagnosing and scoring several types of lesions simultaneously. J. Dent. Res. **37:**1088–1099.

13. **Maniatis, T., E. F. Fritsch, and J. Sambrook.** 1982. Molecular cloning. A laboratory manual. Cold Spring Harbor Laboratory, Cold Spring Harbor, N.Y.

14. **Murchison, H., S. Larrimore, and R. Curtiss III.** 1985. In vitro inhibition of adherence of *Streptococcus mutans* strains by nonadherent mutants of *S. mutans* 6715. Infect. Immun. **50:**826–832.

15. **Nelson, N.** 1944. A photometric adaptation of the Somogyi method for the determination of glucose. J. Biol. Chem. **153:**375–380.

16. **Russell, M. W.** 1986. Protein antigens of *Streptococcus mutans*, p. 51–59. *In* S. Hamada, S. M. Michalek, H. Kiyono, L. Menaker, and J. R. McGhee (ed.), Molecular microbiology and immunology of *Streptococcus mutans*. Elsevier Science Publishers, New York.

17. **Russell, M. W., L. A. Bergmeier, E. D. Zanders, and T. Lehner.** 1980. Protein antigens of *Streptococcus mutans*: purification and properties of a double antigen and its protease-resistant component. Infect. Immun. **28:**486–493.

18. **Russell, R. R. B.** 1979. Wall associated protein antigens of *Streptococcus mutans*. J. Gen. Microbiol. **114:**109–115.

19. **Takada, K., T. Shiota, R. Curtiss III, and S. M. Michalek.** 1985. Inhibition of plaque and caries formation by a glucan produced by *Streptococcus mutans* mutant UAB108. Infect. Immun. **50:**833–843.

Molecular Cloning in *Escherichia coli* of the Gene for a *Streptococcus sobrinus* Surface Protein Containing Two Antigenic Determinants

ROBERT G. HOLT AND JOHN O. OGUNDIPE

Department of Microbiology, Meharry Medical College, Nashville, Tennessee 37208

Streptococcus sobrinus organisms are nonmotile, catalase-negative bacteria found primarily in the oral cavity. This group of gram-positive bacteria, as well as other related species, has been associated with the etiology of dental caries in humans and in experimental animals (4). The virulence of these organisms has been associated with their ability to colonize tooth surfaces and to produce acid from the metabolism of sucrose (4). In recent years, a number of genes for enzymes and other extracellular proteins, including (i) sucrose-metabolizing enzymes (1, 3, 5, 10), (ii) fructosyltransferase (14), (iii) glucan-binding protein (12), and (iv) surface protein antigens (8), that might be involved in the cariogenicity of the organism have been cloned by using recombinant DNA techniques. The surface protein antigens have been prominent in studies involved with the development of an anticaries vaccine (9, 13). Previously, Holt et al. (8) reported the cloning of a gene, designated *spaA,* for an *S. sobrinus* surface protein antigen that lacked at least one antigenic determinant and differed in molecular size from the protein produced in *S. sobrinus.* The product of the cloned gene had a molecular weight of 180,000, a size smaller than the 210,000 molecular weight determined for the *S. sobrinus* protein. At that time, explanations for the differences in electrophoretic mobility and lack of the antigenic determinant between the two proteins were (i) that the *spaA* gene contained on the previously cloned 8.5-kilobase (kb) *Bam*HI fragment was slightly truncated or (ii) that *S. sobrinus* and *Escherichia coli* differed in modification of the *spaA* gene product during or after the translation process. To determine the accuracy of the former explanation, we used here a cloning strategy that increased the probability of the formation of chimeric molecules having large inserts of contiguous stretches of *S. sobrinus* DNA. This allowed us to isolate recombinant plasmids possessing the DNA sequences that flanked the *spaA* gene. Our data presented below support the conclusion that the entire coding sequence for the *S. sobrinus spaA* gene was not cloned previously.

Cosmid Cloning: Identification and Characterization of Recombinant Clones Expressing SpaA Protein Determinants

Cosmid cloning is a system that allows the construction of gene banks of recombinant molecules containing relatively large pieces of contiguous DNA (2). Since we wanted to generate recombinant molecules possessing this trait, a gene bank of *S. sobrinus* 6715 genomic DNA was constructed by using the cosmid system. We performed controlled digestions of *S. sobrinus* chromosomal DNA with the restriction endonuclease *Bam*HI to generate DNA fragments 15 kb or larger in size. The partially digested DNA was further size fractionated by sucrose gradient centrifugation. Fragments of about 20 to 30 kb in size as determined by agarose gel electrophoresis were used in the gene bank construction. To prevent self-ligation of the vector molecules, the cosmid vector pJC74 was digested to completion with *Bam*HI restriction endonuclease and treated with calf intestinal alkaline phosphatase. This step was necessary since this cosmid vector is efficiently packaged in vitro without inserted DNA. Using the approach outlined above, 23 of 1,500 ampicillin-resistant recombinants examined were positive for the production of *S. sobrinus* SpaA protein antigenic determinants when screened by a horseradish peroxidase immunoassay (6). Immunodiffusion analysis of cell-free extracts of these clones and a *S. sobrinus* extracellular protein fraction with a polyclonal antiserum against the *S. sobrinus* SpaA protein showed that four of the extracts contained a protein that formed precipitin lines of complete identity with the *S. sobrinus* SpaA protein. Restriction map analyses of plasmid DNA and product analyses of each of the positive clones allowed us to localize to two *Bam*HI DNA fragments the region encoding the additional antigenic determinant found on the *S. sobrinus* SpaA protein. These fragments, 0.8 and 0.5 kb in size, were contiguous with 8.5-kb *Bam*HI fragment contained in plasmid pYA726 (8). A plasmid, pXI302, containing a DNA insert that encompassed the region having the *Bam*HI fragments

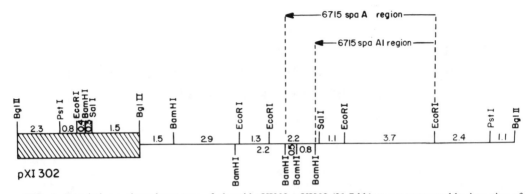

FIG. 1. Restriction endonuclease map of plasmid pXI302. pXI302 (21.7 kb) was constructed by insertion of *S. sobrinus* DNA that was originally contained in the chimeric cosmid pXI300 into the *Bgl*II restriction sites of cosmid pHC79. Cross-hatched area, DNA from pHC79; solid line, DNA from *S. sobrinus*. Distances between sites are given in kilobases.

was constructed by insertion of a 16.2-kb *Bgl*II fragment into the *Bgl*II sites of the cosmid vector pHC79 (7). Figure 1 shows a restriction map of plasmid pXI302 and also shows the approximate coding region of the *S. sobrinus* *spaA* gene.

Characterization of the Product of the Cloned *S. sobrinus spaA* Gene

We have used a variety of immunological techniques and sodium dodecyl sulfate-polyacrylamide gel electrophoresis to demonstrate that the cloned SpaA protein is identical immunologically and in molecular size to the SpaA protein produced by *S. sobrinus* cells. Immunodiffusion analysis of an extract of an *E. coli* strain harboring plasmid pXI302 and an *S. sobrinus* extracellular protein fraction with antibody against the *S. sobrinus* SpaA protein showed continuous precipitin lines, demonstrating the immunological identity between the two protein products (Fig. 2). Furthermore, tandem crossed immunoelectrophoresis of the two fractions also revealed the immunological identity of the SpaA proteins (data not shown).

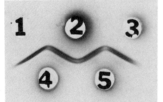

FIG. 2. Immunodiffusion analysis of the reaction between various protein fractions and antisera against *S. sobrinus* SpaA protein. The wells contained the following: 1, purified *S. sobrinus* SpaA protein; 2, a protein extract of HB101(pXI302); 3, a *S. sobrinus* extracellular protein fraction; and 4 and 5, *S. sobrinus* anti-SpaA protein sera.

Sodium dodecyl sulfate-polyacrylamide gel electrophoresis of *E. coli* extracts containing the cloned SpaA protein showed a protein band with a molecular mass of about 210,000 daltons (Fig. 3, lane 2) which migrated with the same mobility as the *S. sobrinus* SpaA protein (Fig. 3, lane 1). This band was absent in an extract prepared from cells containing the plasmid vector (Fig. 3, lane 4). We have also obtained data that suggest that *E. coli* cells harboring the *spaA* gene produced, in addition to the *spaA* gene product, a product that lacked an antigenic determinant. This product is antigenically similar to the truncated gene product produced by the incomplete *spaA* gene previously described (8) and is now referred to as *spaA1*. The presence of the *spaA1*-like gene product also can be demonstrated in the extracellular protein fraction of *S. sobrinus*. Reactions of protein from both sources yielded two closely spaced precipitin bands that exhibit immunological identity (Fig. 2). Using extracts of *E. coli* cells producing the SpaA1 product, we have been able to show that one of two precipitin bands resulted from the reaction of the SpaA1 determinant with *S. sobrinus* SpaA protein antibodies. Presently, we do not know whether these findings resulted from proteolytic breakdown of the SpaA protein in *E. coli* or whether this is a true reflection of how the *spaA* locus is expressed in *E. coli*. However, Russell et al. found that the antigen I/II protein of *S. mutans* strains, which is immunologically related to the SpaA protein, can also exist in a form lacking an antigenic determinant (11).

Conclusions

Evidence in this paper supports the conclusion that the entire coding sequence for the *spaA* gene of *S. sobrinus* has been cloned in *E. coli*. This work necessitates that the gene described in this report be genetically designated *spaA*

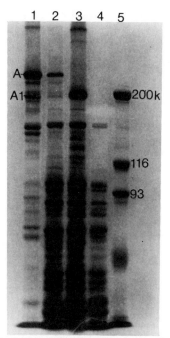

FIG. 3. Sodium dodecyl sulfate-polyacrylamide gel electrophoresis of various protein fractions. A 150-μg amount of protein was loaded in each lane as follows. Lanes: 1, an *S. sobrinus* extracellular protein fraction; 2, a protein extract of HB101 (pXI302); 3, a protein extract of HB101 harboring a pBR322-derived plasmid containing an 8.5-kb *Bam*HI fragment of *S. sobrinus* DNA encoding the SpaA1 protein; 4, a protein extract of HB101 (pHC79); and 5, molecular weight standards. A and A1 indicate protein bands corresponding to the SpaA and SpaA1 proteins, respectively.

since it probably contains the total wild-type coding sequence for the gene. Since the gene described previously and labeled *spaA* contains a deletion of its carboxyl-terminal sequences, we have designated this shortened form of the *spaA* gene as *spaA1* (Fig. 1). Now that the entire *spaA* gene has been cloned, additional work on the genetics of this gene and on the role of the SpaA protein in the caries process can be pursued. In addition, the cloning of the complete *spaA* gene in *E. coli* should allow the production of large quantities of extremely pure antigen for studies involving the production of an effective immunity against dental caries.

This work was supported by Public Health Service grant SO6RR08037-15 from the National Institutes of Health. J.O.O. was supported by the Vivian Beaumont Allen Fund.

Carol McClendon is gratefully acknowledged for typing of the manuscript.

LITERATURE CITED

1. **Aoki, H., T. Shiroza, M. Hayakawa, S. Sato, and H. K. Kuramitsu.** 1986. Cloning of a *Streptococcus mutans* glucosyltransferase coding for insoluble glucan synthesis. Infect. Immun. **53:**587–594.
2. **Collin, J.** 1979. *Escherichia coli* plasmids packageable *in-vitro* in bacteriophage particles. Methods Enzymol. **68:** 309–326.
3. **Gilpin, M. L., R. R. B. Russell, and P. Morrissey.** 1985. Cloning and expression of two *Streptococcus mutans* glucosyltransferases in *Escherichia coli*. Infect. Immun. **49:**414–416.
4. **Hamada, S., and H. D. Slade.** 1980. Biology, immunology, and cariogenicity of *Streptococcus mutans*. Microbiol. Rev. **44:**331–384.
5. **Hayakawa, M., H. Aoki, and H. K. Kuramitsu.** 1986. Isolation and characterization of the sucrose 6-phosphate hydrolase gene from *Streptococcus mutans*. Infect. Immun. **53:**582–586.
6. **Helfman, D. M., J. R. Feramisco, J. C. Fiddes, G. P. Thomas, and S. H. Hughes.** 1983. Identification of clones that encodes chicken tropomyosin by direct immunological screening of a cDNA expression library. Proc. Natl. Acad. Sci. USA **80:**31–35.
7. **Hohn, B., and J. Collins.** 1980. A small cosmid for efficient cloning of large DNA fragments. Gene **11:**291–298.
8. **Holt, R. G., Y. Abiko, S. Saito, M. Smorawinska, J. B. Hansen, and R. Curtiss III.** 1982. *Streptococcus mutans* genes that code for extracellular proteins in *Escherichia coli* K-12. Infect. Immun. **38:**147–156.
9. **Lehner, T., M. W. Russell, J. Caldwell, and R. Smith.** 1981. Immunization with purified protein antigens from *Streptococcus mutans* against dental caries in rhesus monkeys. Infect. Immun. **34:**407–415.
10. **Robeson, J. P., R. G. Barletta, and R. Curtiss III.** 1983. Expression of a *Streptococcus mutans* glucosyltransferase gene in *Escherichia coli*. J. Bacteriol. **153:**211–221.
11. **Russell, M. W., E. D. Zanders, L. A. Bergmeier, and T. Lehner.** 1980. Affinity purification and characterization of protease-susceptible antigen I of *Streptococcus mutans*. Infect. Immun. **29:**999–1006.
12. **Russell, R. R. B., D. Coleman, and G. Dougan.** 1985. Expression of a gene for glucan-binding protein from *Streptococcus mutans* in *Escherichia coli*. J. Gen. Microbiol. **131:**295–299.
13. **Russell, R. R. B., S. L. Peach, G. Coleman, and B. Cohen.** 1983. Antibody responses to antigens of *Streptococcus mutans* in monkeys (*Macaca fascicularis*) immunized against dental caries. J. Gen. Microbiol. **129:**865–875.
14. **Sato, S., and H. K. Kuramitsu.** 1986. Isolation and characterization of a fructosyltransferase gene from *Streptococcus mutans* GS-5. Infect. Immun. **52:**166–170.

Cloning and Partial Characterization of a β-D-Fructosidase of *Streptococcus mutans* GS-5

R. A. BURNE, K. SCHILLING, W. H. BOWEN, and R. E. YASBIN

Cariology Center and Department of Microbiology and Immunology, University of Rochester School of Medicine and Dentistry, Rochester, New York 14642

Many oral streptococci have the ability to use sucrose as a substrate to produce β(2→6)-linked (levan) and β(2→1)-linked (inulin) polymers of D-fructose through the action of fructosyltransferases (FTFs) (8). FTF can represent a significant portion of the total extracellular polymer synthetic activity, depending on the species and the growth conditions. Rolla and co-workers (15) have demonstrated the presence of FTF activity in newly formed pellicle on the tooth surface, while Gold et al. (7) and Higuchi et al. (9) have shown that fructan polymers accumulate in vivo in plaque of human subjects fed a diet rich in sucrose. Studies on the physical and rheological properties of the fructans produced by the oral streptococci indicate that these polymers may not diffuse from plaque and, consequently, they could serve as a storage polysaccharide (5). Thus, the ability to hydrolyze fructan polymers would provide an organism with an additional source of fermentable carbohydrate when extraneous sources are lacking.

Levan hydrolase activity in plaque of human volunteers has been described (12), and DaCosta and Gibbons (4) have identified levan hydrolase activity in culture supernatants from human plaque streptococci. This activity(s) has been shown to be inducible by substrate and catabolite repressed in the presence of glucose (4, 10, 19). Additionally, this activity(s) is produced maximally in continuous culture at low growth rates with fructose as the carbohydrate source (19). This type of regulation would be consistent with the role of fructans as storage polysaccharides (12). Presumably, during a sucrose intake, fructans would accumulate in plaque. Fructan hydrolase activity would be repressed by the presence of glucose, but at a later time when glucose was exhausted, fructanase would be induced, providing the cell with a new source of fermentable hexoses. Utilization of fructans in this manner would have the effect of prolonging the exposure of the tooth surface to organic acids and, therefore, fructanase could contribute to the virulence of the organism.

Investigations of *Streptococcus mutans* virulence determinants have focused primarily on adherence and glucan polymers. In contrast, little attention has been devoted to the role of fructans in the initiation and progression of dental disease. Recombinant DNA technology has been utilized effectively to study the proposed virulence factors of oral streptococci in this laboratory (1) and others (3). In this report, the cloning of a β-D-fructosidase (fructanase) gene from *S. mutans* GS-5 and a preliminary description of the enzyme are presented.

(This report is in partial fulfillment of the requirements for the Ph.D. degree, University of Rochester, for R.A.B.)

Cloning and Selection of the Fructanase

S. mutans GS-5 chromosomal DNA was partially digested with *Sau*3A1, and 4- to 6-kilobase-pair fragments were enriched for by sucrose gradient centrifugation as described (11). The DNA fragments thus generated were cloned into a unique *Bam*HI site in the plasmid vector pMK5. This vector is a chimera of pUC9 (18) and pCl94 (6) which replicates only in *Escherichia coli*, conferring ampicillin resistance (Ap^r) and chloramphenicol resistance (Cm^r). Cloning into the *Bam*HI site destroys β-galactosidase activity, and colonies containing recombinant plasmids appear colorless on MacConkey agar containing lactose and ampicillin (30 μg/ml). Five thousand recombinant clones with an average insert size of 4.5 kilobase pairs were gridded onto a minimal medium containing ampicillin (30 μg/ml) and sucrose as the sole carbohydrate source. After 48 h of incubation, nine clones grew on this medium. Toluenized cells from an overnight culture of each clone were tested for the ability to hydrolyze sucrose. The clone pFRU1, which contains the 4.1-kbp fragment of *S. mutans* GS-5 DNA shown in Fig. 1, was chosen for further study because cells harboring this recombinant plasmid yielded the most sucrolytic activity. Interestingly, mutants of *E. coli* defective in the ability to utilize galactose transformed with pFRU1 were able to grow with raffinose as the sole carbohydrate source as described for the *gtfA* gene of *S. mutants* PS14 (14).

Biochemical Characteristics

Sonic lysates of overnight cultures of *E. coli* JM83 carrying pFRU1 were centrifuged at

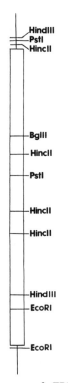

HindIII
PstI
HincII

BglII

HincII

PstI

HincII

HincII

HindIII
EcoRI

EcoRI

FIG. 1. Restriction map of pFRUI, which contains a 4.1-kilobase-pair insert of *S. mutans* GS-5 DNA in the plasmid vector pMK5.

100,000 × *g* for 30 min, and the supernatant fluid (100,000 × *g* soluble fraction) was used for enzymatic assays. When this fraction was incubated with 100 mM sucrose in 50 mM potassium phosphate buffer (pH 6.5), equimolar quantities of glucose and fructose were liberated, as was determined by biochemical means as well as high-pressure liquid chromatography using a Bio-Rad HPX-87H column. There have been reports of two invertase activities from the Bratthall serotype c *S. mutans*: an extracellular invertase (1a), which has never been purified, and an intracellular invertase which has optimal activity on sucrose 6-phosphate (2). When 100,000 × *g* soluble fractions prepared as above were tested for the ability to hydrolyze sucrose 6-phosphate (a gift from B. Chassy, National Institute of Dental Research), no activity was detected.

Prompted by the aforementioned reports of levan hydrolase activity and a report of an exo-β-D-fructosidase from *S. salivarius* (17) that hydrolyzed levan, inulin, sucrose, and raffinose, 100,000 × *g* soluble fractions were assayed using a number of different substrates (Table 1). Under the conditions tested (substrate excess), the enzyme was most active on the levan prepared

from *Aerobacter levanicum* (Sigma), which is predominantly β(2→6)-linked fructans, but contains a significant proportion of β(2→1) linkages (P. Z. Allen, personal communication). The enzyme was able to hydrolyze the β(2→1)-linked fructan, inulin, with approximately 34% of the activity observed for levan. In each instance free fructose was released as determined by high-performance liquid chromatography. In a complete digest, virtually all of the levan or inulin was hydrolyzed to free fructose. Sucrose and raffinose could serve as substrates but were hydrolyzed with only 21 or 12%, respectively, of the activity observed on levan. Under all conditions tested, the enzyme had no detectable activity on α-glucosides, β-glucosides, lactose, or melibiose. The substrate specificities of the cloned gene product indicated that it is a β-D-fructosidase.

pH optimum

The pH optimum for the fructanase from the serotype c *S. mutans* described by Walker et al. (19) was between 5.5 and 6.0, and the enzyme was not active below pH 4.5, suggesting that, at relatively low plaque pH values, the enzyme, even if induced, could not hydrolyze fructans present in the plaque matrix. The pH optimum of the cloned gene product was determined using a potassium phosphate or potassium citrate buffer system (Fig. 2). It was found that the optimum pH for the enzyme was 5.5, consistent with that previously observed (19), but approximately 40% of the activity still remained at a pH as low as 4.0. The enzyme was 50% less active at a pH of 7.0 or above. Interestingly, the pH optimum for activity on sucrose is a full pH unit lower (pH 4.5) than the pH optimum for levan (R. A. Burne, manuscript in preparation), sug-

TABLE 1. Substrate specificities[a]

Substrate	Relative activity
Levan	1.00
Inulin	0.34
Sucrose	0.21
Raffinose	0.12
Dextran	ND[b]
Mutan	ND
Starch	ND
Trehalose	ND
Melibiose	ND
Lactose	ND
Melizitose	ND

[a] All assays were carried out in 50 mM potassium citrate buffer at pH 5.5 at 37°C. All substrates were present in excess. In the case of polymer, the final concentration was 2.0 mg/ml. All sugars were at a final concentration of 50 mM. In no case was product inhibition observed.

[b] ND, None detected.

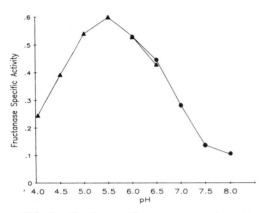

FIG. 2. pH optimum. All assays were carried out in 50 mM potassium citrate (▲) or 50 mM potassium phosphate (●) at the pH indicated for 1 h at 37°C. The reaction was terminated by immersing the tubes in boiling water for 5 min, after which no residual activity remained. One unit of activity was defined as the amount of enzyme necessary to liberate 1 μmol of reducing power in 1 min, with fructose as a standard.

gesting that the enzyme may play a significant role in sucrose dissimilation at very low plaque pH.

Physical Characteristics

The question arises whether a single protein is produced from the cloned DNA fragment which is capable of degrading β(2→6) linkages as well as β(2→1) inulin-type fructans or whether two genes have been cloned which encode polypeptides with specificity for either levan or inulin. To address this issue, 100,000 × g soluble fractions from E. coli JM83 harboring pFRU1 were applied to a Bio-Gel A1.5m gel filtration column, and fractions with activity were concentrated by ammonium sulfate precipitation, dialyzed, and applied to a Bio-Gel DEAE ion-exchange column. Levan, inulin, sucrose, and raffinose hydrolase activity chromatographed in a single peak. Reapplication of fractions from ion exchange onto a Bio-Gel A1.5m column yielded a single peak containing all activities. When this sample was electrophoresed in sodium dodecyl sulfate-polyacrylamide gels, breakdown of the protein was observed. High-performance liquid chromatography has since been used to purify the cloned gene product as well as the β-D-fructosidase from culture supernatant fluids of S. mutans GS-5 (Burne, in preparation). Both the cloned gene product and the S. mutans enzyme purify as a single polypeptide with an approximate molecular weight of 140,000. Levan, inulin, sucrose, and raffinose hydrolase activity are associated with this polypeptide. These data

indicate that S. mutans can synthesize one enzyme, which has previously been shown to be subjected to several levels of genetic control, that can break down both types of fructan polymers synthesized by organisms commonly found in the human mouth.

Conclusions

The cloning and expression of the S. mutans β-D-fructofuranosidase has been demonstrated. This clone allows E. coli JM83 to grow with sucrose as the sole carbohydrate source despite the fact that E. coli JM83 has no sucrose transport system. No enzyme can be detected extracellularly when expressed in E. coli, whereas >95% of the activity produced by S. mutans GS-5 is extracellular (Burne, unpublished data). Growth of JM83 on the minimal sucrose medium probably arises from release of enzyme due to spontaneous lysis of cells as well as cleavage of sucrose by enzyme which has been extruded into the periplasmic space.

The substrate specificity indicated that the enzyme is most active on β(2→6)-linked fructans, which is rather unusual because the serotype c organisms do not produce fructans with this linkage; they make exclusively inulin-type fructans. Other organisms in the mouth such as S. salivarius make large quantities of FTF, and this FTF synthesizes predominantly β(2→6)-linked fructans (13). S. salivarius is found in large numbers on the tongue and in saliva (8) and could possibly serve as a reservoir for FTF which could synthesize levan. S. salivarius FTF may also constitute a significant portion of the total FTF activity which has been found in salivary pellicle (15), although this has not been determined. It appears, though, that S. mutans is able to hydrolyze the predominant types of fructans found in the human mouth.

The description of an extracellular invertase of S. mutans by Chassy et al. (1a) demonstrated that this organism synthesizes an enzyme which could be classified as a β-fructosidase. To date, an extracellular invertase has not been purified from S. mutans. It is quite reasonable to speculate that the activity described in this report could represent the extracellular invertase, the preferred substrate of which is fructan polymer.

The optimum pH for this enzyme was 5.5, and activity remained at a pH as low as 4.0. This is in conflict with previous reports that the enzyme was inactive below pH 4.5. One concern is that the cloned gene product is not modified, or is modified in such a way as to alter the optimal pH for activity. Preliminary results from this laboratory suggest that this is not the case because the fructanase from S. mutans GS-5 has the same pH optimum profile as the cloned gene product. A more likely explanation, therefore, for the difference in pH optimum in this report is differences between the serotype c strains that were examined.

The cloned gene product described here is very similar, from the standpoint of substrate specificities, to the exo-β-D-fructosidase described for *S. salivarius* (17). Antibody has been raised to the cloned gene product, and a nucleic acid probe is now available. It will now be possible to determine whether there is antigenic relatedness or DNA sequence homology between the two gene products and their genes. The molecular weight of the *S. mutans* enzyme is significantly greater than that reported for the *S. salivarius* enzyme. There is preliminary evidence that the cloned *S. mutans* enzyme has the ability to bind glucans (Burne, in preparation), which would be consistent with its proposed role in the plaque environment. The enzyme would bind glucans and remain in plaque, and could then act on fructans when conditions were appropriate. Recently, R. R. B. Russell and M. L. Gilpin (this volume) proposed that there may be a conserved glucan-binding domain in the polymer metabolism enzymes of oral streptococci. It will be interesting to determine whether such a domain exists in the *S. mutans* fructanase and whether the lack of such a domain could account for the lower molecular weight of the *S. salivarius* enzyme.

The genetic regulation of the *S. mutans* fructanase is a point of major interest, and this gene contains a number of features which make it very attractive as a model for studying gene regulation in the oral streptococci. It appears from work by DaCosta and Gibbons (4) and Walker et al. (19), and from preliminary results from this laboratory, that the enzyme is induced by fructose or some cleavage product of a fructan and that the gene is repressed in the presence of glucose. In addition to this level of regulation, the growth rate of the organism plays a profound role in the level of expression of the gene (19). Having cloned the gene, it will now be possible to study the regulation of the gene in detail. Two strategies are being used to explore the regulation of this gene: (i) Northern hybridization and (2) the construction of gene fusions for introduction, by recombination, into the *S. mutans* chromosome. Analysis of the promoter sequence should contribute to the understanding of regulation in streptococci and will also facilitate the comparison of catabolite-repressed promoters from *S. mutans* with catabolite-sensitive promoters from more thoroughly characterized gram-positive organisms, such as *Bacillus subtilis*.

Finally, the role of fructans in the initiation and progression of dental disease is poorly de-

fined. Using strategies for the inactivation of chromosomal genes in oral streptococci, it will be possible to specifically inactivate the fructanase gene of *S. mutans*. Once accomplished, the cariogenic potential of these strains can be assessed in vivo in animal models, and the relative contribution of the fructanase to virulence can be evaluated.

We thank Bonnee Rubinfeld for excellent technical assistance and Rob Quivey and John Keily for providing the mutan used in this study.

This investigation was supported by Public Health Service grant 1 P50 DEO 7003-01 from the National Institute for Dental Research.

LITERATURE CITED

1. **Burne, R. A., B. Rubinfeld, W. H. Bowen, and R. E. Yasbin.** 1986.Tight genetic linkage of a glucosyltransferase and dextranase of *Streptococcus mutans* GS-5. J. Dent. Res. **65:**1392–1401.
1a.**Chassy, B. M., R. M. Bielawski, J. R. Beall, E. V. Porter, M. I. Krichevsky, and J. A. Donkersloot.** 1974. Extracellular invertase in *Streptococcus mutans* and *Streptococcus salivarius*. Life Sci. **15:**1173–1180.
2. **Chassy, B. M., and E. V. Porter.** 1979. Initial characterization of sucrose-6-phosphate hydrolase from *Streptococcus mutans* and its apparent identity with intracellular invertase. Biochim. Biophys. Acta **89:**307–314.
3. **Curtiss, R., III.** 1985. Genetic analysis of *Streptococcus mutans* virulence. Curr. Top. Microbiol. **118:**253–277.
4. **DaCosta, T., and R. J. Gibbons.** 1968. Hydrolysis of levan by human plaque streptococci. Arch. Oral Biol. **13:**609–617.
5. **Ehrlich, J., S. S. Stivala, W. S. Bahary, S. K. Garg, L. W. Long, and E. Newbrun.** 1975. Levans: fractionation, solution viscosity, and chemical analysis of levans produced by *Streptococcus salivarius*. J. Dent. Res. **54:**290–297.
6. **Ehrlich, S. D.** 1977. Replication and expression of plasmids from *Staphylococcus aureus* in *Bacillus subtilis*. Proc. Natl. Acad. Sci. USA **75:**3664–3668.
7. **Gold, W., F. B. Preston, M. C. Lache, and H. Blechman.** 1974. Production of levan and dextran in plaque *in vivo*. J. Dent. Res. **53:**442–449.
8. **Hamada, S., and H. D. Slade.** 1980. Biology, immunology, and cariogenicity of *Streptococcus mutans*. Microbiol. Rev. **44:**331–384.
9. **Higuchi, M., Y. Iwani, T. Yamada, and S. Araya.** 1970. Levan synthesis and accumulation by human dental plaque. Arch. Oral Biol. **15:**563–567.
10. **Jacques, N. J., J. G. Morrey-Jones, and G. J. Walker.** 1985. Inducible and constitutive formation of fructanase in batch and continuous cultures of *Streptococcus mutans*. J. Gen. Microbiol. **131:**1625–1633.
11. **Maniatas, T., E. F. Fritsch, and J. Sambrook.** 1982. Molecular cloning: a laboratory manual. Cold Spring Harbor Laboratory, Cold Spring Harbor, N.Y.
12. **Manly, R. S., and D. T. Richardson.** 1968. Metabolism of levan by oral samples. J. Dent. Res. **47:**1080–1086.
13. **Marshall, K., and H. Weigel.** 1980. Evidence of multiple branching in the levan elaborated by *Streptococcus salivarius* strain 51. Carbohydr. Res. **83:**321–326.
14. **Robeson, J. P., R. G. Barletta, and R. Curtiss, III.** 1983. Expression of a *Streptococcus mutans* glucosyltransferase in *Escherichia coli*. J. Bacteriol. **153:**211–221.
15. **Rolla, G., J. E. Ciardi, K. H. Eggen, W. H. Bowen, and J. Afseth.** 1983. Free glucosyl- and fructosyltransferase in human saliva and adsorption of these enzymes to teeth *in vivo*, p. 21–29. *In* R. J. Doyle and J. E. Ciardi (ed.),

Proceedings: glucosyltransferases, glucans, sucrose and dental caries (a special supplement to Chemical Senses). Information Retrieval, Inc., Washington, D.C.

16. **Sato, S., T. Koga, and M. Inoue.** 1984. Isolation and some properties of extracellular D-glucosyltransferases and D-fructosyltransferases from *Streptococcus mutans* serotypes c, e, and f. Carbohydr. Res. **134:**293–304.

17. **Takahashi, N., F. Mizuno, and K. Talamori.** 1985. Purification and preliminary characterization of exo β-D-fructosidase in *Streptococcus salivarius* KTA-19. Infect. Immun. **47:**271–276.

18. **Viera, J., and J. Messing.** 1982. The pUC plasmids, an M13mp7-derived system for insertion mutagenesis and sequencing with synthetic universal primers. Gene **19:**259–268.

19. **Walker, G. J., M. D. Hare, and J. G. Morrey-Jones.** 1983. Activity of fructanase in batch cultures of oral streptococci. Carbohydr. Res. **113:**101–112.

V. Lactic Acid Streptococci

Characterization of Plasmids and Cloning of the β-Galactosidase Gene from *Streptococcus thermophilus*

RICHARD E. HERMAN,† CRAIG J. SCHROEDER, AND LARRY L. McKAY

Department of Food Science and Nutrition, University of Minnesota, St. Paul, Minnesota 55108

Streptococcus cremoris, S. lactis, S. lactis subsp. *diacetylactis*, and *S. thermophilus* are economically important bacteria which characteristically produce large amounts of lactic acid and are utilized extensively in dairy fermentations as starter cultures. (These bacteria are often referred to as the dairy streptococci or the lactic acid streptococci.) To date, the dairy industry has relied upon the isolation of natural variants to obtain strains suitable for these fermentation processes. The isolation of natural variants is tedious and limited by selection techniques. Mutagenesis, gene transfer systems, and recombinant DNA technology offer more direct approaches to the isolation of desirable strains.

The potential role of recombinant DNA technology in strain improvement programs has resulted in great interest in the development of plasmid cloning vectors for the dairy streptococci. Since cloning vectors would be introduced into microorganisms present in fermented foods, they must be acceptable as food-grade cloning vectors. We are interested in developing a food-grade cloning vector for *S. thermophilus*, as well as other dairy streptococci.

The plasmid cloning vectors that are currently available for the dairy streptococci are derived, at least in part, from *Escherichia coli, Streptococcus agalactiae*, or *Staphylococcus aureus* plasmids (1–3, 7). These plasmid cloning vectors depend upon resistance to antibiotics as selection markers. Alternative selection markers, possibly carbohydrate utilization genes, will have to be developed for a food-grade cloning vector because antibiotic resistance is considered unsuitable for incorporation into starter cultures utilized in dairy and food fermentations.

The development of a food-grade vector from native plasmid DNA and selection markers derived from the dairy streptococci, or other bac-

teria already used in food fermentations, would be highly compatible with the dairy streptococcal hosts. Moreover, a food-grade cloning vector constructed from native DNA would tend to alleviate restrictions by regulatory agencies. These considerations have placed an importance on understanding the plasmid biology of these streptococci.

To develop a suitable plasmid cloning vector for *S. thermophilus* we isolated and characterized five plasmids that were indigenous to *S. thermophilus*. In addition, the β-galactosidase gene of *S. thermophilus* was isolated by cloning in *E. coli* for potential incorporation as a selection marker in an *S. thermophilus* plasmid cloning vector.

S. thermophilus Plasmids

The plasmid biology of the group N dairy streptococci (*S. lactis, S. lactis* subsp. *diacetylactis*, and *S. cremoris*) is currently under intensive study. They have been shown to characteristically contain four to seven plasmids in a wide size range, and while most of these are cryptic, some have been linked to phenotypic traits which are important for dairy fermentations (9). Gene transfer systems and plasmid transformation have also been developed for the genetic manipulation of this group of bacteria (8).

S. thermophilus, a non-group N dairy streptococcus, is utilized in dairy fermentations (e.g., yogurt and Swiss cheese) where survival of the bacteria at high temperatures is required. Very few data are available concerning plasmids in *S. thermophilus*. We examined 23 strains of *S. thermophilus* for plasmid DNA and found only 5 that contained plasmids. A detailed account of the isolation and characterization of these plasmids has been published (5).

A single cryptic plasmid was found in each of the five plasmid-containing strains. The plasmids were designated pHM1 through pHM5. They ranged from 1.4 to 2.2 megadaltons in size

†Present address: Bio Techniques Laboratories, Inc., Redmond, WA 98052.

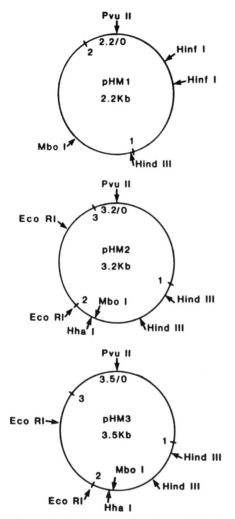

FIG. 1. Restriction maps of *S. thermophilus* plasmids. The maps of pHM1 and pHM5 are identical, as are the maps of pHM2 and pHM4. Restriction sites of the enzymes which cleaved the plasmids one or two times are shown. The *Pvu*II site on each plasmid was arbitrarily chosen as the zero reference point, and distances from it are indicated in kb (5).

and 4 to 18 in copy number. Restriction endonuclease mapping and DNA-DNA hybridization indicated that all five plasmids were closely related, probably having a common ancestor. Plasmids pHM1 and pHM5 were found to be identical, as were pHM2 and pHM4. The net result was the identification of three unique *S. thermophilus* plasmids.

Each plasmid had a single restriction endonuclease site for *Pvu*II and *Mbo*I. In addition, pHM2 and pHM3 had single sites for *Hha*I, and pHM1 had a single site for *Hin*dIII (Fig. 1). The high copy number, small size, and single restriction sites make these plasmids suitable for de-

velopment into cloning vectors for *S. thermophilus*.

Cloning of the *S. thermophilus* β-Galactosidase Gene

One gene involved in carbohydrate utilization which could be used as a selection marker in a food-grade cloning vector is the β-galactosidase gene. The *E. coli* β-galactosidase gene (*lacZ*) has been successfully utilized as a selection marker in several cloning vectors. Its success is partially due to the ease and sensitivity with which it can be detected by incorporating the chromogenic substrate 5-bromo-4-chloro-3-indolyl-β-D-galactoside (X-Gal) into the medium (10). In the presence of X-Gal, β-galactosidase-positive colonies are blue while β-galactosidase-negative colonies remain white. Since *S. thermophilus* also utilizes a β-galactosidase, we have cloned the β-galactosidase gene of *S. thermophilus* in *E. coli*, as reported in detail elsewhere (6), for potential use as a selection marker.

The β-galactosidase gene of *S. thermophilus* was presumed to be chromosomally linked since strains in which plasmids were undetectable contained β-galactosidase activity. Therefore, the isolation of the *S. thermophilus* β-galactosidase gene was approached by shotgun cloning in an *E. coli* host. Chromosomal DNA extracted from the *S. thermophilus* type strain 19258 was cleaved with *Pst*I and ligated to *Pst*I-cleaved pBR322. The ligated DNA was transformed into *E. coli* JM108 (13). JM108 contains a deletion (Δ*lac-pro*) which spans the *E. coli lacZ* gene and renders JM108 β-galactosidase negative.

Transformants were selected on a medium containing tetracycline and X-Gal. A single tetracycline-resistant (Tc^r) colony that was blue, and therefore β-galactosidase positive, was isolated from a background of β-galactosidase-negative transformants. This clone was ampicillin sensitive (Ap^s), indicating that the Ap^r marker of pBR322 had been insertionally inactivated. The plasmid extracted from this clone was designated pRH116, and by cleavage with *Pst*I it was shown to contain a single insertion fragment of 7.0 kilobase pairs (kb).

The product of the cloned β-galactosidase gene was characterized by its electrophoretic mobility in a nondenaturing polyacrylamide gel stained in *O*-nitrophenyl-β-D-galactopyranoside. The β-galactosidase of JM108(pRH116) comigrated with the β-galactosidase of *S. thermophilus* 19258 and was significantly different from the β-galactosidase of *E. coli* HB101 (Fig. 2). No β-galactosidase activity was detected in JM108-(pBR322).

The β-galactosidase of *S. thermophilus* 19258 has been shown by sodium dodecyl sulfate-polyacrylamide gel electrophoresis to have a

1 2 3 4

FIG. 2. Locations of β-galactosidase activity after electrophoresis of cell extracts in a 6% polyacrylamide gel. Lanes: 1, *S. thermophilus* 19258; 2, *E. coli* JM108(pBR322); 3, *E. coli* JM108(pRH116); and 4, *E. coli* HB101. Enzyme activity was detected as described in the text (6).

molecular weight of 105,000 (12). Since about 2.3 kb of DNA would be required to code for a protein of this size, we sought to localize the β-galactosidase gene on the insertion fragment. Restriction endonuclease mapping of the insertion fragment revealed a pair of *Bgl*II and *Bst*EII sites. Since neither of these sites is found in pBR322, they were used to construct deletion derivatives of pRH116. Both of these derivatives, pRH120 and pRH140, respectively, were β-galactosidase negative, indicating the whereabouts of the β-galactosidase gene. By subcloning *Hind*III partial digests of pRH116 we identified two deletion derivatives, pRH172 and pRH173, which were β-galactosidase positive and localized the β-galactosidase gene to a 3.85-kb region consisting of a 0.45-kb *Pst*I-*Hind*III fragment and its adjoining 3.4-kb *Hind*III fragment (Fig. 3).

The proteins encoded by pRH116 were examined by using the *E. coli* Lac⁻ maxicell strain SS5062 (*lacI22 lacZ pro-48 met-90 trpA trpR his rpsL gyrA recA56 srl-2 PI*ˢ) (11). Our results

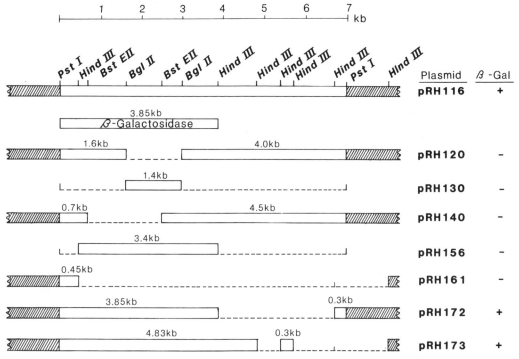

FIG. 3. Linear representation of the physical maps of pRH116 and a series of β-galactosidase-positive and -negative recombinant plasmids used to localize the β-galactosidase gene on the 7.0-kb *Pst*I fragment of *S. thermophilus* 19258 DNA. The construction and characterization of these plasmids are described in the text. The 7.0-kb *Pst*I insert fragment of pRH116 is represented by the open bar flanked by pBR322 vector DNA (closed bar). Deletions occurring in the recombinant plasmids pRH120, pRH140, pRH161, pRH172, and pRH173 are indicated by the dashed line. Plasmid pRH130 was obtained by cloning the 1.4-kb *Bgl*II fragment into the *Bam*HI site of pBR322, and pRH156 was constructed by cloning the 3.4-kb *Hind*III fragment into the *Hind*III site of pBR322. The β-galactosidase activity of each derivative is indicated on the right side as either positive (+) or negative (−). The location of the β-galactosidase gene in the 7.0-kb *Pst*I insert was determined to be within the 3.85-kb region indicated by the bar below pRH116 (6).

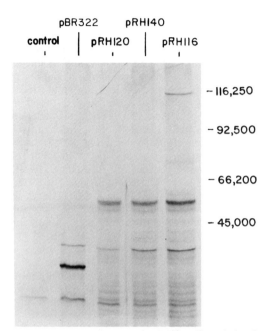

FIG. 4. Autoradiogram of a 7.5% sodium dodecyl sulfate-polyacrylamide gel used to detect plasmid-encoded proteins of maxicell strain SS5062 radiolabeled with [³⁵S]methionine. Extracted cells contained no plasmid (control), pBR322, pRH120, pRH140, or pRH116. The pBR322 lane contains two plasmid-encoded proteins which correspond to the *amp* and *tet* gene products.

(Fig. 4) indicate that, in addition to the *S. thermophilus* β-galactosidase (105,000 daltons), the insertion fragment of pRH116 encodes two proteins of 72,000 and 50,000 daltons. Both the β-galactosidase and the 72,000-dalton protein were absent from extracts of maxicells containing pRH120 and pRH140. We are in the process of determining the role, if any, of these additional proteins in *S. thermophilus* lactose metabolism.

Discussion

The characterization of *S. thermophilus* plasmids and the cloning of the *S. thermophilus* β-galactosidase gene represent preliminary steps toward constructing a food-grade cloning vector for the dairy streptococci. The β-galactosidase gene is being further characterized by subcloning into M13 for DNA sequencing studies. Besides aiding the incorporation of the β-galactosidase gene into a cloning vector, these studies will begin to provide an understanding of the molecular organization of the genes involved in the lactose metabolism of *S. thermophilus*.

In addition to a plasmid cloning vector, the application of recombinant DNA technology to *S. thermophilus* will require the development of gene transfer systems. While gene transfer by conjugation of plasmid DNA from *S. lactis* to *S. thermophilus* has been reported (4), there are no published accounts of plasmid transformation in *S. thermophilus*. However, we have isolated β-galactosidase-negative mutants of *S. thermophilus* (R. E. Herman and L. L. McKay, unpublished data) which could serve as transformation recipients for a plasmid cloning vector which utilizes β-galactosidase as a selection marker.

We appreciate the technical assistance of Nancy Maxfield during these studies.

This research was supported by General Mills, Inc.

LITERATURE CITED

1. **Behnke, D., and M. S. Gilmore.** 1981. Location of antibiotic resistance determinants, copy control, and replication functions on the double-selective streptococcal cloning vector pGB301. Mol. Gen. Genet. **184:**115–120.
2. **Dao, M. L., and J. J. Ferretti.** 1985. *Streptococcus-Escherichia coli* shuttle vector pSA3 and its use in the cloning of streptococcal genes. Appl. Environ. Microbiol. **49:**115–119.
3. **Gasson, M. J., and P. H. Anderson.** 1985. High copy number plasmid vectors for use in lactic streptococci. FEMS Microbiol. Lett. **30:**193–196.
4. **Gasson, M. J., and F. L. Davies.** 1980. Conjugal transfer of the drug resistance plasmid pAMβ in the lactic streptococci. FEMS Microbiol. Lett. **7:**51–53.
5. **Herman, R. E., and L. L. McKay.** 1985. Isolation and partial characterization of plasmid DNA from *Streptococcus thermophilus*. Appl. Environ. Microbiol. **50:**1103–1106.
6. **Herman, R. E., and L. L. McKay.** 1986. Cloning and expression of the β-D-galactosidase gene from *Streptococcus thermophilus* in *Escherichia coli*. Appl. Environ. Microbiol. **52:**45–50.
7. **Kok, J., J. M. B. M. van der Vossen, and G. Venema.** 1984. Construction of plasmid cloning vectors for lactic streptococci which also replicate in *Bacillus subtilis* and *Escherichia coli*. Appl. Environ. Microbiol. **48:**726–731.
8. **Kondo, J. K., and L. L. McKay.** 1985. Gene transfer systems and molecular cloning in group N streptococci: a review. J. Dairy Sci. **68:**2143–2159.
9. **McKay, L. L.** 1983. Functional properties of plasmids in lactic streptococci. Antonie van Leewenhoek J. Microbiol. Serol. **49:**259–274.
10. **Miller, J. H.** 1972. Experiments in molecular genetics, p. 47–55. Cold Spring Harbor Laboratory, Cold Spring Harbor, N.Y.
11. **Singer, J. T., C. S. Barbier, and S. A. Short.** 1985. Identification of the *Escherichia coli deoR* and *cytR* gene products. J. Bacteriol. **163:**1095–1100.
12. **Somkuti, G. A., and D. H. Steinberg.** 1979. β-D-Galactoside galactohydrolase of *Streptococcus thermophilus*: induction, purification, and properties. J. Appl. Biochem. **1:**357–368.
13. **Yanish-Perron, C., J. Vieira, and J. Messing.** 1985. Improved M13 phage cloning vectors and host strains: nucleotide sequences of the M13mp18 and pUC19 vectors. Gene **33:**103–119.

Transformation of *Streptococcus lactis* by Electroporation

SUSAN K. HARLANDER

Department of Food Science and Nutrition, University of Minnesota, St. Paul, Minnesota 55108

The group N streptococci, including *Streptococcus lactis*, *S. cremoris*, and *S. lactis* subsp. *diacetylactis*, are used extensively by the dairy industry for the production of fermented milk products. Recent advances in genetic engineering offer possibilities for strain improvement in this group of industrially important microorganisms. However, to effectively apply recombinant DNA techniques for genetic manipulation, an efficient and reliable high-frequency DNA transformation system should be available.

Protoplast Transformation in *S. lactis*

Kondo and McKay (13, 14) demonstrated polyethylene glycol (PEG)-induced transformation of *S. lactis* protoplasts by plasmid DNA. Briefly, washed cells were suspended in buffer containing an osmotic stabilizer and treated with mutanolysin (12) to digest the cell wall. Plasmid DNA was added to the protoplast suspension, followed immediately by the addition of PEG. The PEG facilitates transformation by an unknown mechanism which could involve making the cell membrane more permeable to the DNA or altering the conformation of the DNA molecule to allow penetration into the protoplast, or both. At high concentrations, PEG causes protoplast fusion. Protoplasts were then washed to remove excess PEG, allowed to express phenotypic determinants under nonselective conditions, and plated in osmotically stabilized soft agar overlays on selective media. Transformants were scored after 5 to 10 days of growth.

Many studies with streptococci (7, 14, 22–24) have shown that there are a number of parameters which can influence the formation of protoplasts and alter the frequency of transformation and regeneration of transformed protoplasts. The phase of growth of the bacterial strain, the composition of the wash buffer, and the number of washes prior to mutanolysin treatment influence the efficiency of protoplast formation. In addition, variability in composition of the commercial mutanolysin preparations and the conditions employed during protoplasting, including initial cell concentration, buffer composition, selection of osmotic stabilizing agent, treatment time, and posttreatment handling, affect protoplast formation. PEG treatment is essential for protoplast transformation, but it has an adverse effect on regeneration of protoplasts (14). PEG concentration, molecular weight, commercial source, method of preparation (autoclaved versus filter sterilized), treatment time, and posttreatment handling all influence transformation frequencies and the efficiency of protoplast regeneration. Other factors affecting regeneration include the composition of the liquid nonselective regeneration medium, the time and temperature of incubation prior to plating, and the use of soft agar overlays. Finally, transformation procedures developed for one strain are not necessarily applicable to other related strains. Despite efforts to optimize and control parameters that influence transformation and regeneration, frequencies remain highly variable, indicating that some critical parameter(s) is yet to be identified and that there are subtle operator-dependent variables in executing the procedures.

Kondo and McKay (14) were able to obtain transformation frequencies as high as 4.1×10^4 transformants per μg of DNA. However, when recombinant DNA was used, the frequency of transformation was 10- to 100-fold lower. This lowering of transformation frequency limits the usefulness of this technique in molecular cloning studies. In this report an experimental approach designed to overcome many of the current limitations with transformation of streptococci and possibly other bacteria is presented.

High-Voltage Electric-Field-Induced Effects

Zimmermann and co-workers have reported extensively on the effects of high-voltage electric field pulses on the membranes of mammalian, plant, and bacterial cells (19, 25–31). High-voltage electric pulses in a very narrow range of intensity and duration result in a reversible, experimentally controllable increase in cell membrane permeability. The electrically induced increase in permeability leads to a transient exchange of matter across the perturbed membrane structures and can cause membrane-membrane fusion when two membranes are in close contact (19).

Electric-field-induced cellular effects have been exploited in a variety of technological applications, including selective and continuous recovery of cell products (8) and generation of hybridomas for monoclonal antibody production (3). In addition, electrical techniques have been successfully used to encapsulate drugs inside erythrocytes and lymphocytes which can then be guided to selected sites via the circulatory system for controlled release of the active agent (30). On the basis of past successes with high-voltage-induced alterations it can be projected that an exciting application of this technology would be the development of high-frequency eucaryotic and procaryotic gene transfer systems.

A variety of biochemical methods have been developed to transfer genes into eucaryotic cells, including calcium phosphate/DNA coprecipitation (9), direct injection of DNA into the nucleus of recipient cells (5), use of viral vectors (10), and application of liposomes as vehicles for gene transfer (6). Neumann et al. (15) recently demonstrated that high-voltage electric pulses enhance the uptake of linear and circular DNA into muse lymphoma cells. Similarly, Potter and co-workers (18) successfully used electroporation to transfect mouse and human B lymphocytes, mouse T lymphocytes, and fibroblast lines with DNA, demonstrating that this physical method is generally applicable to cell lines recalcitrant to traditional biochemical gene transfer methods.

An exciting innovation in plant transformation is the recent demonstration of chemically induced transfer of specific DNAs into dicot and monocot protoplasts (16, 17). However, the frequency obtained by these methods is extremely low (16). Using electroporation, Shillito et al. (20) demonstrated a 1,000-fold increase in the efficiency of direct gene transfer in mesophyll protoplasts of *Nicotiana tabacum*.

Hashimoto et al. (11) used electroporation to transform intact yeast cells with an *Escherichia coli* plasmid, YEp13, which codes for ampicillin resistance. This technique eliminates the need to use yeast protoplasts, which can be difficult to prepare and are often unstable in the presence of cell wall-degrading enzymes.

Electroporation of *S. lactis* Protoplasts

Although electroporation has been demonstrated to be extremely effective in enhancing gene transfer in various mammalian and plant systems, there are few references on the application of this technology to bacterial systems. Shivarova et al. (21) used electroporation to transform *Bacillus cereus* protoplasts with plasmid DNA from *Bacillus thuringiensis* and, in comparison with PEG-induced protoplast transformation, demonstrated an increase in transformation efficiency of one order of magnitude.

To determine whether high-voltage pulses would enhance transformation frequency in *S. lactis*, protoplasts were prepared by the procedure of Kondo and McKay (14). After addition

of pSA3 plasmid DNA and PEG, samples were divided into three 1-ml aliquots and subjected to zero, three, or eight high-voltage pulses generated by a Cober model 615 LabPulser (Cober Electronics, Stamford, Conn.). The duration of each pulse was 5 μs (15 s between pulses) at an initial field strength of 4 kV/cm. Chambers were constructed by the design of Potter et al. (18) and were 5 mm in diameter with a total capacity of 1 ml.

The results of two independent trials indicated that, at the voltage employed, electric pulses did not enhance the frequency of transformation of the *S. lactis* protoplasts (see Table 1). It is important that Shivarova et al. (21) had to subject *B. cereus* protoplasts to three 5-μs pulses of 14 kV/cm to achieve a 10-fold increase in transformation frequency. With the chamber configuration available for use with the Cober LabPulser, the maximum voltage that could be generated was 4 kV/cm. Since Shivarova et al. (21) had demonstrated that *B. cereus* protoplasts were very resistant to high-voltage electric pulses and *S. lactis* protoplasts seemed to have similar properties, alternative equipment, capable of generating higher-voltage electric pulses, was obtained for further studies.

The primary goal of this investigation was to develop a gene transfer system to overcome many of the current limitations with transformation of streptococci, including problems associated with formation and regeneration of protoplasts and the potentially toxic effect of PEG on *S. lactis* protoplasts. Electroporation of *S. lactis* protoplasts offered little advantage over the currently available methods. The successful transformation of intact yeast cells by using electroporation (11) indicated that this might be a viable alternative worthy of investigation. There did not appear to be any reports in the literature of high-voltage electric-pulse-induced transformation of intact bacterial cells.

TABLE 1. Electroporation of *S. lactis* protoplasts[a]

Sample	No. of pulses	Transformants per μg of DNA	
		6 days	10 days
1		1.3×10^3	1.9×10^3
2	3	4.4×10^2	9.7×10^2
3	8	4.4×10^2	9.5×10^2
4		3.4×10^2	8.7×10^2
5	3	4.1×10^2	7.9×10^2
6	8	4.6×10^2	7.9×10^2

[a] Number of input cells was approximately 2×10^9/ml, and 1 μg of pSA3 DNA was used per trial. Pulse voltage was 4 kV/cm, and pulse duration was 5 μs, with 15 s between pulses. Transformants were selected on M17-G agar plates containing 2.5 μg of erythromycin per ml after incubation at 32°C for 6 and 10 days.

Electroporation of Intact Cells of *S. lactis*

Preliminary evidence demonstrating the applicability of electroporation for enhancement of transformation of intact cells of *S. lactis* LM0230 was obtained with a high-voltage cell processor used in conjunction with an electronic cell-processing centrifuge (DEP Systems, Inc., Troy, Mich.). This equipment was selected because it has the capability of delivering an electrical pulse to solutions during centrifugation. With this equipment plasmid DNA and cells can be concentrated on the surface of an electrode prior to electroporation, rather than being dispersed in a large volume of sucrose as in traditional stationary electroporation chambers.

Theory of electroporation. The DEP electroporation chamber is illustrated in Fig. 1. The chamber is filled with a sucrose solution and underlayered with a sucrose/plasmid DNA mixture on the surface of the electrode, E1. Cells to be transformed are layered on top of the sucrose solution. Once loaded and capped, the chamber is placed horizontally in a centrifuge rotor that spins with a vertical axis of rotation. As the chamber spins, the cells sediment through the sucrose and are pelleted against the bottom electrode. While spinning, electric pulses up to 3,500 V (11.6 kV/cm) are applied. If the pulse is of sufficient voltage to perforate the cell wall and membrane, cytoplasm begins to leak out of the cell. In a stationary chamber the pore would close very rapidly, allowing very little uptake of DNA. To keep the pores open, the velocity is increased immediately after the pulse, creating

internal pressure which forces the cytoplasm to flow out of the pore. The amount of cytoplasm leaking from the cell is detected by an ion monitor which measures the conductivity of the solution between E1 and E2. Once the desired amount of cytoplasm has been released, the velocity of the centrifuge is rapidly reduced. As the internal pressure of the cell normalizes, the surrounding liquid, enriched with plasmid DNA, is aspirated into the cell. Transformed cells can be collected from the chamber and detected by plating on a selective medium.

Experimental conditions. A plasmid-cured derivative of *S. lactis* LM0230 was grown to an optical density at 600 nm of 0.15, harvested, washed twice with cold distilled water, and suspended to a final concentration of 10^9 to 3×10^9 cells per ml in 0.5 M sucrose–0.01 M Tris hydrochloride, pH 7.0. Approximately 10^7 to 6×10^7 cells were used for each trial. The sucrose/plasmid DNA mixture consisted of equal volumes of pSA3 (4) or pGB301 (2) plasmid DNA (250 ng/µl) and 1 M sucrose; 125 to 250 ng of plasmid DNA was used for each trial. Chambers were routinely centrifuged for 5 min at 5,000 rpm and subjected to one or two 30-µs pulses of 6 to 8 kV/cm. Velocity was increased to approximately 7,000 rpm for extraction of cytoplasm. Chambers were removed from the centrifuge and allowed to sit at room temperature for 10 min. Excess sucrose was removed, and cells were gently suspended and transferred to a tube containing M17-G broth in 0.5 M sucrose. After incubation at 30°C for 1 to 2 h, cells were plated in soft agar overlays onto M17-G agar plates containing 2.5 µg of erythromycin per ml. Transformants were scored after 36 h of incubation at 32°C.

Results

The results outlined in Table 2 indicate that about 10^4 transformants per µg of DNA were obtained when the lowest-voltage pulse tested, 6 kV/cm, was applied. These results are equivalent to the maximum frequency obtained by Kondo and McKay with PEG-induced protoplast transformation (14). Transformation frequencies are less than 10^4 transformants per µg of DNA when either higher voltages or multiple pulses are employed. Thirty representative erythromycin-resistant clones were analyzed for the presence of plasmid DNA (1), and in all cases the appropriate plasmid was present (see Fig. 2).

Discussion

There are a number of important advantages to an electric-pulse-induced transformation system. Because high-voltage pulses are capable of penetrating bacterial cell walls and membranes,

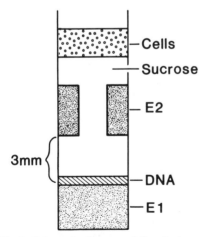

FIG. 1. Schematic diagram of the electroporation chamber. E1, Lower electrode; E2, upper electrode; distance between electrodes, 3 mm. Chamber is filled with a solution of 0.5 M sucrose. Plasmid DNA (in 0.5 M sucrose) is underlayered next to E1 with a syringe. Washed cells (in 0.5 M sucrose–10 mM Tris, pH 7.0) are layered on top of the sucrose.

TABLE 2. Electroporation of intact cells of
S. lactis

| Sample | Plasmid | Pulse | | Transformants per μg of DNA[c] |
		No.	Voltage (kV/cm)	
1	pSA3[a]	1	6.0	1.2×10^4
2	pSA3	1	6.6	1.8×10^3
3	pSA3	1	7.3	3.5×10^2
4	pSA3	1	8.0	2.3×10^2
5	pGB301[b]	1	7.3	6.1×10^2
6	pGB301	2	7.3	2.3×10^2

[a] 250 ng of pSA3 DNA was used per trial.
[b] 125 ng of pGB301 DNA was used per trial.
[c] Number of input cells was approximately 2×10^7. Transformants were selected on M17-G agar plates containing 2.5 μg of erythromycin per ml after incubation at 32°C for 36 h.

it is not necessary to prepare protoplasts. Therefore, osmotic stabilizing agents are not required, and cells transformed by electroporation regenerate in 36 to 48 h, rather than 5 to 10 days as required for protoplast regeneration. Electroporation is a physical method for introducing DNA; therefore, toxic chemical agents such as PEG are not required. Cells and DNA are concentrated in a small volume and kept in close association during centrifugation; thus, nanogram quantities of DNA are required for transformation experiments. In addition, procedures for preparing the cells and carrying out the electroporation are relatively simple, thus minimizing operator-dependent variables.

It is proposed that the successful development

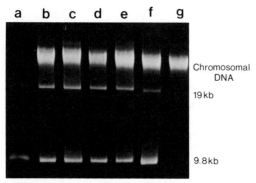

FIG. 2. Agarose gel electrophoresis of plasmids isolated from transformant clones. Lane a, pSA3 plasmid DNA isolated from *S. lactis* and purified via CsCl (monomer molecular weight, 9.8 kilobases; dimer molecular weight, 19 kilobases); lanes b–f, total DNA from five randomly selected erythromycin-resistant clones; lane g, total DNA from *S. lactis* recipient. Electrophoresis conditions: 0.6% agarose gel in TAE buffer (0.04 M Tris-acetate, 0.002 M EDTA, pH 8.0); 2.5 h at 60 V.

of an electroporation transformation system applicable to the lactic streptococci would have broad application in strain improvement programs and serve as the foundation for the application of this technology to other industrially important microorganisms.

I acknowledge the technical assistance of John Biscar, DEP Systems, Inc., Troy, Mich.

This work was supported in part by the University of Minnesota Graduate School and the Agricultural Experiment Station Project no. 18-017.

LITERATURE CITED

1. **Anderson, D., and L. L. McKay.** 1983. Simple and rapid method for isolating large plasmid DNA from lactic streptococci. Appl. Environ. Microbiol. **46:**549–552.
2. **Behnke, D., M. S. Gilmore, and J. J. Ferretti.** 1981. Plasmid pGB301, a new multiple resistance streptococcal cloning vehicle and its use in cloning of the gentamicin/kanamycin resistance determinant. Mol. Gen. Genet. **182:**414–421.
3. **Bischoff, R., R. Eisert, I. Schedel, J. Vienken, and U. Zimmermann.** 1982. Human hybridoma cells produced by electrofusion. FEBS Lett. **147:**64–68.
4. **Dao, M. L., and J. J. Ferretti.** 1985. *Streptococcus-Escherichia coli* shuttle vector pSA3 and its use in the cloning of streptococcal genes. Appl. Environ. Microbiol. **49:**115–119.
5. **Diacumakos, E. G.** 1973. Methods for micromanipulation of human somatic cells in culture, p. 287–311. *In* D. M. Prescott (ed.), Methods in cell biology. Academic Press, Inc., New York.
6. **Fraley, R., S. Subramani, P. Berg, and D. Papahadjopoulos.** 1980. Introduction of liposome-encapsulated SV40 DNA into cells. J. Biol. Chem. **255:**10431–10435.
7. **Gasson, M. J.** 1980. Production, regeneration and fusion of protoplasts in lactic streptococci. FEMS Microbiol. Lett. **9:**99–102.
8. **Gordon, P. B., and P. O. Seglen.** 1982. Autophagic sequestration of ^{14}C-sucrose introduced into rat hepatocytes by reversible electro-permeabilisation. Exp. Cell Res. **142:**1–14.
9. **Graham, F. C., and A. J. van der Eb.** 1973. A new technique for the assay of infectivity of human adenovirus 5 DNA. Virology **52:**456–467.
10. **Hamer, D. H., and P. Leder.** 1979. Splicing and the formation of stable RNA. Cell **18:**1299–1302.
11. **Hashimoto, H., H. Morikawa, Y. Yamada, and A. Kimuara.** 1985. A novel method for transformation of intact yeast cells by electroinjection of plasmid DNA. Appl. Microbiol. Biotechnol. **21:**336–339.
12. **Kondo, J. K., and L. L. McKay.** 1982. Mutanolysin for improved lysis and rapid protoplast formation in dairy streptococci. J. Dairy Sci. **65:**1428–1431.
13. **Kondo, J. K., and L. L. McKay.** 1982. Transformation of *Streptococcus lactis* protoplasts by plasmid DNA. Appl. Environ. Microbiol. **43:**1213–1215.
14. **Kondo, J. K., and L. L. McKay.** 1984. Plasmid transformation of *Streptococcus lactis* protoplasts: optimization and use in molecular cloning. Appl. Environ. Microbiol. **48:**252–259.
15. **Neumann, E., M. Schaefer-Ridder, Y. Wang, and P. Hofschneider.** 1982. Gene transfer into mouse lyoma cells by electroporation in high electric fields. EMBO J. **1:**841–845.
16. **Paszkowski, J., R. Shillito, M. Saul, V. Mandak, T. Hohn, B. Hohn, and I. Potrykus.** 1984. Direct gene transfer to plants. EMBO J. **3:**2717–2722.
17. **Potrykus, I., M. Saul, J. Petruska, J. Paszkowski, and R. Shillito.** 1985. Direct gene transfer to cells of a graminaceous monocot. Mol. Gen. Genet. **199:**183–188.
18. **Potter, H., L. Weir, and P. Leder.** 1984. Enhancer-

dependent expression of human κ immunoglobulin genes introduced into mouse pre-β lymphocytes by electroporation. Proc. Natl. Acad. Sci. USA **81**:7161–7165.

19. **Scheurich, P., and U. Zimmerman.** 1981. Giant human erythrocytes by electric field induced cell-to-cell fusion. Naturwissenschaften **68**:45–46.

20. **Shillito, R., M. Saul, M. Paszkowski, M. Muller, and I. Potrykus.** 1985. High efficiency direct gene transfer to plants. Bio/Technol. **3**:1099–1103.

21. **Shivarova, N., W. Forster, H.-E. Jacob, and R. Grigorova.** 1983. Microbiological implications of electric field effects. VII. Stimulation of plasmid transformation of *Bacillus cereus* protoplasts by electric field pulses. Z. Allg. Mikrobiol. **23**:595–599.

22. **Simon, D., A. Rouault, and M.-C. Chopin.** 1985. Protoplast transformation of group N streptococci with cryptic plasmids. FEMS Microbiol. Lett. **26**:239–241.

23. **von Wright, A., A.-M. Taimisto, and S. Sivela.** 1985. Effect of Ca^{2+} ions on plasmid transformation of *Streptococcus lactis* protoplasts. Appl. Environ. Microbiol. **50**:1100–1102.

24. **Yu, R. S.-T., W. S. A. Kyle, A. A. Azad, and T. V. Hung.** 1984. Conditions for the production and regeneration of protoplasts from lactic streptococci. Milchwissenschaft **39**:136–139.

25. **Zimmermann, U.** 1983. Electrofusion of cells: principles and industrial potential. Trends Biotechnol. **1**:149–155.

26. **Zimmermann, U., G. Kuppers, and N. Salhami.** 1982. Electric field induced release of chloroplasts from plant protoplasts. Naturwissenschaften **69**:451–452.

27. **Zimmermann, U., J. Schultz, and G. Pilwat.** 1973. Transcellular ion flow in *E. coli* B and electrical sizing of bacteria. Biophys. J. **13**:1005–1013.

28. **Zimmermann, U., and J. Vienken.** 1982. Electric field induced cell-to-cell fusion. J. Membr. Biol. **67**:165–182.

29. **Zimmermann, U., J. Vienken, J. Halfmann, and C. Emeis.** 1985. Electrofusion: a novel hybridization technique. Adv. Biotechnol. Processes **4**:79–150.

30. **Zimmermann, U., J. Vienken, and G. Pilwat.** 1980. Development of drug carrier systems: electrical field induced effects in cell membranes. Bioelectrochem. Bioenerg. **7**:553–574.

31. **Zimmermann, U., J. Vienken, and G. Pilwat.** 1984. Electrofusion of cells, p. 89–167. *In* D. Chayen and J. Bitensky (ed.), Investigative microtechniques in medicine and biology. Marcel Dekker, Inc., New York.

Transfection of *Streptococcus thermophilus* Spheroplasts

A. MERCENIER, C. ROBERT, D. A. ROMERO, P. SLOS, AND Y. LEMOINE

Transgene S.A., 67000 Strasbourg, France

Different modes of in vivo gene transfer are known to occur in lactic acid bacteria. Transduction and conjugation of antibiotic markers and metabolic traits are well documented (3, 6). DNA uptake by whole cells of *Streptococcus pneumoniae, S. mutans*, and *S. sanguis* has also been described (10). However, for the dairy streptococci, transformation via natural competence has never been demonstrated.

The use of protoplasts in place of intact cells has opened the way for applying recombinant DNA technology to several gram-positive bacteria of industrial importance. Working with *Streptomyces*, Bibb and co-workers (1) observed first that polyethylene glycol (PEG) was able to promote the uptake of DNA by protoplasts. In 1979, Chang and Cohen (2) developed a protoplast-PEG transformation procedure for *Bacillus subtilis*. This technique has since been successfully adapted to a few strains of *Streptococcus lactis* (8, 9, 11, 13, 15) and *S. faecalis* (12, 14).

For the dairy streptococci, work has focused primarily on the mesophilic strains *S. lactis, S. lactis* subsp. *diacetylactis*, and *S. cremoris*. Despite the economic importance of *S. thermophilus*, which is largely used as a yogurt and hard cheese starter, very little information concerning gene transfer in this species is available. To our knowledge, there are no published reports on the introduction of purified DNA into this organism.

Here we describe a transfection system allowing efficient uptake of DNA by spheroplasts of industrial *S. thermophilus* strains used for yogurt making. Transfection was found to be a powerful tool for developing a protoplast transformation procedure. Detailed results will be published elsewhere (manuscript in preparation).

Restriction Systems in *S. thermophilus*

The *S. thermophilus* strains used in this study were obtained from a yogurt factory and were tested for their sensitivity to virulent phages found in contaminated milk samples. The initial phage lysates formed plaques on strains A023, A032, A050, A054, A078, and A092. From these lysates, two phages, $\phi\alpha32$ and ϕB21, were purified twice onto each host, and high-titer lysates ($>10^9$ PFU/ml) were prepared for each phage/

host pair. Plaquing efficiency of the different $\phi\alpha32$ lysates on the six hosts showed that significant restriction/modification barriers exist between some strains of *S. thermophilus* (Table 1). In subsequent transfection assays, purified DNA from phage propagated on the homologous host was used.

Transfection Assay

Overnight cultures of *S. thermophilus* strains were inoculated at 2% (vol/vol) into rich medium (Elliker medium [6] supplemented with 1% beef extract) and propagated at 42°C. Cells were harvested at an optical density (600 nm) of 0.5, washed once with ice-cold water, and suspended in half the original volume in protoplasting buffer (PB: 20 mM HEPES [*N*-2-hydroxyethylpiperazine-N'-2-ethanesulfonic acid], pH 7.0; 0.3 M raffinose; 0.5% [wt/vol] gelatin; 1 mM MgCl₂). Cell wall digestion was performed using mutanolysin M-1 (Miles Laboratories, Inc.) at 37°C. Both the mutanolysin concentration and the incubation time at 37°C needed to be adjusted for each strain. After the desired time the lytic treatment was interrupted by diluting the cell suspension with two volumes of cold PBMC (PB containing 12.5 mM MgCl₂ and 12.5 mM CaCl₂). The spheroplasts were centrifuged at $2,420 \times g$ for 10 min at 4°C (4,500 rpm in a SS34 Sorvall rotor) and suspended to 1/40 the original volume in PBMC.

Ninety microliters of the spheroplast suspension was mixed with 10 µl of liposomes containing 100 ng of purified phage DNA, followed immediately by the addition of the desired PEG solution. The liposomes used in this work were prepared by the method of B. Chassy (personal communication). PEG solutions were prepared by dissolving PEG 6000 (Fluka) in SMM buffer (0.5 M sucrose, 20 mM sodium maleate, pH 6.5, and 20 mM MgCl₂) and filter sterilizing. The transfection mixture was incubated at 37°C for 10 min (PEG shock) and was then diluted by adding 1.4 ml of rich medium mixed volume to volume with twice-concentrated PBMC. This suspension was then incubated at 37°C to allow expression of the phage genome. After 90 min, a sample of the transfection mixture was added to 3 ml of MRS (5) soft agar supplemented with 12.5 mM MgCl₂ and 12.5 mM CaCl₂ and seeded with the appropriate indicator strain. This was

TABLE 1. Titration of φα32 on different *S. thermophilus* strains

Phage lysate	Efficiency of plaquing on strain					
	A023	A032	A050	A054	A078	A092
A023(φα32)	1	NP[a]	1	1	1	1
A050(φα32)	1	NP	1	1	1	1
A054(φα32)	1.1×10^{-3}	NP	1.6×10^{-4}	1	1.5×10^{-3}	1.0×10^{-3}
A078(φα32)	1	NP	1	1	1	1
A092(φα32)	1	NP	1	1	1	1
A032(φα32*)[b]	NP	1	NP	NP	NP	NP

[a] NP, No plaques detected by spotting 10 μl (three times) of the concentrated phage lysate.

[b] Subsequent DNA analysis showed that the initial φα32 lysate actually contained two phages. One phage (φα32*) plaques only on strain A032; the other phage (φα32) plaques on all strains tested except A032.

overlaid onto MRS agar containing 10 mM CaCl$_2$. MRS agar was used rather than Elliker agar because phage plaques appeared more clearly on the former medium. The number of plaques was determined after overnight incubation at 42°C. The transfection efficiencies were calculated as PFU per microgram of phage DNA.

Transfection of *S. thermophilus* A032 spheroplasts by φα32* DNA

The extent of the lytic treatment and the conditions of PEG shock affect the efficiency of DNA uptake by spheroplasts. Several parameters in these two steps were examined for *S. thermophilus* A032. Spheroplasts were formed by treating a 1-ml cell suspension (at an optical density equal to 1.0) with 10 μg of mutanolysin at 37°C for different times. As shown in Fig. 1A there was a narrow optimum during which the DNA was efficiently taken up. In routine experiments, an incubation time of 10 min was chosen. At this stage, round, refringent spheroplasts were observed under a microscope.

The final concentration of PEG was found to be another critical parameter (Fig. 1B). The optimal concentration of 12% for strain A032 is unusually low when compared with those previously reported for other gram-positive bacteria. Figure 1C shows the effect of incubation time in the presence of 12% PEG on the transfection efficiency. In current assays, a PEG exposure time of 10 min was used. The curves represented in Fig. 1 correspond to an expression time equal to 90 min, which is longer than the eclipse time of phage φα32* on strain A032.

The effect of the chain length of the PEG was examined as well: PEG 4000, 6000, and 12000 led to an effective DNA uptake, while PEG 1000 was totally ineffective. PEGs obtained from Fluka, BDH Laboratories, and Merck were also

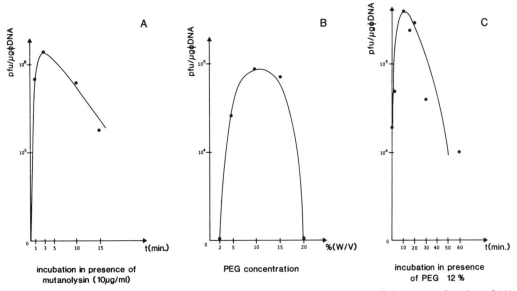

FIG. 1. Transfection of A032 spheroplasts with φα32* DNA. Transfection efficiency as a function of (A) incubation time in the presence of mutanolysin (10 μg/ml), (B) final PEG concentration, and (C) incubation time in the presence of PEG (12%).

TABLE 2. Optimized transfection conditions[a] for the pairs A023/φB1, A054/φB21, and A032/φα32*

| Phage | Strain | Lytic treatment | | PEG shock (final PEG concn, %) | Transfection efficiency (PFU/μg) |
		Mutanolysin concn (μg/ml)	Incubation time (min)		
φB1	A023	7.5	15	30	6×10^4
φB1	A054	2.5	10	15–27	3.2×10^6
φα32*	A032	10.0	10	12	1.4×10^6

[a] Incubation time in the presence of PEG was equal to 10 min, and the expression time was equal to 90 min.

compared, and all were found to promote transfection with efficiencies ranging from 2.5×10^4 to 4.0×10^5 PFU/μg of DNA. PEG 6000 from Fluka gave slightly better results and was chosen for routine assays. The use of purified PEG (7) did not seem to improve transfection efficiencies.

Transfection of *S. thermophilus* A032 was found to be markedly enhanced by the use of phage DNA entrapped into liposomes (1.0×10^5 PFU/μg versus 1.8×10^2 PFU/μg for naked DNA). However, empty liposomes added to the spheroplast/DNA mixture also gave rise to 2.4×10^4 PFU/μg.

Optimal Transfection Conditions Vary from Strain to Strain

In view of the narrow optima shown in Fig. 1, we wanted to determine whether the A032 transfection conditions would work for other strains of *S. thermophilus*. Standard transfection conditions were applied to A023, A054, A078, and A092. Although homologous DNA was used in each assay, no plaques could be detected with any of the mentioned strains. Since PEG concentration was found critical for successful transfection of A032, we first tested this parameter. As shown in Table 2, the optimal concentrations for strains A032 and A054 were found to be higher than 12%. The cell wall digestion conditions were also strain specific. The mutanolysin concentration and the incubation time with the lytic enzyme needed to be adapted for each strain (Table 2). Additionally, for strain A054, no plaques were observed when liposomes were omitted from the transfection mixture. Despite repeated attempts, we were unable to transfect spheroplasts of strains A078 and A092 with φB21 DNA.

When nonhomologous DNA was used, the host restriction system was observed to decrease the transfection efficiency. For strain A023, transfecting with DNA purified from φB21 lysates prepared on strains A023, A078, and A054 gave rise to 1.0×10^5, 2.0×10^4, and 8.0×10^3 PFU/μg, respectively.

Development of a Regeneration Medium for *S. thermophilus* A054

Another requirement for successful transformation is the regeneration of spheroplasts back to whole cells. For the *S. thermophilus* strains we work on, it was not a major problem to form spheroplasts which could be efficiently protected in liquid medium with commonly used osmotic stabilizers (e.g., 0.5 M sucrose, 0.5 M sorbitol, 0.5 M sodium succinate, and 0.3 M raffinose). However, intact cells of our strains did not grow efficiently on the corresponding agar medium (manuscript in preparation). From the numerous protecting agents we tested, only 0.3 M raffinose allowed the growth of most of our strains, provided the medium was supplemented with $MgCl_2$ and $CaCl_2$.

The strain dependence observed for optimal DNA uptake prompted us to focus on a single strain, *S. thermophilus* A054, for developing a protoplast transformation procedure. We took advantage of the transfection assay to define a solid medium which protected A054 spheroplasts. Various media were tested by looking for infective centers appearing immediately after the PEG treatment. On a protective medium, the transfection efficiency reflects the number of viable spheroplasts which have taken up DNA. The enumeration of plaques which appeared after an overnight incubation presented several advantages over the standard regeneration assay. Using the transfection, we were able to screen a large variety of potential protective media. Regeneration and transformation trials are currently under way.

We thank B. Chassy and M. Faelen for having introduced us to different aspects of phage work. Their stimulating discussions were very helpful during this work. We are particularly grateful to J. P. Lecocq for his constant and enthusiastic support. We also acknowledge P. Kourilsky, P. Chambon, and A. Fazel for their scientific interest in this project.

LITERATURE CITED

1. **Bibb, M. J., J. M. Ward, and D. A. Hopwood.** 1978. Transformation of plasmid DNA into streptomyces at a high frequency. Nature (London) **274:**398–400.
2. **Chang, S., and S. N. Cohen.** 1979. High frequency transformation of Bacillus subtilis protoplasts by plasmid DNA. Mol. Gen. Genet. **168:**111–115.
3. **Clewell, D. B.** 1981. Plasmids, drug resistance, and gene transfer in the genus *Streptococcus*. Microbiol. Rev. **45:**409–436.
4. **De Man, J. C., M. Rogosa, and M. E. Sharpe.** 1960. A medium for the cultivation of lactobacilli. J. Appl. Bacteriol. **23:**130–135.

5. **Elliker, P. R., A. W. Anderson, and G. Hannesson.** 1956. An agar culture medium for lactic acid streptococci and lactobacilli. J. Dairy Sci. **89:**1611–1612.

6. **Gasson, M. J.** 1983. Genetic transfer systems in lactic acid bacteria. Antonie van Leeuwenhoek J. Microbiol. Serol. **49:**275–282.

7. **Klebe, R. J., J. V. Harriss, Z. D. Sharpe, and M. G. Douglas.** 1983. A general method for polyethylene-glycol-induced genetic transformation of bacteria and yeast. Gene **25:**333–341.

8. **Kondo, J. K., and L. L. McKay.** 1982. Transformation of *Streptococcus lactis* protoplasts by plasmid DNA. Appl. Environ. Microbiol. **43:**1213–1215.

9. **Kondo, J. K., and L. L. McKay.** 1984. Plasmid transformation of *Streptococcus lactis* protoplasts: optimization and use in molecular cloning. Appl. Environ. Microbiol. **48:**252–254.

10. **Saunders, J. R., A. Docherty, and G. O. Humphreys.** 1984. Transformation of bacteria by plasmid DNA, p. 64–66. *In* P. M. Bennett and J. Grinsted (ed.), Methods in microbiology, vol. 17. Academic Press, Inc., New York.

11. **Simon, D., A. Rouault, and M. C. Chopin.** 1985. Protoplast transformation of group N streptococci with cryptic plasmids. FEMS Microbiol. Lett. **26:**239–241.

12. **Smith, M. D.** 1985. Transformation and fusion of *Streptococcus faecalis* protoplasts. J. Bacteriol. **162:**92–97.

13. **Von Wright, A., A. M. Taimisto, and S. Sivelä.** 1985. Effect of Ca^{2+} ions on plasmid transformation of *Streptococcus lactis* protoplasts. Appl. Environ. Microbiol. **50:**1100–1102.

14. **Wirth, R., F. Y. An, and D. B. Clewell.** 1986. Highly efficient protoplasts transformation system for *Streptococcus faecalis* and a new *Escherichia coli-Streptococcus faecalis* shuttle vector. J. Bacteriol. **165:**831–836.

15. **Yu, R. S.-T., W. S. A. Kyle, T. V. Hung, and A. A. Azad.** 1984. Aspects of genetic transformation involving protoplasts and purified lac plasmid of *S. lactis.* Milchwissenschaft **39:**476–479.

Tn*919* in Lactic Streptococci

GERALD F. FITZGERALD, COLIN HILL, ELAINE VAUGHAN, AND CHARLES DALY

Department of Dairy and Food Microbiology, University College, Cork, Ireland

The lactic streptococci used in dairy fermentations have been the subject of intense genetic study in recent years (see this volume and, for reviews, references 5 and 10). A feature of this work has been the association of plasmid DNA with key functions of lactic streptococci, e.g., lactose and citrate utilization, proteinase activity, and mechanisms of bacteriophage insensitivity. In contrast to the focus on plasmid biology, there has been little genetic analysis of the bacterial chromosome. Transposon mutagenesis may offer a useful approach in this regard. In this report we describe the introduction of the transposon Tn*919* into lactic streptococci with a view to using transposon technology to advance our genetic understanding of these bacteria.

Properties of Tn*919*

Tn*919*, originally isolated from *S. sanguis* FC1, is one of several transposons which have been identified in the genus *Streptococcus* (1–3, 6, 8, 11). It is similar but not identical to Tn*916* from *S. faecalis* DS16 (2, 3, 6). These transposons are between 15 and 17 kilobases in size, encode tetracycline resistance, and are capable of conjugative transfer in the absence of plasmid DNA at frequencies ranging between 10^{-5} and 10^{-8} per donor (2, 3, 6). Tn*919* and Tn*916* have been cloned in *Escherichia coli* where tetracycline resistance is expressed (2, 7). Significantly, in this background, growth of the clones in the absence of tetracycline results in high-frequency excision of the elements from the host DNA. This has led to the development by Gawron-Burke and Clewell of a strategy for cloning streptococcal genes in *E. coli* (7).

Tn*919* has been conjugatively transferred in filter matings, but not on agar surfaces, to all three species of group N streptococci and one species each of *Lactobacillus* and *Leuconostoc* (9). However, before Tn*919* can be exploited to target and clone specific genes from these bacteria in *E. coli*, it is important that the transposon-host system fulfill two requirements. First, insertion in recipient DNA must occur in a nonspecific manner, thereby allowing insertional inactivation at any locus on chromosomal or plasmid DNA. Second, the frequency of conjugative transfer of the transposon must be high enough to yield sufficient numbers of transconjugants for practical mutant-screening purposes.

In this report we describe the manner of insertion of Tn*919* in selected lactic streptococci and also the development of an improved delivery system which exploits the high-frequency conjugative properties of pMG600. This plasmid is derived from pLP712 (the Lac/Prt plasmid of *S. lactis* 712; 4) and confers the Lac$^+$ Lax$^-$ phenotype. (Lax$^-$ refers to the capacity of a strain to clump in broth.)

Nature of Insertion of Tn*919*

Insertion of Tn*916* in the *S. faecalis* chromosome and plasmid DNA has been reported to be random while the same element displays site-specific insertion in *S. mutans* (6; G. F. Fitzgerald and D. B. Clewell, unpublished data). We have also shown that the insertion of Tn*919* in the chromosome of *S. lactis* MG1363 was site specific (9). The nature of insertion of Tn*919* in a further three strains, *S. lactis* SK3S, *S. lactis* subsp. *diacetylactis* 18–16S, and *S. cremoris* 17S, was examined here. Tetracycline-resistant transconjugants of each strain were generated in filter matings with either an *S. lactis* CH919 (i.e., *S. lactis* MG1363 harboring Tn*919* on its chromosome) or an *S. faecalis* GF590 donor, and the transfer frequencies are shown in Table 1. All transconjugants were obtained from independent mating experiments to eliminate the possibility of transconjugant relatedness. The nature of Tn*919* insertion was then determined as summarized here. Chromosomal DNA was isolated from tetracycline-resistant transconjugants and digested with *Hin*dIII (for which Tn*919* has only one recognition site). Two transposon-chromosome junction fragments should be evident when this DNA is probed with the plasmid pAM554, a recombinant molecule containing Tn*919* (2). The size of the junction fragments provides an indication of the insertion specificity; i.e., similarly sized fragments in all transconjugants indicate site-specific insertion.

For the 18–16S transconjugants, probing of the *Hin*dIII-digested chromosome with biotin-labeled pAM554 showed that transposon-chromosome fragments of different sizes were present in four of the five transconjugants tested (Fig. 1A). In one instance (Fig. 1A, lane 6) four bands were evident, indicating the presence of two copies of the transposon on the chromosome of this transconjugant. In addition, no

TABLE 1. Transfer of Tn919 by conjugation to selected lactic streptococci

Donor	Recipient	Frequency per recipient
S. lactis CH919	S. lactis subsp. diacetylactis 18–16S	7.0×10^{-8}
S. faecalis GF590	S. lactis SK3S	5.8×10^{-7}
S. faecalis GF590	S. cremoris 17S	5.3×10^{-5}

specific hybridization was detected with the chromosome of a fifth transconjugant (Fig. 1A, lane 7), designated 18–16SA (which retained tretracycline resistance; see below). Similar results, indicating random insertion, were also obtained with tetracycline-resistant transconjugants of S. lactis SK3S examined in a similar manner. In the case of S. cremoris 17S, the hybridization pattern showed multiple insertion sites with a large degree of coincidence in size, which may indicate a preference for insertion into favored sites.

In the case of the strain 18–16S-derived transconjugants, no insertion could be detected in the chromosome of one isolate (18–16SA; Fig. 1A, lane 7) even though the isolate was tetracycline resistant. This suggested that Tn919 may have inserted in one of the five resident plasmids of 18–16S. Evidence for this was obtained when the plasmid profiles of 18–16SA and 18–16S were compared (Fig. 1B). The results show that the 46-megadalton plasmid (encoding lactose metab-

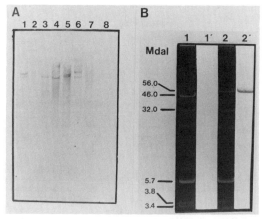

FIG. 1. (A) Color reaction obtained after probing chromosomal DNA from S. lactis subsp. diacetylactis 18–16S tetracycline-resistant transconjugants with biotin-11-dUTP-labeled pAM554. Lanes 3–7 show Hind-III-digested DNA from five independently isolated transconjugants. Lane 8 contains DNA from 18–16S parent strain not harboring Tn919. Lane 1 contains pAM554 DNA. Lane 2 contains λ DNA. (B) Plasmid DNA profiles of strain 18–16S and 18–16SA (lanes 1 and 2, respectively) after agarose gel (0.7%) electrophoresis. Lanes 1' and 2' show the color reaction obtained after hybridization with the biotin-11-dUTP-labeled pAM554 probe.

olism) present in the parent (lane 1) was no longer detectable in 18–16SA (lane 2). However, a larger plasmid of 56 megadaltons was evident. This size difference corresponds to the known size of Tn919 and suggested insertion in the 46-megadalton plasmid. This was confirmed when the plasmid profiles were probed with pAM554 and the novel 56-megadalton molecule showed homology (Fig. 1B; lane 2') while no hybridization with the 46-megadalton plasmid of the parent was detected (lane 1'). Although insertion in the Lac plasmid of 18–16S was observed, expression of the lactose utilization genes in the transconjugant was unaltered, suggesting that the site of insertion was not located in a structural gene or in a region governing its expression. The plasmid profiles of the four other 18–16S tranconjugants were identical to that of the 18–16S parent culture. In the case of S. lactis 17S and S. cremoris SK3S, insertion in plasmid DNA occurred in two of five and in one of five transconjugants, respectively (data not shown). Interestingly, insertion in the chromosome of these isolates was also observed, suggesting that, in some cases at least, Tn919 can insert in multiple sites in the same strain.

Development of a High-Frequency Delivery System for Tn919

Conjugative transfer of Tn919 to lactic acid bacteria, including the lactic streptococci, typically occurred at a frequency of 10^{-7} to 10^{-8} per recipient when the filter mating technique was used (Table 1). However, under these conditions the actual number of transconjugants was low, ranging between 50 and 500 per ml of mating mix. The alternative agar surface mating protocol allows the recovery of a greater number of recipients and consequently a higher yield of transconjugants, provided the transfer frequency remains unaltered. However, as previously reported, Tn919 will not transfer at detectable frequencies when this technique is used (9). When the high-frequency conjugative plasmid pMG600, conferring the Lac⁺ Lax⁻ phenotype, was introduced into S. lactis CH919 (an S. lactis MG1363 tranconjugant harboring Tn919 on its chromosome) from an S. lactis MG4600 donor, a selected transconjugant, designated CH001, was capable of donating Tn919 to S. lactis CK50 (a streptomycin-resistant, plasmid-free derivative of S. lactis 712) in agar surface matings. The

TABLE 2. Comparison of the ability of *S. lactis* CH001 and *S. lactis* CH919 to donate Tn*919* to *S. lactis* and *S. lactis* subsp. *diacetylactis* recipients

S. lactis strain	Recipient	DNA transferred	Frequency per recipient
MG4600	*S. lactis* CH919	pMG600	5.0×10^{-1a}
CH001	*S. lactis* CK50	pMG600	5.0×10^{-1b}
		pMG600, Tn*919*	1.3×10^{-4b}
		Tn*919*	2.0×10^{-5b}
CH919	*S. lactis* MG4600	Tn*919*	1.0×10^{-4b}
CH919	*S. lactis* CK50	Tn*919*	7.0×10^{-8c}
CH001	*S. lactis* subsp. *diacetylactis* 18–16S	Tn*919*	3.0×10^{-6b}
CH919	*S. lactis* subsp. *diacetylactis* 18–16S	Tn*919*	7.0×10^{-8c}

[a] Broth matings.
[b] Agar surface matings.
[c] Filter matings.

transfer frequency was also significantly higher than that obtained with filter matings (Table 2). In addition to this increased transfer frequency, a greater recovery of recipients from plate matings compared with filter matings led to an overall 10,000-fold increase in transconjugant numbers. Analysis of these transconjugants revealed that 80% exhibited the Lac$^+$ Lax$^-$ Tc^r phenotype, suggesting transfer of both pMG600 and Tn*919*. The remaining 20% were Lac$^-$ Lax$^+$ Tc^r, indicating the transfer of Tn*919* only. These data may imply that pMG600 is supplying conjugative functions for Tn*919* which are otherwise poorly expressed by the transposon. This is supported by the fact that the presence of the plasmid allows Tn*919* to transfer on agar surfaces and also the observation that similar high-frequency transfer of the transposon occurs when pMG600 is present in recipient strains (Table 2).

The pMG600-aided transfer of Tn*919* described above occurred between strains derived from *S. lactis* 712 (pMG600 is also derived from the strain 712 Lac/Prt plasmid, pLP712). It was of interest to determine whether this system could also be applied to other strains. When strain 18–16S was used as a recipient in a mating experiment with the CH001 donor, transfer of Tn*919* was observed at frequencies significantly greater than those obtained from a CH919 donor with the filter mating technique (Table 2).

Significantly, CH001 was unable to transfer Tn*919* to *S. cremoris, Lactobacillus*, and *Leuconostoc* strains in agar surface matings. This suggests that the high-frequency delivery system is strain specific and that the conjugational functions specified by pMG600 in the CH001 donor are incompatible with some recipient cell types. The failure to detect transfer of Tn*919* alone probably reflects the inability of strains with the transposon to conjugate on agar surfaces since low-frequency transfer to these recipients was

demonstrated when the filter mating technique was used (9).

Thus, this study has shown that the combination of random insertion and high-frequency transfer of Tn*919* can be achieved with members of the lactic streptococci. This will now enable the targeting and cloning of specific genes in these bacteria, and such experiments are currently under way in this laboratory.

This material is based in part upon contract research supported by the National Board for Science and Technology, by the National Enterprise Agency, and by E.E.C. Contract no. GBI–2–055–EIR.

LITERATURE CITED

1. Clewell, D. B., F. Y. An, B. A. White, and C. Gawron-Burke. 1985. *Streptococcus faecalis* sex pheromones (cAM373) also produced by *Staphylococcus aureus* and identification of a conjugative transposon (Tn*918*). J. Bacteriol. **162:**1212–1220.

2. Fitzgerald, G. F., and D. B. Clewell. 1985. A conjugative transposon (Tn*919*) in *Streptococcus sanguis*. Infect. Immun. **47:**415–420.

3. Franke, A. E., and D. B. Clewell. 1981. Evidence for a chromosome-borne resistance transposon (Tn*916*) in *Streptococcus faecalis* that is capable of "conjugal" transfer in the absence of a conjugative plasmid. J. Bacteriol. **145:**494–502.

4. Gasson, M. J., and F. L. Davies. 1980. High-frequency conjugation associated with *Streptococcus lactis* donor cell aggregation. J. Bacteriol. **143:**1260–1264.

5. Gasson, M. J., and F. L. Davies. 1984. The genetics of dairy lactic acid bacteria, p. 99–126. *In* F. L. Davies and B. A. Law (ed.), Advances in the microbiology and biochemistry of cheese and fermented milk. Elsevier Applied Science Publishers, London.

6. Gawron-Burke, C., and D. B. Clewell. 1982. A transposon in *Streptococcus faecalis* with fertility properties. Nature (London) **300:**281–284.

7. Gawron-Burke, C., and D. B. Clewell. 1984. Regeneration of insertionally inactivated streptococcal DNA fragments after excision of Tn*919* in *Escherichia coli*: a strategy for targeting and cloning genes from gram-positive bacteria. J. Bacteriol. **159:**214–221

8. Hartley, D. L., K. R. Jones, J. A. Tobian, D. J. LeBlanc, and F. L. Macrina. 1984. Disseminated tetracycline resist-

ance in oral streptococci: implication of a conjugative transposon. Infect. Immun. **45**:13–17.

9. **Hill, C., C. Daly, and G. F. Fitzgerald.** 1985. Conjugative transfer of the transposon Tn*919* to lactic acid bacteria. FEMS Microbiol. Lett. **30**:115–119.

10. **Kondo, J. K., and L. L. McKay.** 1985. Gene transfer systems and molecular cloning in group N streptococci: a review. J. Dairy Sci. **68**:2143–2159.

11. **Vijayakumar, M. N., S. D Priebe, G. Pozzi, J. M. Hageman, and W. R. Guild.** 1986. Cloning and physical characterization of chromosomal conjugative elements in streptococci. J. Bacteriol. **166**:972–997.

Molecular Genetics of Metabolic Traits in Lactic Streptococci

M. J. GASSON, S. H. A. HILL, AND P. H. ANDERSON

AFRC Institute of Food Research, Reading Laboratory, Shinfield, Reading RG2 9AT, England

The lactic streptococci have three well-documented plasmid-encoded traits that are important for their growth in milk: the catabolism of lactose with lactic acid production, the presence of a potent cell wall-bound proteinase, and the utilization of citrate with the generation of flavor compounds such as diacetyl.

Genes for lactose catabolism and the proteinase are encoded by pLP712, a 56.5-kilobase plasmid from *Streptococcus lactis* 712 (1, 2). This plasmid has been extensively characterized by restriction endonuclease mapping and by the isolation and mapping of a large number of deleted derivatives of the plasmid. The restriction map for pLP712 and the plasmid deletion map are presented in Fig. 1 and 2, respectively. By correlating the position of the various deletions with their effect on the lactose and proteinase phenotypes, it was possible to locate the genetic determinants for these traits on the pLP712 restriction map (Fig. 2).

The further characterization of these genetic determinants has involved their isolation by cloning onto plasmid vectors. The lactose region from a deleted pLP712 derivative, plasmid pMG820, was cloned by using the vector pAT153 in *Escherichia coli* (8). The gene for phospho-β-D-galactosidase was precisely located within the cloned DNA by isolating and characterizing a series of in vitro-generated deletions (Fig. 3). The gene encoded a 60-kilodalton protein that has been purified from *S. lactis* 712. A molecule of that size was also identified by fluorography of sodium dodecyl sulfate-polyacrylamide gel electrophoresis (SDS-PAGE)-separated ^{35}S-labeled proteins that were produced by the pAT153 clones both in *E. coli* minicells and by in vitro transcription and translation (Fig. 3). A direction of transcription for the gene was determined by the observation of a truncated polypeptide in one deleted derivative (Fig. 3). The phospho-β-D-galactosidase gene has been subcloned onto bacteriophage M13 vectors and is being sequenced in collaboration with W. de Vos, Ede, The Netherlands. Thus far, a promoterlike sequence with −35 and −10 domains and an open reading frame have been determined, and this confirms the direction of transcription for the gene.

The proteinase gene of pLP712 has also been cloned by using lactic streptococcal vectors.

These vectors have been constructed for use in the lactic streptococci by combining antibiotic resistance genes from established *Bacillus subtilis* vectors with the replication regions of small multicopy cryptic plasmids which are common in the lactic streptococci. One such vector, pCK1, consists of the pSH71 (2) replication region and the chloramphenicol and kanamycin resistance genes of *Bacillus* vector pBD64 (3; Fig. 4). This vector and others based on the replication regions of pSH72 (2) and the citrate plasmid pCT176 can be maintained by *S. lactis*, *B. subtilis*, and certain strains of *E. coli*. This promiscuous replication property, first described by Kok et al. for the vector pGK12 (7), makes the lactic streptococcal vectors especially valuable and versatile shuttle vectors. The restriction and deletion maps of pLP712 located the proteinase region on a 4.85-kilobase *Bgl*II D fragment. This was cloned into the *Bgl*II site of pCK1 by using insertional inactivation of the kanamycin resistance gene. This experiment was performed in *B. subtilis* and proved to be difficult. Instability problems were encountered, and only two clones with intact *Bgl*II D fragments were isolated from over 2,000 potential clones that were analyzed by plasmid DNA isolation. These two *Bgl*II D clones were transformed into the plasmid-free *S. lactis* MG1363 (2), and proteinase activity was detected, but the level was less than that of a strain carrying the parent plasmid pLP712. Proteinase activity was assayed by an SDS-PAGE technique which detected the specific breakdown products of β-casein (Fig. 5; S. H. A. Hill and M. J. Gasson, J. Dairy Res., in press). The low activity of the clones was due to truncation of a very large proteinase gene which is consistent with an estimated molecular size of 180 kilodaltons for the proteinase enzyme.

Strains of *S. lactis* subsp. *diacetylactis* ferment citrate, and this unstable property is a well-established plasmid-encoded trait. The citrate permease gene is encoded by a 5.5-megadalton plasmid in a variety of different *S. lactis* subsp. *diacetylactis* strains (5, 6). A detailed restriction map of the citrate plasmid from *S. lactis* subsp. *diacetylactis* 176 has been determined (4), and various fragments of the plasmid have been cloned onto vector pAT153 in *E. coli*.

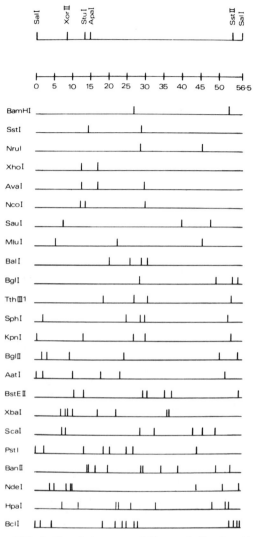

FIG. 1. Restriction map of the metabolic plasmid pLP712. The map was constructed by analysis of single and double digests. Single sites are shown at the top, and increasingly more frequent cutting endonucleases are shown below the kilobase coordinates. The map is oriented about the single site for restriction endonuclease SalI.

These clones have been used for in vitro transcription/translation studies. Two gene products encoded by this pCT176 plasmid were identified by fluorography of SDS-PAGE-separated ^{35}S-labeled proteins (Fig. 6). Thus far, the region of pCT176 that encodes the larger 48-kilodalton protein has been cloned onto the lactic streptococcal vector pCK1 and transformed into a plasmid-free derivative of S. lactis subsp. diacetylactis 176. This did not restore the citrate-fermenting ability of the strain, making it likely that the smaller 32-kilodalton protein

alone or in combination with the larger molecule represents the citrate permease.

This work was supported in part by the Biomolecular Engineering Programme of the Commission of the European Communities.

LITERATURE CITED

1. **Gasson, M. J.** 1982. Identification of the lactose plasmid in *Streptococcus lactis* 712, p. 217–220. *In* D. Schlessinger

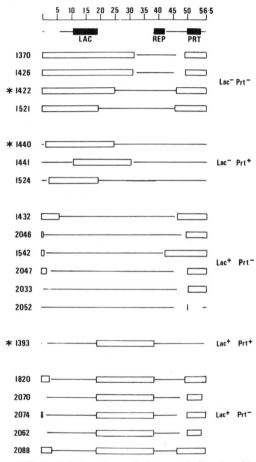

FIG. 2. Deletion map of the metabolic plasmid pLP712. Deleted derivatives of pLP712 were isolated after analysis of lactose-negative (Lac$^-$) or proteinase-negative (Prt$^-$) variants and after transductional shortening (1393). Double deletions (1820–2088) were isolated in two stages as Prt$^-$ variants of the transductionally shortened plasmid. The deletions were mapped by restriction endonuclease analysis. Deleted DNA is shown as an open box, DNA present is a solid line, and gaps show small regions of ambiguity due to the limitation of restriction endonuclease analysis. The asterisks indicate highly preferential deletion events. The location of genetic determinants for lactose utilization (LAC), proteinase production (PRT), and replication (REP) are represented by solid bars below the kilobase coordinates.

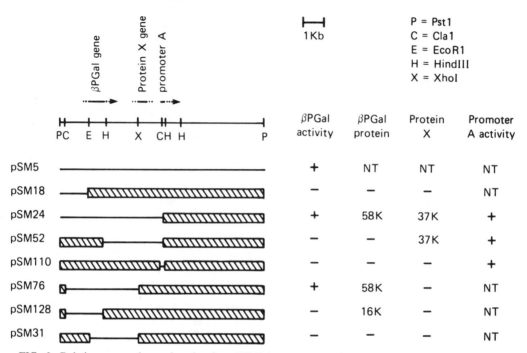

	βPGal activity	βPGal protein	Protein X	Promoter A activity
pSM5	+	NT	NT	NT
pSM18	−	−	−	NT
pSM24	+	58K	37K	+
pSM52	−	−	37K	+
pSM110	−	−	−	+
pSM76	+	58K	−	NT
pSM128	−	16K	−	NT
pSM31	−	−	−	NT

1 Kb

P = Pst1
C = Cla1
E = EcoR1
H = HindIII
X = XhoI

FIG. 3. Deletion map and gene location for pLP712 lactose genes cloned in *E. coli* using vector pAT153. Hatched areas indicate regions deleted in vitro by cleavage with restriction endonucleases and religation. The presence of phospho-β-D-galactosidase (βPgal) activity is indicated. Radiolabeled proteins synthesized in vivo by minicells and in vitro by coupled transcription and translation are shown by quoting the protein sizes in kilodaltons. Promoter A activity is the presence of a tetracycline-resistant phenotype in clones with streptococcal DNA adjacent to the *Hind*III site of pAT153. P, *Pst*I; C, *Cla*I; E, *Eco*RI; H, *Hind*III; X, *Xho*I; NT, not tested. (Reproduced from reference 8.)

(ed.), Microbiology—1982. American Society for Microbiology, Washington, D.C.

2. **Gasson, M. J.** 1983. Plasmid complements of *Streptococcus lactis* NCDO 712 and other lactic streptococci after protoplast-induced curing. J. Bacteriol. **154:**1–9.

3. **Gasson, M. J., and P. H. Anderson.** 1985. High copy number plasmid vectors for use in lactic streptococci. FEMS Microbiol. Lett. **30:**193–196.

4. **Gasson, M. J., and F. L. Davies.** 1984. The genetics of dairy lactic acid bacteria, p. 99–126. *In* F. L. Davies and B. A. Law (ed.), Advances in the microbiology and biochemistry of cheese and fermented milk. Elsevier Applied Science Publishers Ltd., London.

5. **Kemplar, G. M., and L. L. McKay.** 1979. Characterization of plasmid deoxyribonucleic acid in *Streptococcus lactis* subsp. *diacetylactis*: evidence for plasmid-linked citrate

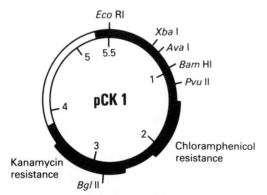

FIG. 4. Restriction map of lactic streptococcal vector pCK1. Solid areas are derived from *B. subtilis* vector pBD64, and open areas are derived from the cryptic streptococcal plasmid pSH71 (2, 3).

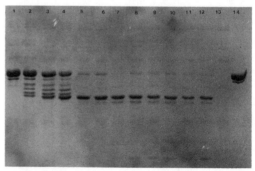

FIG. 5. Detection of *S. lactis* proteinase activity. Hydrolysis of β-casein with time by *S. lactis* strain MG1299 (Prt⁺). Lanes 1 and 14, β-casein standards (no cells); lanes 2–12, MG1299 cells mixed with β-casein (30°C) for 0, 5, 10, 15, 30, 60, 120, 180, 240, 360, 420, and 480 min, respectively; lane 13, MG1299 cells (no β-casein).

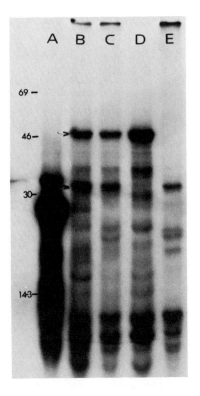

utilization. Appl. Environ. Microbiol. **37**:316–323.

6. **Kemplar, G. M., and L. L. McKay.** 1981. Biochemistry and genetics of citrate utilization in *Streptococcus lactis* ssp. *diacetylactis.* J. Dairy Sci. **64**:1527–1539.

7. **Kok, J., J. M. B. M. van der Vosen, and G. Venema.** 1984. Construction of plasmid cloning vectors for lactic streptococci which also replicate in *Bacillus subtilis* and *Escherichia coli.* Appl. Environ. Microbiol. **48**:726–731.

8. **Maeda, S., and M. J. Gasson.** 1985. Cloning, expression and location of the *Streptococcus lactis* NCDO 712 gene for phospho-β-D-galactosidase. J. Gen. Microbiol. **132**:331–340.

FIG. 6. Fluorography of an SDS-PAGE gel showing ³⁵S-labeled proteins encoded by the citrate plasmid pCT176. The tracks carry the *E. coli* vector pAT153 (A), pAT153 clones with the entire citrate plasmid in two orientations (B and C), and the plasmid cloned as two separate halves (D and E). Two pCT176-encoded proteins of 48 and 32 kilodaltons, one located on each half of the plasmid, are visible.

Molecular and Genetic Analysis of Plasmid-Encoded Structural Genes for Lactose Metabolism from *Streptococcus cremoris*

JULIA M. INAMINE,[1] LINDA N. LEE,[2] AND DONALD J. LeBLANC[2]

Department of Microbiology and Immunology, University of North Carolina at Chapel Hill, Chapel Hill, North Carolina 27514,[1] and Bacterial Virulence Section, Laboratory of Molecular Microbiology, National Institute of Allergy and Infectious Diseases, Fort Detrick, Frederick, Maryland 21701-1013[2]

The group N streptococci (*Streptococcus cremoris*, *S. lactis*, and *S. lactis* subsp. *diacetylactis*) are used extensively in the dairy industry as starter cultures for making cheese (10). The usefulness of these species in this regard is due, in part, to the production of lactic acid from the homolactic fermentation of lactose by the following pathway (10). Two lactose-specific components of the phosphoenolpyruvate-dependent phosphotransferase system (PTS), enzyme II-lac and factor III-lac, are required for the transport of lactose into the cell as lactose phosphate. This intermediate is then hydrolyzed to glucose and galactose 6-phosphate by the enzyme phospho-β-D-galactosidase (p-β-gal). Glucose is further metabolized via the Embden-Meyerhof pathway, and the galactose 6-phosphate is further metabolized via enzymes of the D-tagatose 6-phosphate pathway (galactose-6-phosphate isomerase, D-tagatose-6-phosphate kinase and tagatose-1,6-disphosphate aldolase) to the triose phosphates.

The results of experiments utilizing plasmid curing techniques, transductional analyses, or conjugation indicate that lactose metabolism is associated with the presence of specific plasmids among numerous strains of group N streptococci (9). McKay and co-workers (1, 16) provided evidence correlating the presence of such plasmids in these organisms with the synthesis of the two components of the lac-PTS, as well as p-β-gal activity. Crow et al. (4) provided evidence for plasmid linkage of the D-tagatose 6-phosphate pathway enzymes in *S. lactis*. However, none of these studies confirmed that the specific structural genes for lactose metabolism were plasmid borne. Recent attempts to identify such structural genes on lactose-metabolic plasmids from group N streptococci have incorporated transformation and recombinant DNA technologies.

The Challis strain of *S. sanguis* can express a natural competence for transformation (17), was the first streptococcal strain used in plasmid transformation experiments (11), and metabolizes lactose by the same pathway as the group N streptococci (5). Thus, it seemed appropriate to use this strain as a recipient in transformation experiments designed to establish the presence of lactose-specific structural genes on plasmid DNA from the dairy streptococci. Lac⁻ mutants of *S. sanguis* Challis were obtained after treatment with the mutagen N-methyl-N'-nitro-N-nitrosoguanidine. The *lac-8* isolate was deficient in the p-β-gal activity but continued to synthesize lac-PTS activity and was thus sensitive to the presence of lactose in the growth medium (E. J. St. Martin, personal communication). A second mutant, obtained after further treatment of the *lac-8* strain with N-methyl-N'-nitro-N-nitrosoguanidine, was lactose insensitive and defective in the synthesis of both lac-PTS and p-β-gal activities (19). This new strain (*lac-83*) was used in transformation experiments with purified plasmid DNA from *S. lactis* ATCC 11454, and Lac⁺ transformants were selected on a medium containing lactose as the sole source of carbon and energy (19). The transformants had acquired a lac-PTS activity with the same substrate specific activity (galactose/lactose PTS activity ratio) as the *S. lactis* donor strain (0.30), rather than that of the wild-type Challis strain (0.05), and synthesized a p-β-gal protein with the same size as that produced by the *S. lactis* donor (40,000 daltons) rather than that of *S. sanguis* (52,000 daltons). These results provided the first evidence that a group N lactose-metabolic plasmid actually encoded structural genes for p-β-gal activity and at least one component of the lac-PTS. Because of the stability of the lactose-metabolic trait and the lack of detectable plasmids in the transformants of the *lac-83* strain, it was suggested that the *S. lactis* plasmid-encoded genes had integrated into the Challis chromosome (19). Similar results were obtained by Harlander and McKay (6) following transformation of the *lac-8* and *lac-83* strains with purified pLM2001 DNA, a lactose-metabolic plasmid from *S. lactis* LM0232. These investigators could not detect any Lac⁻ derivatives of the Challis transformants after treatment with acriflavine under conditions that cured greater than 60% of Lac$^+$ *S. lactis* C2 and LM0232. Additional evidence for the integration of lactose-metabolic genes into the chromosome of *S. sanguis* Lac$^+$ transformants was obtained when it was shown that restriction endonuclease-digested lactose-metabolic plasmid DNA, or purified fragments thereof, could also transform the *lac-8* and *lac-83* strains (7; D. J. LeBlanc et al., unpublished data).

Several attempts have been made to clone *S. lactis* lactose-metabolic plasmid DNA directly into *S. sanguis* or *S. lactis* using streptococcal plasmid vectors. Harlander et al. (7) inserted a 23-kilobase-pair (kbp) *Kpn*I fragment from pLM2001 into the *S. sanguis* vector pDB101 (2), which contains an erythromycin (Em) resistance marker, and obtained Lac$^+$ transformants of the *lac-8* and *lac-83* strains. A recombinant plasmid from one of these transformants was used to retransform the Lac$^-$ Challis strains to a Lac$^+$ Em^r phenotype. A second *Kpn*I fragment from pLM2001 was also cloned with pDB101, and it too restored the Lac$^+$ phenotype to the *lac-83* strain. These results suggested that pLM2001 contains duplicate copies of the genes for p-β-gal and lac-PTS activity. The erythromycin resistance trait, but not the Lac$^+$ phenotype of the recombinant clones, could be cured, suggesting that the genes for the latter readily integrated into the Challis chromosome. No further delineation of the lactose-metabolic genes was attempted. Kondo and McKay (8) then used a second streptococcal plasmid, pGB301 (3), and an improved protoplast transformation system (8) to clone lactose-metabolic plasmid DNA into a plasmid-free strain of *S. lactis*. By a series of restriction endonuclease analyses and Southern blot hybridizations, they were able to localize the lactose-metabolic genes within a 17.9-kbp *Bgl*II fragment and a 19.4-kbp *Bcl*I fragment of pLM2001. Again, the delineation of individual structural genes was not reported. In this latter case, the requirement for at least six plasmid-mediated enzyme activities for the metabolism of lactose (two components of the lac-PTS system, p-β-gal, and three enzymes of the D-tagatose 6-phosphate pathway) may have precluded the direct cloning of individual lactose-metabolic genes in a plasmid-free strain of *S. lactis*.

Recently, the problems associated with the cloning of individual lactose-metabolic structural genes in *S. sanguis* and *S. lactis*, presumably due to the tendency of these genes to integrate into the host chromosome of the former and the requirement for several plasmid-mediated genetic determinants by the latter, were circumvented by cloning a plasmid-borne p-β-gal gene from *S. lactis* in *Escherichia coli*. Maeda and Gasson (14) were able to clone and express the p-β-gal gene of pLP712, an *S. lactis* lactose-metabolic plasmid, in *E. coli* using the *E. coli* vector pAT153. By examining a series of in vitro-constructed deletions of a hybrid plasmid originally containing a 10.4-kbp *Pst*I fragment from pLP712, they were able to locate the p-β-gal gene within a 3.5-kbp segment of that fragment. They were also able to correlate the synthesis of p-β-gal activity with the presence of a 58,000-dalton protein, either in minicells or by the use of an in vitro system for coupled transcription and translation.

In this report, we describe the use of an *E. coli* host-vector system to clone the genes for p-β-gal activity and at least one component of the lac-PTS system from a lactose-metabolic plasmid of *S. cremoris*. The data have been presented elsewhere (7a). These studies were begun by transferring the Lac$^+$ phenotype of *S. cremoris* H2 to a plasmid-free derivative of *S. lactis* H1. A single plasmid approximately 65 kbp in size, pJI70, was purified from one of the Lac$^+$ transconjugants and used to transform the *S. sanguis* *lac-8* and *lac-83* strains to a Lac$^+$ phenotype. As with similar Lac$^+$ transformants obtained with lactose-metabolic plasmids from *S. lactis* strains (6, 19), no plasmid DNA could be detected. *Ava*I- and *Pst*I-digested pJI70 DNA also transformed both the *lac-8* and *lac-83* strains to a Lac$^+$ phenotype, while *Eco*RI and *Sst*I digests of the plasmid could transform the *lac-8* strain, but not the *lac-83* strain. Subsequently, *Pst*I and *Sst*I fragments from pJI70 were cloned in *E. coli* using pUC19 (20) as the vector. Clones containing a 10.9-kbp *Pst*I fragment or a 10.1-kbp *Sst*I fragment were shown to express p-β-gal activity in the *E. coli* host. Because the *E. coli* phosphoenolpyruvate-dependent PTS components do not complement the corresponding systems in gram-positive bacteria (18), it was not possible to determine, in *E. coli*, whether any of the lac-PTS genes of plasmid pJI70 were also present on either of the cloned fragments. Therefore, the *Pst*I and *Sst*I fragments were separated from the vector by agarose gel electrophoresis and purified by electroelution for use in transformation with the *lac-8* and *lac-83* strains. As expected from the results obtained with the restriction endonuclease digests of pJI70, the cloned 10.9-kbp *Pst*I fragment transformed the *lac-8* and *lac-83* strains to a Lac$^+$ phenotype, but the cloned 10.1 *Sst*I fragment transformed only the *lac-8* strain (Fig. 1).

Also illustrated in Fig. 1 are the results of subcloning and transformation experiments that led to the delineation of the structural genes for p-β-gal and lac-PTS activity on the cloned fragments. A 3.5-kbp *Pst*I-*Ava*I fragment was subcloned from the *Pst*I fragment, and this also transformed both the *lac-8* and *lac-83* strains to a Lac$^+$ phenotype. Since the nature of the

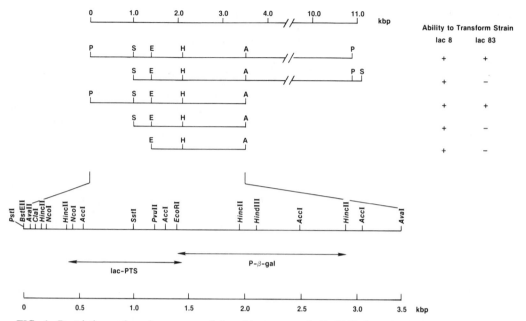

FIG. 1. Restriction endonuclease maps of cloned lactose-metabolic DNA from *S. cremoris* plasmid pJI70. The ability of each cloned fragment to transform *S. sanguis* Challis Lac⁻ mutants, *lac-8* and *lac-83*, to a Lac⁺ phenotype is indicated at the right of the figure. Letters on the linear maps indicate sites for cleavage by restriction endonucleases: A, *Ava*I; E, *Eco*RI; H, *Hin*dIII; P, *Pst*I; S, *Sst*I. The legend at the top of the figure corresponds to the top five fragments. The legend at the bottom of the figure corresponds to the expanded map of the 3.5-kbp *Pst*I-*Ava*I fragment encoding lac-PTS and p-β-gal activities. The two bidirectional arrows on the map of the 3.5-kbp *Pst*I-*Ava*I fragment indicate the proposed limits of the structural genes for lac-PTS and p-β-gal activities.

lac-PTS defect in the *lac-83* strain is not known, it was not possible to determine whether the cloned 3.5-kbp *Pst*I-*Ava*I fragment contained a gene for factor III-lac or enzyme II-lac, or both. The lac-PTS and p-β-gal sequences were further delineated by transformation experiments with smaller fragments obtained from the 3.5-kbp *Pst*I-*Ava*I fragment. Purified restriction endonuclease fragments from the hybrid plasmid containing this fragment were used to transform the *lac-8* and *lac-83* strains. A 2.5-kbp *Sst*I-*Ava*I fragment and a 2.1-kbp *Eco*RI-*Ava*I fragment transformed the *lac-8* strain to a Lac⁺ phenotype. However, the *lac-83* strain could be transformed to Lac⁺ only with the 3.5-kbp *Pst*I-*Ava*I fragment. These results permitted the mapping of the p-β-gal region within a 1.6-kbp segment of the 3.5-kbp *Pst*I-*Ava*I fragment, and at least one component of the lac-PTS system adjacent to the p-β-gal gene, within a 0.75- to 1.0-kbp segment of this fragment (bottom of Fig. 1). In addition to the lac-PTS and p-β-gal genes, this 3.5-kbp *Pst*I-*Ava*I fragment contained all that was necessary and sufficient for integration into the *S. sanguis* Challis chromosome.

Cell-free extracts from several independently isolated Lac⁺ *S. sanguis* transformants, derived by transformation of the *lac-8* and *lac-83* strains with either whole pJI70 DNA or restriction fragments from this plasmid, as well as *E. coli* strains harboring hybrid plasmids derived from pUC19 and appropriate restriction endonuclease fragments from pJI70, all contained a 40,000-dalton p-β-gal activity band in polyacrylamide gels typical of that found in *S. cremoris* H2. None of the transformants of the *lac-8* and *lac-83* strains contained the 52,000-dalton activity band associated with the wild-type *S. sanguis* Challis. These results, in addition to providing evidence for integration of the *S. cremoris* Lac⁺ DNA into the *S. sanguis* chromosome, also demonstrated that gene replacement and not gene repair was occurring in the *S. sanguis* transformants. A series of Southern blot hybridizations, in which the 3.5-kbp *Pst*I-*Ava*I fragment was used as a probe, showed that the wild-type Challis strain and the *lac-8* and *lac-83* strains each contained a single 3.3-kbp *Hinc*II DNA fragment with homology to the probe. This fragment was not among the hybridizing fragments found in the DNA of any of the Lac⁺ transformants of either the *lac-8* or *lac-83* strain, indicating that in each case the lactose-metabolic sequences from pJI70 had integrated into a common region of the Challis chromosome.

The cloned Lac[+] DNA from pJI70 was also used as a probe to determine whether homologous sequences could be found in the DNA of other gram-positive bacteria which metabolize lactose by the same pathway. Homology was detected with DNA from Lac[+] strains of *S. sanguis* and *S. faecalis*, which presumably carry the lactose-metabolic genes on their respective chromosomes, and *S. lactis* ATCC 11454, which has been shown to harbor a lactose-metabolic plasmid (12). There was also faint but detectable homology to DNA from a strain of *S. mutans*. There was no detectable hybridization to DNAs from Lac[+] strains of *S. salivarius, Staphylococcus aureus, Lactobacillus casei,* or a Lac[-] derivative of *S. lactis* ATCC 11454. The hybridization pattern obtained with DNA from the Lac[+] strain of *S. lactis* ATCC 11454 contained several restriction endonuclease fragments in common with that of the *S. cremoris* plasmid, pJI70.

The problems associated with the cloning of lactose-metabolic sequences in *S. sanguis* and group N streptococci have been circumvented by the cloning of a p-β-gal gene from *S. lactis* (14) and *S. cremoris* (this report) using *E. coli* host-vector systems. In both instances the p-β-gal gene was expressed by *E. coli* recombinant clones. In the latter case we were also able to show that plasmid DNA sequences from *S. cremoris*, cloned in *E. coli*, could complement the lac-PTS and p-β-gal mutations in Lac[-] derivatives of *S. sanguis* Challis. More recently, one of the genes of the D-tagatose 6-phosphate pathway, tagatose-1,6-diphosphate aldolase, was also cloned in *E. coli* (13). Certainly, it is very likely that the genes for all six plasmid-mediated enzyme activities associated with the metabolism of lactose by the group N streptococci will be cloned in *E. coli*. Further studies on the regulation of these enzymes and their possible arrangement in one or more operons will depend on complementation of streptococcal mutants defective in the synthesis of each of the six enzyme components. In this regard, Park and McKay (16) have obtained a derivative of *S. lactis* in which portions of the lactose-metabolic plasmid appear to be integrated into the host chromosome. The application of an improved protoplast transformation system (8) to similar derivatives, containing point mutations in the individual genes associated with lactose metabolism, may be useful for such complementation studies.

J.M.I. was the recipient of National Research Service Award DE05346 from the National Institute of Dental Research.

LITERATURE CITED

1. **Anderson, D. G., and L. L. McKay.** 1977. Plasmids, loss of lactose metabolism, and appearance of partial and full lactose-fermenting revertants in *Streptococcus cremoris* B1. J. Bacteriol. **129:**367–377.
2. **Behnke, D., and J. J. Ferretti.** 1980. Physical mapping of plasmid pDB101: a potential vector plasmid for molecular cloning in streptococci. Plasmid **4:**130–138.
3. **Behnke, D., M. S. Gilmore, and J. J. Ferretti.** 1981. Plasmid pGB301, a new multiple resistance streptococcal cloning vehicle and its use in cloning the gentamicin/kanamycin resistance determinant. Mol. Gen. Genet. **182:**414–421.
4. **Crow, V. L., G. P. Davey, L. E. Pearce, and T. D. Thomas.** 1983. Plasmid linkage of the D-tagatose 6-phosphate pathway in *Streptococcus lactis:* effect on lactose and galactose metabolism. J. Bacteriol. **153:**76–83.
5. **Hamilton, I. R., and G. C. Y. Lo.** 1978. Co-induction of β-galactosidase and the lactose-P-enolpyruvate phosphotransferase system in *Streptococcus salivarius* and *Streptococcus mutans.* J. Bacteriol. **136:**900–908.
6. **Harlander, S. K., and L. L. McKay.** 1984. Transformation of *Streptococcus sanguis* Challis with *Streptococcus lactis* plasmid DNA. Appl. Environ. Microbiol. **48:**342–346.
7. **Harlander, S. K., L. L. McKay, and C. F. Schachtele.** 1984. Molecular cloning of lactose-metabolizing genes from *Streptococcus lactis.* Appl. Environ. Microbiol. **48:**347–351.
7a. **Inamine, J. M., L. N. Lee, and D. J. LeBlanc.** 1986. Molecular and genetic characterization of lactose metabolic genes of *Streptococcus cremoris.* J. Bacteriol. **167:**855–862.
8. **Kondo, J. K., and L. L. McKay.** 1984. Plasmid transformation of *Streptococcus lactis* protoplasts: optimization and use in molecular cloning. Appl. Environ. Microbiol. **48:**252–259.
9. **Kondo, J. K., and L. L. McKay.** 1985. Gene transfer systems and molecular cloning in group N streptococci: a review. J. Dairy Sci. **68:**2143–2159.
10. **Lawrence, R. C., T. D. Thomas, and B. E. Terzaghi.** 1976. Reviews in the progress of dairy science: cheese starters. J. Dairy Res. **43:**141–193.
11. **LeBlanc, D. J., and F. P. Hassell.** 1976. Transformation of *Streptococcus sanguis* Challis by plasmid deoxyribonucleic acid from *Streptococcus faecalis.* J. Bacteriol. **128:**347–355.
12. **LeBlanc, D. J., V. L. Crow, L. N. Lee, and C. F. Garon.** 1979. Influence of the lactose plasmid on the metabolism of galactose by *Streptococcus lactis.* J. Bacteriol. **137:**878–884.
13. **Limsowtin, G. K. Y., V. L. Crow, and L. E. Pearce.** 1986. Molecular cloning and expression of the Streptococcus lactis tagatose-1,6-bisphosphate aldolase gene in Escherichia coli. FEMS Microbiol. Lett. **33:**79–83.
14. **Maeda, S., and M. J. Gasson.** 1986. Cloning, expression and location of the Streptococcus lactis gene for phosphobeta-D-galactosidase. J. Gen. Microbiol. **132:**331–340.
15. **Molskness, T. A., W. E. Sandine, and L. R. Brown.** 1974. Characterization of Lac[+] transductants of *Streptococcus lactis.* Appl. Microbiol. **28:**753–758.
16. **Park, Y. H., and L. L. McKay.** 1982. Distinct galactose phosphoenolpyruvate-dependent phosphotransferase system in *Streptococcus lactis.* J. Bacteriol. **149:**420–425.
17. **Perry, D., and H. D. Slade.** 1962. Transformation of streptococci to streptomycin resistance. J. Bacteriol. **83:**443–449.
18. **Postma, P. W., and S. Roseman.** 1976. The bacterial phosphoenolpyruvate: sugar phosphotransferase system in Staphylococcus aureus. Biochim. Biophys. Acta **457:**213–257.
19. **St. Martin, E. J., L. N. Lee, and D. J. LeBlanc.** 1982. Genetic analysis of carbohydrate metabolism in streptococci, p. 232–233. *In* D. Schlessinger (ed.), Microbiology—1982. American Society for Microbiology, Washington, D.C.
20. **Yanisch-Perron, J., J. Vieira, and J. Messing.** 1985. Improved M13 phage cloning vectors and host strains: nucleotide sequences of the M13mp18 and pUC19 vectors. Gene **33:**103–119.

Genetic Characterization of Lactic Streptococcal Bacteriophages

MICHAEL TEUBER AND MARTIN LOOF

Institute of Microbiology, Federal Dairy Research Center, D-2300 Kiel, Federal Republic of Germany

The mesophilic group N lactic acid streptococci, *Streptococcus lactis, S. lactis* subsp. *diacetylactis*, and *Streptococcus cremoris*, are used as starter cultures for the manufacture of fermented dairy products like cheese, sour cream, and sour milk. Their indispensable functions include acid production, aroma formation (e.g., diacetyl), and not yet completely elucidated contributions to cheese ripening. The total world output for cheese is now well over 10 million tons per year. Since exact numbers for the other fermented milk derivatives, including lactic butter, are not available, their volume can only be roughly estimated to amount to another 5 to 10 million tons per year. This adds up to about 2×10^{10} kg of fermented dairy products for which a milk supply of about 10^{11} kg or 25% of the total available milk is needed. Since the lactic flora grows to more than 10^9 viable cells per ml of fermented milk, the total number of viable lactic acid bacteria involved in dairy fermentations is in the order of 10^{23} per year. This underlines their enormous economic and nutritional impact.

Cheese cannot be made from heat-sterilized milk since denaturation of whey proteins onto the surface of the casein micelles would prevent a proper development of cheese texture. The only reasonable measure to lower microbial numbers in milk intended for cheese production is pasteurization for 15 to 30 s at between 70 and 74°C. However, certain thermoduric bacteria, and also bacteriophages, survive this heat treatment.

In addition, since cheese making has been traditionally performed in open vats, infection of cheese fermentations with bacteriophages is a common and serious problem aggravated by the ever increasing fermentation volumes, which have reached up to 50,000 liters in a single tank in the case of fresh cheese manufacture.

These phage problems were recognized for the first time in 1934 by Whitehead in New Zealand. The measures developed to handle the problems include the following: (i) external hygienic conditions during manufacture, with special emphasis on the prevention of recontamination of cheese milk with whey, aerosols, or particulate remains from previous or parallel fermentations; (ii) rotation of cultures of different phage types; and (iii) application of phage-resistant cultures (20, 21).

Spontaneous highly phage-resistant cultures have been selected by challenging the cultures with cocktails of all available virulent phages. These cultures are presently being used with some success in New Zealand, the United States, Ireland, and Denmark. A mixture of two to six different strains is used. If a phage against one of the strains is detected in the whey, this strain is replaced by another resistant strain. In the light of modern developments in genetics and genetic engineering of lactic acid streptococci, it appears timely to use this knowledge to construct bacteriophage-resistant strains and cultures. As a preliminary to this task, it is necessary to identify all the existing phages and know their relationships on a broad basis.

Ultrastructure of Virulent Phages (11, 21)

The ultrastructural analysis of the virulent bacteriophages of group N lactic acid streptococci provides a suitable basis for further microbiological, biochemical, and genetic characterization. The body of knowledge accumulated by extensive investigations in many laboratories all over the world is summarized in Fig. 1 and 2 and Tables 1 to 3. On the basis of head shapes and sizes, the most common phages are classified into four groups: (i) phages with prolate heads (type P001; Fig. 1A, B, and C), DNA size about 13 megadaltons; (ii) phages with small isometric heads, DNA size 18 to 20 megadaltons (type P008; Fig. 1G); (iii) phages with small isometric heads, DNA size 22 to 25 megadaltons (type P335; Fig. 1D, E, and I); and (iv) phages with large isometric heads (type P026; Fig. 2E), DNA size about 34 megadaltons.

Less common phages characterized by morphological peculiarities regarding collars, base plates, and tail fibers are listed in Table 3 and shown in Fig. 2. Further differentiation of morphologically identical phages is necessitated by the observation of different host-range patterns of otherwise morphologically identical phages.

Ultrastructure of Prophages (21)

Many strains of mesophilic lactic acid streptococci can be induced by mitomycin C or UV irradiation to show prophage-dependent lysis of the cells. However, only a few indicator strains

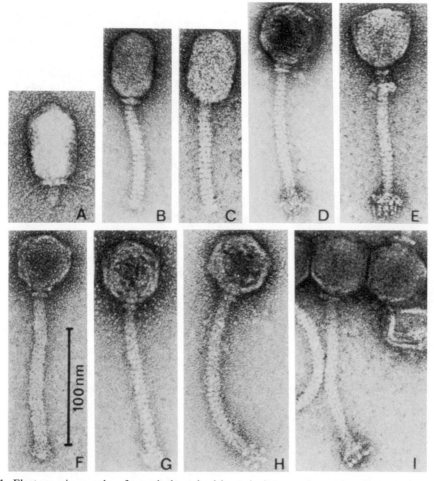

FIG. 1. Electron micrographs of negatively stained bacteriophages active against lactic streptococci. (A) Phage P034; (B) P109; (C) P001; (D) P013; (E) P059; (F) P191; (G) P008; (H) P219; (I) P142. The magnification is the same for all pictures (see bar). (From reference 11.)

are available which are able to propagate some of the temperate phages (e.g., *S. cremoris* Wg2). Therefore, we know only the ultrastructural characteristics of most prophages (see Fig. 3 and 4). Comparison of these data with those for previously recognized virulent phages (Fig. 1 and 2) indicates that some of the phages isolated as virulent phages on the basis of a plaque test may actually be temperate phages. In summary, there seems to be a broad diversity of morphologically distinct forms within the prophages of group N streptococci.

Host-Range Pattern

Determination of host-range pattern depends on the availability of genetically distinct bacterial host strains. There is no clear indication of how many different strains of *S. cremoris* and *S. lactis* may exist and be in industrial use all over

the world. On the basis of individual plasmid profiles, it may be estimated that at least several hundred distinct strains exist and are employed in the form of undefined, mixed-strain cultures (1). Since phage resistance develops easily and since resistance mechanisms (mutation of adsorption sites, restriction/modification) are not well defined in most strains, the host-range pattern is not a very suitable marker for differentiating bacteriophages.

Immunological Characterization

The reaction of phages with specific antibodies can be observed in an electron microscope or can be measured by inhibition of infectivity. Morphologically identical phages may show different immunological properties (12), and immunologically identical phages may have different host ranges (3).

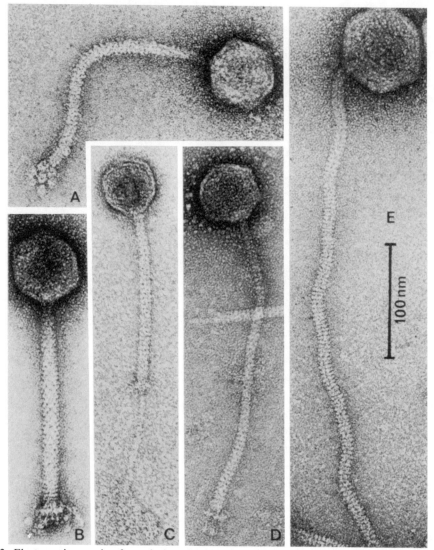

FIG. 2. Electron micrographs of negatively stained bacteriophages active against lactic streptococci (continued from Fig. 1). (A) P047; (B) P087; (C) P107; (D) P204; (E) P026. The magnification is the same for all pictures (see bar). (From reference 11.)

Genetic Characterization

A potent aid to supplement classical methods of phage differentiation could be the use of modern methods of DNA analysis, including moles percent guanine-plus-cytosine (G+C) content, mapping by restriction endonucleases, and electron microscopy of heteroduplex molecules, in addition to analysis of bacteriophage mutants. However, no systematic study has yet been even initiated to characterize the genomes of lactic streptococcal phages by directed mutagenesis.

Moles percent G+C content, genome sizes, and forms. The moles percent G+C contents of

group N streptococcal bacteriophages indicated in Tables 1 and 2 were determined mainly by the melting-point method and in several instances were confirmed by high-performance liquid chromatography of DNA hydrolysates. The values, ranging from 36.7 to 40.7, are in the range of the G+C content of the genome of group N streptococci (38 to 40%). The genome sizes of the different phage types (12.7 to 34) were determined by direct electron microscopy and by electrophoretic measurements of restriction endonuclease fragments. For phage types P001 and P008, cohesive ends of the DNA have been described which are responsible for the circular

TABLE 1. Characteristics of common prolate-head bacteriophages (type P001) of group N streptococci

Head size (nm)	Tail size (nm) Length	Tail size (nm) Width	Proteins (kilodaltons)	Mol wt of DNA[a]	G+C (mol%)[a]	Peculiarities	V/T[b]	No. of phages	No. of subgroups	Origin	Reference
51-62 by 40-45	90-102	5-10	169, 85, 64, 51, 35, 31, 26, 23, 22, 21, 20, 19	14.6×10^6	36.7	Cohesive ends Physical map	V	12		Australia	10 17
56-61 by 46-51	87-96	8-10		13×10^6	41		V	1		Germany	6
59-61 by 42-44	105-111	11		12.7×10^6			V	49	5[c]		11 (type P001)
56-63 by 44-48	82-95			15×10^6	39.9		V	14	2-3[d]	New Zealand	19 (type 4) 7 (type d) 8 (type d)
60 by 46	118 265	10 8					V	2		Norway	2
52-69 by 36-45	102-110	7-10					V	3		Argentina	16
52 by 40	105			14.7×10^6		Cohesive ends	V	1		Ireland	5
							T	1		France	4

[a] G+C content of DNA.
[b] V, Virulent phages; T, temperate phages.
[c] By immune electron microscopy.
[d] By hybridization with labeled DNA and neutralization with antiserum.

TABLE 2. Characteristics of common isometric head phages of group N streptococci

Phage type	Head size (nm)	Tail size (nm)		Base plate (nm)	Proteins (kilodaltons)	Mol wt of DNA
		Length	Width			
P008	52–62	125–153				
	50–61	140–154				17.8×10^6–21.9×10^6
	49–54	150–154	12–14			
	50–54	143–172	10.5–12.5		122, 77, 41.5, 35, 28.5, 22.5	18.0×10^6–19.8×10^6
	60	150	11			
	52	143–152				20.6×10^6
P026	81–88	450–500				
	83–88	450–457				34×10^6
	80	500				
	98	551	12			
P335	53–61	113–134				
	53–54	109–110				22×10^6–25×10^6
	52–55	128–147	11	25		
	51–55	110–146[h]	8–10	14–22 × 18–27		23.9×10^6–24.6×10^6

[a] V, Virulent phages; T, temperate phages.
[b] By presence of collar.
[c] By tail morphology.
[d] By restriction endonuclease patterns and immune electron microscopy.
[e] By DNA hybridization.
[f] By morphology of tail and collar.
[g] By morphology of tail and whisker, and unusual or modified bases.
[h] Length without length of base plate.

appearance of the phage DNA (13, 17; M. Loof, Ph.D. thesis, University of Kiel, Kiel, Federal Republic of Germany).

Restriction endonuclease mapping of phage DNA. Physical restriction maps have been determined for phages of types P001 and P008 (see Tables 1 and 2).

Figure 5 shows a restriction endonuclease map of phage P008. The position of the cohesive ends has been taken as the zero point in the map, which lies within the fragment C region of EcoRI

digests. By heteroduplex analysis of the DNA of the P008 type, which differed only by the presence or absence of a collar and whiskers, it was possible to establish an absolute orientation of the DNA of phage P008. The stippled area in Fig. 5 corresponds to the position of a single-stranded loop observed by electron microscopy (Fig. 4) in the heteroduplex between DNA of phage P008 (having collar and whiskers) and phage P113 G (lacking collar and whiskers). Heteroduplex molecules between P008 DNA

TABLE 3. Additional, less common phages of group N streptococci

Phage type	Head size (nm)	Tail size (nm)		Details	V/T	No. of isolates	Origin	Reference
		Length	Width					
P107	55	152	11	Long tail fiber	V/T	2	Germany	11, 21
				From S. cremoris TL853, TL925	T	3	France	4
P047	65	260	14	Baseplate 30-nm width	V/T	51	Germany	11
P087	65	200	16	Baseplate 40-nm width	V	1	Germany	11
C10III	70	180	12	Baseplate 30-nm width	V	1	Australia	10
Type II	73	195	20	Baseplate	V		Bulgaria	22
Phage 963	70	170		Baseplate	V	1	New Zealand	19
P204	55	255	10	Baseplate 16-nm width	V/T	1	Germany	11
P034	65 × 44	24	10	Collar with appendages	V	4	Germany	11
KSY1	230 × 50	35		Complex collar structure	V	1	Finland	18

G+C (mol%)	Peculiarities	V/T[a]	No. of		Origin	References
			Phages	Subgroups		
		V	90	2[b]	New Zealand	19 (group 1 and 2)
40.7	Physical map	V	12	2[b]		8, 9 (type a and b)
		V	35	2[c]	Germany	11 (type P008 and P191)
37.5	Cohesive ends		43	4[d]		13; Loof, Ph.D. thesis
	Physical map	V				
		V	14	2[b]	Norway	2
	Cohesive ends Physical map	V	2		Ireland	5
		V	5		New Zealand	19 (group 5)
38.8		V	4	2[e]	New Zealand	8 (type e)
		V	3		Germany	11 (type P026)
					Bulgaria	22 (type III)
		V	17		New Zealand	19 (group 3)
		V	2			8 (type c)
		V/T	9	3[f]	Germany	11 (types P013, P059, P142)
39–40.7	Terminal redundant (12% of the genome) and circular permuted genome; unusual or modified bases	V/T	13	3[g]		Loof, Ph.D. thesis (type P335)

and the DNA of phage P272 of the same type, but again without collar and whiskers, demonstrate a remarkable disparity on the right-hand side of the heteroduplex molecule in addition to the large loop at position 4 of the megadalton scale (14). A similar experiment conducted by Jarvis and Meyer (9) with phages of the P008 type isolated in New Zealand over a period of several years revealed striking parallels to the observations in our laboratory: the heteroduplex molecules had long stretches of undisturbed double-strand areas on the "left-hand side" of the molecule but many single-stranded loops on the "right-hand side." This suggests that natural evolution of the phage DNA by spontaneous mutation might occur mainly in this portion of the DNA.

Comparison of a P008-type phage, am1A, isolated from an Irish dairy, with phage P008 showed a few small fragments of am1A to hybridize with P008 fragments and map in the same position (5). Daly also presented evidence that the cohesive ends of phages am1A and P008 might be similar.

DNA-DNA hybridization by Southern blotting. Southern blotting has been used to compare different virulent and temperate phage types (8, 9). Until now, no DNA homology has been detected between the temperate and virulent phages investigated, whereas strong homology was observed within the four individual main phage types listed above (8; Loof, Ph.D. thesis).

Conclusion

The genetic characterization of bacteriophages of lactic acid streptococci is an important method to gain insight into the evolution of these phages. Together with morphological and immunological methods, it enables a detailed taxonomic characterization which is a prerequisite for further work on the construction of bacteriophage-resistant starter cultures.

Part of this work was supported by grants from the Deutsche Forschungsgemeinschaft (Bonn) and the European Community.

LITERATURE CITED

1. **Andresen, A., A. Geis, U. Krusch, and M. Teuber.** 1984. Plasmidmuster milchwirtschaftlich genutzter Starterkulturen. Milchwissenschaft **39**:140–143.
2. **Brinchmann, E., E. Namork, B. V. Johansen, and T. Langsrud.** 1983. A morphological study of some lactic streptococcal bacteriophages isolated from Norwegian cultured milk. Milchwissenschaft **38**:1–4.
3. **Budde-Niekiel, A., V. Möller, J. Lembke, and M. Teuber.** 1985. Ökologie von Bakteriophagen in einer Frischkäserei. Milchwissenschaft **40**:477–481.
4. **Chopin, M.-C., A. Rouault, and M. Rousseau.** 1983. Elimination d'un prophage dans les souches mono- et multilysogènes de streptocoques de groupe N. Lait **63**:102–115.
5. **Daly, C.** 1986. Exploitation of recombinant DNA technology to provide improved cultures for dairy fermentations, p. 453–463. In E. Magnien (ed.), Biomolecular engineering in the European community: achievements of the research programme (1982–1986)—final report. Martinus Nijhoff Publishers, Dordrecht, The Netherlands.
6. **Engel, G., K. E. v.Milczewski, and A. Lembke.** 1975. Versuche zur Differenzierung von Phagen der Streptococcus lactis und cremoris Gruppe: Bakterienspektrum, serologische und morphologische Kriterien. Kiel. Milchwirtsch. Forschungsber. **27**:25–48.
7. **Jarvis, A. W.** 1977. The serological differentiation of lactic streptococcal bacteriophage. N.Z. J. Dairy Sci. Technol. **12**:176–181.
8. **Jarvis, A. W.** 1984. Differentiation of lactic streptococcal phages into phage species by DNA-DNA homology. Appl.

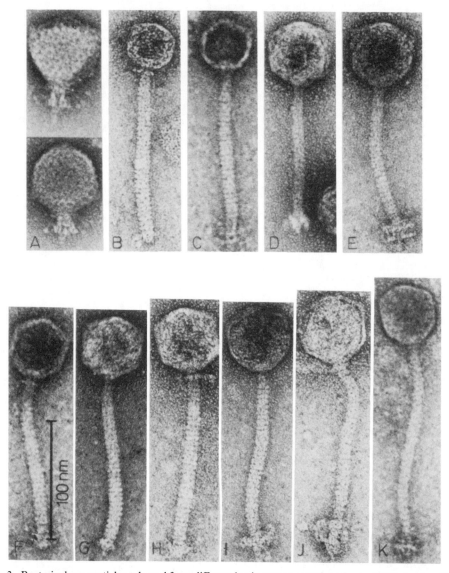

FIG. 3. Bacteriophage particles released from different lactic streptococcal strains by UV induction (21): (A) *S. cremoris* S18/4; (B) *S. lactis* subsp. *diacetylactis* 54296; (C) *S. lactis* subsp. *diacetylactis* 20384; (D) *S. lactis* C10; (E) *S. cremoris* P8/0–51; (F) *S. cremoris* C13; (G) *S. lactis* 51732; (H) *S. cremoris* E8K; (I) *S. cremoris* Wis98; (J) *S. lactis* 47724; (K) *S. cremoris* S18/4; (L) *S. cremoris* P8/2–3; (M) *S. lactis* subsp. *diacetylactis* 54292; (N) *S. cremoris* 3096; (O) *S. lactis* subsp. *diacetylactis* 54292; (P) *S. lactis* 20388; (Q) *S. lactis* 25552; (R) *S. lactis* 42172; (S) *S. lactis* 42200. The magnification is the same for all pictures (see bar). (From reference 21.)

Environ. Microbiol. **47:**343–349.

9. **Jarvis, A. W., and J. Meyer.** 1986. Electron microscopic heteroduplex study and restriction endonuclease cleavage analysis of the DNA genome of three lactic streptococcal bacteriophages. Appl. Environ. Microbiol. **51:**566–571.

10. **Keogh, B. P., and P. D. Shimmin.** 1974. Morphology of the bacteriophages of lactic streptococci. Appl. Microbiol. **27:**411–415.

11. **Lembke, J., U. Krusch, A. Lompe, and M. Teuber.** 1980. Isolation and ultrastructure of bacteriophages of group N (lactic) streptococci. Zentralbl. Bakteriol. Parasitenkd. Infektionskr. Abt. 1 Orig. Reihe C **1:**79–91.

12. **Lembke, J., and M. Teuber.** 1981. Serotyping of morphologically identical bacteriophages of lactic streptococci by immuno electronmicroscopy. Milchwissenschaft **36:**10–12.

13. **Loof, M., J. Lembke, and M. Teuber.** 1983. Characterization of the genome of the Streptococcus lactis subsp. diacetylactis bacteriophage P008 wide-spread in German cheese factories. Syst. Appl. Microbiol. **4:**413–423.

14. **Loof, M., and M. Teuber.** 1986. Heteroduplex analysis of the genomes of Streptococcus lactis "subsp. diacetylactis" bacteriophages of the P008-type isolated from German cheese factories. Syst. Appl. Microbiol. **8:**226–229.

15. **Lyttle, D. J., and G. B. Petersen.** 1984. The DNA of

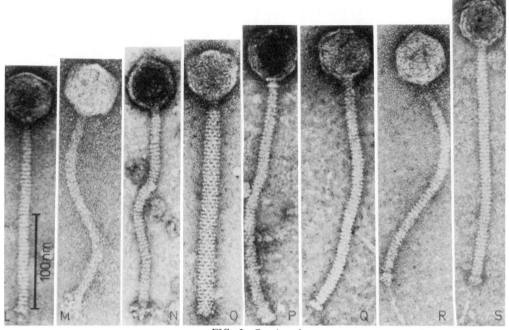

FIG. 3. *Continued*

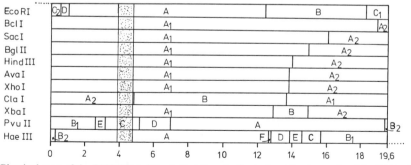

bacteriophage 643: isolation and properties of the DNA of a bacteriophage infecting lactic streptococci. Virology **133:**403–415.

16. **Parada, J. L., M. I. La Via, and A. J. Solari.** 1984. Isolation of *Streptococcus lactis* bacteriophages and their interaction with the host cell. Appl. Environ. Microbiol. **47:**1352–1354.

17. **Powell, I. B., and B. E. Davidson.** 1985. Characterization of streptococcal bacteriophage c6A. J. Gen. Virol. **66:**2737–2741.

FIG. 4. Heteroduplex molecule P008-P113G (interpretative drawing). The arrow points to the left end of the molecule (0-megadalton site; see Fig. 5). (From reference 14.)

	0	2	4	6	8	10	12	14	16	18	19,6
EcoRI	C₂ D			A				B			C₁
Bcl I			A₁								A₂
Sac I			A₁						A₂		
Bgl II			A₁						A₂		
Hind III			A₁					A₂			
Ava I			A₁					A₂			
Xho I			A₁					A₂			
Cla I		A₂		B				A₁			
Xba I			A₁				B		A₂		
Pvu II	B₁	E	C	D		A					B₂
Hae III	B₂		A			F	D E	C	B₁		

FIG. 5. Physical map of the P008 DNA. The dotted lines indicate the cohesive ends of the DNA. The capital letters are assigned in alphabetical order to restriction fragments according to decreasing sizes. A_1 and A_2, B_1 and B_2, and C_1 and C_2 indicate fragments generated by opening of the cohesive ends within an otherwise larger restriction fragment. The stippled area indicates the 0.9-megadalton deletion found in phage P113G-DNA. (From reference 14.)

18. **Saxelin, M.-L., E.-L. Nurmiaho, M. P. Korhola, and V. Sundman.** 1979. Partial characterization of a new C3-type capsule-dissolving phage of Streptococcus cremoris. Can. J. Microbiol. **25:**1182–1187.

19. **Terzaghi, B. E.** 1976. Morphologies and host range sensitivities of lactic streptococcal phages from cheese factories. N.Z. J. Dairy Sci. Technol. **11:**155–163.

20. **Teuber, M.** 1986. Mikrobiologie fermentierter Milchpro-

dukte. Chem. Mikrobiol. Technol. Lebensm. **9:**162–172.

21. **Teuber, M., and J. Lembke.** 1983. The bacteriophages of lactic acid bacteria with emphasis on genetic aspects of group N lactic streptococci. Antonie van Leeuwenhoek J. Microbiol. Serol. **49:**283–295.

22. **Tsaneva, K. P.** 1976. Electron microscopy of virulent phages for *Streptococcus lactis*. Appl. Environ. Microbiol. **31:**590–601.

Mechanisms of Bacteriophage Insensitivity in the Lactic Streptococci

CHARLES DALY AND GERALD FITZGERALD

Department of Dairy and Food Microbiology, University College, Cork, Ireland

The mesophilic lactic streptococci, which comprise *Streptococcus cremoris, S. lactis*, and *S. lactis* subsp. *diacetylactis*, are of major industrial importance as components of starter cultures used in the manufacture of a variety of fermented dairy products, e.g., cheeses, lactic butter, cultured buttermilk, sour cream, and quarg. Bacteriophage attack has been recognized since the 1930s as the most serious cause of starter culture inhibition in commercial practice. The consequence may be significant economic loss due to downgrading of product or, in severe cases, total loss of the fermentation. The vulnerability of dairying, in contrast to other modern industrial fermentations, to phages exists partly because the substrate, milk, cannot be sterilized for biochemical reasons, and in addition, the starter culture must perform in large mechanized and automated units that demand consistent, predictable acid production to ensure high-quality end products with the desired flavor and texture characteristics.

Not surprisingly, in view of their commercial significance, considerable research has been devoted to understanding and controlling lactic streptococcal phages. The heat treatment and physical protection of bulk culture media, the use of phage-inhibitory media, the rotation of starters, the chlorination of fermentation vats between fills, and improved whey disposal procedures have all helped to limit the phage problem. More relevant to this report, however, is the evidence provided by a combination of laboratory research and practical experience that, although the instability of key traits of lactic streptococci is well recognized, a limited number of strains possess a marked insensitivity to phage attack. The recent upsurge in genetic studies of lactic acid bacteria, especially the lactic streptococci, has provided evidence for the role of plasmid DNA in key functions such as lactose and citrate utilization, proteinase activity, and insensitivity to bacteriophage attack. The concomitant rapid progress in the development of gene transfer systems for the lactic streptococci has raised hopes that meaningful genetic manipulation of these bacteria is imminent. The availability of naturally occurring phage-insensitive strains as model systems for studying the mechanisms involved is providing both opportunity and challenge to the starter culture molecular biologist. A number of plasmid-encoded phage defense mechanisms have recently been identified, and the construction of new phage-insensitive strains may offer the first practical application of the present intensive genetic studies. In addition to the papers in this volume, excellent recent reviews are available on bacteriophage (21, 48, 85) and genetic studies (27, 50) of lactic streptococci. Ongoing research on mechanisms of phage insensitivity in our laboratory is summarized in this paper in the context of recent developments elsewhere.

Naturally Occurring Phage-Insensitive Strains

Although the main focus of this report is on genetic aspects of phage-host interactions in the lactic streptococci, it is useful first to summarize studies on the natural selection of phage-insensitive strains. In this regard, the instability of phage insensitivity and variation within strains are well established (10, 12, 13, 41, 48, 57, 74, 75). Nevertheless, starter culture systems based on stable phage-insensitive strains, both as unknown mixtures and as defined strain blends, are proving commercially successful.

The majority of dairy fermentations utilize lactic streptococci as mixtures of unknown composition, often in association with *Leuconostoc* species. Originally, these mixtures were propagated in the dairies without protection against phage attack, and they evolved, as a result of selective pressure, as phage-containing cultures with a dominance of phage-insensitive strains. These natural mixtures work well in small production units (15, 24, 82) and may be used more or less continuously, i.e., without rotation. However, in modern large production units their performance is not adequate and leads to problems such as inadequate moisture control in cheese making (82). The evolution, strain composition, and phage-host interactions of these mixtures have been elegantly studied by Stadhouders and co-workers (24, 81, 82), who have established guidelines for their successful use in large-scale fermentations. The mixtures with optimum balance of phage-insensitive strains and ability to "repair" activity after attack by phage have been selected and stored as frozen stocks. Provided that transfers are minimized, to avoid changes in strain balance, these cultures may be used in fermentations without rotation when additional precautions, including protection of the bulk culture from phage and chlorination of vats between fills, are taken. Notwithstanding their value in commercial fermentations and their role as a source of phage-insensitive strains, these mixtures do not readily lend themselves to detailed analysis or improvement by genetic techniques.

Many mixed-strain starter cultures available to the dairy industry have been propagated under aseptic conditions, i.e., in the absence of phage. These cultures have become quite phage sensitive and can be used only as components of culture rotations (17, 18, 36, 78). This emphasizes the importance of storage and propagation conditions in maintaining the desired characteristics of lactic streptococci.

In contrast to the mixed-strain starter cultures, defined strains of lactic streptococci have traditionally been used in New Zealand (53–55) and Australia (16, 38, 39) and have recently found increasing favor in the United States (73, 88, 89) and Ireland (17, 18). The known identity of the strains facilitates research input, including the assessment of phage-host interactions. As with mixtures, the practical use of these strains has also had to respond to the rationalization of the cheese industry into large-scale production units (55). Although there is considerable evidence that many defined strains of lactic streptococci possess significant insensitivity to phage (7, 35, 54, 56, 73, 89), a test designed by Heap and Lawrence in 1976 (32, 33) is especially useful for selecting strains that remain insensitive under practical cheese-making conditions. A limited number of strains survive successive exposures to all known groups of phages under simulated cheese-making conditions that allow the effects of low-efficiency phage replication, mutations in phage or host, and temperature on phage-defense mechanisms to be manifested. When used as pairs or blends of three to six strains, without rotation, these remarkably stable strains provide consistent performance and high-quality end products under intense manufacturing conditions. Again, it is imperative to retain the integrity of these strains and avoid the accumulation of variants by careful attention to storage and propagation conditions.

Use of Phage-Insensitive Mutants

The idea of isolating phage-insensitive mutants of lactic streptococci by growing cultures in the presence of phage was proposed shortly after phages were first recognized (92, 93). However, most mutants were less active in terms of acid production, reverted to sensitivity to the original phage, or were attacked by new phages.

Likewise, the use of mutagens to select phage-insensitive strains has been attempted with varying results (21, 48, 80; A. V. Gudkov, L. A. Ostroumov, and N. I. Odegov, Proc. 21st Int. Dairy Congr., Moscow, 1982, vol. 1, book 2, p. 311). In some of the earlier studies, the criteria used to define phage insensitivity (or phage resistance, which was the term frequently used) were not clearly presented. Some of the problems encountered may have been due to the failure to differentiate clearly between genotypic resistance and the establishment of phage-carrying cultures which can manifest phenotypic resistance (2).

As mentioned, propagation of mixed-strain cultures in the presence of phage ensures the emergence of phage-insensitive strains (15, 24, 81, 82). With the use of improved isolation techniques, this concept has recently been applied to defined strains with reasonable success. Phage-insensitive mutants arise spontaneously in some strains of lactic streptococci exposed to phages (58, 62). Most of the mutants characterized no longer adsorbed the phage, although in a few cases adsorption was normal but phage DNA injection did not occur. In addition, several reports of the selection of phage-insensitive derivatives by extended growth in the presence of cheese factory wheys have appeared (16, 38, 39, 43, 58, 87, 89), and this procedure is used to select replacement strains when a strain used commercially has succumbed to phage attack. Although there is good evidence that the mutations to phage insensitivity and slow acid production are separate genetic events (47, 62), it is nevertheless essential to carefully characterize phage or whey "adapted" derivatives as suitable for cheese making. While phage- or whey-derived bacteriophage-insensitive mutants are used successfully in commercial cheese making (16, 39, 89), the approach is not successful with all strains (43, 58; C. Daly and T. M. Cogan, unpublished data), and care must be taken to ensure that the mutant is a genuine derivative of the parent and does not contribute any undesirable effects during product manufacture or ripening (14, 39).

Lysogeny and the Phage-Carrier State

The presence of a phage in a bacterial culture may influence the effect of newly encountered phage and thus must be considered as a possible factor in phage insensitivity.

While the lytic cycle of phage proliferation, resulting in lysis of the host bacterium, is the main problem encountered in industrial fermentations, the lysogenic state has also been clearly demonstrated. In fact, most strains of lactic streptococci are lysogenic (as evidenced by electron microscope observation of UV light- or mitomycin C-induced lysates), but success in finding indicator strains has varied (20, 34, 37, 51, 72, 84). The presence of multiple lysogens (9, A. W. Jarvis, Proc. 21st Int. Dairy Congr., Moscow, 1982, vol. 1, book 2, p. 314) and the spontaneous liberation of temperate phages have also been observed (37, 72). Lysogeny has received most attention as a potential source of phages in dairy fermentations, and although Jarvis (44), using DNA-DNA hybridization techniques, found no relationship between lytic and lysogenic phage for selected S. cremoris strains, this aspect is worthy of further consideration.

The superinfection immunity to closely related phage represents a positive dimension to the phenomenon of lysogeny in dairy fermentations. However, the susceptibility of lysogens to phage suggests that superinfection immunity is too specific to be of major value in overall phage insensitivity. The recent availability of procedures for prophage curing (9, 26, 28, 85) provides an opportunity to examine the immunity aspect of lysogeny more closely. Chopin et al. (9) reported that the acid-producing ability of all but one strain tested was unaffected by prophage curing. In addition, the pattern of sensitivity to virulent phages was unchanged in three of four strains after curing; the fourth strain, a bi-lysogen, became sensitive to several prolate-headed virulent phages when one of its prophages was cured.

Chopin et al. (8) have identified a plasmid, pIL7, that increases UV resistance and prophage instability in S. lactis. The subsequent cloning of these traits in S. lactis has been achieved (6).

As mentioned above, defined-strain starter systems incorporating only a small number (two to six) of strains are finding increasing usage and will provide further opportunity to assess the contribution of lysogeny both as a source of phage and to the insensitivity inherent in specific strains. The goal of cloning superimmunity to the limited number of phages that attack some S. cremoris strains, in particular, is attractive. A prerequisite, however, is a concerted study of the biochemistry and molecular biology of phage repressor systems in the lactic streptococci.

The existence of the phage-carrier state has been recognized in the lactic streptococci since 1947 (42), and although its contribution to insensitivity to phage has been observed in defined strains (42, 59) and especially in mixtures (24, 30, 82), it has received little attention. Limsowtin and Terzaghi (59) showed that all of nine strains of lactic streptococci examined could establish the phage-carrier state, although only two strains retained the acid-producing ability of the parent.

The existence of a phage-carrier state in S. cremoris 320 has been observed in our laboratory (J. Lyne, C. Daly, and T. M. Cogan, Irish J. Food Sci. Technol. 8:153, 1984; M. Murphy, J. Lyne, C. Daly, and G. F. Fitzgerald, J. Appl. Bacteriol. 59:x, 1985). While the mechanism involved in the phage-host association has not yet been clarified, a bonus of the study has been the opportunity to assess the possible role of the carried-state phage (designated CS) as a source of phages active against strain 320 that were detected in three cheese factories using the strain as a component of a defined-strain starter culture system.

Although the presence of phage could only rarely be detected (and then only a few plaques) in the supernatant of S. cremoris 320 grown under noncontrolled conditions in milk, a phage was released when the strain was grown in either 2.5 or 10% skim-milk solids medium when the pH was controlled at 6.5 or 6.0. The phage titer, against strain 320 itself, reached 10^{10} PFU/ml after 4 h at 30°C and caused lysis of the culture. When the pH was controlled at 5.5, phage were not detected, although exogenously added CS phage caused lysis of the culture at this pH. Neither was CS phage detected in S. cremoris 320 cultures grown in broth media or in Phase 4, a commercially available phage-inhibitory medium (64).

The phage CS, released from S. cremoris 320 during pH-controlled growth in milk, was compared with five phages, I1, I2, I3, B, and M, which appeared as lytic phages for strain 320 when used in three cheese factories. The data (summarized in Table 1) show that, on the basis of morphology, restriction enzyme digest patterns of the DNA, and DNA-DNA hybridization, the CS phage was very similar to four of the five factory phages. This suggests that the culture itself was a likely source of phage.

TABLE 1. Comparison of phage CS released by S. cremoris 320 and virulent phages I1, I2, I3, B, and M isolated from cheese factory wheys

Phage	Morphology (nm)			Restriction enzyme[a] digest pattern compared with CS	Hybridization with CS DNA[b]
	Tail length	Tail width	Head diam		
CS	168	10.1	58.7		+
I1	271	11.0	63.6	Distinct	−
I2	169	11.6	64.8	Similar	+
I3	171	10.6	63.1	Similar	+
B	168	9.8	65.2	Similar	+
M	164	10.9	59.6	Similar	+

[a] The restriction enzymes used were EcoRI, MboI, Sau3A, and PvuII.

[b] For hybridization analysis the PvuII digest was transferred to a nitrocellulose filter and probed with biotin-dUTP-labeled CS phage DNA.

The phage-carrier state in lactic streptococci is certainly worthy of detailed investigation both as a source of phage and, of direct relevance to this report, as a contributor to phage insensitivity.

Phage Lysins

The production of phage-associated lytic enzymes, lysins, has been observed in phage-host interactions in the lactic streptococci. These lysins can cause lysis of strains that are insensitive to infection by the lysin-producing phage. The properties and broad host spectrum of a lysin from *S. lactis* C2 have recently been highlighted by Mullan and Crawford (65, 66). The idea of isolating lysin-resistant mutants of lactic streptococci in the hope that they would have insensitivity to a wide range of phages has been suggested (21, 71) and should be examined further. However, because of the low levels of phage present, lysin is rarely a problem when only a small number of carefully selected phage-insensitive strains are used for cheese making.

Adsorption

The ideal, most effective defense against phage is to prevent any productive contact between the phage and the bacteria. This could be achieved by mutation of phage receptors in the cell wall or by secretion of a capsule or slime layer that acts as a barrier to approach of the phage. Although the phage, for its part, may adapt to resistant cells, e.g., by a host-range mutation which alters its adsorption specificity, there have been recent demonstrations of potentially useful mechanisms affecting phage adsorption by the lactic streptococci.

At least some of the phage-insensitive mutants isolated by the challenge of phage-sensitive cultures with phage have shown a decreased ability to adsorb the phage (47). This may involve mutations in chromosomal genes that affect adsorption. However, there have recently been two reports of plasmid-encoded adsorption mechanisms. Sanders and Klaenhammer (76) showed that a 30-megadalton (MDa) plasmid, pME0030, genetically determined the ability of *S. lactis* ME2 to prevent the adsorption of phage. The parent adsorbed phage at a level of 20 to 40%, whereas mutants lacking pME0030 adsorbed at levels exceeding 99.8%. The adsorption of four phages by the mutants suggested a broad-spectrum effect of the mechanism examined. In a similar study, deVos et al. (22) showed that variants of *S. cremoris* SK11 which contained a 34-MDa plasmid were resistant to phage sk11g, whereas those that had lost the plasmid were sensitive. Phage adsorption to strains containing the plasmid was 1 to 5% compared with 80 to 90% adsorption to phage-sensitive variants lacking the plasmid. The potential of these plasmids to contribute to phage insensitivity of other strains after their introduction by conjugation or other means and the mechanism by which a plasmid-encoded product(s) prevents phage adsorption warrant investigation. Indeed, these interesting observations regarding the role of plasmids in altering phage receptors highlight how little is known about the nature or location of these receptors in lactic streptococci. The involvement of the plasma membrane and cell wall in the adsorption of phage m13 on *S. lactis* was reported by Oram and Reiter (68), and Oram (67) isolated a receptor from the lipoprotein fraction of the plasma membrane of the same strain. A study of *S. cremoris* EB7 suggested that the same substances were involved in phage adsorption and serological specificity (46). Future research on cell surfaces will benefit our understanding of phage adsorption and also aspects such as the lysozyme sensitivity and tolerance to lyophilization of lactic streptococci.

The occurrence of lactic streptococci that produce a mucoid, ropy texture in milk has long been recognized, and these strains find favor in some fermented products. Evidence for the involvement of a conjugative 18.5-MDa plasmid, pSRQ220, in the expression of the Muc$^+$ (mucoidness) phenotype in *S. cremoris* MS has been presented (90). Transconjugants of *S. lactis* 4T4.2 and *S. lactis* subsp. *diacetylactis* SLA.25 that acquired the Muc$^+$ phenotype also became resistant to phages that were lytic for the parent strains. While the mechanism of phage resistance associated with the Muc plasmid has yet to be clarified, the association of phage resistance with mucoid phenotype offers another potentially useful approach for the derivation of phage-insensitive strains of lactic streptococci.

Restriction-Modification

When phage DNA enters a bacterial cell, it may be replicated, leading to phage propagation. However, following DNA injection, at least two defense mechanisms may still contribute to the survival of the host (52). If some component of the host cell is not compatible, the infection may be abortive, or alternatively, DNA restriction enzymes in the host may destroy the phage DNA. However, at low efficiency, the phage DNA may become chemically modified so that it is not recognized by the host's restriction enzymes and is able to direct lytic phage maturation. The modified phage are able to grow well in further encounters with this host, thus limiting the practical significance of restriction as a phage defense mechanism. The host-controlled modifications are characterized by the reversibility of the phenotypic change by one growth

TABLE 2. Plasmid DNA involvement in restriction-modification in the lactic streptococci

Strain	Plasmid size (MDa)	Reference
S. cremoris KH	10	75
S. lactis IL594	19 and 21	5
S. cremoris F	28	This paper
S. lactis 3085	10	M. Teuber (personal communication)
S. lactis ME2	60	Klaenhammer and Sanozky-Dawes[a]
S. cremoris CN96	20	Klaenhammer and Sanozky-Dawes[a]

[a] Abstr. 2nd ASM Conf. Streptococcal Genetics, 1986, abstr. no. 222, p. 25.

cycle of the phage on another host strain and can, therefore, be distinguished from mutation.

Biological evidence for the presence of restriction-modification in the lactic streptococci was provided by Collins (11) and was confirmed in numerous investigations (4, 5, 19, 57, 69, 70, 74, 75, 77). The temperature sensitivity observed in some systems is of practical significance for the control of phage during the cooking stages of cheese manufacture (69, 75, 77). Also, a restriction endonuclease, ScrFI, which recognizes the sequence CC↓NGG, has been isolated from S. cremoris F, providing the first biochemical confirmation of restriction in the lactic streptococci (19, 23).

The accumulating evidence for the involvement of plasmid DNA in restriction-modification systems (Table 2) helps explain their widespread distribution in the lactic streptococci and the ease with which variants deficient in restriction ability can be accumulated during subculture. Some of the plasmids are conjugative, providing an opportunity to assess their effects in a variety of strain backgrounds. A contribution by T. R. Klaenhammer and R. B. Sanozky-Dawes (Abstr. 2nd ASM Conf. Streptococcal Genetics, 1986, abstr. no. 222, p. 25) to this conference provided an example of the additive effect of two plasmids encoding restriction-modification activities. The plasmid pTN1060 was formed by a natural recombination event in S. lactis ME2 between a 40-MDa plasmid, pTR1040 (encoding lactose-fermenting ability), and DNA sequences which conferred high-frequency conjugal transfer ability and restriction-modification activities. pTN1060 was introduced into S. cremoris CN9S, which itself contained restriction-modification activities (effiency of plating for phage M12.M12R, 10^{-3}) encoded by a second plasmid, pLR1020. The resulting transconjugants exhibited an additive level of phage restriction (efficiency of plating, 10^{-5} to 10^{-8}) due to the combined activity of pLR1020 and pTN1060. This result augurs well for the use of "super-restriction" as a component of maximum phage insensitivity in the lactic streptococci.

Since both biological and biochemical evidence for restriction-modification are available

in S. cremoris F (19, 23), we have attempted to characterize the system with the long-term objective of cloning ScrFI and other restriction enzymes in lactic streptococci. Mutants of S. cremoris F lacking restriction activity were isolated spontaneously and at a level of 2% by treatment with acridine orange (20 μg/ml). The mutants were R⁻M⁻ and were consistently lacking a 28-MDa plasmid present in the parent strain. However, the mutants still showed ScrFI activity, suggesting that the enzyme may not have a role in restriction in vivo but rather serves in some other function such as recombination or repair of pathways. Alternatively, the restriction-deficient mutants of F may lack some cofactor or regulatory substance required for in vivo ScrFI activity. Efforts to clarify the roles of the 28-MDa plasmid and ScrFI in restriction-modification in S. cremoris F are ongoing.

In our initial screening for the presence of type II restriction endonucleases in the lactic streptococci (19), we could not demonstrate activity in several strains maintained in our laboratory collection but did so in the case of S. cremoris F, a strain recently isolated from a mixed-strain starter culture used in Cheddar cheese manufacture. This fact, coupled with the apparent complexity of the restriction-modification system in strain F, tempted us to examine further industrial isolates for enzyme activity. Seven additional S. cremoris strains, each isolated from separate mixed-strain starter cultures, were examined. Six strains, A22, D09, F21, G26, H14, and J11, contained a restriction enzyme which, on preliminary investigation, appears similar to ScrFI. These strains are not identical to S. cremoris F on the basis of plasmid DNA profiles, although several common plasmids are present. While it is interesting to speculate on the implications of the presence of isolates containing ScrFI in separate mixed-strain starter cultures, three of the isolates (D09, G26, and F) show biological evidence of restriction of phage DNA and provide additional systems for investigation of the role of the enzyme and plasmid DNA species in restriction-modification.

Although the potential of restriction-modification as a sole defense mechanism against phage

TABLE 3. Plasmid DNA involvement in the inhibition of phage maturation in the lactic streptococci

Strain	Plasmid size (MDa)	Reference
S. lactis subsp. diacetylactis DRC3−	40	63
S. lactis ME2	30	45, 49, 79, 83
S. cremoris UC653	50	3; this paper
S. lactis 811	29	This paper
S. lactis subsp. diacetylactis BuI	39	M. Teuber (personal communication)

is limited by the possible destructive effect of the modified phage, the concept of constructing strains with super restriction-modification systems is attractive. As mentioned above, the additive effect of plasmids encoding restriction-modification systems is significant in this regard. Advances in the understanding of restriction-modification systems in other bacteria (52), including the successful cloning of several restriction enzymes (29, 61, 86, 91), should stimulate further studies in lactic streptococci, especially in strains demonstrating marked insensitivity to phage. The ability to manipulate the DNA modification systems of these strains may be useful to counteract phages that overcome the restriction-modification systems present.

Inhibition of Phage Maturation

The current intense genetic studies on aspects of phage insensitivity have been rewarded by the discovery of several plasmids (Table 3) that prevent phage replication by an, as yet, unknown mechanism. The main aspects of the systems involved are examined here, with emphasis on those discovered in our laboratory (3).

We used conjugation on agar surfaces to screen lactic streptococci for the ability to trans-

fer determinants of phage insensitivity to plasmid-free strains of S. lactis used as recipients. The primary selection was for Lac transfer, and transconjugants were screened for insensitivity to phage. This approach permitted the identification of phage-insensitivity plasmids in S. cremoris UC653 and S. lactis 811. Initial matings between these donors and S. lactis MG1363 demonstrated that Lac$^+$ could be conjugationally transferred at frequencies of $\sim 10^{-7}$ and 10^{-4} per recipient in the case of UC653 and 811, respectively, and there was a very high cotransfer of insensitivity to phage 712 (Table 4).

When S. cremoris UC653 was used as donor, the Lac$^+$ phage-insensitive transconjugants obtained were of three types (Fig. 1): (i) containing 50- and 26-MDa plasmids, both present in the donor (example of this type designated AB001); (ii) containing a large plasmid of 77 MDa not present in the donor (examples of this type designated AB002 and MM8); and (iii) containing a large plasmid of 83 MDa not present in the donor (examples of this type designated AB003 and MM7). Curing experiments showed that the 50-MDa plasmid, CI750, encoded phage insensitivity while the 26-MDa plasmid coded for Lac.

When selected transconjugants were used as donors in subsequent mating experiments, those containing the large 77-MDa plasmid, e.g., AB002 or MM8, transferred Lac$^+$ and phage insensitivity at a frequency four log cycles greater than that achieved with the AB001 donor, containing separate Lac and phage-insensitivity plasmids (Table 4). In addition, a Lac$^+$ phage-sensitive transconjugant containing only the 26-MDa Lac plasmid pCI726 was incapable of transferring the Lac$^+$ phenotype. These results suggest that both phage insensitivity and transfer (tra) genes are encoded by pCI750 and that this plasmid mobilizes the Lac plasmid. The nature of the recombination events leading to

TABLE 4. Conjugative transfer of Lac and phage-insensitivity plasmids to plasmid-free S. lactis strains

Donor	Recipient	Lac$^+$ transconjugants		Representative transconjugants[a]
		Frequency/recipient	% Phage 712 insensitive	
S. cremoris UC653	S. lactis MG1363	3.3×10^{-7}	70	AB001, AB002, MM8, MM8E
S. lactis 811	S. lactis MG1363	1.4×10^{-4}	11	AC001
S. lactis AB001	S. lactis LM0230	1.5×10^{-8}	15	
S. lactis AB002	S. lactis LM0230	5.9×10^{-4}	98	
S. lactis AB001	S. lactis LM2306	4.0×10^{-7}	70	
S. lactis MM8	S. lactis LM2306	3.7×10^{-3}	92	
S. lactis MM8E	S. lactis LM2306	0	0	
S. lactis AC001	S. lactis LM0230	1.8×10^{-5}	60	

[a] Transconjugant AB001 was Lac$^+$ and phage insensitive and contained two plasmids of 26 and 50 MDa. Transconjugants AB002 and MM8 were Lac$^+$ and phage insensitive and contained one large 77-MDa plasmid. Transconjugant MM8E was Lac$^+$ and phage sensitive and contained a 26-MDa plasmid. Transconjugant AC001 was Lac$^+$ and phage insensitive and contained two plasmids of 42 and 29 MDa.

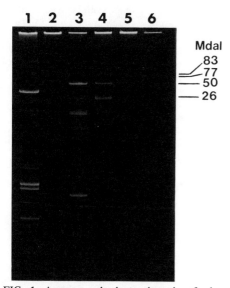

FIG. 1. Agarose gel electrophoresis of plasmid DNA from the *S. cremoris* UC653 parent (lane 3) and three Lac⁺ phage 712-insensitive transconjugants (lanes 4–6) after conjugation with a plasmid-free *S. lactis* MG1363 recipient (lane 2). Lane 1 contains the size reference plasmids from *Escherichia coli* V517 (reprinted from Baumgartner et al. [3]).

the formation of the 77-MDa Lac⁺ Tra⁺ phage-insensitivity plasmid has not yet been determined, but recombinant plasmids have previously been observed among transconjugants of lactic streptococci (1, 25). The linkage of the easily scored marker, Lac, in addition to transfer and phage-insensitivity genes on the same plasmid is proving a very useful tool for the transfer of phage insensitivity to other strains in our laboratory.

Five phages lytic for *S. lactis* MG1363 were examined for their ability to plaque at 21, 30, and 37°C on the Lac⁺ phage-insensitive transconjugant AB001. No plaques were detected for phages 712 or sk1 at any temperature. Phages c2, eb1, and stl5 were capable of forming plaques on AB001 but at a reduced titer and with a considerably reduced plaque size. Further analysis revealed no evidence for altered phage adsorption or a restriction-modification system, but one-step growth experiments revealed a drastic reduction in the burst size of phage c2 when propagated on AB001 compared with *S. lactis* MG1363. When pCI750 was transferred to *S. lactis* subsp. *diacetylactis* 18-16, it conferred total insensitivity to phage 18-16. This insensitivity was also evident in milk cultures of strain 18-16 exposed to successive additions of the phage in a simulated cheese-making activity test, thus confirming the temperature tolerance of the phage-insensitivity mechanism present.

Findings similar to those described for *S. cremoris* UC653 were obtained when *S. lactis* 811 was used as a donor in conjugation experiments. Phage insensitivity of transconjugants was associated with a 29-MDa plasmid which conferred total and partial insensitivity to phages 712 and c2. As with pCI750, there was no effect on the adsorption of the phage or of temperatures in the range 21 to 37°C.

Thus, the two plasmids pCI750 and pCI829 confer a similar phage-insensitive phenotype, total insensitivity to phage 712, and partial insensitivity to phage c2. It was of interest to determine whether the two plasmids could coexist in the same strain and provide effective barriers to phage. When the two plasmids were introduced into *S. lactis* MG1363 (Fig. 2), together they conferred total insensitivity to phage c2, whereas this strain with either plasmid alone was only partially insensitive to the phage. The ability to totally exclude a prolate-headed phage, c2, is significant in view of the wide host range of prolate phages (31) and the inability of another plasmid, pTR2030 (45), or indeed either pCI750 or pCI829 on their own, to totally inhibit prolate phages. The additive effect of the plasmids pCI750 and pCI829 is encouraging for the concept of constructing strains of lactic streptococci with enhanced insensitivity to phage attack.

The mechanism of phage insensitivity encoded by plasmids pCI750 and pCI829 described here and, indeed, that of similar plasmids reported (Table 3) need to be examined in detail. The mechanism associated with pNP40 (63) and pT2030 (49) is temperature sensitive in *S. lactis* strains. However, the phage insensitivity conferred by pT2030 was clearly shown to be unaffected by cheese-making temperatures in the industrially more useful *S. cremoris* strains (79, 83). The pTR2030 system has been studied in most detail and confers stable phage insensitiv-

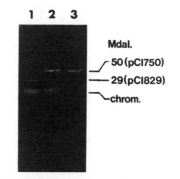

FIG. 2. Agarose gel electrophoresis patterns of plasmid DNA of transconjugants of *S. lactis* MG1363 containing the phage-insensitivity plasmids, pCI829 (lane 1) and pCI750 (lane 2), and the two plasmids together (lane 3).

ity against a range of small isometric-headed phages that attack *S. cremoris* and *S. lactis* strains (45). This is a very significant finding in view of the widespread occurrence of these phages.

The conjugative nature of the plasmids listed in Table 3 and their ability to confer total insensitivity to at least some phages makes them very useful for the construction of valuable phage-insensitive strains in advance of more complex gene transfer systems, e.g., transformation, becoming routinely available for the lactic streptococci. The availability of alternative methods, instead of antibiotic selection, for the screening of phage-insensitive derivatives should remove any objection to the use of genetically manipulated strains, containing only DNA of lactic streptococcal origin, for the commercial production of dairy products (79, 90).

Conclusion

This report has highlighted the advances in the genetic studies of phage-insensitivity mechanisms in the lactic streptococci. The rapid progress to date is such that manipulated strains containing conjugative plasmids are now available to extend the limited number of naturally occurring, stable phage-insensitive strains useful for commercial dairy fermentations. A continued major research effort is needed to elucidate the mechanisms associated with the plasmids conferring phage insensitivity at the levels of adsorption, restriction-modification, and phage maturation. These plasmids are now being characterized in detail, and strategies are being developed for the cloning of phage-insensitivity genes in suitable hosts. The search for further mechanisms contributing to phage insensitivity must be continued. In particular, mechanisms encoded by chromosomal genes are likely to be of major significance in some very stable strains and need to be investigated. The elegant studies of Klaenhammer's group in unraveling the complex mechanisms associated with phage insensitivity in *S. lactis* ME2 have shown the additive effect of systems operating at different stages of the phage lytic cycle (49, 76, 77, 79, 83; Klaenhammer and Sanozky-Dawes, Abstr. 2nd ASM Conf. Streptococcal Genetics, 1986, abstr. 222, p. 25). This approach should be extended to other strains to increase our understanding of mechanisms present in naturally occurring insensitive strains and to aid in the construction of "super strains" with multiple additive defense mechanisms. It is essential that newly constructed strains retain their insensitivity under the conditions pertaining in practical dairy fermentations, e.g., in pasteurized milk as the growth medium and at cheese-making temperatures (40, 48). The achievement of these goals will be aided by closer collaboration among molecular biologists, biochemists, and microbial physiologists. Our knowledge of the protagonist, the phage, must also increase. Studies of lactic streptococcal phages at the genomic level (60) may clarify the way in which they overcome the host defense mechanisms that they encounter. A variety of survival techniques have been described for phages attacking other bacteria (52). Increased understanding of both phage and host will maximize our potential to eliminate the phage problem in dairy fermentations.

This material is based in part upon contract research supported by the National Board for Science and Technology, by the National Enterprise Agency, and by E.E.C. Contract no. GBI-2-055-EIR.

LITERATURE CITED

1. **Anderson, D. G., and L. L. McKay.** 1984. Genetic and physical characterization of recombinant plasmids associated with cell aggregation and conjugal transfer in *Streptococcus lactis* ML3. J. Bacteriol. **47:**245–249.
2. **Barksdale, L., and S. B. Arden.** 1974. Persisting bacteriophage infections, lysogeny and phage conversions. Annu. Rev. Microbiol. **28:**265–299.
3. **Baumgartner, A., M. Murphy, C. Daly, and G. F. Fitzgerald.** 1986. Conjugative co-transfer of lactose and bacteriophage resistance plasmids from *Streptococcus cremoris* UC653. FEMS Microbiol. Lett. **35:**233–237.
4. **Boussemaer, J. P., P. P. Schrauwen, J. L. Sourrouille, and P. Guy.** 1980. Multiple modification-restriction systems in lactic streptococci and their significance in defining a phage-typing system. J. Dairy Res. **47:**401–409.
5. **Chopin, A., M. C. Chopin, A. Moillo-Batt, and P. Langella.** 1984. Two plasmid determined restriction and modification systems in *Streptococcus lactis*. Plasmid **11:**260–263.
6. **Chopin, M. C., A. Chopin, A. Ronault, and D. Simon.** 1986. Cloning in *Streptococcus lactis* of plasmid-mediated UV resistance and effect on prophage stability. Appl. Environ. Microbiol. **51:**233–237.
7. **Chopin, M. C., A. Chopin, and C. Roux.** 1976. Definition of bacteriophage groups according to their lytic action on mesophilic lactic streptococci. Appl. Environ. Microbiol. **31:**741–746.
8. **Chopin, M. C., A. Moillo-Batt, and A. Ronault.** 1985. Plasmid mediated UV-protection in *Streptococcus lactis*. FEMS Microbiol. Lett. **26:**243–245.
9. **Chopin, M. C., A. Ronault, and M. Rousseau.** 1983. Elimination d'un prophage dans des souches mono- et multilysogenes de streptocoques du groupe N. Lait **63:**102–115.
10. **Collins, E. B.** 1955. Action of bacteriophage on mixed strain cultures. III. Strain dominance due to the action of bacteriophage and variations in the acid production of secondary growth bacteria. Appl. Microbiol. **3:**137–140.
11. **Collins, E. B.** 1956. Host-controlled variations in bacteriophages active against lactic streptococci. Virology **2:**261–271.
12. **Collins, E. B.** 1958. Changes in the bacteriophage sensitivity of lactic streptococci. J. Dairy Sci. **41:**41–48.
13. **Collins, E. B.** 1962. Behaviour and use of lactic streptococci and their bacteriophages. J. Dairy Sci. **45:**552–558.
14. **Coventry, M. J., A. J. Hillier, and G. R. Jago.** 1984. Changes in the metabolism of factory-derived bacteriophage resistant derivatives of *Streptococcus cremoris*. Aust. J. Dairy Technol. **39:**154–159.
15. **Crawford, R. J. M., and J. H. Galloway.** 1962. Bacteriophage contamination of mixed-strain cultures in cheese-

making. Proceedings of the 16th International Dairy Congress, vol. B, p. 785–791.

16. **Czulak, J., D. H. Bant, S. C. Blyth, and J. B. Grace.** 1979. A new cheese starter system. Dairy Ind. Int. **44:**17–19.

17. **Daly, C.** 1983. Starter culture developments in Ireland. Irish J. Food Sci. Technol. **7:**39–48.

18. **Daly, C.** 1983. The use of mesophilic cultures in the dairy industry. Antonie van Leeuwenhoek J. Microbiol. Serol. **49:**297–312.

19. **Daly, C., and G. F. Fitzgerald.** 1982. Bacteriophage DNA restriction and the lactic streptococci, p. 213–216. *In* D. Schlessinger (ed.), Microbiology—1982. American Society for Microbiology, Washington, D.C.

20. **Davies, F. L., and M. J. Gasson.** 1981. Reviews of the progress of dairy science: genetics of lactic acid bacteria. J. Dairy Res. **48:**363–376.

21. **Davies, F. L., and M. J. Gasson.** 1984. Bacteriophages of lactic-acid bacteria, p. 127–151. *In* F. L. Davies and B. A. Law (ed.), Advances in the microbiology and biochemistry of cheese and fermented milk. Elsevier Applied Science Publishers, London.

22. **deVos, W. M., H. M. Underwood, and F. L. Davies.** 1984. Plasmid encoded bacteriophage resistance in *Streptococcus cremoris* SK11. FEMS Microbiol. Lett. **23:**175–178.

23. **Fitzgerald, G. F., C. Daly, L. R. Brown, and T. R. Gingeras.** 1982. ScrFI: a new sequence-specific endonuclease from *Streptococcus cremoris*. Nucleic Acids Res. **10:**8171–8179.

24. **Galesloot, T. E., F. Hassing, and J. Stadhouders.** 1966. Differences in phage sensitivity of starters propagated in practice and in a dairy research laboratory. Proceedings of the 17th International Dairy Congress, vol. D2, p. 491–498.

25. **Gasson, M. J.** 1983. Genetic transfer systems in lactic acid bacteria. Antonie van Leeuwenhoek J. Microbiol. Serol. **49:**275–282.

26. **Gasson, M. J., and F. L. Davies.** 1980. Prophage-cured derivatives of *Streptococcus lactis* and *Streptococcus cremoris*. Appl. Environ. Microbiol. **40:**964–966.

27. **Gasson, M. J., and F. L. Davies.** 1984. The genetics of lactic-acid bacteria, p. 99–126. *In* F. L. Davies and B. A. Law (ed.), Advances in the microbiology and biochemistry of cheese and fermented milk. Elsevier Applied Science Publishers, London.

28. **Georghiou, D., S. H. Phua, and E. Terzaghi.** 1981. Curing of a lysogenic strain of *Streptococcus cremoris* and characterization of the temperate bacteriophage. J. Gen. Microbiol. **122:**295–303.

29. **Gingeras, T. R., and J. E. Brooks.** 1983. Cloned restriction/modification system from *Pseudomonas aeruginosa*. Proc. Natl. Acad. Sci. U.S.A. **80:**402–406.

30. **Graham, D. M., C. E. Parmelee, and F. E. Nelson.** 1952. The carrier-state of lactic streptococcus bacteriophage. J. Dairy Sci. **35:**813–822.

31. **Heap, H. A., and A. W. Jarvis.** 1980. A comparison of prolate- and isometric-headed lactic streptococcal bacteriophages. N.Z. J. Dairy Sci. Technol. **15:**75–81.

32. **Heap, H. A., and R. C. Lawrence.** 1976. The selection of starter strains for cheesemaking. N.Z. J. Dairy Sci. Technol. **11:**16–20.

33. **Heap, H. A., and R. C. Lawrence.** 1981. Recent modifications of the New Zealand activity test for Cheddar cheese starters. N.Z. J. Dairy Sci. Technol. **15:**91–94.

34. **Heap, H. A., G. K. Y. Limsowtin, and R. C. Lawrence.** 1978. Contribution of *Streptococcus lactis* strains in raw milk to phage infection in commercial cheese factories. N.Z. J. Dairy Sci. Technol. **13:**16–22.

35. **Henning, D. R., C. H. Black, W. E. Sandine, and P. R. Elliker.** 1968. Host-range studies of lactic streptococci bacteriophages. J. Dairy Sci. **51:**16–21.

36. **Huggins, A. R.** 1984. Progress in dairy starter culture technology. Food Technol. **38:**41–50.

37. **Huggins, A. R., and W. E. Sandine.** 1977. Incidence and properties of temperate bacteriophages induced from lactic streptococci. Appl. Environ. Microbiol. **33:**184–191.

38. **Hull, R. R.** 1977. Control of bacteriophage in cheese factories. Aust. J. Dairy Technol. **32:**65–66.

39. **Hull, R. R.** 1983. Factory-derived starter cultures for the control of bacteriophage in cheese manufacture. Aust. J. Dairy Technol. **38:**149–154.

40. **Hull, R. R., and A. R. Brooke.** 1982. Bacteriophages more active against Cheddar cheese starters in unheated milk. Aust. J. Dairy Technol. **37:**143–146.

41. **Hunter, G. J. E.** 1939. Examples of variation within pure cultures of *Streptococcus cremoris*. J. Dairy Res. **10:**464–470.

42. **Hunter, G. J. E.** 1947. Phage-resistant and phage-carrying strains of lactic streptococci. J. Hyg. **45:**307–312.

43. **Jarvis, A. W.** 1981. The use of whey derived phage resistant starter strains in New Zealand cheese plants. N.Z. J. Dairy Sci. Technol. **16:**25–31.

44. **Jarvis, A. W.** 1984. DNA-DNA homology between lactic streptococci and their temperate and lytic phages. Appl. Environ. Microbiol. **47:**1031–1038.

45. **Jarvis, A. W., and T. R. Klaenhammer.** 1986. Bacteriophage resistance conferred on lactic streptococci by the conjugative plasmid pTR2030: effects on small isometric-, large isometric-, and prolate-headed phages. Appl. Environ. Microbiol. **51:**1272–1277.

46. **Keogh, B. P., and G. Pettingall.** 1983. Adsorption of bacteriophage eb7 on *Streptococcus cremoris* EB7. Appl. Environ. Microbiol. **45:**1946–1948.

47. **King, W. R., E. B. Collins, and E. L. Barrett.** 1983. Frequencies of bacteriophage-resistant and slow acid-producing variants of *Streptococcus cremoris*. Appl. Environ. Microbiol. **45:**1481–1485.

48. **Klaenhammer, T. R.** 1984. Interactions of bacteriophages with lactic streptococci. Adv. Appl. Microbiol. **30:**1–29.

49. **Klaenhammer, T. R., and R. B. Sanozky.** 1985. Conjugal transfer from *Streptococcus lactis* ME2 of plasmids encoding phage resistance, nisin-resistance and lactose-fermenting ability: evidence for a high frequency conjugative plasmid responsible for abortive infection of virulent bacteriophage. J. Gen. Microbiol. **131:**1531–1541.

50. **Kondo, J. K., and L. L. McKay.** 1985. Gene transfer systems and molecular cloning in group N streptococci: a review. J. Dairy Sci. **68:**2143–2159.

51. **Kozak, W., M. Rajchert-Trzpil, J. Zajdel, and W. T. Dobrzanski.** 1973. Lysogeny in lactic streptococci producing and not producing nisin. Appl. Microbiol. **25:**305–308.

52. **Kruger, D. H., and T. A. Bickle.** 1983. Bacteriophage survival: multiple mechanisms for avoiding the deoxyribonucleic acid restriction systems of their hosts. Microbiol. Rev. **47:**345–360.

53. **Lawrence, R. C., H. A. Heap, and J. Gilles.** 1984. A controlled approach to cheese technology. J. Dairy Sci. **67:**1632–1645.

54. **Lawrence, R. C., H. A. Heap, G. K. Y. Limsowtin, and A. W. Jarvis.** 1978. Cheddar cheese starters: current knowledge and practices of phage characterization and strain selection. J. Dairy Sci. **61:**1181–1191.

55. **Lawrence, R. C., and L. E. Pearce.** 1972. Cheese starters under control. Dairy Ind. **37:**73–78.

56. **Lawrence, R. C., T. D. Thomas, and B. E. Terzaghi.** 1976. Reviews of the progress of dairy science: cheese starters. J. Dairy Res. **43:**141–193.

57. **Limsowtin, G. K. Y., H. A. Heap, and R. C. Lawrence.** 1978. Heterogeneity among strains of lactic streptococci. N.Z. J. Dairy Sci. Technol. **13:**1–8.

58. **Limsowtin, G. K. Y., and B. E. Terzaghi.** 1976. Phage resistant mutants: their selection and use in cheese factories. N.Z. J. Dairy Sci. Technol. **11:**251–266.

59. **Limsowtin, G. K. Y., and B. E. Terzaghi.** 1977. Characterisation of bacterial isolates from a phage-carrying culture of *Streptococcus cremoris*. N.Z. J. Dairy Sci. Technol. **12:**22–28.

60. **Loof, M., J. Lembke, and M. Teuber.** 1983. Characterization of the genome of *Streptococcus lactis* subsp. *diacetylactis* bacteriophage P008 wide-spread in German cheese factories. Syst. Appl. Microbiol. **4:**413–424.

61. **Mann, M. B., R. Nagaraja-Rao, and H. O. Smith.** 1978. Cloning of restriction and modification genes in *E. coli*: the HhaII system from *Haemophilus haemolyticus*. Gene **3**:97–112.

62. **Marschall, R. J., and N. J. Berridge.** 1976. Selection and some properties of phage-resistant starters for cheesemaking. J. Dairy Res. **43**:449–458.

63. **McKay, L. L., and K. A. Baldwin.** 1984. Conjugative 40-megadalton plasmid in *Streptococcus lactis* subsp. *diacetylactis* DRC3 is associated with resistance to nisin and bacteriophage. Appl. Environ. Microbiol. **47**:68–74.

64. **Mermelstein, N. H.** 1982. Advanced bulk starter medium improves fermentation process. Food Technol. **36**:69–76.

65. **Mullan, W. M. A., and R. J. M. Crawford.** 1985. Lysin production by phage c²(w), a prolate phage for *Streptococcus lactis* C2. J. Dairy Res. **52**:113–121.

66. **Mullan, W. M. A., and R. J. M. Crawford.** 1985. Partial purification and some properties of phage c²(w) lysin, a lytic enzyme produced by phage-infected cells of *Streptococcus lactis* C2. J. Dairy Res. **52**:123–128.

67. **Oram, J. D.** 1971. Isolation and properties of a phage receptor substance from the plasma membrane of *Streptococcus lactis* ML3. J. Gen. Virol. **13**:59–71.

68. **Oram, J. D., and B. Reiter.** 1968. The adsorption of phage to group N streptococci. The specificity of adsorption and location of phage receptor substances in cell-wall and plasma-membrane fractions. J. Gen. Virol. **3**:103–119.

69. **Pearce, L. E.** 1978. The effect of host-controlled modification on the replication rate of a lactic streptococcal bacteriophage. N.Z. J. Dairy Sci. Technol. **13**:166–171.

70. **Potter, N. N.** 1970. Host-induced changes in lactic streptococcal bacteriophages. J. Dairy Sci. **53**:1358–1362.

71. **Reiter, B.** 1973. Some thoughts on cheese starters. J. Soc. Dairy Technol. **26**:3–15.

72. **Reyrolle, J., M. C. Chopin, F. Letellier, and G. Novel.** 1982. Lysogenic strains of lactic streptococci and lytic spectra of their temperate bacteriophages. Appl. Environ. Microbiol. **43**:349–356.

73. **Richardson, G. H., G. L. Hong, and C. A. Ernstrom.** 1980. Defined strains of lactic streptococci in bulk culture for Cheddar and Monterey cheese manufacture. J. Dairy Sci. **63**:1981–1986.

74. **Sanders, M. E., and T. R. Klaenhammer.** 1980. Restriction and modification in group N streptococci: effect of heat on development of modified lytic bacteriophage. Appl. Environ. Microbiol. **40**:500–506.

75. **Sanders, M. E., and T. R. Klaenhammer.** 1981. Evidence for plasmid linkage of restriction and modification in *Streptococcus cremoris* KH. Appl. Environ. Microbiol. **42**:944–950.

76. **Sanders, M. E., and T. R. Klaenhammer.** 1983. Characterization of phage-sensitive mutants from a phage-insensitive strain of *Streptococcus lactis*: evidence for a plasmid determinant that prevents phage adsorption. Appl. Environ. Microbiol. **46**:1125–1133.

77. **Sanders, M. E., and T. R. Klaenhammer.** 1984. Phage resistance in a phage-insensitive strain of *Streptococcus*

lactis: temperature-dependent phage development and host-controlled phage replication. Appl. Environ. Microbiol. **47**:979–985.

78. **Sandine, W. E.** 1977. New techniques in handling lactic cultures to enhance their performance. J. Dairy Sci. **60**: 822–828.

79. **Sing, W. D., and T. R. Klaenhammer.** 1986. Conjugal transfer of bacteriophage resistance determinants on pTR2030 into *Streptococcus cremoris* strains. Appl. Environ. Microbiol. **51**:1264–1271.

80. **Sinha, R. P.** 1980. Alteration of host specificity to lytic bacteriophages in *Streptococcus cremoris*. Appl. Environ. Microbiol. **40**:326–332.

81. **Stadhouders, J.** 1975. Microbes in milk and dairy products. An ecological approach. Neth. Milk Dairy J. **29**:104–126.

82. **Stadhouders, J., and G. J. M. Leenders.** 1984. Spontaneously developed mixed-strain starters. Their behaviour towards phages and their use in the Dutch cheese industry. Neth. Milk Dairy J. **38**:157–181.

83. **Steenson, L. R., and T. R. Klaenhammer.** 1985. *Streptococcus cremoris* M12R transconjugants carrying the conjugal plasmid pTR2030 are insensitive to attack by lytic bacteriophages. Appl. Environ. Microbiol. **50**:851–858.

84. **Terzaghi, B. E., and W. E. Sandine.** 1981. Bacteriophage production following exposure of lactic streptococci to ultraviolet radiation. J. Gen. Microbiol. **122**:305–311.

85. **Teuber, M., and J. Lembke.** 1983. The bacteriophages of lactic acid bacteria with emphasis on genetic aspects of group N lactic streptococci. Antonie van Leeuwenhoek J. Microbiol. Serol. **49**:283–295.

86. **Theriault, G., and P. H. Roy.** 1982. Cloning of *Pseudomonas* plasmid pMG7 and its restriction-modification system in *Escherichia coli*. Gene **19**:355–359.

87. **Thomas, T. D., and R. J. Lowrie.** 1975. Starters and bacteriophages in lactic acid casein manufacture. II. Development of a controlled starter system. J. Milk Food Technol. **38**:275–278.

88. **Thunnell, R. K., F. W. Bodyfelt, and W. E. Sandine.** 1984. Economic comparisons of Cheddar cheese manufactured with defined strain and commercial mixed strain cultures. J. Dairy Sci. **67**:1061–1068.

89. **Thunell, R. K., W. E. Sandine, and F. W. Bodyfelt.** 1981. Phage-insensitive, multiple-strain starter approach to Cheddar cheesemaking. J. Dairy Sci. **64**:2270–2277.

90. **Vedamuthu, E. R., and J. M. Neville.** 1986. Involvement of a plasmid in production of ropiness (mucoidness) in milk cultures by *Streptococcus cremoris* MS. Appl. Environ. Microbiol. **51**:677–682.

91. **Walder, R. Y., J. L. Hartley, J. E. Donelson, and J. A. Walder.** 1981. Cloning and expression of the PstI restriction-modification system in *Escherichia coli*. Proc. Natl. Acad. Sci. USA **78**:1503–1507.

92. **Whitehead, H. R.** 1953. Bacteriophage in cheese manufacture. Bacteriol. Rev. **17**:109–123.

93. **Whitehead, H. R., and G. A. Cox.** 1936. Bacteriophage phenomena in cultures of lactic streptococci. J. Dairy Res. **7**:55–62.

Appendix A

Streptococcal Cloning Vectors

Vector: pMV158
Size: 5.4 kilobases (kb)
Single sites: *Eco*RI
Reference: V. Burdett, Antimicrob. Agents Chemother. **18:** 753–760, 1980

Vector: pMP5
Size: 11.9 kb
Single sites for gene inactivation: Kan^r: *Bgl*II
Other single sites: *Bam*HI
Reference: M. Espinosa, P. Lopez, M. T. Perez-Urena, and S. A. Lacks, Mol. Gen. Genet. **188:**195–201, 1982

Vector: pIP501
Size: 30 kb
Single sites for gene inactivation: Cm^r: *Bst*EII
Other single sites: *Ava*I, *Bcl*I, *Hae*III, *Hpa*II, *Pvu*II
Reference: R. P. Evans and F. L. Macrina, J. Bacteriol. **154:**1347–1355, 1983

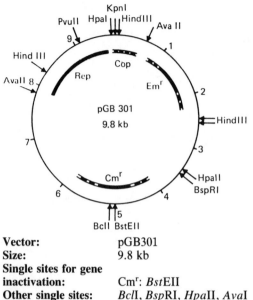

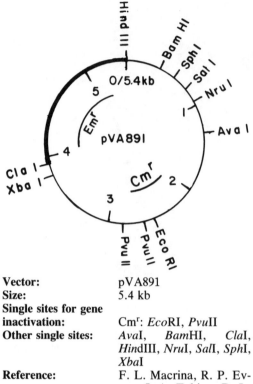

Vector: pGB301
Size: 9.8 kb
Single sites for gene inactivation: Cm^r: *Bst*EII
Other single sites: *Bcl*I, *Bsp*RI, *Hpa*II, *Ava*I
Other hosts: *Staphylococcus* spp., *Bacillus* spp.
Reference: D. Behnke, M. S. Gilmore, and J. J. Ferretti, Mol. Gen. Genet. **182**:414–421, 1981

Vector: pVA891
Size: 5.4 kb
Single sites for gene inactivation: Cm^r: *Eco*RI, *Pvu*II
Other single sites: *Ava*I, *Bam*HI, *Cla*I, *Hin*dIII, *Nru*I, *Sal*I, *Sph*I, *Xba*I
Reference: F. L. Macrina, R. P. Evans, J. A. Tobian, D. L. Hartley, D. B. Clewell, and K. R. Jones, Gene **25**:145–150, 1983

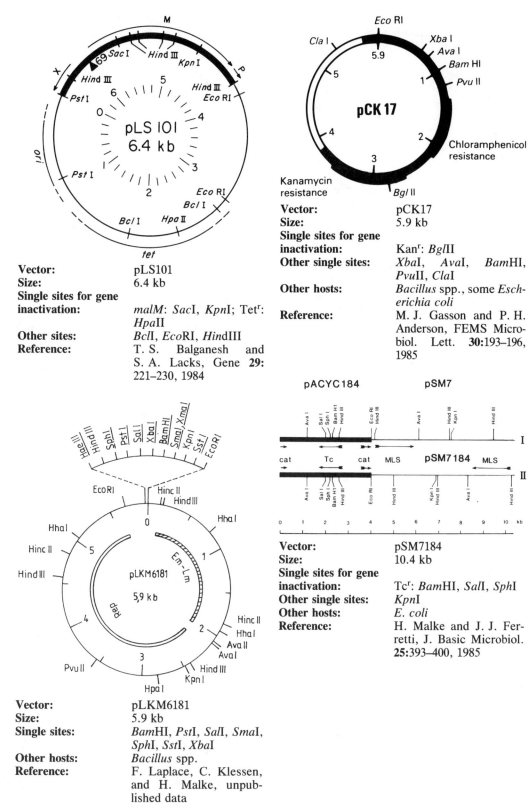

Vector: pLS101
Size: 6.4 kb
**Single sites for gene
inactivation:** *malM*: *Sac*I, *Kpn*I; Tet^r: *Hpa*II
Other sites: *Bcl*I, *Eco*RI, *Hind*III
Reference: T. S. Balganesh and S. A. Lacks, Gene **29**: 221–230, 1984

Vector: pCK17
Size: 5.9 kb
**Single sites for gene
inactivation:** Kan^r: *Bgl*II
Other single sites: *Xba*I, *Ava*I, *Bam*HI, *Pvu*II, *Cla*I
Other hosts: *Bacillus* spp., some *Escherichia coli*
Reference: M. J. Gasson and P. H. Anderson, FEMS Microbiol. Lett. **30**:193–196, 1985

Vector: pLKM6181
Size: 5.9 kb
Single sites: *Bam*HI, *Pst*I, *Sal*I, *Sma*I, *Sph*I, *Sst*I, *Xba*I
Other hosts: *Bacillus* spp.
Reference: F. Laplace, C. Klessen, and H. Malke, unpublished data

Vector: pSM7184
Size: 10.4 kb
**Single sites for gene
inactivation:** Tc^r: *Bam*HI, *Sal*I, *Sph*I
Other single sites: *Kpn*I
Other hosts: *E. coli*
Reference: H. Malke and J. J. Ferretti, J. Basic Microbiol. **25**:393–400, 1985

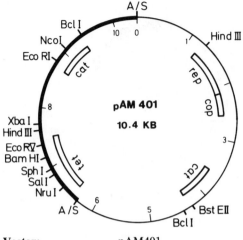

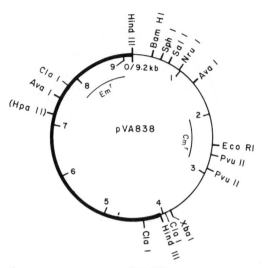

Vector: pAM401
Size: 10.4 kb
Single sites for gene inactivation: Cm^r (pGB354): *Bst*EII; Cm^r (pACYC184): *Eco*RI, *Nco*I; Tc^r: *Bam*HI, *Eco*RV, *Sal*I, *Sph*I
Other hosts: *E. coli*
Reference: R. Wirth, F. Y. An, and D. B. Clewell, J. Bacteriol. **165:**831–836, 1986

Vector: pVA838
Size: 9.2 kb
Single sites for gene inactivation: Cm^r: *Eco*RI, *Pvu*II
Other single sites: *Bam*HI, *Hpa*II, *Nru*I, *Sal*I, *Sph*I, *Xba*I
Other hosts: *E. coli*
Reference: F. L. Macrina, J. A. Tobian, K. R. Jones, R. P. Evans, and D. B. Clewell, Gene **19:**345–353, 1982

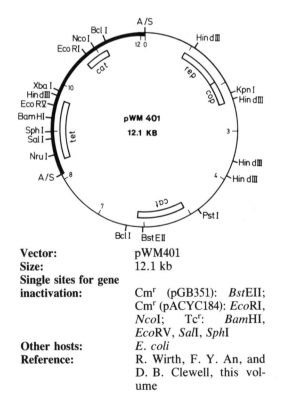

Vector: pWM401
Size: 12.1 kb
Single sites for gene inactivation: Cm^r (pGB351): *Bst*EII; Cm^r (pACYC184): *Eco*RI, *Nco*I; Tc^r: *Bam*HI, *Eco*RV, *Sal*I, *Sph*I
Other hosts: *E. coli*
Reference: R. Wirth, F. Y. An, and D. B. Clewell, this volume

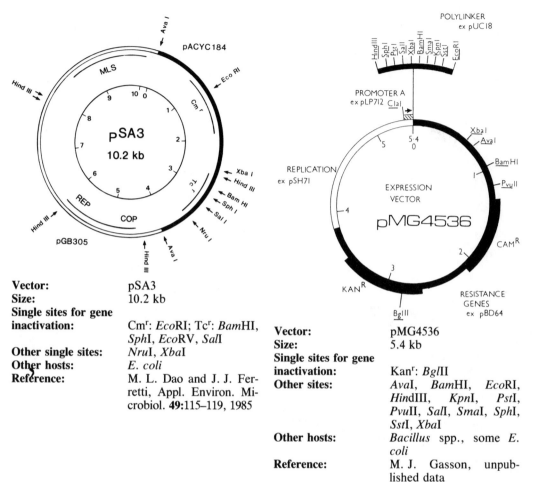

Vector: pSA3
Size: 10.2 kb
Single sites for gene inactivation: Cm^r: *Eco*RI; Tc^r: *Bam*HI, *Sph*I, *Eco*RV, *Sal*I
Other single sites: *Nru*I, *Xba*I
Other hosts: *E. coli*
Reference: M. L. Dao and J. J. Ferretti, Appl. Environ. Microbiol. **49**:115–119, 1985

Vector: pMG4536
Size: 5.4 kb
Single sites for gene inactivation: Kan^r: *Bgl*II
Other sites: *Ava*I, *Bam*HI, *Eco*RI, *Hind*III, *Kpn*I, *Pst*I, *Pvu*II, *Sal*I, *Sma*I, *Sph*I, *Sst*I, *Xba*I
Other hosts: *Bacillus* spp., some *E. coli*
Reference: M. J. Gasson, unpublished data

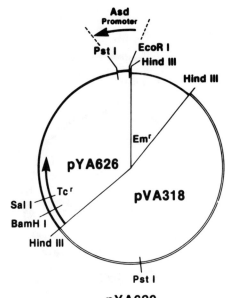

pYA629

Vector:	pYA629
Size:	11.2 kb
Single sites for gene inactivation:	Tc^r: *Bam*HI, *Sal*I
Other hosts:	*E. coli*
Reference:	H. H. Murchison, J. F. Barrett, G. A. Cardineau, and R. Curtiss III, Infect. Immun. **54**:273–282, 1986

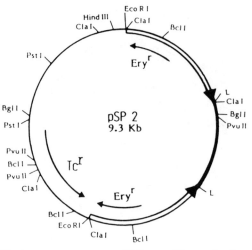

Vector:	pSP2, positive selection vector
Size:	9.3 kb/7.9 kb
Single sites for positive selection:	*Bam*HI, *Sal*I (L on map)
Reference:	H. Prats, B. Martin, P. Pognonec, A. C. Burger, and J. P. Claverys, Gene **39**:41–48, 1985

Appendix B

Nucleotide Sequences of Streptococcal Genes

```
GAATTCTCTT TAATGTTTGT AAAAGCAAAT GCGGGGGTAG AGGATAAATA TTATATTGCA GGCCATGTTT TTCGTATTAT TTCATGCTTA      90
                                                                              -35                -10
AATCAAGTAC TATTTGCATG TAATAATGCT TATTGCATTA ACGAAAAGAA AGCTATAAAA CTGCTTGAAA CTTTTGAATA TAAACCTGAA     180

AAATATGCCG AAAGAGTTAA TCATATTTTT GAAGTACTCG GTCTTTCTCT TTTTGAATGC TACGATATGA CCGAGAAACT TTATAAAGAA     270
                                                RBS
                                                             MET LYS GLU ASN LYS TYR ASP ASP ASN ILE PHE
GTGAATGAAA TTGTATCGGA GATAAATAACTTTTTAAACGAGGAGAGTTCAG ATG AAA GAA AAC AAA TAT GAT GAT AAT ATA TTT     355

PHE GLN LYS TYR SER GLN MET SER ARG SER GLN LYS GLY LEU ALA GLY ALA GLY GLU TRP GLU THR LEU LYS LYS
TTT CAA AAA TAC AGT CAA ATG AGT CGC TCG CAG AAA GGA CTG GCT GGT GCG GGA GAA TGG GAG ACT TTG AAA AAG     430

MET LEU PRO ASP PHE LYS GLY LYS ARG VAL LEU ASP LEU GLY CYS GLY TYR GLY TRP HIS CYS ILE TYR ALA MET
ATG CTA CCT GAT TTT AAG GGT AAG CGT GTG CTT GAT TTA GGA TGC GGC TAT GGA TGG CAC TGT ATA TAT GCG ATG     505

GLU ASN GLY ALA SER SER VAL VAL GLY VAL ASP ILE SER HIS LYS MET LEU GLU VAL ALA LYS GLY LYS THR HIS
GAA AAC GGT GCT TCC TCT GTA GTA GGT GTT GAT ATT TCT CAT AAA ATG CTC GAA GTA GCA AAA GGA AAA ACC CAT     580

PHE PRO GLN ILE GLU TYR GLU CYS CYS ALA ILE GLU ILE ASP VAL ASP PHE PRO GLU GLU SER PHE ASP VAL ILE LEU
TTT CCA CAG ATT GAA TAT GAA TGC TGT GCC ATA GAA GAT GTG GAT TTC CCA GAG GAG AGC TTT GAT GTA ATA CTA     655

SER SER LEU ALA PHE HIS TYR VAL ALA ASP TYR GLU ASN LEU ILE LYS LYS ILE TYR ARG MET LEU LYS ALA GLY
AGT TCG CTT GCG TTT CAT TAT GTA GCA GAC TAT GAG AAT TTA ATA AAA AAG ATA TAT AGG ATG CTG AAG GCT GGT     730

GLY ASN LEU VAL PHE THR VAL GLU HIS PRO VAL PHE THR ALA HIS GLY THR GLN ASP TRP TYR TYR ASN GLU LYS
GGC AAT TTA GTT TTT ACA GTT GAA CAT CCT GTT TTT ACT GCT CAT GGA ACA CAA GAC TGG TAT TAT AAC GAA AAA     805

GLY GLU ILE LEU HIS PHE PRO VAL ASP ASN TYR TYR TYR GLU GLY LYS ARG THR ALA MET PHE LEU GLU GLU LYS
GGA GAA ATA CTG CAT TTC CCG GTG GAC AAT TAT TAT TAT GAG GGC AAA CGG ACA GCT ATG TTT TTG GAA GAA AAG     880

VAL THR LYS TYR HIS ARG THR LEU THR TYR LEU ASN THR LEU LEU SER ASN SER PHE ILE ILE ASN GLN ILE
GTT ACA AAA TAT CAT AGA ACA CTG ACC ACA TAT CTA AAT ACA CTG CTT TCA AAT AGT TTT ATA ATA AAT CAG ATT     955

VAL GLU PRO GLN PRO PRO GLU ASN MET MET ASP ILE PRO GLY MET ALA ASP GLU MET ARG ARG PRO MET MET LEU
GTG GAG CCA CAG CCG CCA GAG AAC ATG ATG GAT ATT CCG GGG ATG GCG GAT GAA ATG CGA CGC CCA ATG ATG CTG    1030

ILE VAL SER ALA LYS LYS LYS MET *** ***                        RBS                    MET ARG SER GLU LYS
ATT GTA TCG GCA AAA AAG AAG ATG TAA TAA   TATAGAAAAA ATAAACGAGG AGTATGTAA   ATG AGA TCA GAA AAA    1104
        -35           -10
GLU MET MET ASP LEU VAL LEU SER LEU ALA GLU GLN ASP GLU ARG ILE ARG ILE VAL THR LEU GLU GLY SER ARG
GAA ATG ATG GAT TTA GTA CTT TCT TTA GCA GAA CAG GAT GAA CGT ATT CGA ATT GTG ACC CTT GAG GGG TCA CGC    1179

ALA ASN ILE ASN ILE PRO LYS ASP GLU PHE GLN ASP TYR ASP ILE THR TYR PHE VAL SER ASP ILE GLU PRO PHE
GCA AAT ATT AAT ATA CCT AAA GAT GAA TTT CAG GAT TAT GAT ATT ACA TAT TTT GTA AGT GAT ATA GAA CCG TTT    1254

ILE SER ASN ASP ASP TRP LEU ASN GLN PHE GLY ASN ILE ILE MET MET GLN LYS PRO GLU ASP MET GLU LEU PHE
ATA TCT AAT GAT GAC TGG CTT AAT CAA TTT GGG AAT ATA ATA ATG ATG CAA AAG CCG GAG GAT ATG GAA TTA TTC    1329

PRO PRO GLU GLU LYS GLY PHE SER TYR LEU MET LEU PHE ASP ASP TYR ASN LYS ILE ASP LEU THR LEU LEU PRO
CCA CCT GAA GAA AAG GGA TTT TCC TAT CTT ATG CTA TTT GAT GAT TAC AAT AAA ATT GAT CTT ACC TTA TTG CCC    1404

LEU GLU GLU LEU ASP ASN TYR LEU LYS GLY ASP LYS LEU ILE LYS VAL LEU ILE ASP LYS ASP CYS ARG ILE LYS
TTG GAA GAG TTA GAT AAT TAC CTA AAG GGC GAT AAA TTA ATA AAG GTT CTA ATT GAT AAA GAT TGT AGA ATT AAA    1479

ARG ASP ILE VAL PRO THR ASP ILE ASP TYR HIS VAL ARG LYS PRO SER ALA ARG GLU TYR ASP ASP CYS CYS ASN
AGG GAC ATA GTT CCG ACT GAT ATA GAT TAT CAT GTA AGA AAG CCA AGC GCA AGG GAG TAT GAT GAT TGC TGC AAT    1554

GLU PHE TRP ASN VAL THR PRO TYR VAL ILE LYS GLY LEU CYS ARG LYS GLU ILE LEU PHE ALA ILE ASP HIS PHE
GAA TTT TGG AAT GTA ACA CCT TAT GTT ATT AAA GGA TTG TGC CGT AAG GAA ATT TTA TTT GCT ATT GAT CAT TTT    1629

ASN GLN ILE VAL ARG HIS GLU LEU LEU ARG MET ILE SER TRP LYS VAL GLY ILE GLU THR GLY PHE LYS LEU SER
AAT CAG ATT GTT CGC CAT GAG CTG CTG AGA ATG ATA TCA TGG AAG GTC GGC ATC GAA ACA GGC TTT AAA TTA AGT    1704

VAL GLY LYS ASN TYR LYS PHE ILE GLU ARG TYR ILE SER GLU ASP LEU TRP GLU LYS LEU LEU SER THR TYR ARG
GTA GGC AAG AAC TAT AAG TTT ATT GAA AGG TAT ATA TCC GAG GAT TTG TGG GAG AAA CTT TTG TCC ACC TAC CGG    1779

MET ASP SER TYR GLU ASN ILE TRP GLU ALA LEU PHE LEU CYS HIS GLN LEU PHE ARG ALA VAL SER GLY GLU VAL
ATG GAT TCC TAT GAA AAC ATA TGG GAA GCA TTA TTT CTA TGC CAT CAA TTG TTC AGG GCG GTA TCC GGT GAG GTG    1854

ALA GLU ARG LEU HIS TYR ALA TYR PRO GLU TYR ASP ARG ASN ILE THR LYS TYR THR ARG ASP MET TYR LYS LYS
GCG GAA AGG CTT CAT TAT GCC TAT CCG GAG TAT GAT AGG AAT ATA ACA AAA TAT ACC AGG GAC ATG TAT AAA AAA    1929

TYR THR GLY LYS THR GLY CYS LEU ASP SER THR TYR ALA ALA ASP ILE GLU GLU ARG ARG GLU GLN ***
TAC ACT GGT AAA ACC GGC TGC CTG GAT AGC ACA TAT GCC GCT GAT ATA GAA GAG AGG CGG GAA CAG TGA TTACAGA    2005

AATGAAAGCA GGGCACCTGA AAGATATCGAT                                                                    2036
```

Gene: *aadE*
Protein: 6-Streptomycin adenyltransferase
Organism: *S. faecalis*
Reference: H. Ounissi and P. Courvalin, unpublished data

```
                                                      50
GATAAACCCA GCGAACCATT TGAGGTGATA GGTAAGATTA TACCGAGGTA TGAAAACGAG AATTGGACCT TTACAGAATT

              100                                                        150
ACTCTATGAA GCGCCATATT TAAAAAGCTA CCAAGACGAA GAGGATGAAG AGGATGAGGA GGCAGATTGC CTTGAATATA

                              200
TTGACAATAC TGATAAGATA ATATATCTTT TATATAGAAG ATATCGCCGT ATGTAAGGAT TTCAGGGGGC AAGGCATAGG

       250                                           300
CAGCGCGATT ATCAATATAT CTATAGAATG GGCAAAGCAT AAAAACTTGC ATGGACTAAT GCTTGAAACC CAGGACAATA

                      350                                         400
ACCTTATAGC TTGTAAATTC TATCATAATT GTGGTTTCAA AATCGGCTCC GTCGATACTA TGTTATACGC CAACTTTCAA

                                 450
AACAACTTTG AAAAAGCTGT TTTCTGGTAT TTAAGGTTTT AGAATGCAAG GAACAGTGAA TTGGAGTTCG TCTTGTTATA

              500                                                        552
ATTAGCTTCT TGGGGTATCT TTAAATACTG TAGAAAAGAG GAAGGAAATA ATAA ATG GCT AAA ATG AGA ATA TCA
                                                         M   A   K   M   R   I   S

                                                   600
CCG GAA TTG AAA AAA CTG ATC GAA AAA TAC CGC TGC GTA AAA GAT ACG GAA GGA ATG TCT CCT GCT
 P   E   L   K   K   L   I   E   K   Y   R   C   D   K   D   T   E   G   M   S   P   A

                                        651
AAG GTA TAT AAG CTG GTG GGA GAA AAT GAA AAC CTA TAT TTA AAA ATG ACG GAC AGC CGG TAT AAA
 K   V   Y   K   L   V   G   E   N   E   N   L   Y   L   K   M   T   D   S   R   Y   K

              699                                                      750
GGG ACC ACC TAT GAT GTG GAA CGG GAA AAG GAC ATG ATG CTA TGG CTG GAA GGA AAG CTG CCT GTT
 G   T   T   Y   D   D   E   R   E   K   D   M   M   L   W   L   E   G   K   L   P   V

                                                801
CCA AAG GTC CTG CAC TTT GAA CGG CAT GAT GGC TGG AGC AAT CTG CTC ATG AGT GAG GCC GAT GGC
 P   K   V   L   H   F   E   R   H   D   G   W   S   N   L   L   M   S   E   A   D   G

                            849
GTC CTT TGC TCG GAA GAG TAT GAA GAT GAA CAA AGC CCT GAA AAG ATT ATC GAG CTG TAT GCG GAG
 V   L   C   S   E   E   Y   E   D   E   Q   S   P   E   K   I   I   E   L   Y   A   E

                 900                                                     951
TGC ATC AGG CTC TTT CAC TCC ATC GAC ATA TCG GAT TGT CCC TAT ACG AAT AGC TTA GAC AGC CGC
 C   I   R   L   F   H   S   I   D   I   S   D   C   P   Y   T   N   S   L   D   S   R

                                     999
TTA GCC GAA TTG GAT TAC TTA CTG AAT AAC GAT CTG GCC GAT GTG GAT TGC GAA AAC TGG GAA GAA
 L   A   E   L   D   Y   L   L   N   N   D   L   A   D   V   D   C   E   N   W   E   E

                            1050
GAC ACT CCA TTT AAA GAT CCG CGC GAG CTG TAT GAT TTT TTA AAG ACG GAA AAG CCC GAA GAG GAA
 D   T   P   F   K   D   P   R   E   L   Y   D   F   L   K   T   E   K   P   E   E   E

            1098                                                          1149
CTT GTC TTT TCC CAC GGC GAC CTG GGA GAC AGC AAC ATC TTT ATG AAA GAT GGC AAA GTA AGT GGC
 L   V   F   S   H   G   D   L   G   D   S   N   I   F   M   K   D   G   K   V   S   G

                                              1200
TTT ATT GAT CTT GGG AGA AGC GGC AGG GCG GAC AAG TGG TAT GAC ATT GCC TTC TGC GTC CGG TCG
 F   I   D   L   G   R   S   G   R   A   D   K   W   Y   D   I   A   F   C   V   R   S

                        1248
ATC AGG GAG GAT ATC GGG GAA GAA CAG TAT GTC GAG CTA TTT TTT GAC TTA CTG GGG ATC AAG CCT
 I   R   E   D   I   G   E   E   Q   Y   V   E   L   F   F   D   L   L   G   I   K   P

              1299                                       1329            1349
GAT TGG GAG AAA ATA AAA TAT TAT ATT TTA CTG GAT GAA TTG TTT TAG  TACCTAGATT TAGATGTCTA
 D   W   E   K   I   K   Y   Y   I   L   L   D   E   L   F   -

                                     1399
AAAAGCTTTA ACTACAAGCT TTTTAGACAT CTAATCTTTT CTGAAGTACA TCCGCAACTG TCCATACTCT GATGTTTTAT

              1449                                       1489
ATCTTTTCTA AAAGTTCGCT AGATAGGGGT CCCGAGCGCC TACGAGGAAT TTGTATCGAT
```

Gene: *aph3*
Protein: 3'5"-Aminoglycoside phosphotransferase type III
Organism: *S. faecalis*
Reference: P. Trieu-Cuot and P. Courvalin, Gene **23**:331–341, 1983

5'

AGCTAAAGAAAATAAATAAAATTATAGGATTTCCATATTGCTATACACTTTAAGACCTGATGGAAAAACAATGTTTTATTTACACTCAATAGGAATGTTAC
                    50   100

CTAACTATCAAGACAAAGGTTATGGTTCAAAATTATTATCTTTTATTAAGGAATATTCTAAAGAGATTGGTTGTTCTGAAATGTTTTTAATAACTGATAA
                    150   200

AGGTAATCCTAGAGCCTTGCCATGTATATGAAAAATTAGGTCGGTAAAAAATGATTATAAAGATGAAATAGTATATGTATATGATTATGAAAAAGGTGATAA
                    250   300

```
                         10                                          20
TAA ATG AAT ATA GTT GAA AAT GAA ATA TGT ATA AGA ACT TTA ATA GAT GAT GAT TTT CCT TTG ATG TTA AAA TGG TTA
    Met Asn Ile Val Glu Asn Glu Ile Cys Ile Arg Thr Leu Ile Asp Asp Asp Phe Pro Leu Met Leu Lys Trp Leu

                         30                                          40                                     50
ACT GAT GAA AGA GTA TTA GAA TTT TAT GGT GGT AGA GAT AAA AAA TAT ACA TTA GAA TCA TTA AAA AAA CAT TAT ACA
Thr Asp Glu Arg Val Leu Glu Phe Tyr Gly Gly Arg Asp Lys Lys Tyr Thr Leu Glu Ser Leu Lys Lys His Tyr Thr

                         60                                          70
GAG CCT TGG GAA GAT GAA GTT TTT AGA GTA ATT ATT GAA TAT AAC AAT GTT CCT ATT GGA TAT GGA CAA ATA TAT AAA
Glu Pro Trp Glu Asp Glu Val Phe Arg Val Ile Ile Glu Thr Asn Asn Val Pro Ile Gly Tyr Gly Gln Ile Tyr Lys

                         80                                          90                                     100
ATG TAT GAT GAG TTA TAT ACT GAT TAT CAT TAT CCA AAA ACT GAT GAG ATA GTC TAT GGT ATG GAT CAA TTT ATA GGA
Met Tyr Asp Glu Leu Tyr The Asp Tyr His Tyr Pro Lys Thr Asp Glu Ile Val Tyr Gly Met Asp Gln Phe Ile Gly

                         110                                         120
GAG CCA AAT TAT TGG AGT AAA GGA ATT GGT ACA AGA TAT ATT AAA TTG ATT TTT GAA TTT TTG AAA AAA GAA AGA AAT
Glu Pro Asn Tyr Trp Ser Lys Gly Ile Gly Thr Arg Tyr Ile Lys Leu Ile Phe Glu Phe Leu Lys Lys Glu Arg Asn

                         130                                         140                                     150
CCT AAT GCA GTT ATT TTA GAC CCT CAT AAA AAT AAT CCA AGA GCA ATA AGG GCA TAC CAA AAA TCT GGT TTT AGA ATT
Ala Asn Ala Val Ile Leu Asp Pro His Lys Asn Asn Pro Arg Ala Ile Arg Ala Tyr Gln Lys Ser Gly Phe Arg Ile

                         160                                         170                                     180
ATT GAA GAT TTG CCA GAA CAT GAA TTA CAC GAG GGC AAA AAA GAA GAT TGT TAT TTA ATG GAA TAT AGA TAT GAT GAT
Ile Glu Asp Leu Pro Glu His Glu Leu His Glu Gly Lys Lys Glu Asp Cys Tyr Leu Met Glu Tyr Arg Tyr Asp Asp

                         190                                         200
AAT GCC ACA AAT GTT AAG GCA ATG AAA TAT TTA ATT GAG CAT TAC TTT GAT AAT TTC AAA GTA GAT AGT ATT GAA ATA
Asn Ala Thr Asn Val Lys Ala Met Lys Tyr Leu Ile Glu His Tyr Phe Asp Asn Phe Lys Val Asp Ser Ile Glu Ile

                         210                                         220                                     230
ATC GGT AGT GGT TAT GAT AGT GTG GCA TAT TTA GTT AAT AAT GAA TAC ATT TTT AAA ACA AAA TTT AGT ACT AAT AAG
Ile Gly Ser Gly Tyr Asp Ser Val Ala Tyr Leu Val Asn Asn Glu Tyr Ile Phe Lys Thr Lys Phe Ser Thr Asn Lys

                         240                                         250
AAA AAA GGT TAT CCA AAA GAA AAA GCA ATA TAT AAT TTT TTA AAT ACA AAT TTA GAA ACT AAT GTA AAA ATT CCT AAT
Lys Lys Gly Try Ala Lys Glu Lys Ala Ile Tyr Asn Phe Leu Asn Thr Asn Leu Glu Thr Asn Val Lys Ile Pro Asn
```

Gene: *aac6/aph2*
Protein: 6′-Aminoglycoside acetyltransferase-2″-aminoglycoside phosphotransferase
Organism: *S. faecalis*
Reference: J. J. Ferretti, K. S. Gilmore, and P. Courvalin, J. Bacteriol. **167**:631–638, 1986

aac6/aph2 sequence, continued.

```
260                          270                          280
ATT GAA TAT TCG TAT ATT AGT GAT GAA TTA TCT ATA CTA GGT TAT AAA GAA ATT AAA GGA ACT TTT TTA ACA CCA GAA
Ile Glu Tyr Ser Tyr Ile Ser Asp Glu Leu Ser Ile Leu Gly Tyr Lys Glu Ile Lys Gly Thr Phe Leu Thr Pro Glu

             290                          300                          310
ATT TAT TCT ACT ATG TCA GAA GAA GAA CAA AAT TTG TTA AAA CGA GAT ATT GCC AGT TTT TTA AGA CAA ATG CAC GGT
Ile Try Ser Thr Met Ser Glu Glu Glu Gln Asn Leu Leu Lys Arg Asp Ile Ala Ser Phe Leu Arg Gln Met His Gly

             320                          330
TTA GAT TAT ACA GAT ATT AGT GAA TGT ACT ATT GAT AAT AAA CAA AAT GTA TTA GAA GAG TAT ATA TTG TTG CGT GAA
Leu Asp Try Thr Asp Ile Ser Glu Cys Thr Ile Asp Asn Lys Gln Asn Val Leu Glu Glu Tyr Ile Leu Leu Arg Glu

        340                          350                          360
ACT ATT TAT AAT GAT TTA ACT GAT ATA GAA AAA GAT TAT ATA GAA AGT TTT ATG GAA AGA CTA AAT GCA ACA ACA GTT
Thr Ile Tyr Asn Asp Leu Thr Asp Ile Glu Lys Asp Tyr Ile Glu Ser Phe Met Glu Arg Leu Asn Ala Thr Thr Val

             370                          380
TTT GAG GGT AAA AAG TGT TTA TGC CAT AAT GAT TTT AGT TGT AAT CAT CTA TTG TTA GAT GGC AAT AAT AGA TTA ACT
Phe Glu Glu Lys Lys Cys Leu Cys His Asn Asp Phe Ser Cys Asn His Leu Leu Leu Asp Gly Asn Asn Arg Leu Thr

390                          400                          410
GGA ATA ATT GAT TTT GGA GAT TCT GGA ATT ATA GAT GAA TAT TGT GAT TTT ATA TAC TTA CTT GAA GAT AGT GAA GAA
Gly Ile Ile Asp Phe Gly Asp Ser Gly Ile Ile Asp Glu Tyr Cys Asp Phe Ile Tyr Leu Leu Glu Asp Ser Glu Glu

             420                          430                          440
GAA ATA GGA ACA AAT TTT GGA GAA GAT ATA TTA AGA ATG TAT GGA AAT ATA GAT ATT GAG AAA CCA AAA GAA TAT CAA
Glu Ile Gly Thr Asn Phe Gly Glu Asp Ile Leu Arg Met Tyr Gly Asn Ile Asp Ile Glu Lys Ala Lys Glu Tyr Gln

             450                          460
GAT ATA GTT GAA GAA TAT TAT CCT ATT GAA ACT ATT GTT TAT GGA ATT AAA AAT ATT AAA CAG GAA TTT ATC GAA AAT
Asp Ile Val Glu Glu Tyr Tyr Pro Ile Glu Thr Ile Val Tyr Gly Ile Lys Asn Ile Lys Gln Glu Phe Ile Glu Asn

        470
GGT AGA AAA GAA ATT TAT AAA ACG ACT TAT AAA GAT TGA TTATATAATATATGAAAGCTATTATAAAAGACATTAGTATTAAATAGTTT
Gly Arg Lys Glu Ile Tyr Lys Arg Thr Tyr Lys Asp  *

AAAAAAATGAAAAATAATAAAGGAAGTGAGTCAAGTCCAGACTCCTGTGTAAAATGCTATACAATGTTTTTACCATTTCTACTTATCAAAATTGATGTAT
   1800                          1850

TTTCTTGAAGAATAAATCCATTCATCATGTAGGTCCATAAGAACGGCTCCAATTAAGCGATTGGCTGATGTTTGATTGGGGAAGATGCGAATAATCTTTT
   1900                          1950

CTCTTCTGGGTACTTCTTGATTCAGTCGTTCAATTAGATTGGTACTCTTTAGTCGATTGTGGGGAATTCCTTGTACGGTATATTGAAAGCCGTCTTCGAA
   2000                          2050

TCCATCATCCAATGATGCCCAAGCTT 3'
   2100          2120
```

```
GATTCTAAAG TATCCGGACA ATATCTGTAT GCTTTGTATG CCTATGGTTA TGCATAAAAA TCCCAGTGAT AAGAGTATTT ATCACTGGGA TTTTTATGCC   100

                              MET LYS ILE ILE ASN ILE GLY VAL LEU ALA HIS VAL ASP ALA GLY LYS THR THR LEU THR
            RBS
       CTTTTGGGCT TTTgaaTgga ggAAAATCAC ATG AAA ATT ATT AAT ATT GGA GTT TTA GCT CAT GTT GAT GCG GGA AAA ACT ACC TTA ACA   190

GLU SER LEU LEU TYR ASN SER GLY ALA ILE THR GLU LEU GLY SER VAL ASP ARG GLY THR THR LYS THR ASP ASN THR LEU LEU GLU ARG GLN ARG GLY
GAA AGC TTA TTA TAT AAC AGT GGA GCG ATT ACA GAA TTA GGA AGC GTG GAC AGA GGT ACA ACG AAA ACG GAT AAT ACG CTT TTA GAA CGT CAG AGA GGA   289

ILE THR ILE GLN THR ALA ILE THR SER PHE GLN TRP LYS ASN THR LYS VAL ASN ILE ILE ASP THR PRO GLY HIS MET ASP PHE LEU ALA GLU VAL TYR
ATT ACA ATT CAG ACG GCG ATA ACC TCT TTT CAG TGG AAA AAT ACT AAG GTG AAC ATC ATA GAC ACG CCA GGA CAT ATG GAT TTT TTA GCA GAA GTA TAT   388

ARG SER LEU SER VAL LEU ASP GLY ALA ILE LEU LEU ILE SER ALA LYS ASP GLY VAL GLN ALA GLN THR ARG ILE LEU PHE HIS ALA LEU ARG LYS ILE
CGT TCA TTA TCA GTA TTA GAT GGG GCA ATT CTA CTG ATT TCT GCA AAA GAT GGC GTA CAA GCA CAA ACT CGT ATA TTG TTT CAT GCA CTT AGG AAA ATA   487

GLY ILE PRO THR ILE PHE PHE ILE ASN LYS ILE ASP GLN ASN GLY ILE ASP LEU SER THR VAL TYR GLN ASP ILE LYS GLU LYS LEU SER ALA GLU ILE
GGT ATT CCC ACA ATC TTT TTT ATC AAT AAG ATT GAC CAA AAT GGA ATT GAT TTA TCA ACG GTT TAT CAG GAT ATT AAA GAG AAA CTT TCT GCG GAA ATT   586

VAL ILE LYS GLN LYS VAL GLU LEU HIS PRO ASN MET ARG VAL MET ASN PHE THR GLU SER GLU GLN TRP ASP MET VAL ILE GLU GLY ASN ASP TYR LEU
GTA ATC AAA CAG AAG GTA GAA CTG CAT CCT AAT ATG CGT GTA ATG AAC TTT ACC GAA TCT GAA CAA TGG GAT ATG GTA ATA GAA GGA AAT GAT TAC CTT   685

LEU GLU LYS TYR THR SER GLY LYS LEU LEU GLU ALA LEU GLU LEU GLU GLN GLU GLU SER ILE ARG PHE HIS ASN CYS SER LEU PHE PRO VAL TYR HIS
TTG GAG AAA TAT ACG TCT GGG AAA TTA TTG GAA GCA TTA GAA CTC GAA CAA GAG GAA AGC ATA AGA TTT CAT AAT TGT TCC CTG TTC CCT GTT TAT CAC   784

GLY SER ALA LYS ASN ASN ILE GLY ILE ASP ASN LEU ILE GLU VAL ILE THR ASN LYS PHE TYR SER SER THR HIS ARG GLY GLN SER GLU LEU CYS GLY
GGA AGT GCA AAA AAC AAT ATA GGG ATT GAT AAC CTT ATA GAA GTG ATT ACG AAT AAA TTT TAT TCA TCA ACA CAT CGA GGT CAG TCT GAA CTT TGC GGA   883

LYS VAL PHE LYS ILE GLU TYR SER GLU LYS ARG GLN ARG LEU ALA TYR ILE ARG LEU TYR SER GLY VAL LEU HIS LEU ARG ASP PRO VAL ARG ILE SER
AAA GTT TTC AAA ATT GAG TAT TCG GAA AAA AGA CAG CGT CTT GCA TAT ATA CGT CTT TAT AGT GGC GTA CTG CAT TTG CGA GAT CCG GTT AGA ATA TCG   982

GLU LYS GLU LYS ILE LYS ILE THR GLU MET TYR THR SER ILE ASN GLY GLU LEU CYS LYS ILE ASP LYS ALA TYR SER GLY GLU ILE VAL ILE LEU GLN
GAA AAG GAA AAA ATA AAA ATT ACA GAA ATG TAT ACT TCA ATA AAT GGT GAA TTA TGT AAA ATC GAT AAG GCT TAT TCC GGG GAA ATT GTT ATT TTG CAG   1081

ASN GLU PHE LEU LYS LEU ASN SER VAL LEU GLY ASP THR LYS LEU LEU PRO GLN ARG GLU ARG ILE GLU ASN PRO LEU PRO LEU LEU GLN THR THR VAL
AAT GAG TTT TTG AAG TTA AAT AGT GTT CTT GGA GAT ACA AAG CTA TTG CCA CAG AGA GAG AGA ATT GAA AAT CCC CTC CCT CTG CTG CAA ACG ACT GTT   1180

GLU PRO SER LYS PRO GLN GLN ARG GLU MET LEU LEU ASP ALA LEU LEU GLU ILE SER ASP SER ASP PRO LEU LEU ARG TYR TYR VAL ASP SER ALA THR
GAA CCG AGC AAA CCT CAA CAA AGG GAA ATG TTA CTT GAT GCA CTT TTA GAA ATC TCC GAC AGT GAC CCG CTT CTG CGA TAT TAT GTG GAT TCT GCA ACA   1279

HIS GLU ILE ILE LEU SER PHE LEU GLY LYS VAL GLN MET GLU VAL THR CYS ALA LEU LEU GLN GLU LYS TYR HIS VAL GLU ILE GLU ILE LYS GLU PRO
CAT GAA ATC ATA CTT TCT TTC TTA GGG AAA GTA CAA ATG GAA GTG ACT TGT GCT CTG CTG CAA GAA AAG TAT CAT GTG GAG ATA GAA ATA AAA GAG CCT   1378

THR VAL ILE TYR MET GLU ARG PRO LEU LYS LYS ALA GLU TYR THR ILE HIS ILE GLU VAL PRO PRO ASN PRO PHE TRP ALA SER ILE GLY LEU SER VAL
ACA GTC ATT TAT ATG GAA AGA CCG TTA AAA AAA GCA GAG TAT ACC ATT CAC ATC GAA GTT CCA CCG AAT CCT TTC TGG GCT TCC ATT GGT CTA TCT GTA   1477

ALA PRO LEU PRO LEU GLY SER GLY VAL GLN TYR GLU SER SER VAL SER LEU GLY TYR LEU ASN GLN SER PHE GLN ASN ALA VAL MET GLU GLY ILE ARG
GCA CCG CTT CCA TTA GGG AGC GGA GTA CAG TAT GAG AGC TCG GTT TCT CTT GGA TAC TTA AAT CAA TCG TTT CAA AAT GCA GTT ATG GAG GGG ATA CGC   1576

TYR GLY CYS GLU GLN GLY LEU TYR GLY TRP ASN VAL THR ASP CYS LYS ILE CYS PHE LYS TYR GLY LEU TYR TYR SER PRO VAL SER THR PRO ALA ASP
TAT GGC TGT GAA CAA GGA TTG TAT GGT TGG AAT GTG ACG GAC TGT AAA ATC TGT TTT AAG TAT GGC TTA TAC TAT AGC CCT GTT AGT ACC CCA GCA GAT   1675

PHE ARG MET LEU ALA PRO ILE VAL LEU GLU GLN VAL LEU LYS LYS ALA GLY THR GLU LEU LEU GLU PRO TYR LEU SER PHE LYS ILE TYR ALA PRO GLN
TTT CGG ATG CTT GCT CCT ATT GTA TTG GAA CAA GTC TTA AAA AAA GCT GGA ACA GAA TTG TTA GAG CCA TAT CTT AGT TTT AAA ATT TAT GCG CCA CAG   1774

GLU TYR LEU SER ARG ALA TYR ASN ASP ALA PRO LYS TYR CYS ALA ASN ILE VAL ASP THR GLN LEU LYS ASN ASN GLU VAL ILE LEU SER GLY GLU ILE
GAA TAT CTT TCA CGA GCA TAC AAC GAT GCT CCT AAA TAT TGT GCC AAC ATC GTA GAC ACT CAA TTG AAA AAT AAT GAG GTC ATT CTT AGT GGA GAA ATC   1873

PRO ALA ARG CYS ILE GLN GLU TYR ARG SER ASP LEU THR PHE PHE THR ASN GLY ARG SER VAL CYS LEU THR GLU LEU LYS GLY TYR HIS VAL THR THR
CCT GCT CGG TGT ATT CAA GAA TAT CGT AGT GAT TTA ACT TTC TTT ACA AAT GGA CGT AGT GTT TGT TTA ACA GAG TTA AAA GGG TAC CAT GTT ACT ACC   1972

GLY GLU PRO VAL CYS GLN PRO ARG ARG PRO ASN SER ARG ILE ASP LYS VAL ARG TYR MET PHE ASN LYS ILE THR ***
GGT GAA CCT GTT TGC CAG CCC CGT CGT CCA AAT AGT CGG ATA GAT AAA GTA CGA TAT ATG TTC AAT AAA ATA ACT TAG TGTATTTTAT GTTGTTATAT   2070

       AAATATGGTT TCTTGTTAAA TAAGCAAACA AAAAAACCAC CGCTACCAGC GGTGGTTTGT TTGCCGGATC AAGAGCTACC AACTCTTTTT CCGAAGGTAA   2170
```

Gene: *tetM*
Protein: Tetracycline resistance protein of transposon Tn*1545*
Organism: *Streptococcus* spp.
Reference: P. Martin, P. Trieu-Cuot, and P. Courvalin, Nucleic Acids Res. **14:**7047–7058, 1986

```
DdeI                    -35                      -10        +1
CTTAGAAGCAAACTTAAGAGTGTGTTGATAGTGCATTATCTTAAAATTTTGTATAATAGGAATTGAAGTTAAATTAGATGCTAAAAATTTGTAATTAAGA
---------+---------+---------+---------+---------+---------+---------+---------+---------+---------+    100

  SD-1 HinfI    MetLeuValPheGlnMetArgAsnValAspLysThrSerThrIluLeuLysGlnThrLysAsnSerAspTyrValAspLysTyrVa
AGGAGGGATTCGTCATGTTGGTATTCCAAATGCGTAATGTAGATAAAACATCTACTATTTTGAAACAGACTAAAAACAGTGATTACGTAGATAAATACGT
---------+---------+---------+---------+------●--+---------+---------+---------+---------+---------+    200
                         (icrs 1)

lArgLeuIluProThrSerAspEND
TAGATTAATTCCTACCAGTGACTAATCTTATGACTTTTTAAACAGATAACTAAAATTACAAACAAATCGTTTAACTTCGTGTATTTGTTTATAGATGTATC
---------+---------●-----+---------+---------+---------+---------+---------+---------+---------+---------+    300
       (icrs 6)

                 MetLysLysAsnIluLysTyrSerGlnAsnPheLeuThrAsnGluLysValLeuAsnGlnIluIluLysGlnLeuAsnLeuLy
         SD-2                                  RsaI
ACTTCAGGAGTGATTACATGAAAAAAAATATAAAATATTCTCAAAACTTTTTAACGAATGAAAAGGTACTCAACCAAATAATAAAACAATTGAATTTAAA
----●---+---------+---------+---------+---------+---------+---------+---------+---------+---------+    400
(icrs 14)
sGluThrAspThrValTyrGluIluGlyThrGlyLysGlyHisLeuThrThrLysLeuAlaLysIluSerLysGlnValThrSerIluGluLeuAspSer
AGAAACCGATACCGTTTACGAAATTGGAACAGGTAAAGGGCATTTAACGACGAAACTGGCTAAAATAAGTAAACAGGTAACGTCTATTGAATTAGACAGT
---------+---------+---------+---------+---------+---------+---------+---------+---------+---------+    500

HisLeuPheAsnLeuSerSerGluLysLeuLysLeuAsnIluArgValThrLeuIluHisGlnAspIluLeuGlnPheGlnPheProAsnLysGlnArgT
CATCTATTCAACTTATCTTCAGAAAAATTAAAACTGAACATTCGTGTCACTTTAATTCACCAAGATATTCTACAGTTTCAATTCCCTAACAAACAGAGGT
---------+---------+---------+---------+---------+---------+---------+---------+---------+---------+    600

yrLysIluValGlySerIluProTyrHisLeuSerThrGlnIluIluLysLysValValPheGluSerHisAlaSerAspIluTyrLeuIluValGluGl
ATAAAATTGTTGGGAGTATTCCTTACCATTTAAGCACACAAATTATTAAAAAAGTGGTTTTTGAAAGCCATGCGTCTGACATCTATCTGATTGTTGAAGA
---------+---------+---------+---------+---------+---------+---------+---------+---------+---------+    700

uGlyPheTyrLysArgThrLeuAspIluHisArgSerLeuGlyLeuLeuLeuHisThrGlnValSerIluGlnGlnLeuLeuLysLeuProAlaGluCys
  HinfI        RsaI                                          TaqIHinfI          Fnu4HI
AGGATTCTACAAGCGTACCTTGGATATTCACCGTTCACTAGGGTTGCTCTTGCACACTCAAGTCTCGATTCAGCAATTGCTTAAGCTGCCAGCGGAATGC
---------+---------+---------+---------+---------+---------+---------+---------+---------+---------+    800

PheHisProLysProLysValAsnSerValLeuIluLysLeuThrArgHisThrThrAspValProAspLysTyrTrpLysLeuTyrThrTyrPheValS
                                                                                            RsaI
TTTCATCCTAAACCAAAAGTAAACAGTGTCTTAATAAAACTTACCCGCCATACCACAGATGTTCCAGATAAATATTGGAAGCTATATACGTACTTTGTTT
---------+---------+---------+---------+---------+---------+---------+---------+---------+---------+    900

erLysTrpValAsnArgGluTyrArgGlnLeuPheThrLysAsnGlnPheHisGlnAlaMetLysHisAlaLysValAsnAsnLeuSerThrValThrTy
TaqI                                                                              RsaI
CAAAATGGGTCAATCGAGAATATCGTCAACTGTTTACTAAAAATCAGTTTCATCAAGCAATGAAACACGCCAAAGTAAACAATTTAAGTACCGTTACTTA
---------+---------+---------+---------+---------+---------+---------+---------+---------+---------+    1000

rGluGlnValLeuSerIluPheAsnSerTyrLeuLeuPheAsnGlyArgLysEND    MetSerArgPheCysLysPheGlyLysLeuHisValThrLy
                                 SD-3                      HinfI
TGAGCAAGTATTGTCTATTTTTAATAGTTATCTATTATTTAACGGGAGGAAATAATTCTATGAGTCGCTTTTGTAAATTTGGAAAGTTACACGTTACTAA
---------+---------+---------+---------+----●----+---------+---------+---------+---------+---------+    1100

sGlyAsnValAspLysLeuLeuGlyIluLeuLeuThrAlaSerLysLysLeuLysArgSerLeuAlaProThrGlyAsnLeuTyrArgEND
                                                        AvaII               ClaI        RsaI
AGGGAATGTAGATAAATTATTAGGTATACTACTGACAGCTTCCAAGAAGCTAAAGAGGTCCCTAGCGCCTACGGGGAATTTGTATCGATAAGGGGTACAA
---------+---------+---------+---------+---------+---------+---------+---------+---------+---------●---+    1200

   DdeI      AvaII                DdeI                  RsaI
ATTCCCACTAAGCGCTCGGGACCCCTTGTAGGAAATTGTCCTAAGTGTGGCAACAATATTGGGGTACATGGAACATAGAAAACAAAGTGAAGTATCTTTC
---------+---------+---------+---------+---------+---------+---------+---------+--------●----+---------+    1300

                                                 DdeI
TAGGGTAAATATACCACCCCCACCTTTAAGGCTCTTAAAATACGTTTTAGAGCCTTAGAAAACGACACAAAAAAGCAAGAGCATTTTTGACCTTGCTTTT
---------+---------+---------+---------+●●------+---------+---------+---------+----●●---+---------+    1400

TTTATTGTCGTGCAATCCGATACACTTTATACGAAGCTCTAAA
---------+---------+---------+--- 1443
```

Gene: pAM77/MLS resistance
Protein: N^6-Methylase of adenine in 23S rRNA
Organism: *S. sanguis*
Reference: S. Horinouchi, W. H. Byeon, and B. Weisblum, J. Bacteriol. **154**:1252–1262, 1983

```
                                                                              *  AV
AGAATTAGTGTTTAGAGCGAACGATTAAATATTTATATTAGGCAAATATTGTCTCAAAAGGTAATAACAACGATTTAATAGATAGTCTTTTTGGGGTCC
         20            40            60            80           100

A 1                  CLA 1                              RSA 1
CGAGCGCCTACGAGGAATTTGTATCGATAAGAAATAGATTTAAAAATTTCGCTGTTATTTTGTACATTTAACTTGACGGTGACATCTCTCTATTGTGAGT
        120           140           160           180           200

       RSA 1
TATTAGTGGTACAGTTTTCAACCGTTTTAATTATAAAAAAGTGGTGCATTTTTAAATTGGCACAAACAGGTAACGGTTATTGCAGGTGTATTTCTTATCT
        220           240           260           280           300

                                                       HPA 1
ATGGGTTTAACATGGATTTTATCATTAAAATCATGAGTATTGTCCGAGAGTGATTGGTCTTGCGTATGGTTAACCCTAAAGTTATGGAAATAAGACTTAG
        320           340           360           380           400

                 -35               -10                                   S.D.
AAGCAAACTTAAGAGTGTGTTGATAGTGCATTATCTTAAAATTTTGTATAATAGGAATTGAAGTTAAATTAGATGCTAAAAATTTGTAATTAAGAAGGAG
        420           440           460           480           500
     ORF1
     MetLeuValPheGlnMetArgAsnValAspLysThrSerThrValLeuLysGlnThrLysAsnSerAspTyrAlaAspLysTyrValArgL

GGATTCGTCATGTTGGTATTCCAAATGCGTAATGTAGATAAAACATCTACTGTTTTGAAACAGACTAAAAACAGTGATTACGCAGATAAATACGTTAGAT
        520           540           560           580           600

euIleProThrSerAsp

TAATTCCTACCAGTGACTAATCTTATGACTTTTTAAACAGATAACTAAAATTACAAACAAATCGTTTAACTTCTGTATTTATTTATAGATGTATCACTTC
        620           640           660           680           700
     ORF2
     MetAsnLysAsnIleLysTyrSerGlnAsnPheLeuThrAsnGluLysValLeuAsnGlnIleIleLysGlnLeuAsnLeuLysGluT
 S.D.                                           RSA 1
AGGAGTGATTACATGAACAAAAATATAAAATATTCTCAAAACTTTTTAACGAATGAAAAGGTACTCAACCAAATAATAAAACAATTGAATTTAAAAGAAA
        720           740           760           780           800

hrAspThrValTyrGluIleGlyThrGlyLysGlyHisLeuThrThrLysLeuAlaLysIleSerLysGlnValThrSerIleGluLeuAspSerHisLe

CCGATACCGTTTACGAAATTGGAACAGGTAAAGGGCATTTAACGACGAAACTGGCTAAAATAAGTAAACAGGTAACGTCTATTGAATTAGACAGTCATCT
        820           840           860           880           900

uPheAsnLeuSerSerGluLysLeuLysLeuAsnIleArgValThrLeuIleHisGlnAspIleLeuGlnPheGlnPheProAsnLysGlnArgTyrLys

ATTCAACTTATCGTCAGAAAAATTAAAACTGAACATTCGTGTCACTTTAATTCACCAAGATATTCTACAGTTTCAATTCCCTAACAAACAGAGGTATAAA
        920           940           960           980          1000

IleValGlyAsnIleProTyrHisLeuSerThrGlnIleIleLysLysValValPheGluSerHisAlaSerAspIleTyrLeuIleValGluGluGlyP

ATTGTTGGGAATATTCCTTACCATTTAAGCACACAAATTATTAAAAAAGTGGTTTTTGAAAGCCATGCGTCTGACATCTATCTGATTGTTGAAGAAGGAT
       1020          1040          1060          1080          1100

heTyrLysArgThrLeuAspIleHisArgThrLeuGlyLeuLeuLeuHisThrGlnValSerIleGlnGlnLeuLeuLysLeuProAlaGluCysPheHi
          RSA 1                                      TAQ 1           ALU 1
TCTACAAGCGTACCTTGGATATTCACCGAACACTAGGGTTGCTCTTGCACACTCAAGTCTCGATTCAGCAATTGCTTAAGCTGCCAGCGGAATGCTTTCA
       1120          1140          1160          1180          1200

sProLysProLysValAsnSerValLeuIleLysLeuThrArgHisThrThrAspValProAspLysTyrTrpLysLeuTyrThrTyrPheValSerLys
                                                            ALU 1        RSA 1
TCCTAAACCAAAAGTAAACAGTGTCTTAATAAAACTTACCCGCCATACCACAGATGTTCCAGATAAATATTGGAAGCTATATACGTACTTTGTTTCAAAA
       1220          1240          1260          1280          1300

TrpValAsnArgGluTyrArgGlnLeuPheThrLysAsnGlnPheHisGlnAlaMetLysHisAlaLysValAsnAsnLeuSerThrValThrTyrGluG
   TAQ 1                                                                   RSA 1
TGGGTCAATCGAGAATATCGTCAACTGTTTACTAAAAATCAGTTTCATCAAGCAATGAAACACGCCAAAGTAAACAATTTAAGTACCGTTACTTATGAGC
       1320          1340          1360          1380          1400
                                              ORF3
 lnValLeuSerIlePheAsnSerTyrLeuLeuPheAsnGlyArgLys       MetSerArgPheCysLysPheGlyLysLeuHisValThrLysGlyA
                      S.D.
AAGTATTGTCTATTTTTAATAGTTATCTATTATTTAACGGGAGGAAATAATTCTATGAGTCGCTTTTGTAAATTTGGAAAGTTACACGTTACTAAAGGGA
       1420          1440          1460          1480          1500
```

Gene: Tn*917* entire sequence
Protein: N^6-Methylase of adenine in 23S rRNA and five other open reading frames
Organism: *S. faecalis*
Reference: J. H. Shaw, and D. B. Clewell, J. Bacteriol. **164**:782–796, 1985

281

Tn917 sequence, continued.

```
snValAspLysLeuLeuGlyIleLeuLeuThrAlaSerLysGluLeuLysArgSerLeuAlaProThrGlyAsnLeuTyrArg
                    ALU 1        ALU 1                                    CLA 1
ATGTAGATAAATTATTAGGTATACTACTGACAGCTTCCAAGGAGCTAAAGAGGTCCCTAGCGCCTACGGGGAATTTGTATCGATAAGGAATAGATTTAAA
        1520             1540             1560             1580             1600

              RSA 1                  KPN 1                  RSA 1
AATTTCGCTGTTATTTTGTACAATAAGGATAAATTTGAATGGTACCATAAACGACCGTTTATGGTACTTTTCATTTTCCTGCTTTTTCTAAATGTTTTTT
        1620             1640             1660             1680             1700

        RSA 1                                      -35          RSA 1     -10        S.D.
AAGTAAATCAAGTACCAAAATCCGTTCCTTTTTCATAGTTCCTATATAGTATACTTAATGAGTTATGGTACATTTAAATTATAAAATTAAGGAGGTTTTT
        1720             1740             1760             1780             1800
ORF4
   MetIlePheGlyTyrAlaArgValSerThrAspAspGlnAsnLeuSerLeuGlnIleAspAlaLeuThrHisTyrGlyIleAspLysLeuPheGlnGl
                 AVA 1    RSA 1 BCL 1
TTATGATTTTTGGCTATGCTCGAGTGAGTACGGATGATCAAAATCTTAGTTTACAAATTGATGCACTTACTCATTATGGAATTGATAAATTATTTCAAGA
        1820             1840             1860             1880             1900

uLysValThrGlyAlaLysLysAspArgProGlnLeuGluGluMetIleAsnLeuLeuArgGluGlyAspSerValValIleTyrLysLeuAspArgIle
                            BCL 1                                              TAQ 1
AAAAGTAACTGGTGCGAAAAAAGACCGACCGCAATTAGAAGAAATGATCAACCTACTACGTGAAGGAGATTCTGTTGTCATTTACAAGTTAGATCGAATT
        1920             1940             1960             1980             2000

SerArgSerThyLysHisLeuIleGluLeuSerGluLeuPheGluGluLeuSerValAsnPheIleSerIleGlnAspAsnValAspThrSerThrSerM
TCACGATCAACTAAACATTTGATTGAACTTTCTGAATTATTTGAAGAACTTAGTGTCAATTTTATATCTATTCAAGATAACGTAGATACTTCAACGTCTA
        2020             2040             2060             2080             2100

etGlyArgPhePhePheArgValMetAlaSerLeuAlaGluLeuGluArgAspIleIleIleGluArgThrAsnSerGlyLeuLysAlaAlaArgValAr
TGGGAAGATTCTTTTTTCCGAGTTATGGCTAGTTTAGCAGAACTGGAACGGGATATTATTATTGAACGAACTAACTCTGGTCTTAAGGCAGCCAGAGTCCG
        2120             2140             2160             2180             2200

gGlyLysLysGlyGlyArgProSerLysGlyLysLeuSerIleAspLeuAlaLeuLysMetTyrAspSerLysGluTyrSerIleArgGlnIleLeuAsp
                     ALU 1            ALU 1
AGGAAAAAAAGGGGGCCGTCCAAGTAAAGGTAAGCTATCAATTGATTTAGCTTTAAAAAATGTATGACAGCAAAGAGTATTCTATTCGTCAAATTCTTGAT
        2220             2240             2260             2280             2300
                                           ORF5
AlaSerLysLeuLysAsnAsnLeuLeuProLeuProGln          MetAlaMetLysArgIleLeuThrThrSerGlnArgGluGln
GCCTCTAAATTAAAAAACAACCTTTTACCGTTACCTCAATAAAAGGTATGCTTAAGATATGGCTATGAAAAGAATTTTAACTACTTCACAGCGTGAACAA
        2320             2340             2360             2380             2400

LeuLeuSerValAspHisLeuSerGluGluAspPheLysAlaTyrPheSerPheSerAspTyrAspLeuGluValIleAsnGlnHisArgGlyLysValA
                                                                                           HPA 11
CTTCTTTCTGTAGACCACTTATCAGAAGAGGATTTTAAAGCGTATTTTAGTTTTTCTGATTATGATCTGGAGGTTATTAATCAACACCGTGGAAAGGTCA
        2420             2440             2460             2480             2500

snLysLeuGlyPheAlaIleGlnLeuCysLeuAlaArgTyrProGlyCysSerLeuSerAsnTrpProIleLysSerThrArgLeuThrSerTyrValSe
                                                                                               S
ATAAACTAGGATTTGCGATACAACTTTGTTTGGCCCGGTATCCTGGGTGTTCTTTAAGTAATTGGCCGATTAAATCAACCAGACTAACTTCTTATGTGAG
        2520             2540             2560             2580             2600

rArgGlnLeuHisLeuAspAlaIleAspLeuAsnSerTyrAspHisArgAsnThrArgAlaAsnHisPheAsnGluIleLeuLeuGluValPheAsnTyrHis
AL 1                                                                          XHO 11              CL
TCGACAGCTCCATCTTGATGCAATTGATTTAAATTCATATGATCATAGAAATACACGTGCAAATCACTTCAACGAGATCTTAGAAGTATTCAACTATCAT
        2620             2640             2660             2680             2700

ArgPheGlySerAlaAsnThrGlnLysGlnLeuIleGluTyrLeuIleGluLeuAlaLeuGluAsnAspAspSerIleTyrLeuMetLysLysThrIleA
A 1                                                      ALU 1
CGATTCGGTAGTGCTAATACACAAAAACAGTTAATAGAATATTTAATTGAACTAGCTTTAGAAAATGATGACTCTATCTATCTAATGAAAAAAACAATTG
        2720             2740             2760             2780             2800

spPheLeuThrArgLysArgIleIlePheProSerIleAlaThrLeuGluAspIleIleSerArgCysArgAspLysAlaGluAsnAsnLeuPheSerIl
        TAQ 1                       ALU 1                         TAQ 1
ATTTCTTAACTCGAAAAAGAATTATTTTTTCCATCTATAGCTACACTTGAAGACATTATAAGCCGCTGTCGAGATAAAGCAGAAAAGAACTTATTTTCAAT
        2820             2840             2860             2880             2900
```

Tn917 sequence, continued.

eLeuLeuCysSerLeuThrAspIleGlnIleGluLysLeuGluSerLeuPheGlnIleTyrGluGluThrLysIleThrLysLeuAlaTrpLeuLysAsp

ATTACTCTGTTCATTAACAGATATACAAATTGAAAAACTAGAGAGTTTGTTTCAAATTTATGAAGAGACGAAAATAACTAAACTCGCTTGGCTAAAAGAC
　　　　　2920　　　　　　　　　2940　　　　　　　　　2960　　　　　　　　　2980　　　　　　　　　3000

IleProGlyLysAlaAsnProGluSerPheMetSerIleCysLysLysValGluValIleAlaSerMetGlyLeuGlyThrIleAsnValSerHisIleA
　　　　　　　　　HIND 111　　　　　　　　　　　　　　　　　　　　　　　　AVA 1
ATTCCAGGTAAGGCAAATCCAGAAAGCTTTATGAGTATTTGTAAAAAAGTGGAAGTGATTGCTTCCATGGGACTCGGGACAATTAATGTCTCCCATATTA
　　　　　3020　　　　　　　　　3040　　　　　　　　　3060　　　　　　　　　3080　　　　　　　　　3100

snArgAsnArgPheLeuGlnLeuAlaArgLeuGlyGluAsnTyrAspAlaTyrAspPheSerArgPheGluLeuGluLysArgTyrSerLeuLeuIleAl
　　　　　ALU1ALU1　　　　　　　　　　　　　　　　　　　　　　　　　　ALU 1
ATCGGAACAGGTTTCTTCAGCTAGCTAGACTAGGGGAAAATTATGATGCATATGACTTCTCCCGTTTTGAGCTTGAAAAAAGATACTCTTTACTTATTGC
　　　　　3120　　　　　　　　　3140　　　　　　　　　3160　　　　　　　　　3180　　　　　　　　　3200

aPheLeuValAsnHisHisGlnTyrLeuIleAspGlnLeuIleGluIleAsnAspArgIleLeuAlaSerIleLysArgLysGlyThrArgAspSerGln
　　　　　　　　　CLA 1
TTTTTTAGTCAATCATCATCAATATCTGATCGATCAACTGATTGAGATTAATGACCGCATTTTAGCAAGTATTAAACGCAAAGGGACACGTGATTCACAA
　　　　　3220　　　　　　　　　3240　　　　　　　　　3260　　　　　　　　　3280　　　　　　　　　3300

GluGlnLeuLysGluLysGlyLysLeuAlaThrLysLysLeuGluHisTyrAlaSerLeuIleAspAlaLeuHisPheAlaLysAspAsnAspSerAsnP

GAACAGTTAAAAGAAAAAGGAAAATTGGCTACTAAAAAATTGGAACATTATGCTTCTTTAATTGATGCTCTTCACTTTGCAAAAGATAATGATAGTAATC
　　　　　3320　　　　　　　　　3340　　　　　　　　　3360　　　　　　　　　3380　　　　　　　　　3400

roPheAspGluIleGluArgIleMetProTrpGluAspLeuValGlnAspGlyGluGluAlaLysLysAlaIleThrGlyAsnLysAsnHisGlyTyrLe
　　　　　　　　　　　　　　　RSA 1　　　　　　　　　ALU 1　　　　　ALU 1
CTTTTGACGAAATTGAACGAATCATGCCTTGGGAAGATTTAGTACAAGATGGAGAAGAAGCTAAAAAAGCTATTACAGGTAATAAAAATCATGGCTATTT
　　　　　3420　　　　　　　　　3440　　　　　　　　　3460　　　　　　　　　3480　　　　　　　　　3500

uGluMetValArgAsnLysAlaAsnTyrLeuArgArgTyrThrProMetLeuLeuArgThrLeuSerPheLysAlaThrProAlaAlaAsnProValLeu
　　　　　ALU 1　　　　　　　　　　　　　　　　　　　　　　　　　　　　　　HPA 11
AGAAATGGTTCGAAATAAAGCTAATTACCTCCGAAGATACACGCCAATGTTATTGAGGACCCTTTCGTTCAAAGCAACTCCGGCAGCAAATCCAGTCCTC
　　　　　3520　　　　　　　　　3540　　　　　　　　　3560　　　　　　　　　3580　　　　　　　　　3600

MetAlaLeuThrGlnLeuThrAspLeuHisAsnSerGlyLysArgLysIleProAlaAspThrSerThrAspPheValSerLysLysTrpLysSerLeuV
　　　　　　　　　　　　　　HPA 11
ATGGCCCTAACTCAACTAACTGATTTACACAATAGTGGTAAAAGAAAAATACCGGCAGATACTTCTACTGATTTTGTGAGTAAAAAATGGAAAAGCCTTG
　　　　　3620　　　　　　　　　3640　　　　　　　　　3660　　　　　　　　　3680　　　　　　　　　3700

alArgProGluGluGlyLysIleAspArgSerTyrTyrGluLeuValAlaPheThrGluLeuLysAsnAsnIleArgSerGlyAspIleSerValGluGl
　　　　　　　　　　　　　　　　　ALU 1　　　　　ALU　　　　　　　　TAQ 1
TTCGGCCAGAAGAGGGGAAAATAGATCGGTCTTACTATGAGTTAGTAGCTTTCACCGAGCTAAAGAACAATATTCGATCAGGAGATATTTCAGTTGAAGG
　　　　　3720　　　　　　　　　3740　　　　　　　　　3760　　　　　　　　　3780　　　　　　　　　3800

ySerMetIleHisArgAsnIleAspAspTyrLeuValAspLeuSerAlaCysIleAspSerGluThrIleProAspThrPheGluAspTyrLeuLysAsp
　　　　　TAQ 1
AAGTATGATCCATCGAAATATTGATGATTACTTAGTTGATTTATCTGCTTGTATTGATTCAGAAACTATTCCAGACACGTTTGAGGACTATTTAAAGGAT
　　　　　3820　　　　　　　　　3840　　　　　　　　　3860　　　　　　　　　3880　　　　　　　　　3900

ArgGluIleIleLeuAspLeuGlnLeuGlnPheTyrSerThrValAspLysArgIleSerArgAlaAsnLeuLysLysLeuGluLysValThrProSerA
　　　　　ALU 1　　　　　　　　TAQ 1
CGGGAAATAATTTTAGATTTACAGCTTCAATTTTATTCGACAGTTGATAAGAGAATTTCAAGAGCAAACCTTAAAAAGTTGGAAAAAGTTACACCTAGCG
　　　　　3920　　　　　　　　　3940　　　　　　　　　3960　　　　　　　　　3980　　　　　　　　　4000

spArgLysTyrIleGluLysAsnPheIleGln

ACAGGAAATATATAGAAAAAAACTTTATTCAATAATTCCTAAGATAAGGCTTAGTGATCTTTTAATTGAGGTGGACAGTTGGACCAACTTTTCACAAGAA
　　　　　4020　　　　　　　　　4040　　　　　　　　　4060　　　　　　　　　4080　　　　　　　　　4100

TTTTAGTCATGATTCTACAGGGAAACCGCCGAGTGAACAAGAAAGAAAAATTATTTTTGCTGCTTTGCTGGGTTTAGGGATGAATATTGGTCTTGAAAAA
　　　　　4120　　　　　　　　　4140　　　　　　　　　4160　　　　　　　　　4180　　　　　　　　　4200

　　　　　　　　　　　　　　　　　　　-35　　　　　　　　　-10　　　ALU 1
ATGGCCCAATCAACTCCTGGAATTTCTTATTCTCAGTTAGCCAATGCCAAACAATGGCGCTTTTATAAAGAAGCTCTGACTCGTGCTCAATCTGTTTTGG
　　　　　4220　　　　　　　　　4240　　　　　　　　　4260　　　　　　　　　4280　　　　　　　　　4300

Tn917 sequence, continued.

```
          HIND 111                                                                              ALU 1
TTAATTATCAGTTAAAGCTTCCTGTTGCAGACTTTTGGGGTGAAGGAAAAACCACTGCTTCAGACGGAATGCGCGTCCCAGTGGCGTCTCAGCTCTAAAA
          4320            4340            4360            4380            4400

                                      ORF6
                           S.D.        MetIleArgSerIleAsnAspArgHisThrThrHisHisIleGluValAlaSerT
                           ALU 1                                                TAQ 1
TCCGATGTTAATCCACATTACAAAAGTATGGAAAAAGGAGCTACAATGATTCGATCAATAAATGATAGGCATACGACTCATCATATCGAGGTTGCTTCAA
          4420            4440            4460            4480            4500

  hrAsnThrArgGluAlaThrHisThrLeuAspGlyLeuLeuTyrHisGluThrAspLeuAspIleGluGluHisPheThrAspThrAsnGlyTyrSerAs
          ALU 1                                 XBA 1                                       BCL
CTAATACAAGGGAAGCTACTCATACCCTTGATGGCCTACTTTATCATGAAACAGATCTAGATATTGAGGAACATTTTACTGATACAAATGGGTATTCTGA
          4520            4540            4560            4580            4600

  pGlnValPheGlyMetThrAlaLeuLeuGlyPheAspPheGluProArgIleArgAsnIleLysLysSerGlnLeuPheSerIleLysSerProSerTyr
   1
TCAGGTGTTTGGAATGACCGCATTACTAGGCTTTGATTTTGAACCTCGCATCAGAAATATAAAAAAATCACAATTATTTTCTATCAAATCACCTTCCTAC
          4620            4640            4660            4680            4700

  TyrProAsnLeuSerGluAspIleSerGlyLysIleAsnValLysIleIleGluGluAsnTyrAspGluIleLysArgIleAlaTyrSerIleGlnThrG
                                                                                            TAQ 1
TACCCTAACTTATCAGAAGATATAAGCGGAAAAATCAATGTAAAAATTATTGAAGAAAACTATGATGAAATTAAACGAATCGCCTATTCGATTCAAACAG
          4720            4740            4760            4780            4800

  lyLysValSerSerSerLeuLeuLeuGlyLysLeuGlySerTyrAlaArgLysAsnArgValAlaLeuAlaLeuArgGluLeuGlyArgIleGluLysSe
          ALU 1                                          ALU 1
GAAAAGTATCTAGTTCTTTACTATTAGGAAAGCTAGGCTCATACGCACGTAAGAATAGAGTAGCTCTTGCACTGAGAGAACTAGGTCGCATTGAAAAGAG
          4820            4840            4860            4880            4900

  rIlePheMetIleAspTyrIleThrAspSerGluLeuArgArgArgIleThrHisGlyLeuAsnLysThrGluAlaIleAsnAlaLeuArgArgGluLeu
          ALU 1
CATTTTTATGATAGATTATATTACAGATAGTGAGCTACGGCGAAGGATCACTCATGGACTAAATAAGACAGAAGCGATTAATGCTTTACGTAGAGAACTA
          4920            4940            4960            4980            5000

  PhePheGlyAspAlaTluAsnLeuTrpSerAlaIlePheAlaAspAsnPheLysValLeuValArgLeuMetCys
TTTTTTGGCGACGCGGAAAATTTATGGAGCGCGATATTCGCCGACAACTTCAAAGTGCTAGTGCGCTTAATGTGTTAATAAATGCAATAAGTATATGGAA
          5020            5040            5060            5080            5100

                AL
CGCCGTCTACTTACAAGCAGCTTATAATTATCTCGTCAAAATAGATCCCGAAGTAACTAAGTATATGAAGCATGTATCTCCTATTAATTGGGAGCATATC
          5120            5140            5160            5180            5200

ACTTTTCTTGGAGAGTATAAAATTTGACTTGTTATCTATTCCTAAACACTTAAGAGAATTGAATATAAAAAATAAAAGGCCTTGAAACATTGGTTTAGTGG
          5220            5240            5260            5280            5300

                CLA 1             AVA 1      *                     HINC 11
GAATTTGTACCCCTTATCGATACAAATTCCCACTAAGCGCTCGGGACCCCTTTTTTAGGATATATTTGTTTTTAATGGTTAACTATTCTATTTTACTGAC
          5320            5340            5360            5380            5400

                ALU 1
AATAATAGCTCTTTTCTAATCTCTTTAATAGCTTTTTTAAGTATTATAAATTCGCATACAATAAAAAGATTTGTAGATAAAGAAATAATGGAACAAGGAA
          5420            5440            5460            5480            5500

ACGTCCAGAGAATTATTACAGAAGCAATTGAAGGAATTGAAACCATTAAATCTGAATGCAGAAAAGAGTTTTTTGTTAAATTGGAAAAACATGTTTACGT
          5520            5540            5560            5580            5600

CTCAA
```

5' TCATGTTTGACAGCTTATCATCGATAAGCTTACTTTTCGAATCAGGTCTATCCTTGAAACAGGTGCAACATAGATTAGGGCATGGAGATTTACCAGACAA
                        50   100

CTATGAACGTATATACTCACATCACGCAATCGGCAATTGATGACATTGGAACTAAATTCAATCAATTTGTTACTAACAAGCAACTAGATTGACAACTAAT
                        150   200

TCTCAACAAACGTTAATTTAACAACATTCAAGTAACTCCCACCAGCTCCATCAATGCTTACCGTAAGTAATCATAACTTACTAAAACCTTGTTACATCAA
                        250   300

GGTTTTTTCTTTTTGTCTTGTTCATGAGTTACCATAACTTTCTATATTATTGACAACTAAATTGACAACTCTTCAATTATTTTTCTGTCTACTCAAAGTT
                        350   400

TTCTTCATTTGATATAGTCTAATTCCACCATCACTTCTTCCACTCTCTCTACCGTCACAACTTCATCATCTCTCACTTTTTCGTGTGGTAACACATAATC
                        450   500

AAATATCTTTCCGTTTTTACGCACTATCGCTACTGTGTCACCTAAAATATACCCCTTATCAATCGCTTCTTTAAACTCATCTATATATAACATATTTCAT
                        550   600

CCTCCTACCTATCTATTCGTAAAAAGATAAAAATAACTATTGTTTTTTTTGTTATTTTATAATAAAATTATTAATATAAGTTAATGTTTTTTAAAAATAT
                        650   700

ACAATTTTATTCTATTTATAGTTAGCTATTTTTTCATTGTTAGTAATATTGGTGAATTGTAATAACCTTTTTAAATCTAGAGGAGAACCCAGATATAAAA
                        750   800

                            M   E   N   N   K   K   V   L   K   K   M   V   F   F   V   L   V   T   F   L   G   L
TGGAGGAATATTA ATG GAA AAC AAT AAA AAA GTA TTG AAG AAA ATG GTA TTT TTT GTT TTA GTG ACA TTT CTT GGA CTA
     RBS   10   20

     T   I   S   Q   E   V   F   A   Q   Q   D   P   D   P   S   Q   L   H   R   S   S   L   V   K   N   L
     ACA ATC TCG CAA GAG GTA TTT GCT CAA CAA GAC CCC GAT CCA AGC CAA CTT CAC AGA TCT AGT TTA GTT AAA AAC CTT
                            30   40

     Q   N   I   Y   F   L   Y   E   G   D   P   V   T   H   E   N   V   K   S   V   D   Q   L   L   S   H
     CAA AAT ATA TAT TTT CTT TAT GAG GGT GAC CCT GTT ACT CAC GAG AAT GTG AAA TCT GTT GAT CAA CTT TTA TCT CAC
        50   60   70

     D   L   I   Y   N   V   S   G   P   N   Y   D   K   L   K   T   E   L   K   N   Q   E   M   A   T   L
     GAT TTA ATA TAT AAT GTT TCA GGG CCA AAT TAT GAT AAA TTA AAA ACT GAA CTT AAG AAC CAA GAG ATG GCA ACT TTA
        80   90   100

     F   K   D   K   N   V   D   I   Y   G   V   E   Y   Y   H   L   C   Y   L   C   E   N   A   E   R   S
     TTT AAG GAT AAA AAC GTT GAT ATT TAT GGT GTA GAA TAT TAC CAT CTC TGT TAT TTA TGT GAA AAT GCA GAA AGG AGT
                            110   120

     A   C   I   Y   G   G   V   T   N   H   E   G   N   H   L   E   I   P   K   K   I   V   V   K   V   S
     GCA TGT ATC TAC GGA GGG GTA ACA AAT CAT GAA GGG AAT CAT TTA GAA ATT CCT AAA AAG ATA GTC GTT AAA GTA TCA
                            130   140   150

     I   D   G   I   Q   S   L   S   F   D   I   E   T   N   K   K   M   V   T   A   Q   E   L   D   Y   K
     ATC GAT GGT ATC CAA AGC CTA TCA TTT GAT ATT GAA ACA AAT AAA AAA ATG GTA ACT GCT CAA GAA TTA GAC TAT AAA
                            160   170

     V   R   K   Y   L   T   D   N   K   Q   L   Y   T   N   G   P   S   K   Y   E   T   G   Y   I   K   F
     GTT AGA AAA TAT CTT ACA GAT AAT AAG CAA CTA TAT ACT AAT GGA CCT TCT AAA TAT GAA ACT GGA TAT ATA AAG TTC
        180   190   200

     I   P   K   N   K   E   S   F   W   F   D   F   F   P   E   P   E   F   T   Q   S   K   Y   L   M   I
     ATA CCT AAG AAT AAA GAA AGT TTT TGG TTT GAT TTT TTC CCT GAA CCA GAA TTT ACT CAA TCT AAA TAT CTT ATG ATA
        210   220   230

     Y   K   D   N   E   T   L   D   S   N   T   S   Q   I   E   V   Y   L   T   T   K   *
     TAT AAA GAT AAT GAA ACG CTT GAC TCA AAC ACA AGC CAA ATT GAA GTC TAC CTA ACA ACC AAG TAA CTTTTTGCTTTTGGC
                            240   250

AACCTTACCTACTGCTGGATTTAGAAATTTTATTGCAATTCTTTTATTAATGTAAAAACCGCTCATTTGATGAGCGGTTTGTCTTATCTAAAGGAGCTTTAC
                1600   1650

CTCCTAATGCTGCAAAATTTTAAATGTTGGATTTTTGTATTTGTCTATTGTATTTGATGGGTAATCCCATTTTTCGACAGACATCGTCGTGCCACCTCTAACA
                1700   1750

CCAAAATCATAGACAGGAGCTTGTAGCTTAGCAACTATTTTATCGTC 3'   EcoRI
                1800   1837

Gene: *speA*
Protein: Streptococcal exotoxin type A (erythrogenic toxin)
Organism: *S. pyogenes* bacteriophage T12
Reference: C. R. Weeks and J. J. Ferretti, Infect. Immun. **52**:144–150, 1986

```
        PstI
5'  CTGCAGCTACCTGATACCAGGCATTTCCAACAAACATGGTTAAGGGCCAAACCAAAATCACTTTCTAGCGTTGGCAAGAGACCTTCAAGCGAGCGCAAGACCTTTATTGAAGTTGCTTGTC
                                 60                                                 120

    GACATAAAAATGCTGTTTGGGTTGTGCTGATAGGCAAAATGACCTCAAGCCCTGCAATCATCTGCTGGAGCAACTCAACTAAGTCAGCTGGTAAAACCTGCTGATGATTGAGGTAAATAA
                                 180                                                240

    ACTGAGAAGTCTCAAACAGCTGAGGGGGATTGCCCTGATGATCAAGCAAATACCGCTGCCAAGGTGACCCTAGCGGCTGCAAGACCTCATATTGACCCAACCCCACCTCAAGTAATAAGC
                                 300                                                360

    GCTCTTTTTCGGATAAACATGATTTGGGAAAATGCACATATTGGTCCCCTTCTTTGACACTCACCCACTCTTTATCTCCTAACGGATGAGGGCCTACTTGCATCTCTGGAAAATAGTCTT
                                 420                                                480

    TTAGCTCCATAGCCATTCCTTTCATGACGGTCTTTAAACCATTATAACACATGACTCTTTATCACACAGTTCAGTTTGTTGTCAGCACGATTTTGTATTTTCTGCCTTTTTAATCATTAA
                                 540                                                600

    AACTAAATAAGGGTTATTCATTTTTAGCAAGAACATTCAATTAAATAGCTATTTATCGGAATATTAATTTATGTTTATGCTAAAAAAGGTATTATTTACCTTTTTTCATTGTCATTAAAA
                                 660                                                720

                                                                                        M  K  N  Y  L
    TATCATTTTAAAAAAATCATTAGGTTTTTATTTGTGTCTTTAAAACCATTATGTTATTCTAATAATGGGGATTGAAACTTAACTTTTAGGAGGTTTCT ATG AAA AAT TAC TTA
                                 780                                      SD                    30

     S   F   G   M   F   A   L   L   F   A   L   T   F   G   T   V   N   S   V   Q   A│I   A   G   P   E   W   L   L   D
    TCT TTT GGG ATG TTT GCA CTG CTG TTT GCA CTA ACA TTT GGA ACA GTC AAT TCT GTC CAA GCT│ATT GCT GGA CCT GAG TGG CTG CTA GAC
              10                                20                                60

     R   P   S   V   N   N   S   Q   L   V   V   S   V   A   G   T   V   E   G   T   N   Q   D   I   S   L   K   F   F   E
    CGT CCA TCT GTC AAC AAC AGC CAA TTA GTT GTT AGC GTT GCT GGT ACT GTT GAG GGG ACG AAT CAA GAC ATT AGT CTT AAA TTT TTT GAA
              40                                50                                60

     I   D   L   T   S   R   P   A   H   G   G   K   T   E   Q   G   L   S   P   K   S   K   P   F   A   T   D   S   G   A
    ATC GAT CTA ACA TCA CGA CCT GCT CAT GGA GGA AAG ACA GAG CAA GGC TTA AGT CCA AAA TCA AAA CCA TTT GCT ACT GAT AGT GGC GCG
              70                                80                                90

     M   S   H   K   L   E   K   A   D   L   L   K   A   I   Q   E   Q   L   I   A   N   V   H   S   N   D   D   Y   F   E
    ATG TCA CAT AAA CTT GAG AAA GCT GAC TTA CTA AAG GCT ATT CAA GAA CAA TTG ATC GCT AAC GTC CAC AGT AAC GAC GAC TAC TTT GAG
              100                               110                               120

     V   I   D   F   A   S   D   A   T   I   T   D   R   N   G   K   V   Y   F   A   D   K   D   G   S   V   T   L   P   T
    GTC ATT GAT TTT GCA AGC GAT GCA ACC ATT ACT GAT CGA AAC GGC AAG GTC TAC TTT GCT GAC AAA GAT GGT TCG GTA ACC TTG CCG ACC
              130                               140                               150

     Q   P   V   Q   E   F   L   L   S   G   H   V   R   V   R   P   Y   K   E   K   P   I   Q   N   Q   A   K   S   V   D
    CAA CCT GTC CAA GAA TTT TTG CTA AGC GGA CAT GTG CGC GTT AGA CCA TAT AAA GAA AAA CCA ATA CAA AAC CAA GCG AAA TCT GTT GAT
              160                               170                               180

     V   E   Y   T   V   Q   F   T   P   L   N   P   D   D   D   F   R   P   G   L   K   D   T   K   L   L   L   K   T   L   A
    GTG GAA TAT ACT GTA CAG TTT ACT CCC TTA AAC CCT GAT GAC GAT TTC AGA CCA GGT CTC AAA GAT ACT AAG CTA TTG AAA ACA CTA GCT
              190                               200                               210

     I   G   D   T   I   T   S   Q   E   L   L   A   Q   A   Q   S   I   L   N   K   N   H   P   G   Y   T   I   Y   E   R
    ATC GGT GAC ACC ATC ACA TCT CAA GAA TTA CTA GCT CAA GCA CAA AGC ATT TTA AAC AAA AAC CAC CCA GGC TAT ACG ATT TAT GAA CGT
              220                               230                               240

     D   S   S   I   V   T   H   D   N   D   I   F   R   T   I   L   P   M   D   Q   E   F   T   Y   R   V   K   N   R   E
    GAC TCC TCA ATC GTC ACT CAT GAC AAT GAC ATT TTC CGT ACG ATT TTA CCA ATG GAT CAA GAG TTT ACT TAC CGT GTT AAA AAT CGG GAA
              250                               260                               270

     Q   A   Y   R   I   N   K   K   S   G   L   N   E   E   I   N   N   T   D   L   I   S   E   K   Y   Y   V   L   K   K
    CAA GCT TAT AGG ATC AAT AAA AAA TCT GGT CTG AAT GAA GAA ATA AAC AAC ACT GAC CTG ATC TCT GAG AAA TAT TAC GTC CTT AAA AAA
              280                               290                               300

     G   E   K   P   Y   D   P   F   D   R   S   H   L   K   L   F   T   I   K   Y   V   D   V   D   T   N   E   L   L   K
    GGG GAA AAG CCG TAT GAT CCC TTT GAT CGC AGT CAC TTG AAA CTG TTC ACC ATC AAA TAC GTT GAT GTC GAT ACC AAC GAA TTG CTA AAA
              310                               320                               330

     S   E   Q   L   L   T   A   S   E   R   N   L   D   F   R   D   L   Y   D   P   R   D   K   A   K   L   L   Y   N   N
    AGT GAG CAG CTC TTA ACA GCT AGC GAA CGT AAC TTA GAC TTC AGA GAT TTA TAC GAT CCT CGT GAT AAG GCT AAA CTA CTC TAC AAC AAT
              340                               350                               360

     L   D   A   F   G   I   M   D   Y   T   L   T   G   K   V   E   D   N   H   D   D   T   N   R   I   I   T   V   Y   M
    CTC GAT GCT TTT GGT ATT ATG GAC TAT ACC TTA ACT GGA AAA GTA GAG GAT AAT CAC GAT GAC ACC AAC CGT ATC ATA ACC GTT TAT ATG
              370                               380                               390

     G   K   R   P   E   G   E   N   A   S   Y   H   L   A   Y   D   K   D   R   Y   T   E   E   E   R   E   V   Y   S   Y
    GGC AAG CGA CCC GAA GGA GAG AAT GCT AGC TAT CAT TTA GCC TAT GAT AAA GAT CGT TAT ACC GAA GAA GAA CGA GAA GTT TAC AGC TAC
              400                               410                               420

     L   R   Y   T   G   T   P   I   P   D   N   P   N   D   K   *
    CTG CGT TAT ACA GGG ACA CCT ATA CCT GAT AAC CCT AAC GAC AAA TAA CCACGGTCTTCTAAAACGATGAGATTAACTGACAAAAAAAGCAAGCAACATGCTAT
              430                               440                                       2160

    CAACAGTTGCTTGCTTTTTTCTAACCTCTTAGTTGTAGAGACTAGTGACATTTCGTGTCTAAAATAATCGTAACTGGTCCATCATTGATGAGACTAACCTGCATATCTGCCCCAAAAACG
                           2220                                                2280

    CCACGCTCAACTGGCACAAAATCTGCCAATTGTTCATTAAAGCGATCATAAAACTGGCTAGCCATATCAGCTTTGCAGCTCCTGTAAAGGCTGGGCGATTTCCCTTTTTGGTGTCAGCAT
                           2340                                                2400

    AAAGGGTAAATTGCGACACAGATAAGATACTACCCTTGATGTCTTGGATAGACTGATTCATCTTGCCATCAGCATCTGAAAAAATGCGCATGTTGACTATTTTTGCACAGCGTAAGCCAA
                           2460                                                2520

           PstI
    ATCTTCTGCAG  3'
           2568
```

Gene: *skc*
Protein: Streptokinase
Organism: *S. equisimilis*
Reference: H. Malke, B. Roe, and J. J. Ferretti, Gene **34**:357–362, 1985

```
        10        20        30        40        50        60        70        80        90       100
GATCTAGAAG AGATTGAAAA ACAGTATGAT GTGATCGTGA CAGATGTTAT GGTAGGAAAA AGCGATGAGT TAGAAATTTT CTTTTTCTAC AAAATGATTC

       110       120       130       140       150       160       170       180       190       200
CAGAAGCGAT TATTGACAAG CTCAATGTGT TTTTAAACAT CAGCTTTGCA GACAGCTTGC CACTAGACAA ACCCATCAAG AACCCCTTGG ACTTTCATCG

       210       220       230       240       250       260       270       280       290       300
CAAAGAGCTT ACCTTACCCA CTCCCCCCAA CAAGTTGCAC GCCCCCCCCT CCACAACTTA GACAGCCTAG CCGCAGAAAC TCAAAAACAG ATTCATCATT

       310       320       330       340       350       360       370       380       390       400
AATAGCATTT AGGTCAAAAA GGTGGCAAAA GCTAAAAAAG CTGGTCTTTA CCTTTTTGGCT TTTATTATTT ACAATACAAT TATTAGAGTT AAACCCTGAA

       410       420       430       440       450       459              474            489
AATGAGGGTT TTTCCTAAAA AATGATAACA TAAGGAGCAT AAAA     ATG GCT AAA AAT AAC ACG AAT AGA CAC TAT TCG CTT AGA AAA TTA
                                                     Met Ala Lys Asn Asn Thr Asn Arg His Tyr Ser Leu Arg Lys Leu

           504              519              534              549              564
AAA AAA GGT ACT GCA TCA GTA GCA GTG GCT TTG AGT GTA ATA GGG GCA GGA TTA GTT GTC AAT ACT AAT GAA GTT AGT GCA AGA
Lys Lys Gly Thr Ala Ser Val Ala Val Ala Leu Ser Val Ile Gly Ala Gly Leu Val Val Asn Thr Asn Glu Val Ser Ala Arg

       579              594              609              624              639              654
GTG TTT CCT AGG GGG ACG GTA GAA AAC CCG GAC AAA GCA CGA GAA CTT CTT AAC AAG TAT GAC GTA GAG AAC TCT ATG TTA CAA
Val Phe Pro Arg Gly Thr Val Glu Asn Pro Asp Lys Ala Arg Glu Leu Leu Asn Lys Tyr Asp Val Glu Asn Ser Met Leu Gln

           669              684              699              714              729
GCT AAT AAT GAC AAG TTA ACA ACT GAG AAT AAT AAC TTA ACA GAT CAG AAT AAA AAC TTA ACA ACT GAG AAT AAA AAC TTA ACA
Ala Asn Asn Asp Lys Leu Thr Thr Glu Asn Asn Asn Leu Thr Asp Gln Asn Lys Asn Leu Thr Thr Glu Asn Lys Asn Leu Thr

       744              759              774              789              804              819
GAT CAG AAT AAA AAC TTA ACA ACT GAG AAT AAA AAC TTA ACA GAT CAG AAT AAA AAC TTA ACA ACT GAG AAT AAG GAG TTA AAA
Asp Gln Asn Lys Asn Leu Thr Thr Glu Asn Lys Asn Leu Thr Asp Gln Asn Lys Asn Leu Thr Thr Glu Asn Lys Glu Leu Lys

           834              849              864              879              894              909
GCT GAG GAG AAT AGG TTA ACA ACT GAG AAT AAA GGG TTA ACT AAA AAG TTG AGT GAA GCT GAA GAA GCA GCA AAT AAA GAG
Ala Glu Glu Asn Arg Leu Thr Thr Glu Asn Lys Gly Leu Thr Lys Lys Leu Ser Glu Ala Glu Glu Ala Ala Asn Lys Glu

           924              939              954              969              984
CGA GAA AAT AAA GAA GCC ATT GGT ACC CTT AAA AAA ACC TTG GAT GAG ACA GTA AAA GAT AAA ATT GCT AAG GAG CAA GAA AGT
Arg Glu Asn Lys Glu Ala Ile Gly Thr Leu Lys Lys Thr Leu Asp Glu Thr Val Lys Asp Lys Ile Ala Lys Glu Gln Glu Ser

       999             1014             1029             1044             1059             1074
AAA GAA ACC ATT GGT ACC CTT AAA AAA ACC TTG GAT GAG ACA GTA AAA GAT AAA ATT GCT AAG GAG CAA GAA AGT AAA GAA ACC
Lys Glu Thr Ile Gly Thr Leu Lys Lys Thr Leu Asp Glu Thr Val Lys Asp Lys Ile Ala Lys Glu Gln Glu Ser Lys Glu Thr

              1089             1104             1119             1134             1149
AAT GGT ACC CTT AAA AAA ACC TTG GAT GAG ACA GTA AAA GAT AAA ATT GCT AAG GAG CAA GAA AGT AAA GAA ACC ATT GGT ACC
Ile Gly Thr Leu Lys Lys Thr Leu Asp Glu Thr Val Lys Asp Lys Ile Ala Lys Glu Gln Glu Ser Lys Glu Thr Ile Gly Thr

1164             1179             1194             1209             1224             1239
CTT AAA AAA ATC TTG GAT GAG ACA GTA AAA GAT AAA ATT GCG AGA GAG CAA AAA AGT AAA CAA GAC ATT GGT GCC CTT AAA CAA
Leu Lys Lys Ile Leu Asp Glu Thr Val Lys Asp Lys Ile Ala Arg Glu Gln Lys Ser Lys Gln Asp Ile Gly Ala Leu Lys Gln

       1254             1269             1284             1299             1314             1329
GAA TTA GCT AAA AAA GAT GAA GGA AAC AAA GTT TCA GAA GCA AGC CGT AAG GGT CTT CGC CGT GAC TTG GAC GCA TCA CGT GAA
Glu Leu Ala Lys Lys Asp Glu Gly Asn Lys Val Ser Glu Ala Ser Arg Lys Gly Leu Arg Arg Asp Leu Asp Ala Ser Arg Glu

              1344             1359             1374             1389             1404
GCT AAG AAA CAG GTT GAA AAA GAT TTA GCA AAC TTG ACT GCT GAA CTT GAT AAG GTT AAA GAA GAA AAA CAA ATC TCA GAC GCA
Ala Lys Lys Gln Val Glu Lys Asp Leu Ala Asn Leu Thr Ala Glu Leu Asp Lys Val Lys Glu Glu Lys Gln Ile Ser Asp Ala

1419             1434             1449             1464             1479             1494
AGC CGT CAA GGT CTT CGC CGT GAC TTG GAC GCA TCA CGT GAA GCT AAG AAA CAA GTT GAA AAA GCT TTA GAA GAA GCA AAC AGC
Ser Arg Gln Gly Leu Arg Arg Asp Leu Asp Ala Ser Arg Gln Ala Lys Lys Gln Val Glu Lys Ala Leu Glu Glu Ala Asn Ser

       1509             1524             1539             1554             1569
AAA TTA GCT GCT CTT GAA AAA CTT AAC AAA GAG CTT GAA GAA AGC AAG AAA TTA ACA GAA AAA GAA AAA GCT GAG CTA CAA GCA
Lys Leu Ala Ala Leu Glu Lys Leu Asn Lys Glu Leu Glu Glu Ser Lys Lys Leu Thr Glu Lys Glu Lys Ala Glu Leu Gln Ala

1584             1599             1614             1629             1644             1659
AAA CTT GAA GCA GAA GCA AAA GCA CTC AAA GAA CAA TTA GCG AAA CAA GCT GAA GAA CTT GCA AAA CTA AGA GCT GGA AAA GCA
Lys Leu Glu Ala Glu Ala Lys Ala Leu Lys Glu Gln Leu Ala Lys Gln Ala Glu Glu Leu Ala Lys Leu Arg Ala Gly Lys Ala

       1674             1689             1704             1719             1734             1749
TCA GAC TCA CAA ACC CCT GAT GCA AAA CCA GGA AAC AAA GTT GTT CCA AAA GGT CAA GCA CCA CAA GCA GGT ACA AAA CCT
Ser Asp Ser Gln Thr Pro Asp Ala Lys Pro Gly Asn Lys Val Val Pro Lys Gly Gln Ala Pro Gln Ala Gly Thr Lys Pro

          1764             1779             1794             1809             1824
AAC CAA AAC AAA GCA CCA ATG AAG GAA ACT AAG AGA CAG TTA CCA TCA ACA GGT GAA ACA GCT AAC CCA TTC TTC ACA GCG GCA
Asn Gln Asn Lys Ala Pro Met Lys Glu Thr Lys Arg Gln Leu Pro Ser Thr Gly Glu Thr Ala Asn Pro Phe Phe Thr Ala Ala

1839             1854             1869             1884                           1903      1913
GCC CTT ACT GTT ATG GCA ACA GCT GGA GTA GCA GCA GTT GTA AAA CGC AAA GAA GAA AAC     TAAGCTATCA CTTTGTAATA
Ala Leu Thr Val Met Ala Thr Ala Gly Val Ala Ala Val Val Lys Arg Lys Glu Glu Asn

      1923      1933      1943      1953      1963      1973      1983      1993      2003      2013
CTGAGTGAAC ATCAAGAGAC AAACCAGTCGG TTCTCTCTTT TATGTATAGA AGAATGAGAT TAAGGAGGTC ACAAACTAAA CAACTCTTAA AAAGCTGACC

      2023      2033      2043      2053      2063      2073      2083      2093      2103
TTTACTAATA ATCGTCTTTT TTTTATAATA AAGATGTTAA TAATATAATT GATAAATGAG ATACATTTAA TGATTATGAC AAAAGCAAGA AAA
```

Gene: *emm6*
Protein: Type 6 M protein
Organism: *S. pyogenes*
Reference: S. K. Hollingshead, V. A. Fischetti, and J. R. Scott, J. Biol. Chem. **261**:1677–1686, 1986

```
        .                  .              .            .              .
AAGCTTTTTAGTCTGGGGTGTTATTGTAGATAGAATGCAGACCTTGTCAGTCCTATTTAC 60

        .              .            .              .              .
AGTGTCAAAATAGTGCGTTTTGAAGTTCTATCTACAAGCCTAATCGTGACTAAGATTGTC 120

          .              .            .              .            .
TTCTTTGTAAGGTAGAAATAAAGGAGTTTCTGGTTCTGGATTGTAAAAAATGAGTTGTTT 180

            .                    .Met Glu Ile Asn Val Ser Lys Leu Arg Thr Asp
TAATTGATAAGGAGTAGAATATG GAA ATT AAT GTG AGT AAA TTA AGA ACA GAT 233
                        -----

Leu Pro Gln Val Gly Val Gln Pro Tyr Arg Gln Val His Ala His Ser
TTG CCT CAA GTC GGC GTG CAA CCA TAT AGG CAA GTA CAC GCA CAC TCA 281

Thr Gly Asn Pro His Ser Thr Val Gln Asn Glu Ala Asp Tyr His Trp
ACT GGG AAT CCG CAT TCA ACC GTA CAG AAT GAA GCG GAT TAT CAC TGG 329

Arg Lys Asp Pro Glu Leu Gly Phe Phe Ser His Ile Val Gly Asn Gly
CGG AAA GAC CCA GAA TTA GGT TTT TTC TCG CAC ATT GTT GGG AAC GGT 377

Cys Ile Met Gln Val Gly Pro Val Asp Asn Gly Ala Trp Asp Val Gly
TGC ATC ATG CAG GTA GGA CCT GTT GAT AAT GGT GCC TGG GAC GTT GGG 425

Gly Gly Trp Asn Ala Glu Thr Tyr Ala Ala Val Glu Leu Ile Glu Ser
GGC GGT TGG AAT GCT GAG ACC TAT GCA GCG GTT GAA CTG ATT GAA AGC 473

His Ser Thr Lys Glu Glu Phe Met Thr Asp Tyr Arg Leu Tyr Ile Glu
CAT TCA ACC AAA GAA GAG TTC ATG ACG GAC TAC CGC CTT TAT ATC GAA 521

Leu Leu Arg Asn Leu Ala Asp Glu Ala Gly Leu Pro Lys Thr Leu Asp
CTC TTA CGC AAT CTA GCA GAT GAA GCA GGT TTG CCG AAA ACG CTT GAT 569

Thr Gly Ser Leu Ala Gly Ile Lys Thr His Glu Tyr Cys Thr Asn Asn
ACA GGG AGT TTA GCT GGA ATT AAA ACG CAC GAG TAT TGC ACG AAT AAC 617

Gln Pro Asn Asn His Ser Asp His Val Asp Pro Tyr Pro Tyr Leu Ala
CAA CCA AAC AAC CAC TCA GAC CAC GTT GAC CCT TAT CCA TAT CTT GCT 665

Lys Trp Gly Ile Ser Arg Glu Gln Phe Lys His Asp Ile Glu Asn Gly
AAA TGG GGC ATT AGC CGT GAG CAG TTT AAG CAT GAT ATT GAG AAC GGC 713

Leu Thr Ile Glu Thr Gly Trp Gln Lys Asn Asp Thr Gly Tyr Trp Tyr
TTG ACG ATT GAA ACA GGC TGG CAG AAG AAT GAC ACT GGC TAC TGG TAC 761

Val His Ser Asp Gly Ser Tyr Pro Lys Asp Lys Phe Glu Lys Ile Asn
GTA CAT TCA GAC GGC TCT TAT CCA AAA GAC AAG TTT GAG AAA ATC AAT 809

Gly Thr Trp Tyr Tyr Phe Asp Ser Ser Gly Tyr Met Leu Ala Asp Arg
GGC ACT TGG TAC TAC TTT GAC AGT TCA GGC TAT ATG CTT GCA GAC CGC 857

Trp Arg Lys His Thr Asp Gly Asn Trp Tyr Trp Phe Asp Asn Ser Gly
TGG AGG AAG CAC ACA GAC GGC AAC TGG TAC TGG TTC GAC AAC TCA GGC 905

Glu Met Ala Thr Gly Trp Lys Lys Ile Ala Asp Lys Trp Tyr Tyr Phe
GAA ATG GCT ACA GGC TGG AAG AAA ATC GCT GAT AAG TGG TAC TAT TTC 953

Asn Glu Glu Gly Ala Met Lys Thr Gly Trp Val Lys Tyr Lys Asp Thr
AAC GAA GAA GGT GCC ATG AAG ACA GGC TGG GTC AAG TAC AAG GAC ACT 1001

Trp Tyr Tyr Leu Asp Ala Lys Glu Gly Ala Met Val Ser Asn Ala Phe
TGG TAC TAC TTA GAC GCT AAA GAA GGC GCC ATG GTA TCA AAT GCC TTT 1049

Ile Gln Ser Ala Asp Gly Thr Gly Trp Tyr Tyr Leu Lys Pro Asp Gly
ATC CAG TCA GCG GAC GGA ACA GGC TGG TAC TAC CTC AAA CCA GAC GGA 1097

Thr Leu Ala Asp Arg Pro Glu Phe Thr Val Glu Pro Asp Gly Leu Ile
ACA CTG GCA GAC AGG CCA GAA TTC ACA GTA GAG CCA GAT GGC TTG ATT 1145

Thr Val Lys END
ACA GTA AAA TAA TAATGGAATGTCTTTCAAATCAGAACAGCGCATATTATTAGGTCTTG 1204

AAAAAGCTT 1213
```

Gene: *lytA*
Protein: Pneumonococcal autolysin
Organism: *S. pneumoniae*
Reference: P. Garcia, J. L. Garcia, E. Garcia, and R. Lopez, Gene **43**:265–272, 1986

```
3'-GATTTCTGAGTTCGATAAAAACCAGCGGAAACAGAAACTATGCGACTTAATCGGTTAGTCGCTGATGAA   -1

GACGTCTGTAAACTCTCTACAAACCGTCACCAACTCACAAGAACTTGACAAACTATTGTCGACAACAGTTGAGTAGCAAAAATGGAGTCGCATACTTGC  100
ALA SER MET GLN SER ILE ASN PRO LEU PRO GLN THR ASN LYS PHE GLN LYS ILE VAL ASP THR THR LEU GLU ASP ASN LYS ILE GLY GLU ALA TYR SER ARG

TCGGAGTCATAATCGACCCTAAAGCAATCAGAATAGTATATTATGAAAAACAACAAGTCAACGTTGTTCCTTCAGATGTTAAAAAACTCTTCGAAGTTCC  200
ALA GLU THR ASN ALA PRO ILE GLU ASN THR LYS ASP TYR LEU VAL LYS GLN GLN GLU THR ALA VAL LEU PHE ASP VAL PHE LYS GLN SER ALA GLU LEU

AAGAATTGCCCAACTCCTTACTGGGTTCGAAATGGTGGTGGCTTACGTCGTATAAGAAAAGGTAAACCTTCTCAACCCTATCAACGTTGAGGCATCAAAT  300
ASN ASN VAL ALA GLN LEU VAL THR PHE ALA LYS GLY VAL GLY PHE ALA ALA TYR GLU LYS GLY ASN PRO LEU THR HIS ALA VAL GLY TYR LYS ASP VAL

GAAATCGTAGAAATTTCCGAACTCGAAAGGTTCCAGGCAGCTACTATCGTCGACAAAATGGAAGAACCTTAACTCAAACCTAATTCAAAGGTCGTCGAAG  400
LYS ALA ASP LYS PHE ALA GLN ALA LYS TRP PRO GLY ASP ILE ILE ALA ALA THR LYS GLY GLU GLN PHE GLN THR GLN ILE LEU ASN GLY ALA GLY GLU

ACATAGAACGTATGGAAATCCGGTAAAAAGCATGGTTCTAAATCGCATCAACTATGGACGCTATCTTGGCAGCAAACGTTCTGGCTACAGAAATCGCAGA  500
THR ASP GLN MET GLU ILE LYS PRO TYR LYS GLY TYR TRP LYS GLU TYR TRP ASN ILE ALA TYR ASN ILE GLY ALA GLU ASP ASN VAL LEU GLY ILE ASP LYS ALA ASP

AATGGCAAAACCGGTTTCTGCATTCGTGGCAATGGCCGTTCTTCAGGTATACATATCATCTTCAAACAGGTCAGTCGATCCTTCCGTCATCAAAATGGTA  600
LYS GLY ASN GLN GLY PHE VAL TYR ALA GLY ASN GLY ALA LEU LEU GLY TYR THR TYR TYR PHE ASN THR TRP ASP ALA LEU PHE ALA THR THR LYS GLY ASP

GAAGTGGTCGCTTACGCATAAACGATAGAAATCGTTCCAAAGGTTCAGTCGTTTACAAAAAACCTCGTAGAAGTGGTTCAGAAACAACATCATGTGATTG  700
GLU VAL ALA PHE ALA TYR LYS SER ASP LYS ALA LEU ASN GLU LEU ASP ALA PHE THR LYS PRO ALA ASP LYS VAL VAL GLN ASN ASN ILE MET VAL

TTCACTGAGCTATTGCCCGTCCTCGTGGCATTTGAAATGGTAAATCGTCGACAATGTTCTCTAAATCAACACAGCAGACAAAATCGTGGTAGCGAGTTAAAG  800
LEU SER GLU ILE VAL ALA PRO ALA TYR VAL LYS ASN ALA ALA THR VAL LEU SER LYS THR THR ASP ASP THR LYS ALA GLY ASP SER LEU LYS

TGAAGACTTTCAACAGGCAGTCTTGGTTCCGATGGATGTGCCAGCATACCTCGGTAGTATTGTAGTCCCTGTAATGGTCTAACCAACAGTTCTCTTTCAA  900
VAL GLU SER LEU GLN GLY ASP SER GLY LEU SER GLY VAL ARG ASP TYR PRO ALA MET MET VAL ASP PRO VAL ASN GLY SER GLN ASN ASP LEU SER LEU LYS

ATAGTTCTGGAGGATCTCGTAGTGGTCAAATTCTCACTGAAAATGAGGTCGAAGAAAAATGATTCGAAATCGTTGGAGAAGTTATATCGAGAATATAGG  1000
ASP LEU GLY GLY LEU ALA ASP GLY TYR THR LYS SER LEU PHE THR LEU ILE GLU ALA GLU LYS GLU TYR ILE LYS ALA VAL GLU GLU TYR GLU LYS TYR ARG

GAGCAGATGTATATGTCACTCAAGAAACTGAAGTCTACTTGGTCTTAGTCGTCGAATAGTCGTCAAAACGAAGGCGTTCGATGGTTTTCGTTCGATCGT  1100
GLU ASP VAL TYR VAL THR LEU GLU LYS VAL GLU SER SER GLY SER ASP ALA PRO LYS ASP ALA THR LYS SER GLY CYS ALA VAL LEU LEU LEU SER ALA

TCACATTGTCAAGGTTCGTCGGCGTCACAGGCGGTATTTAAATCTACTGTATCACATAAGGAGGGTTTCTTATCGTTCAAAATAACTATTCCTTTGCGTTT  1200
LEU THR VAL THR GLY TYR LEU VAL ALA THR SER LYS MET PHE LYS SER ASP SER SER MET

5'-CGTTTTCCTTTATGAGCTTAGTATAGCACAAATAGAAAACGGTTGCAAGTATTTTTTGTAAAAATTTTAAAAAATTTTTAAGCTTGATTTGATAGCATAA  1300
   GCAAAAGGAAATACTCGAATCATATCGTGTTTTATCTTTTGCCAACGTTCATAAAAAACATTTTTAAAATTTTTTTAAAAATTCGAACTAAACTATCGTATT-5'

TTTTGCATTTTAATAGAAGAAAAGTATGAAATAGATAAGAATACAGTATTTTTTAATAAAAATATATGGAATCTGTCTAAAGAGGGAAACCAGGAGGGGCT  1400

ACCTCCCTGGTTCTAATCCCTATCTATTTTCTATGGACGTTTGTGCTTTGTTACATATAAATTCAAGCTCCCTAAGGAGGTTTTGTACTTTGTAGACTGT  1500

CTCGTGAGAGTCAAACCTTAGGATTAGATAGTTTTAGTTATCTTGCTAAATGAGTTAGTCTCCTAAATATCAAAAGTTTTTAAAAAGCTTTGCAATTATCGCG  1600

                                                                                       MET LYS LYS ARG
TTGAAAAGGAGTATACTTATAAGTAACGCAAACGTTTGCGTCTGCAAAATACGCAACGTTCCATTATTTTAACACACGAGGTGCTATTATGAAAAAACGT  1700
GLN SER GLY VAL LEU MET HIS ILE SER SER SER PRO GLY ALA TYR GLY ILE GLY SER PHE SER ALA TYR ASP PHE VAL LEU ASP ALA
CAAAGTGGTGTGTTTGATGCACATCTCTTCTCTTCCAGGAGCTTACGGAATCGGATCATTTGGTCAAAGTGCTTACGACTTCGTTGATTTCTTGGTCCGTA  1800

THR LYS GLN ARG TYR TRP GLN ILE LEU PRO LEU GLY ALA THR SER TYR GLY ASP SER PRO TYR GLN SER PHE SER ALA PHE ALA GLY ASN THR HIS PHE ILE
CAAAACAACGTTACTGGCAAATCCTTCCATTAGGAGCAACTAGTTACGGGGATTCTCCTTACCAATCTTTCTCAGCCTTCGCAGGAAACACTCATTTTTAT  1900

ASP LEU ASP ILE LEU VAL GLU GLN GLY LEU LEU GLU ALA SER ASP LEU GLY GLY VAL ASP PHE GLY SER ASP ALA SER GLU VAL ASP TYR ALA LYS ILE
CGATTTAGATATCTTGGTGGAGCAAGGTTTGTTGGAAGCAAGTGACCTTGAAGGAGTTGACTTTGGTAGCGATGCGTCTGAAGTTGACTATGCTAAAATC  2000

TYR TYR ALA ARG ARG PRO LEU LEU GLY LYS ALA VAL LYS ARG PHE PHE GLU VAL GLY TYR ASP PHE GLU LYS PHE ALA GLN ASP HIS PHE ASP GLN PHE LYS
TACTATGCACGTCGTCCTCTTTTAGAAAAAGCGGTGAAACGTTTCTTTGAAGTCGGAGATGTTAAAGATTTTTGAGAAATTTGCTCAAGACAACCAATCAT  2100

TRP LEU GLU LEU PHE ALA GLU TYR MET ALA ILE LYS GLU TYR PHE ASP ASN LEU ALA TRP THR GLU TRP PRO ASP ALA ASP ALA ARG ALA ARG LYS ALA SER
GGCTTGAGCTCTTTGCTGAGTATATGGCTATCAAAGAGTATTTTGACAATCTTGCTTGGACTGAATGGCCAGATGCAGATGCTCGTGCTCGTAAAGCTTC  2200

ALA LEU GLU SER TYR ARG GLU GLN LEU ALA ASP LYS LEU VAL TYR HIS VAL THR GLN TYR PHE PHE GLN GLN TRP LEU GLU LEU LYS ALA TYR
AGCACTTGAAAGCTATCGTGAGCAATTGGCAGATAAGCTTGTCTACCACGTGTGACTCAATACTTCTTCTTCCAACAATGGTTGAAATTGAAAGCTTAC  2300

ALA ASN ASP ASN HIS ILE GLU ILE VAL GLY ASP MET PRO ILE TYR VAL ALA GLU ASP SER SER ASP MET TRP ALA ASN PRO HIS LEU PHE LYS THR ASP
GCTAACGACAACCACATCGAAATCGTTGGGGACATGCCAATCTACGTAGCGGGAAGATTCAAGTCGATAATGTGGGCAAATCCACATCTCTTCAAAACAGATG  2400

VAL ASN GLY LYS ALA THR CYS ILE ALA GLY CYS PRO PRO ASP GLU PHE SER VAL THR GLY GLN LEU TRP GLY ASN PRO ILE TYR ASP TRP GLU ALA MET ASP
TCAATGGTAAGGCTACTTGTATCGCAGGATGCCCACCAGATGAGTTTTCTGTAACTGGTCAGCTTTGGGGTAACCCAATCTATGACTGGGAAGCAATGGA  2500

LYS ASP GLY TYR LYS TRP TRP ILE GLU VAL ARG LEU ARG GLU SER PHE LYS ILE TYR ASP VAL ILE ALA ASP PRO HIS GLU PHE GLU LEU GLY SER TYR TRP
CAAAGACGGCTACAAATGGTGGATTGAACGCTTGCGTGAAAGCTTCAAAAATCTACGATATCGTTCGTATCGACCACTTCCGTGGCTTCGAATCTTACTGG  2600

GLU ILE PRO ALA GLY SER ASP THR ALA ALA PRO GLY GLU TRP VAL LYS GLY PRO GLY TYR LYS LEU PHE ALA ALA VAL LYS GLU GLU LEU GLY GLU LEU
GAAATCCCTGCTGGTTCCGATACAGCAGCACCTGGTGAGTGGGTGAAAGGTCCAGGCTACAAGCTTTTTGCAGCCGTTAAGGAAGAACTTGGTGAGCTAA  2700

ASN ILE ILE ALA GLU ASP LEU GLY PHE MET THR ASP GLU VAL ILE GLU LEU ARG GLU ARG THR GLY PHE PRO GLY MET LYS ILE LEU GLN PHE ALA PHE ASN
ACATCATCGCAGAAGACCTTGGCTTCATGACAGATGAAGTGATCGAATTGCGTGAACGTACTGGCTTCCCAGGAATGAAGATTTTGCAATTTGCCTTCAA  2800

PRO GLU ASP GLU SER ILE ASP SER PRO HIS LEU ALA PRO ALA ASN SER VAL MET TYR THR GLY THR HIS GLY ASP ASN ASN THR VAL LEU GLY LYS TRP TYR ARG
CCCAGAAGACGAAAGCATTGATAGCCCACACTTGGCACCTGCTAACTCAGTTATGTACACAGGAACACACGATAACAATACGGTTCTTGGTTGGTACCGT  2900

ASN GLU ILE ASP ASP ALA THR ARG GLU TYR MET ALA ARG TYR THR ASN ARG LYS GLU TYR GLU THR VAL VAL HIS ALA MET LEU ARG THR VAL PHE SER
AATGAGATTGATGATGCGACTCGTGAGTACATGGCTCGTTACACGAACCGTAAAGAATACGAAACAGTGGTACACGCTATGCTTCGTACAGTATTTTCAT  3000

SER VAL SER PHE MET ALA ILE ALA THR MET GLN ASN PHE LEU GLU LEU ASP ASP VAL GLN ALA ALA ASP ARG MET ASN PHE PRO SER THR LEU GLY VAL ASN TRP SER TRP
CAGTTAGCTTTATGGCAATTCAACTATGCAAGATTTACTAGAATTGGATGAGGCAGCTCGTATGAACTTCCCATCTACCCTTGGTGGAAATGGTCTTG  3100

ARG MET THR GLU ASP GLN LEU THR PRO ALA VAL GLU GLU GLY LEU LEU ASP LEU THR THR ILE TYR ARG ARG ILE ASN GLU ASN LEU VAL ASP LEU LYS
GCGTATGACTGAAGATCAATTGACACCAGCTGTCGAGGAAGGTTTGCTTGACTTGACAACAATTTATCGCCGAATTAATGAAAAATTTGGTAGATTTAAAG  3200

LYS
AAATAAGACAATAATCAGGAGACAACTAAACATGTTATCACTACAAGAATTTGTACAAAATCGTTACAATAAAACCATTGCAGAATGTAGCAATGAAGAG  3300

LEU TYR LEU ALA LEU LEU ASN TYR SER LEU SER LEU ALA SER SER GLN LEU GLY LYS LYS LYS VAL TYR TYR LEU ASP ASP
CTTTACCTTGCTCTTCTTAACTACAGCAAGCTTGCAAGCAGCCAAAAACCAGTCAACACTGGTAAGAAAAAAGTTTACTACATCTCAGCTGAGTTCTTGA  3400

ILE GLY LYS LEU LEU SER ASN ASN LEU ILE ASN LEU GLY LEU TYR ASP ASP VAL LYS LYS GLU LEU ALA ALA ALA
TTGGTAAACTCTTTGTCAAACAACTTGATTAACCTTGGTCTTTACGACGATGTTAAAAAGAACTTGCAGCTGCAGCAACGCCAACCAAAAGGGACAAA-3'
```

Gene: *malM*
Protein: Amylomaltase
Organism: *S. pneumoniae*
Reference: S. A. Lacks, J. J. Dunn, and B. Greenberg, Cell **31**:327–336, 1982

```
                                .                   .                        .                   .                   .       450
AAT TCT AAA ATA GAA AAT TTA GTA ATG TGG TAG AAA ATA CAA GAG TTG TGT TTT AAT TTC TAT GGT ATA ATT AAA AGC ATG AAG ATA AAA
                                                                                            MET LYS ILE LYS
                                .                   .                        .                   .                   .       540
GAA ATA AAG AAA GTT ACT TTA CAA CCG TTC ACG AAA TGG ACA GGT GGT AAA AGA CAA TTA TTG CCT GTT ATT AGA GAA TTA ATA CCT AAA
GLU ILE LYS LYS VAL THR LEU GLN PRO PHE THR LYS TRP THR GLY GLY LYS ARG GLN LEU LEU PRO VAL ILE ARG GLU LEU ILE PRO LYS
                                .                   .                        .                   .                   .       630
ACC TAT AAC AGG TAT TTC GAA CCT TTT GTT GGA GGT GGA GCT TTA TTT TTT GAT TTG GCT CCT AAA GAT GCA GTT ATT AAT GAT TTT AAC
THR TYR ASN ARG TYR PHE GLU PRO PHE VAL GLY GLY GLY ALA LEU PHE PHE ASP LEU ALA PRO LYS ASP ALA VAL ILE ASN ASP PHE ASN
                                .                   .                        .                   .                   .       720
GCT GAA CTA ATA AAT TGC TAT CAA CAA ATT AAG GAC AAT CCT CAA GAA TTG ATT GAA ATT TTG AAA GTT CAT CAG GAA TAT AAT TCA AAA
ALA GLU LEU ILE ASN CYS TYR GLN GLN ILE LYS ASP ASN PRO GLN GLU LEU ILE GLU ILE LEU LYS VAL HIS GLN GLU TYR ASN SER LYS
                                .                   .                        .                   .                   .       810
GAA TAT TAT TTA GAT TTA CGT TCT GCA GAT CGT GAT GAA AGA ATA GAT ATG ATG TCC GAA GTA CAA AGA GCT GCA CGT ATT CTA TAT ATG
GLU TYR TYR LEU ASP LEU ARG SER ALA ASP ARG ASP GLU ARG ILE ASP MET MET SER GLU VAL GLN ARG ALA ALA ARG ILE LEU TYR MET
                                .                   .                        .                   .                   .       900
TTG AGA GTG AAC TTT AAT GGT CTA TAT CGT GTG AAT TCT AAG AAT CAA TTT AAT GTT CCA TAT GGA CGT TAT AAG AAT CCT AAA ATT GTT
LEU ARG VAL ASN PHE ASN GLY LEU TYR ARG VAL ASN SER LYS ASN GLN PHE ASN VAL PRO TYR GLY ARG TYR LYS ASN PRO LYS ILE VAL
                                .                   .                        .                   .                   .       990
GAT GAG GAA TTG ATA TCT GCT ATT TCA GTT TAT ATA AAT AAC AAT CAA CTA GAA ATT AAA GTG GGA GAT TTT GAA AAG GCA ATT GTA GAT
ASP GLU GLU LEU ILE SER ALA ILE SER VAL TYR ILE ASN ASN ASN GLN LEU GLU ILE LYS VAL GLY ASP PHE GLU LYS ALA ILE VAL ASP
                                .                   .                        .                   .                   .      1080
GTT CGA ACA GGA GAT TTT GTG TAT TTT GAC CCT CCA TAT ATT CCA TTG TCT GAG ACG AGT GCA TTT ACG TCT TAT ACT CAT GAG GGA TTC
VAL ARG THR GLY ASP PHE VAL TYR PHE ASP PRO PRO TYR ILE PRO LEU SER GLU THR SER ALA PHE THR SER TYR THR HIS GLU GLY PHE
                                .                   .                        .                   .                   .      1170
TCT TTT GCA GAT CAA GTA AGA TTA AGA GAT GCC TTT AAG AGA TTG AGT GAT ACA GGA GCT TAT GTT ATG TTA TCA AAT TCT TCT AGT GCT
SER PHE ALA ASP GLN VAL ARG LEU ARG ASP ALA PHE LYS ARG LEU SER ASP THR GLY ALA TYR VAL MET LEU SER ASN SER SER SER ALA
                                .                   .                        .                   .                   .      1260
TTA GTA GAG GAG TTG TAT AAG GAT TTT AAT ATA CAT TAT GTT GAA GCT ACC CGA ACT AAT GGA GCA AAA TCT TCA AGT CGA GGA AAA ATT
LEU VAL GLU GLU LEU TYR LYS ASP PHE ASN ILE HIS TYR VAL GLU ALA THR ARG THR ASN GLY ALA LYS SER SER SER ARG GLY LYS ILE
                                .                   .                        .                   .                   .      1350
TCT GAA ATT ATA GTC ACA AAT TAT GAA AAA TAA CGA ATA TAA GTA TGG AGG TGT TCT TAT GAC AAA ACC ATA CTA CAA TAA AAA GAT
SER GLU ILE ILE VAL THR ASN TYR GLU LYS ###
```

Gene: *dpnM*
Protein: *Dpn*II DNA adenine methylase
Organism: *S. pneumoniae*
Reference: B. M. Mannarelli, T. S. Balganeshi, B. Greenberg, S. S. Springhorn, and S. A. Lacks, Proc. Natl. Acad. Sci. USA **82**:4468–4472, 1985

```
AAGCTTTGGTGGAGAAATTGGCTGGCGAATCCAGCTTCACCGGTGTTTCACCAGTAGATGCTTTCTGTGGTCTTATTGACACGCACTTGTGGCGAGAGTACTAACAGTCACAGCGACGTT   120

AACTTTATTTTCCTTATGAGAGGTTAAGAAAAAACGTTATTAAATAGCAGAAAAGAATATTATGACTGACGTTAGGAGTTTTCTCCTAACGTTTTTTTTAGTACAAAAAGAGAATTCTCT   240
                                                  -------->    <---------                              EcoRI

ATTATAAATAAAATAAATAGTACTATAGATAGAAAATCTCATTTTTAAAAAGTCTTGTTTTCTTAAAGAAGAAAATAATTGTTGAAAAATTATAGAAAATCATTTTTATACTAATGAAAT   360

                                                                                   -35
AGACATAAAGGCTAAATTGGTGAGGTGATGATAGGAGATTTATTTGTAAGGATTCCTTAATTTTATTAATTCAACAAAAATTGATAGAAAAATTAAATGGAATCCTTTGATTTAATTTTATT   480
                                                                                --------->   <----------
                                                                                            -------

          -10                                                     rbs        fMetGluLysGluLysLysValLys   8
AAAGTTGTATAATAAAAAGTGAAATTATTAAATCGTAGTTTCAAATTTGTCGGCTTTTTAATATGTGCTGGCATATTAAAATTAAAAAAGGAGAAAAAATGGAAAAGAAAAAAGGTAAAA   601
>        <-------                                           --------->   <---------

TyrPheLeuArgLysSerAlaPheGlyLeuAlaSerValSerAlaAlaPheLeuValGlySerThrValPheAlaValAspSerProIleGluAspProIleIleArgAsnGlyGly   48
TACTTTTTTACGTAAATCAGCTTTTGGGTTAGCATCCGTATCAGCTGCATTTTTAGTGGGATCAACGGTATTCGCTGTTGATTCACCAATCGAAGATACCCCAATTATTCGTAATGGTGGT   721

GluLeuThrAsnLeuLeuGlyAsnSerGluThrThrLeuAlaLeuArgAsnGluGluSerAlaThrAlaAspLeuThrAlaAlaAlaValAlaAspThrValAlaAlaAlaAlaAlaGlu   88
GAATTAACTAATCTTCTGGGGAATTCAGAGACAACACTGGCTTTGCGTAATGAAGAGAGTGCTACAGCTGATTTGACAGCAGCAGCGGTAGCCGATACTGTGGCAGCAGCGGCAGCTGAA   841
                EcoRI

                                            A1
AsnAlaGlyAlaAlaAlaTrpGluAlaAlaAlaAlaAlaAlaAspAlaLeuAlaLysAlaLysAlaAspAlaLeuLysGluPheAsnLysTyrGlyValSerAspTyrTyrLysAsnLeuIle   128
AATGCTGGGGCAGCAGCTTGGGAAGCAGCGGCAGCAGCAGATGCTCTAGCAAAAGCCAAAGCAGATGCCCTTAAAGAATTCAACAAATATGGAGTAAGTGACTATTACAAGAATCTAATC   961
                                                                                   EcoRI

AsnAsnAlaLysThrValGluGlyIleLysLysAspLeuGlnAlaGlnValValGluSerAlaLysLysAlaArgIleSerGluAlaThrAspGlyLeuSerAspPheLeuLysSerGlnThr   168
AACAATGCCAAAACTGTTGAAGGCATAAAAGACCTTCAAGCACAAGTTGTTGAATCAGCGAAGAAAGCGCGTATTTCAGAAGCAACAGATGGCTTATCTGATTTCTTGAAATCGCAAACA   1081

                                            A2
ProAlaGluAspThrValLysSerIleLeuAlaGluAlaLysValLeuAlaLeuAsnAsnArgGluLeuAspLysTyrGlyValSerAspTyrHisLysAsnLeuIleAsnAsnAlaLysThr   208
CCTGCTGAAGATACTGTTAAATCAATTGAATTAGCTGAAGCTAAAGTCTTAGCTAACAGAGAACTTGACAAATATGGAGTAAGTGACTATCACAAGAACCTAATCAACAATGCCAAAACT   1201

                                            B1
ValGluGlyValLysGlyLeuIleAspGluIleLeuAlaAlaLeuProLysThrAspThrTyrLysLeuIleLeuAsnGlyLysThrLeuLysGlyGluThrThrThrGluAlaValAsp   248
GTTGAAGGTGTAAAAGAACTGATAGATGAAATTTTAGCTGCATTACCTAAGACTGACACTTACAAATTAATCCTTAATGGTAAAACATTGAAAGGCGAAACAACTACTGAAGCTGTTGAT   1321

                                                                                  AlaAlaThrAlaGluLysValPheLysGlnTyrAlaAsnAspAsnGlyValAspGlyGluTrpThrTyrAspAspAlaThrLysThrPheThrValThrGluLysProGluValIleAsp   288
GCTGCTACTGCAGAAAAAGTCTTCAAACAATACGCTAACGACAACGGTGTTGACGGTGAATGGACTTACGACGATGCGACTAAGACCTTTACAGTTACTGAAAAACCAGAAGTGATCGAT   1441
              PstI

                    B2
AlaSerGluLeuThrProAlaValThrThrTyrLysLeuValIleAsnGlyLysThrLeuLysGlyGluThrThrThrLysAlaValAspAlaGluThrAlaGluLysAlaPheLysGln   328
GCGTCTGAATTAACACCAGCCGTGACAACTTACAAACTTGTTATTAATGGTAAAACATTGAAAGGCGAAACAACTACTAAAGCAGTAGACGCAGAAACTGCAGAAAAAGCCTTCAAACAA   1561
                                                                                           PstI

                                                                      TyrAlaAsnAspAsnGlyValAspGlyValTrpThrTyrAspAspAlaThrLysThrPheThrValThrGluMetValThrGluValProGlyAlaProThrGluProGluLysPro   368
TACGCTAACGACAACGGTGTTGATGGTGTTTGGACTTATGATGATGCGACTAAGACCTTTACGGTAACTGAAATGGTTACAGAGGTTCCTGGTGATGCACCAACTGAACCAGAAAAACCA   1681

                                      C1           C2           C3           C4           C5
GluAlaSerIleProGluValProLeuThrProAlaThrProIleAlaLysAspAspAlaLysLysAspAspThrLysLysGluAspAlaLysLysProGluAlaLysLysAspAspAla   408
GAAGCAAGTATCCCTCTTGTTCCGTTAACTCCTGCAACTCCAATTGCTAAAGATGACGCTAAGAAAGACGATACTAAGAAAGAAGATGCTAAAAAACCAGAAGCTAAGAAAGATGACGCT   1801

      LysLysAlaGluThrLeuProThrThrGlyGluGlySerAsnProPhePheThrAlaAlaAlaLeuAlaValMetAlaGlyAlaGlyAlaLeuAlaValAlaSerLysArgLysGluAsp   448
AAGAAAGCTGAAACTCTTCCTACAACTGGTGAAGGAAGCAACCCATTCTTCACAGCAGCTGCGCTTGCAGTAATGGCTGGTGCGGGTGCTTTGGCGGTCGCTTCAAAACGTAAAGAAGAC   1921
                                                                                                                   --- -

TAATTGTCATTATTTTTGACAAAAAGCTT   1950
- ---->   <----  -----
```

Gene: *spg*
Protein: Streptococcal immunoglobulin G-binding protein
Organism: Group G *Streptococcus* spp.
Reference: S. R. Fahnestock, P. Alexander, J. Nagle, and D. Filpula, J. Bacteriol. **167**:870–880, 1986

```
        |-35|              |-10 |                                                        |  S-D  |            100
TTTTGATGGĊ TTGGCAATTĊ AGACGATTTG TGTTATACTA AAACTATGCG ATGAGTCGAT TGTGGTTTAT ACCACATACG CTAAGGAGAT CATA  ATG CAĠ GAG CAG ATC TTG
                                                                                                         Met Gln Glu Gln Ile Leu

                                                    ←──────  ·  ──────→                                                       200
GCA AGT TĠT GTG AAC CTĠ CTT GCC ACA ĊTT AAT CAA AĂG TGC CTT TTĊ CCT TGT AAT ĂAC AAG GGC TTT TTT ATT GAĂ TTC AAT AGT TTA TGG ACT TTT
Ala Ser Cys Val Asn Leu Leu Ala Thr Leu Asn Gln Lys Cys Leu Phe Pro Cys Asn Asn Lys Gly Phe Phe Ile Glu Phe Asn Ser Leu Trp Thr Phe
  |-35 |         |1  2  3   4 , 5     |                -10                |  S-D  |        ·                                    300
TTT TTG TAT AAA AAG TAG TATAATGTCT ATTATTAGAA TGAATTAGAA TAAAGAGGTA AACT   ATG GGC TAC ACA GTT GĊT ATC GTT GGṪ GCT ACA GGC GĊC GTT
Phe Leu Tyr Lys Lys ***                                                   Met Gly Tyr Thr Val Ala Ile Val Gly Ala Thr Gly Ala Val

                                                                                                         400
GGA ACṪ CGT ATG ATT CAA CAA TTG ĠAA CAA TCG AĊA CTT CCA GTṪ GAT AAG GTA ĊGG CTT TTG TĊA TCT TCA CGṪ TCT GCA GGṪ AAĂ GTT TTG CĂA TAT
Gly Thr Arg Met Ile Gln Gln Leu Glu Gln Ser Thr Leu Pro Val Asp Lys Val Arg Leu Leu Ser Ser Ser Arg Ser Ala Gly Lys Val Leu Gln Tyr

                                                                                              500
AAA GAṪ CAA GAT GTC ĂCG GTT GAA TṪA ACT ACG AAĂ GAT TCC TTT ĠAA GCT GTT GĂT ATT GCG CTṪ TTT TCA GCT ĠGC GGT TCT GṪT TCG GCA AAĂ TTT
Lys Asp Gln Asp Val Thr Val Glu Leu Thr Thr Lys Asp Ser Phe Glu Ala Val Asp Ile Ala Leu Phe Ser Ala Gly Gly Ser Val Ser Ala Lys Phe

                                                                                    600
GCT CCC TĂT GCA GTC AĂA GCT GGT GCĂ GTC GTT GTT ĠAT AAT ACC TĊT CAT TTT CGṪ CAA AAT CCA ĠAT GTG CCT TṪG GTT GTT CCT ĠAA GTC AAT ĠCT
Ala Pro Tyr Ala Val Lys Ala Gly Ala Val Val Val Asp Asn Thr Ser His Phe Arg Gln Asn Pro Asp Val Pro Leu Val Val Pro Glu Val Asn Ala

                                                                              700
TAT GCT AṪG GAT GCT CAṪ AAT GGG ATT ĂTT GCT TGT CĊT AAC TGC TCĂ ACG ATT CAA ĂTG ATG GTA GĊC TTG GAA CCṪ ATT CGT CAA AĂA TGG GGA TṪA
Tyr Ala Met Asp Ala His Asn Gly Ile Ile Ala Cys Pro Asn Cys Ser Thr Ile Gln Met Met Val Ala Leu Glu Pro Ile Arg Gln Lys Trp Gly Leu

                                                                        800
AGT CGT GTṪ ATT GTT TCA AĊC TAT CAA GĊT GTT TCA GGĂ GCA GGT CAA ṪCA GCT ATT AĂT GAA ACT GTṪ CGT GAA ATT ĂAA GAA GTT GṪT AAT GAT GGṪ
Ser Arg Val Ile Val Ser Thr Tyr Gln Ala Val Ser Gly Ala Gly Gln Ser Ala Ile Asn Glu Thr Val Arg Glu Ile Lys Glu Val Val Asn Asp Gly

                                                                  |-35| TTG GCĂ CAG ATṪ GAT GTC TTC          900
GTG GAT CCT ĂAA GCT GTT CĂT GCT GAT ATṪ TTT CCA TCA ĠGT GGT GAT AĂA AAG CAT TAṪ CCA ATT GCT TTC AAT GCT          ATT GAT GTC TTC
Val Asp Pro Lys Ala Val His Ala Asp Ile Phe Pro Ser Gly Gly Asp Lys Lys His Tyr Pro Ile Ala Phe Asn Ala Leu Ala Gln Ile Asp Val Phe

 ·  |-10 | ·                                                                                       1000
ACC ĠAT AAT ĠAT TAT ACT TAṪ GAA GAA ATG ĂAG ATG ACT AĂC GAA ACC AAG AAA ATC ATG ĠAA GAA CCT ĠAA CTT CCC GTṪ TCG GCC CAT TĠT GTT CGT
Thr Asp Asn Asp Tyr Thr Tyr Glu Glu Met Lys Met Thr Asn Glu Thr Lys Lys Ile Met Glu Glu Pro Glu Leu Pro Val Ser Ala His Cys Val Arg

                                                                                       1100
GṪT CCA ATC CTṪ TTT TCA CAT TĊT GAG GCT GṪT TAT ATT GAĂ ACT AAA GAC ĠTT GCT CCA AṪT GAA GAA GTĂ AAA GCA GCT ATṪ GCA GCA TṪT CCA GGT
Val Pro Ile Leu Phe Ser His Ser Glu Ala Val Tyr Ile Glu Thr Lys Asp Val Ala Pro Ile Glu Glu Val Lys Ala Ala Ile Ala Ala Phe Pro Gly

                                                                                1200
GCṪ GTT CTT GAA ĠAT GAT ATT AĂA CAT CAA ATṪ TAC CCA CAA ĠCA GCG AAT GĊT GTT GGC AGṪ CGT ACT TTT ĠTC GGC CGT AṪT CGT AAG GAṪ TTA GAT
Ala Val Leu Glu Asp Asp Ile Lys His Gln Ile Tyr Pro Gln Ala Ala Asn Ala Val Gly Ser Arg Thr Phe Val Gly Arg Ile Arg Lys Asp Leu Asp

                                                                                1300
ATT ĠAA AAT GGT AṪT CAT ATG TGĠ GTC GTT TCA ĠAC AAT CTT CṪT AAA GGT GCṪ GCT TGG AAT TĊA ATC ATC AĊC GCT AAC CGṪ CTA CAT GAA ĊGG GGT
Ile Glu Asn Gly Ile His Met Trp Val Val Ser Asp Asn Leu Leu Lys Gly Ala Ala Trp Asn Ser Ile Ile Thr Ala Asn Arg Leu His Glu Arg Gly

                                                          1400
CTT GṪT CGT TCG ACĂ TCA G^A TTG ĂAG TTT GAA CṪG AAA TAA AAṪAACTAAT AAṪGGTGTAT TAĂAGTGAGA GGĠAGAGAAC ATĠTCAATTC AAG
Leu Val Arg Ser Thr Ser Glu Leu Lys Phe Glu Leu Lys ***
```

Gene: *asd*
Protein: Aspartate semialdehyde dehydrogenase
Organism: *S. mutans*
Reference: G. A. Cardineau and R. Curtiss III, J. Biol. Chem., in press

Appendix C

Compilation of Nucleotide Sequences That Signal the Initiation of Transcription and Translation in Streptococci

```
aadE            TTGAAACTTTTGAATATAAACCTGAAAAATATGCCGAAAGAGTT-100 bp-TTTAAACGAGGAGAGTTCAGATG

aph3            TTGAAAAGCTGTTTTCTGGTATTTAAGGTTTTAGAATGCAAGGA-50 bp-AAATACTGTAGAAAGAGGAAGGAAATAATAAATG

aac6/aph2       ATGAAAAATTAGGTGGTAAAAATGATTATAAAGATGAAATAGTATATGTATATGATTATGAAAAAGGTGATAAATAAATG

tetM            GCTTTGTATGCCTATGGTTATGCATAAAAATCCCAGTGATAAGAGTATTTATCACTGGGATTTTTATGCCCTTTTGGGCTTTTGAATGGAGGAAAATCACATG

pAM77/MLS       TGTTGATAGTGCATTATCTTAAAATTTTGTATAATAGGAATTGAAGTTAAATTAGATGCTAAAAATTTGTAATTAAGAGGAGGGATTCGTCATG

Tn917/MLS ORF1  GTTGATAGTGCATTATCTTAAAATTTTGTATAATAGGAATTGAAGTTAAATTAGATGCTAAAAATTTGTAATTAAGAGGAGGGATTCGTCATG

          ORF4  TTAATGAGTTATGGTACATTTAAATTATAAAATTAAGGAGGTTTTTTTATG

          ORF6  ATGCCAAACAATGGCGCTTTTATAAAGAAGCTCTGACTCGTGCTCAA - 110 bp - TCCGATGTTAATCCACATTACAAAGTATGGAAAAAGGAGCTACAATG

speA            TTGTTAGTAATATTGGTGAATTGTAATAACCTTTTTAAATCTAGAGGAGAACCCAGATATAAAATGGAGGAATATTAATG

skc             TTAAAACCATTATGTTATTCTAATAATGGGGATTGAAACTTAACTTTTAGGAGGTTTCTATG

emm6            TTTACCTTTTGGCTTTTATTATTTACAATACAATTATTAGAGTTAAACCCTGAAAATGAGGGTTTTTCCTAAAAAATGATAACATAAGGAGCATAAAAATG

lytA            TAAGGTAGAAATAAAGGAGTTTCTGGTTCTGGATTGTAAAAAATGAGTTGTTTTAATTGATAAGGAGTAGAATATG

malM            TTGCAATTATGCGTTGAAAAGGAGTATACTTATAAGTAACGCAAACGTTTGCGTCTGCAAAATACGCAACGTTCCATTATTTTAACACACGAGGTGCTATTATG

spg             TTGATTTAATTTTATTAAGTTGTATAATAAAAAAGTGAAATTATTTAAATCGTAGTTTCAAATTTGTCGGCTTTTTAATATGTGCTGGCATATTAAAATTAAAAAGGAGAAAAAATG

asd             TTGTATAAAAAGTAGTATAATGTCTATTATTAGAATGTTAGAATAGAGGTAAACTATG

gtfI            TTGACAATTAGAATTATTTAACTCTTAAAAAAGTTAAAGGTCGTTATTTTTTCAAATTAGGAGGAACTCCATTGATG

M12 protein     AAAAGCTGGTCTTTACCTTTTGGCTTATATTATTTACAATAGAATTATTAGAGTTAAACCCTGAAAAATGAGGGTTTTTTCCTAAAAAAATGATAACATAAGGAGCATAACATGATG
```

The underlined sequences are identified by the original authors as putative or actual sequences for transcription and translation initiation. References are presented in Appendix B of this volume with the complete nucleotide sequence. *gtfI*, Glucosyltransferase, insoluble form, gene from *S. sobrinus* (J. J. Ferretti and R. R. B. Russell, unpublished data). Type 12 M protein gene from *S. pyogenes*, J. Robbins and P. Cleary, this volume.

Appendix D

The Genetic Code and Single-Letter Amino Acid Designations

TTT	Phe	F	TCT	Ser	S	TAT	Tyr	Y	TGT	Cys	C
TTC	Phe	F	TCC	Ser	S	TAC	Tyr	Y	TGC	Cys	C
TTA	Leu	L	TCA	Ser	S	TAA	——	–	TGA	——	–
TTG	Leu	L	TCG	Ser	S	TAG	——	–	TGG	Trp	W
CTT	Leu	L	CCT	Pro	P	CAT	His	H	CGT	Arg	R
CTC	Leu	L	CCC	Pro	P	CAC	His	H	CGC	Arg	R
CTA	Leu	L	CCA	Pro	P	CAA	Gln	Q	CGA	Arg	R
CTG	Leu	L	CCG	Pro	P	CAG	Gln	Q	CGG	Arg	R
ATT	Ile	I	ACT	Thr	T	AAT	Asn	N	AGT	Ser	S
ATC	Ile	I	ACC	Thr	T	AAC	Asn	N	AGC	Ser	S
ATA	Ile	I	ACA	Thr	T	AAA	Lys	K	AGA	Arg	R
ATG	Met	M	ACG	Thr	T	AAG	Lys	K	AGG	Arg	R
GTT	Val	V	GCT	Ala	A	GAT	Asp	D	GGT	Gly	G
GTC	Val	V	GCC	Ala	A	GAC	Asp	D	GGC	Gly	C
GTA	Val	V	GCA	Ala	A	GAA	Glu	E	GGA	Gly	G
GTG	Val	V	GCG	Ala	A	GAG	Glu	E	GGG	Gly	G

Appendix E

Mole Percent Guanosine plus Cytosine (G+C) of Streptococci

Lancefield group	Organism	Mol% G+C[a]	
		Bridge and Sneath (1)	Other references
A	S. pyogenes	35.8	34.5–38.5 (11)
B	S. agalactiae	36.6	35–37 (10)
C	S. equi	39.3	40.7 (7)
C	S. equisimilis	37.3	38.1–40.0 (7)
C	S. dysgalactiae	37.2	38.5–39.8 (7)
C	S. zooepidemicus		41.3–42.7 (7)
D	S. faecalis	38.4	33.5–38.0 (11)
D	S. faecium	39.0	38.3–39.0 (11)
D	S. durans	37.0	
D	S. bovis	39.3	36.3–40.3 (9)
D	S. equinus	35.6	36.2–38.6 (9)
			39.9–41.3 (9)
D	S. faecium subsp. casseliflavus	39.9	42.2 (11)
D	S. alactolyticus		39.9–41.3 (9)
D	S. gallinarum		37.4 (11)
E	S. infrequens		37.1 (7)
F	S. minutis		38.0 (6)
G	S. anginosus		38.8 (7)
G			39.3–40.2 (4)
H	S. sanguis	39.3	40.8–42.8 (3)
			42.7–44.0 (3)
			43.8–46.4 (3)
I			
J			
K	S. salivarius	38.9	39.5–41.0 (8)
L		38.1	39.2–40.2 (7)
M			
N	S. lactis	35.5	38.6 (11)
N	S. cremoris	34.1	
N	S. raffinolactis	39.8	40.3–41.5 (11)
N	S. lactis subsp. diacetylactis	34.8	
O			
P			
Q	S. avium	39.9	39.9 (11)
R			
S		40.2	
Ungrouped	S. acidominimus		39.7 (11)
	S. constellatus		38.8 (7)
	S. iniae		32.9 (11)
	S. intermedius		37.7–38.4 (7)
	S. pneumoniae	35.1	38.5 (11)
	S. saccharolyticus		37.6–38.3 (9)
	S. suis		40.9–42.0 (7, 9)
	S. thermophilus		37.2–40.3 (8)
	S. uberis	36.4	36.3–37.5 (7)

Lancefield group	Organism	Mol% G+C[a]	
		Bridge and Sneath (1)	Other references
Oral	*S. cricetus*		42–44 (2, 5)
	S. ferus		43–45 (5)
	S. macacae		35–36 (5)
	S. milleri	34.1	39.5 (7)
	S. mitior		39.9–41.0 (4)
			41.3–42.6 (4)
	S. mitis	38.9	39.5 (8)
	S. mutans		36–38 (2, 5)
	S. oralis	39.9	41.3 (8)
	S. rattus	38.8	41–43 (2)
	S. sobrinus		44–46 (2, 5)
Anaerobic	*S. hansenii*		37–38 (11)
	S. morbillorum		
	S. parvulis		46 (11)
	S. pleomorphus		39 (11)

[a] References:
1. Bridge, P. D., and P. H. A. Sneath, J. Gen. Microbiol. **129**:565–596, 1983.
2. Coykendall, A. L., J. Gen. Microbiol. **83**:327–338, 1974.
3. Coykendall, A. L., and P. A. Specht, J. Gen. Microbiol. **91**:92–98, 1975.
4. Coykendall, A. L., and A. J. Munzenmaier, Int. J. Syst. Bacteriol. **28**:511–515, 1987.
5. Coykendall, A. L., and K. B. Gustafson, *in* S. Hamada, S. M. Michalek, H. Kiyono, L. Menaker, and J. R. McGhee (ed.), *Molecular Microbiology and Immunology of S. mutans*, Elsevier, New York, 1986.
6. Ezaki, T., R. Facklam, N. Takeuchi, and E. Yabuuchi, Int. J. Syst. Bacteriol. **36**:345–347, 1986.
7. Farrow, J. A. E., and M. D. Collins, Syst. Appl. Microbiol. **5**:483–493, 1984.
8. Farrow, J. A. E., and M. D. Collins, J. Gen. Microbiol. **130**:357–362, 1984.
9. Farrow, J. A. E., J. Kruze, B. A. Phillips, A. J. Bromley, and M. D. Collins, Syst. Appl. Microbiol. **5**:467–482, 1984.
10. Forbes, B. A., and J. J. Ferretti, Infect. Immun. **18**:866–867, 1977.
11. Hardie, J. M., p. 1043–1071, *in* J. G. Holt, P. H. A. Sneath, N. S. Mair, and M. E. Sharpe (ed.), *Bergey's Manual of Systematic Bacteriology*, vol. 2, The Williams & Wilkins Co., Baltimore, 1986.

Author Index

Subject Index

†Initial page of chapter.